Advances in

ORGANOMETALLIC CHEMISTRY

VOLUME 26

Advances in Organometallic Chemistry

EDITED BY

F. G. A. STONE

DEPARTMENT OF INORGANIC CHEMISTRY
THE UNIVERSITY
BRISTOL, ENGLAND

ROBERT WEST

DEPARTMENT OF CHEMISTRY
UNIVERSITY OF WISCONSIN
MADISON, WISCONSIN

VOLUME 26

1986

ACADEMIC PRESS, INC.
Harcourt Brace Jovanovich, Publishers
Orlando   San Diego   New York   Austin
Boston   London   Sydney   Tokyo   Toronto

ACADEMIC PRESS, INC.
Orlando, Florida 32887

United Kingdom Edition published by
ACADEMIC PRESS INC. (LONDON) LTD.
24–28 Oval Road, London NW1 7DX

LIBRARY OF CONGRESS CATALOG CARD NUMBER: 64-16030

ISBN 0–12–031126–7 (alk. paper)

PRINTED IN THE UNITED STATES OF AMERICA

86 87 88 89   9 8 7 6 5 4 3 2 1

Contents

Triosmium Clusters

ANTONY J. DEEMING

Chemistry of 1,3-Ditungstacyclobutadienes

MALCOLM H. CHISHOLM and JOSEPH A. HEPPERT

Acyclic Pentadienyl Metal Complexes

P. POWELL

Computer Aids for Organometallic Chemistry and Catalysis

I. THEODOSIOU, R. BARONE, and M. CHANON

π Bonding to Main-Group Elements

PETER JUTZI

Dienes, Versatile Reactants in the Photochemistry of Group 6 and 7 Metal Carbonyl Complexes

CORNELIUS G. KREITER

Triosmium Clusters

ANTONY J. DEEMING

Department of Chemistry
University College London
London WC1H OAJ, England

I

INTRODUCTION

Dodecacarbonyltriosmium, $Os_3(CO)_{12}$, has been one of the key compounds in the development of transition metal clusters since it can undergo a wide range of reactions giving many new clusters, more often than not without any change of nuclearity. The chemistries of neighboring systems such as $Re_2(CO)_{10}$ or $Ir_4(CO)_{12}$ are much less rich than that of $Os_3(CO)_{12}$, and, while $Ru_3(CO)_{12}$ is a fascinating material, and its chemistry is similar in many ways to that of $Os_3(CO)_{12}$, it is more difficult to control. With clusters larger than trinuclear most interest has been in their skeletal shapes and how these are determined by the outer coordination shell or by encapsulated atoms. Variety of cluster shape for trinuclear clusters is obviously severely limited, but on the other hand triosmium clusters have been synthesized with many different ligands and the coordination chemistry of clusters reaches its most developed form with triosmium. Triosmium clusters are generally easily handled in air, are easily crystallized and usually easy to characterize. The wealth of chemistry and relatively easy handling has led to rapid growth of the subject, moderated only by the expense of the metal which limits the scale of research and makes applications unattractive.

Our knowledge of the chemistry of doubly bridging inorganic and organic ligands in triosmium clusters may be extrapolated to that of ligands in dinuclear compounds. For example, the chemistry of $Os_3H(\mu\text{-}X)(CO)_{10}$ is finding parallels in that of $Re_2H(\mu\text{-}X)(CO)_8$ [X = 2-pyridyl (1), alkynyl (2), alkenyl (3), acyl (4) etc]. Likewise the chemistry of μ_3 ligands in triosmium clusters relates to that of the triangular faces of much larger clusters. Triosmium compounds are a reasonable starting point for understanding much of the ligand chemistry in clusters.

Most of the chemistry of $Os_3(CO)_{12}$ has been learned since 1970, and a review on osmium carbonyls in 1975 contains only a little triosmium chemistry (5). Comprehensive Organometallic Chemistry (1982) includes several times more material and cites 82 X-ray structures of triosmium compounds

1

(6). Other recent accounts of cluster chemistry have emphasized osmium chemistry (7–9). Perhaps a growing interest in dinuclear chemistry, in large clusters, and in heteronuclear clusters is overshadowing trinuclear chemistry at the moment, but even here $Os_3(CO)_{12}$ and its derivatives play the role of useful starting materials. Triosmium systems will continue to be developed and to be used increasingly to establish reaction mechanisms and detailed structural and physical properties and behavior because of their extreme versatility in the coordination of essentially any inorganic and organic ligand. These are ideal systems for testing the viability of newly conceived ligands or known ligands in unusual bonding modes. For example, μ_3-benzene (10), μ_3-C_2O (11,12), semibridging μ_3-CH (13), and μ_3-BCO (14) are examples discovered in triosmium chemistry, but one could well consider the following as synthetic targets in triosmium chemistry: μ_2-C_2O, μ_3-CH_2 (with agostic bonding), μ_3-CO_2, μ_3-N_2 (analogous to μ_3-alkyne), μ_3-$CHCH_2$, and so on.

Triosmium dodecacarbonyl is the best (almost only) starting material for triosmium clusters, and hundreds of compounds have now been formed by reaction of inorganic or organic substrates with $Os_3(CO)_{12}$ directly or more commonly with more reactive derivatives such as $Os_3H_2(CO)_{10}$ (15), $Os_3(CO)_{11}L$ (16–18), or $Os_3(CO)_{10}L_2$ (18–21) (where L is a weakly bound, easily displaced ligand such as acetonitrile or cyclooctene) or with diene complexes such as $Os_3(CO)_{10}$(cyclohexadiene) (22). Mononuclear osmium chemistry is extensive, especially that with tertiary phosphine ligands (6), dinuclear chemistry very limited (6), while high nuclearity clusters of osmium are an important and exciting development to come in the main from the work of Johnson and Lewis (6–9). There are, however, very few routes to triosmium clusters from these sources. $Os_2(CO)_9$ gives $Os_3(CO)_{12}$ at room temperature (23) while $Os(CO)_5$ does so at a significant rate only at 80°C (24). Whereas $K[RuH(CO)_4]$ is readily converted with acid to $K[Ru_3H(CO)_{11}]$, the related osmium compound only gives $K[Os_3H(CO)_{11}]$ in low yield over long periods (25). These are certainly not synthetic routes of any value. Solid $Os_6(CO)_{18}$ reacts with CO to give $Os_6(CO)_{20}$, heptane solutions react with CO (90 atm, 160°C) to give $Os(CO)_5$ and $Os_5(CO)_{19}$, while under slightly more forcing conditions (100 atm, 170°C) a complete range of compounds is obtained: $Os(CO)_5$, $Os_3(CO)_{12}$, $Os_5(CO)_{16}$, $Os_5(CO)_{19}$, $Os_6(CO)_{20}$, and $Os_7(CO)_{21}$ (26,27). The cluster opening of $Os_6(CO)_{18}$ by ligand addition is very interesting but is not an effective route to triosmium compounds. In its reaction with CS_2 and CO, the cluster $Os_6(CO)_{18}$ gives $Os_4(CO)_{12}(CS)(S)$ and the trinuclear compounds $Os_3(CO)_8(CS)(S)_2$ and $Os_3(CO)_9(S)_2$, both made more easily from $Os_3(CO)_{12}$ (28). Other than these few examples, triosmium clusters are prepared from $Os_3(CO)_{12}$.

Virtually all Os_3 species are diamagnetic, air-stable, inert at room temperature, and comply with the 18-electron rule in most cases without straining one's credulity as to the number of Os—Os bonds or the order of these bonds. The easy application of simple elecon counting must mask an underlying complexity (it could not be otherwise), but simple predictions on the basis of the 18-electron rule are extremely powerful in Os_3 chemistry. There are a few systems requiring three-center–two-electron bonding (other than hydride bridges). Bridging phenyl has been found in $Os_3H(CO)_9$-$(\mu_3\text{-}Ph_2PCH_2PPhC_6H_4)$ (29) and in $Os_3(\mu\text{-}Ph)(PPh_2)(PPhC_6H_4)(CO)_8$ (30). Although they do not give unique predictions of structure, simple 18-electron considerations work very well; this is not true for higher clusters of osmium where the polyhedral skeletal electron pair (PSEP) and related approaches work much better (8,9,31–33)

From the above point of view it might seem that Os_3 chemistry is dull. This is not so because the very inertness of the system has allowed the isolation of interesting compounds otherwise unavailable. Their interest has sometimes arisen from their relation to chemisorbed species. The formation of the μ_3-ethylidyne ligand in $Os_3H_3(CCH_3)(CO)_9$ from *ethyl*ene and $Os_3(CO)_{12}$ via $Os_3H_2(C{=}CH_2)(CO)_9$(34–37) parallels the formation of CCH_3 on the Pt(111) surface on treatment with ethylene(38). Similarly the identification of μ_3-CHCH on Pt(111) and Pd(111) surfaces has been made by comparison of the vibrational spectra of the chemisorbed species with that of the ligand in $Os_3(CO)_{10}(\mu_3\text{-}C_2H_2)$ (39) and of CH_2 at W(111) and Ni(111) surfaces by infrared comparison with $Os_3H_2(\mu\text{-}CH_2)(CO)_{10}$ and $Os_3(CO)_{11}(\mu\text{-}CH_2)$ (40). Agostic bonding in $Os_3H(CH_3)(CO)_{10}$ allows the bridging methyl group to be a three-electron donor (41,42). This and the facile interconversion of the methyl in this compound with the methylene in the isomeric species $Os_3H_2(CH_2)(CO)_{10}$ were reactions first identified in the Os_3 area (43,44). Ortho metallation of pyridine (45), formation of $\mu_3\text{-}ortho$-phenylene (benzyne) by direct reaction of $Os_3(CO)_{12}$ with benzene (35,46), formation of $\mu_3\text{-}\eta^2\text{-}C{=}CH_2$ from ethylene (34–36), simple direct conversion of CO to $\mu_2\text{-}CH_2$ (47,48), and formation of $\mu_3\text{-}C_2O(11,12)$ and $\mu_3\text{-}BCO$ (14) have all been observed in triosmium chemistry. Other important areas that have been developed include the chemistry of μ_2-acyls (49), the capping of trinuclear clusters with other metal atoms, and the reactivity of the coordinatively unsaturated cluster hydride, $Os_3H_2(CO)_{10}$.

As far as one can tell, most Os_3 chemistry involves even-electron reactions of diamagnetic compounds, but it may be that radical chemistry will be more important once attention is paid to it; this is usually the case. The esr spectrum of $[Os_3(CO)_{12}]^-$ has been given (50), and this species is presumably involved in the Ph_2CO^--catalyzed substitution of $Os_3(CO)_{12}$. The radical anion is substitutionally labile.

II

TYPES OF SYSTEMS

A. *Number of Metal–Metal Bonds*

It is useful to classify compounds so that they may be compared and normal behavior distinguished from the abnormal. The "electron count" of clusters is a useful idea in this respect. This is equal to the number of valence electrons of the metal atoms (24 for Os_3) added to the number of electrons supplied by the ligands for metal–ligand bonding, allowing for overall charge on the molecule. The 18-electron rule is valuable in analyzing and predicting modes of ligand attachment, structure, reaction mechanisms, etc., and superficially applies well to isolable Os_3 clusters if some assumptions are made about the number of metal–metal bonds. The increasing number of non-18-electron systems (radicals and coordinatively nonprecise compounds) in organometallic chemistry in general only emphasizes the value of electron counting for considering isolable compounds. There is no overriding value in applying the PSEP approach to the small osmium clusters considered here.

Triosmium clusters with electron counts of 46, 48, 49, 50, or 52 have been described (see Table I). The 49-electron system $[Os_3(CO)_{12}]^-$ appears unique and has so far only been characterized by its esr spectrum (*50*). The 44-electron compound $Os_3(CO)_{10}$ has been considered as a reaction intermediate but without substantiating evidence (*51*). In established cases the varying

TABLE I

ELECTRON COUNT AND NUMBER OF OS—OS BONDS

Electron count	Os—Os bonds predicted	Examples	Occurrence
46	4	$Os_3H_2(CO)_{10}(15)$, $Os_3H(AuPPh_3)(CO)_{10}(52)$	Rare
48	3	$Os_3(CO)_{12}$, $Os_3H_2(CO)_{11}(53-55)$, and most other triosmium clusters	Very common
49	2.5	$[Os_3(CO)_{12}]^-(50)$	Unique
50	2	$Os_3H_2(CO)_{12}(56,57)$, $Os_3(OMe)_2(CO)_{10}(58)$, $Os_3H(AsMe_2)(C_6H_4)(CO)_9(59)$, $Os_3(S)_2(CO)_9(60)$, $Os_3(NO)_2(CO)_9\{P(OMe)_3\}(61)$, $Os_3H(CO)_{10}(CH{=}CHNC_5H_4)(62)$, $Os_3H(CO)_{10}(PEt_3)(CF_3C{=}CHCF_3)(63)$	Quite common
52	1	$Os_3(S)(CH_2S)(CO)_9(PMe_2Ph)(64)$, $Os_3H(S)(CH_2S)(Cl)(CO)_8(PMe_2Ph)(65)$	Rare

TABLE II

Os—Os Bond Lengths (Å)[a]

	Os$_3$H$_2$(CO)$_{10}$	Os$_3$HX(CO)$_{10}$		Os$_3$X$_2$(CO)$_{10}$	
		X = OMe	X = Cl	X = OMe	X = Cl
a	2.815(1)	2.813(1)	2.829(1)	2.815(3)	2.849(2)
b	2.815(1)	2.822(1)	2.836(1)	2.823(3)	2.855(2)
c	2.682(1)	2.812(1)	2.848(1)	3.078(3)	3.233(2)
Ref.	58,66–68	69	70	58	71

[a] The a and b are nonbridged edges, and c is bridged.

"electron-counts" may be associated with different numbers of Os-Os bonds or different bond multiplicities.

There have been difficulties in defining the number of metal–metal bonds when there are ligand bridges. The 18-electron rule gives bond orders for the bridged Os—Os pair in the compounds Os$_3$H$_2$(CO)$_{10}$, Os$_3$HX(CO)$_{10}$, and Os$_3$X$_2$(CO)$_{10}$ of 2, 1, and 0, respectively, if X is a three-electron donor group or atom. These orders appear to be supported by the Os—Os distances (Table II). It is unlikely, however, that there is any build up of bonding electron density along the Os—Os direction, but rather metal–metal interactions are likely to take place through the bridging atoms. Various treatments have been applied to Os$_3$H$_2$(CO)$_{10}$ and related systems (72–74), but a qualitative approach emphasizing a similarity with bridging in B$_2$H$_6$ is attractive (68). The OsH$_2$Os group involves four-center four-electron bonding employing d^1sp^3 hybrid metal orbitals with occupancy of the orbitals shown in Fig. 1. A more quantitative MO treatment supported this overall view but also identified an additional bonding interaction not found in diborane (74).

Octahedral geometries are closely maintained in the clusters in Table II, and direct metal–metal bonding would need to bisect these octahedral axes. It is a feature of many Os$_3$ clusters that octahedral geometry at the metal atoms is maintained as far as constraints such as ring formation will allow. These

FIG. 1. Occupied bonding orbitals in the dihydride bridges of Os$_3$H$_2$(CO)$_{10}$ in a four-center four-electron approach.

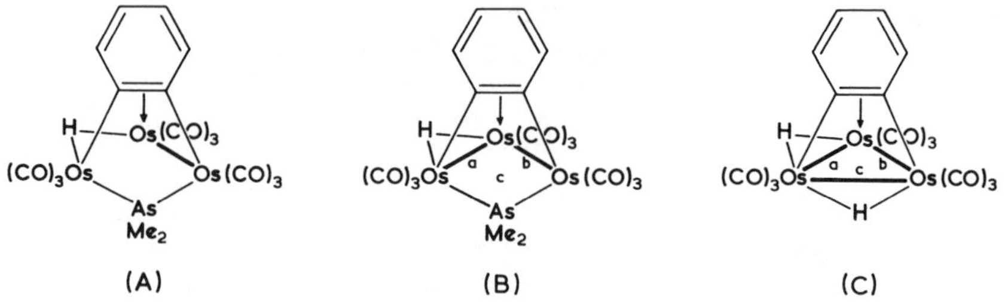

FIG. 2. Different representations of Os_3 clusters. Bond lengths (Å): *molecule B*: a = 2.946(4), b = 2.839(4), c = 3.929(5); *molecule C*: a = 2.866(2), b = 2.751(2), c = 3.026(2).

octahedral directions coincide with metal–metal vectors when there are no bridges across that pair of atoms but coincide with the metal to bridging ligand atom direction when bridges are present. Some authors have chosen to omit the metal–metal vectors in structural diagrams if these do not coincide with the octahedral axes. This leads to representations such as A in Fig. 2, which are aesthetically unappealing and possibly misleading. In this article an Os—Os bond will be drawn if the 18-electron rule requires one and if the Os—Os distance is below about 3.05 Å as it invariably is if metal–metal bonding is appropriate. In general distances greater than this are found where no bond is required by electron-counting considerations. Thus the compound $Os_3H(AsMe_2)(C_6H_4)(CO)_9$ will be represented as B in Fig. 2 (50 valence electrons) (*59*) and $Os_3H_2(C_6H_4)(CO)_9$ as C (48 valence electrons) (*20*). The metal–metal distances (a, b, and c) are clearly consistent with there being two and three metal–metal bonds, respectively, regardless of whether or not one should really consider the OsHOs bridge as involving three-center two-electron bonding.

Although very reliable for Os_3 clusters, the 18-electron rule is of limited use for larger ones. In the case of Os_3 clusters with added transition metal atoms (Section VIII) one cannot simply apply the rule assuming that each short M—M contact involves a two-electron bond, but despite that the rule works well in most cases.

B. *Main Structural Types of Os₃ Clusters*

1. *One Osmium–Osmium Bond*

This situation is found rarely and is only to be expected when strongly bridging ligands maintain the trinuclear cluster intact. Sulfur ligands have this property. The cluster $Os_3H_2(CO)_{10}$ reacts in refluxing CS_2 to give $\{Os_3H(CO)_{10}\}_2CH_2S_2$ (**1**) (Scheme 1) (*75*). The substituted compound

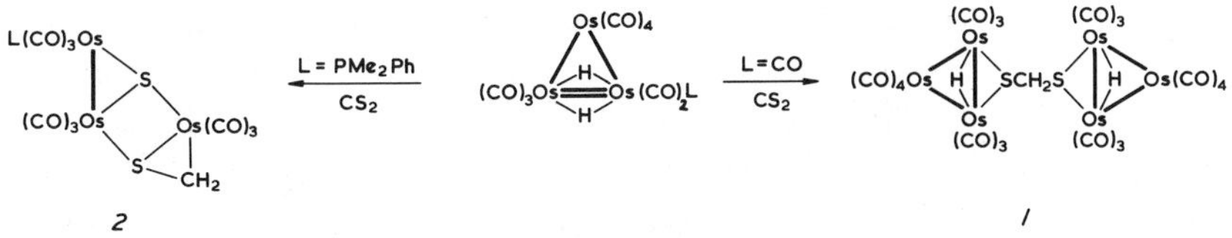

SCHEME 1.

$Os_3H_2(CO)_9(PMe_2Ph)$ reacts with a similar double hydrogen transfer to the CS_2 carbon atom to give some of the analog of **1** but C—S cleavage also occurs to give compound **2** (64,76). The S and the CH_2S (thioformaldehyde) ligands strongly bridge as four-electron donors and the cluster is maintained with only one Os—Os bond (2.855 Å) as required by the 18-electron rule. Compound **2** readily decarbonylates in refluxing hexane to give $Os_3(\mu_3\text{-}S)\text{-}$ $(\mu_3\text{-}CH_2S)(CO)_8(PMe_2Ph)$ with an extra Os—Os bond (64,76). The newly formed Os—Os bond may be cleaved again by forming an HCl adduct (**3**) with only one Os—Os bond (65). The compound $Os_3H(\mu_3\text{-}S)(\mu_3\text{-}CH_2S)\text{-}$ $(CO)_8(PMe_2Ph)(SnMe_3)$ formed by Me_3SnH addition probably also has one Os—Os bond (77).

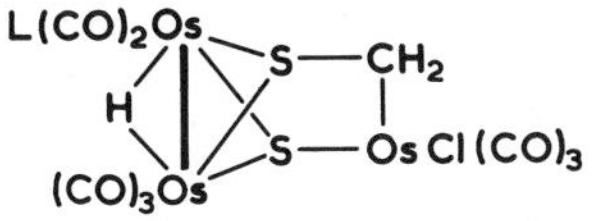

3

2. Two Osmium–Osmium Bonds in Linear Chain

Halogenation of $Os_3(CO)_{12}$ with Cl_2, Br_2, or I_2 (X_2) gives $Os_3X_2(CO)_{12}$ compounds which were originally thought to be of D_{4h} symmetry on the basis of three infrared active $\nu(CO)$ bands (78). The X-ray structure of compound **4** (X = I) shows that, although the Os_3 chain is linear, the I atoms are cis to the metal–metal bonds (79). There is a stereochemical change at the central osmium atom from cis to trans on formation from $Os_3(CO)_{12}$. The kinetics are first order in cluster and zero order in X_2, and a rapid preequilibrium mechanism was considered (80). Perhaps the initial Os—Os cleavage is

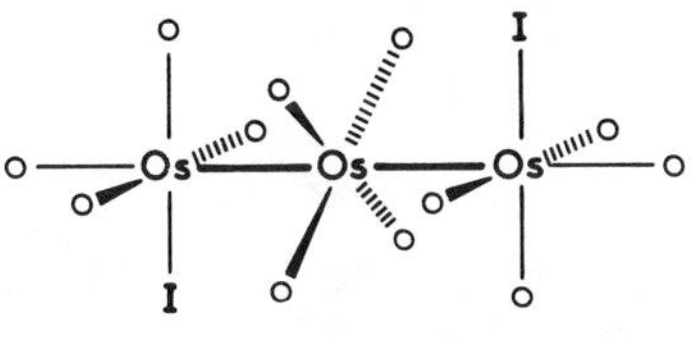

4 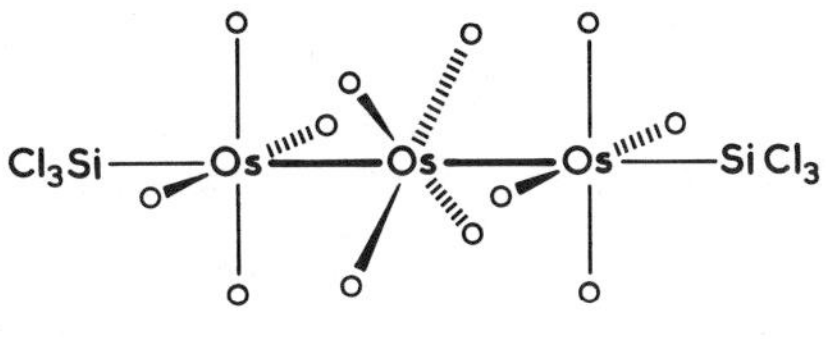   **5**

reversible prior to this cis to trans conversion. $Os_3H_2(CO)_{12}$, probably with a structure like that of 4 (56), reacts with CCl_4 or CBr_4 to give the corresponding dihalides (57). The dihalides $Os_3X_2(CO)_{12}$ decarbonylate thermally to $Os_3X_2(CO)_{10}$, in which the terminal metal atoms are linked by halogen bridges forming a triangular cluster (81).

A linear arrangement (D_{4h} ignoring Cl atoms) has been found for $Os_3(SiCl_3)_2(CO)_{12}$ (5) (82). Other clusters are closely related to 4 and 5 but do not have totally linear metal chains because of distortions caused by bridging ligands. The μ,η^2-vinyl bridge in $Os_3H(CO)_{10}(PEt_3)(CF_3C=CHCF_3)$ (6) distorts the octahedral geometry at the central metal atom leading to an Os—Os—Os angle of 162.3(1)° (63). Similar bends are observed in Os_3H-(μ-CH=CHNC$_5$H$_4$)(CO)$_9$L (7) [160.0(1)° for L = CO and 160.4(1)° for L = PMe$_2$Ph] (62). The cluster $Os_3H(CH=CHCOMe)(CO)_{10}$ (8) is isostructural with 7; their ir spectra in the $\nu(CO)$ region are very similar (83). In these examples the open "linear" structures are derived from closed triangular structures, and rather than having closed structures with three Os—Os bonds the vinyl groups are η^2-coordinated in compounds 6 to 8 as well as the ketonic group in compound 8.

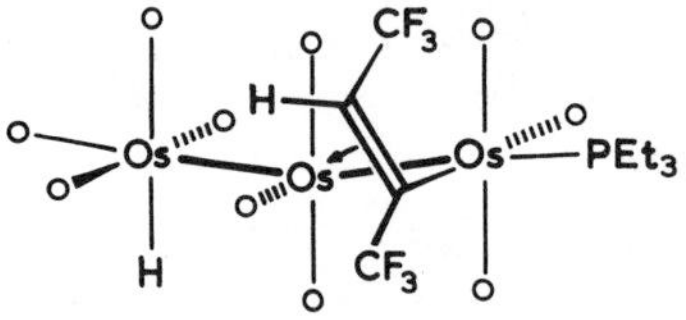

6

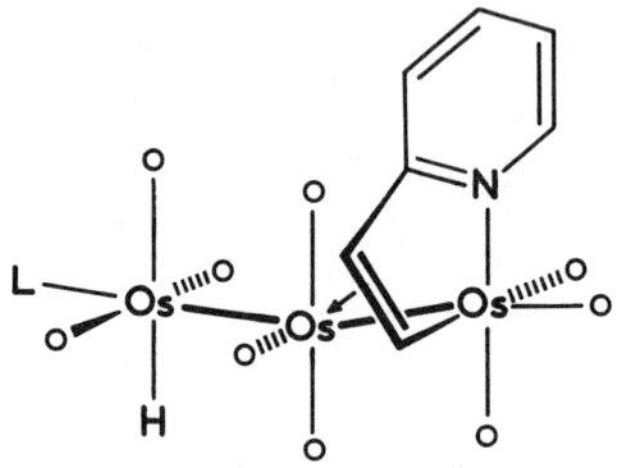

7

(L = CO or PMe$_2$Ph)

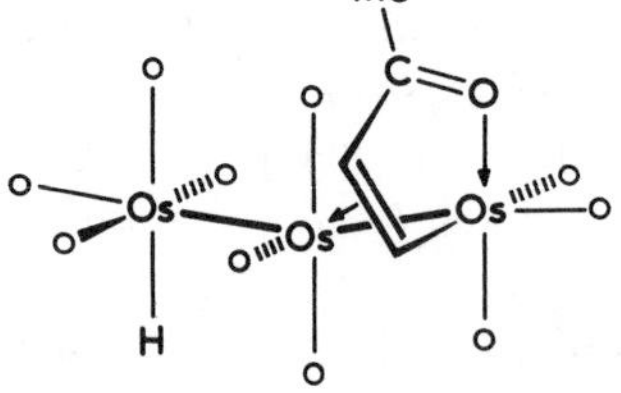

8

3. Two Osmium–Osmium Bonds with Terminal Atoms Bridged

There are four main ways to achieve bridging of terminal atoms in clusters with two Os—Os bonds. Structures A to D in Fig. 3 illustrate major bonding modes, and examples of each will be considered below.

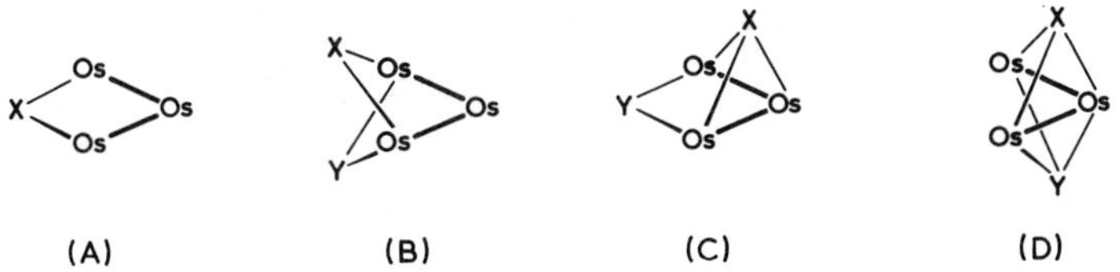

FIG. 3. Ways of bridging Os_3 clusters with two Os—Os bonds.

There are several compounds of type $Os_3(CO)_{12-x}L$ where $x = 0, 1,$ or 2 and L is a two-, four-, or six-electron donor, respectively, bridging two Os atoms which are not directly bonded, type A. The very interesting reaction of $Os_3(CO)_{11}(CH_2)$ with CO to give $Os_3(CO)_{12}(CH_2CO)$ (9), containing the

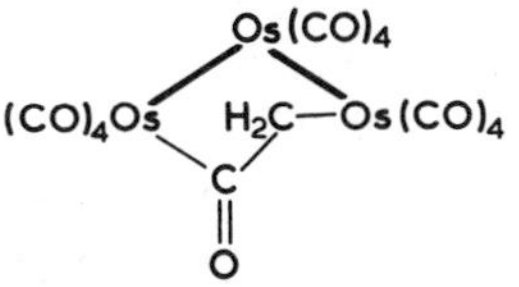

9

two-electron donating ketene ligand, is an extremely important example of the chemistry of μ-CH_2 groups. The incoming CO does not enter the CH_2CO ligand, and an intramolecular preequilibrium involving C—C coupling is likely (84). A closely related compound $Os_3(CO)_{12}L$ (10) (Scheme 2) results from CO addition (1500 psi, 25°C) to $Os_3(CO)_{11}L$ (11), itself formed by the addition of L ($MeO_2CN_2CO_2Me$) to $Os_3(CO)_{11}(MeCN)$. Thermal decarbonylation of 11 gives 12, as shown in Scheme 2; clusters 10 and 12 both have distances exceeding 4 Å between the bridged Os atoms. Coordination of the ester groups rather than construction of metal–metal bonds occurs to compensate for the loss of CO (85,86). A two-electron donor (MeCN) in $Os_3(CO)_{11}(MeCN)$ is displaced by allene, which is a four-electron donor in the product $Os_3(CO)_{11}(C_3H_5)$ (13); the cluster is opened and a long Os—Os distance (3.860 Å) is found (87). Compound 13 relates to 11 with the μ ligand donating one electron to the $Os(CO)_4$ group and three electrons to the Os $(CO)_3$ group.

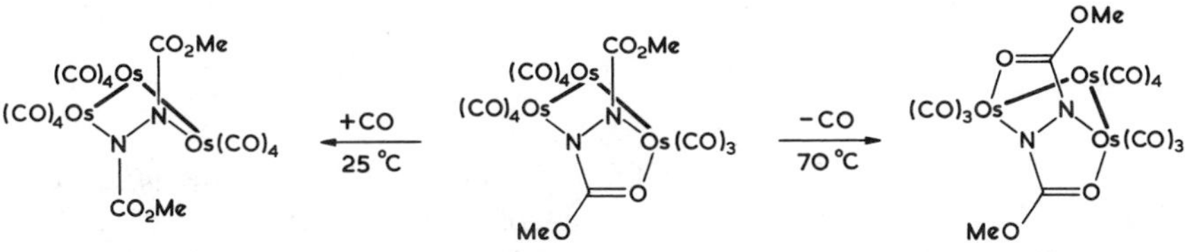

SCHEME 2.

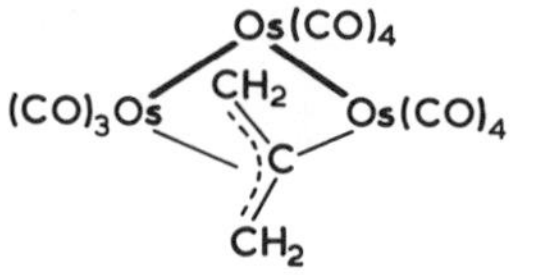

13 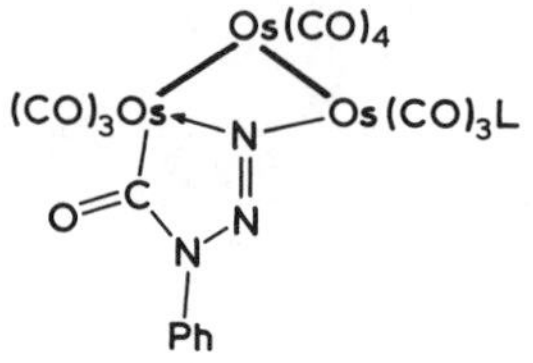   14

The clusters $Os_3(CO)_{10}L(NNNPhCO)$ (**14**) formed by adding PhN_3 to $Os_3(CO)_{11}L$ (L = MeCN or py) are also related to **11** and **13**. Since $Os_3(CO)_{12}$ does not react with PhN_3, the nucleophilic azide does not seem to attack CO directly but only after displacing L which then is recoordinated in product **14** (*88,89*).

Type B systems of general formula $Os_3(X)(Y)(CO)_{10}$ require X and Y to be three-electron donor ligands. $Os_3Cl_2(CO)_{10}$, formed by thermal decarbonylation of $Os_3Cl_2(CO)_{12}$, has a bridged Os—Os distance of 3.233(2) Å (Table II), too long for an Os—Os bond (*71*). Likewise $Os_3(\mu\text{-}NO)_2(CO)_9L$, where L = $P(OMe)_3$ (*61*) or NMe_3 (*90*), have long Os—Os distances of 3.217 and 3.197 Å, respectively. The NO bridges in these and their parent compound $Os_3(\mu\text{-}NO)_2(CO)_{10}$ are three-electron donors. In NO clusters which require three Os—Os bonds such as $Os_3(CO)_9(NO)_2$ (*91*) and $Os_3(CO)_8(NO)_2\{P(OMe)_3\}$ (*92*), the nitrosyl ligands are terminal, but since there are linear M—N—O bonds these are still three-electron donors.

Related SR clusters $Os_3(\mu\text{-}SR)_2(CO)_{10}$, synthesized by oxidative addition of R_2S_2 to $Os_3(CO)_{11}(MeCN)$ exist as a single isomer with nonequivalent SR ligands when first formed. This isomer requires heating to 95°C to give an isomer with equivalent SR groups, which implies that there are much higher barriers to inversion at sulfur than is normal. These isomers are believed to be **15** and **16**, respectively (*93*).

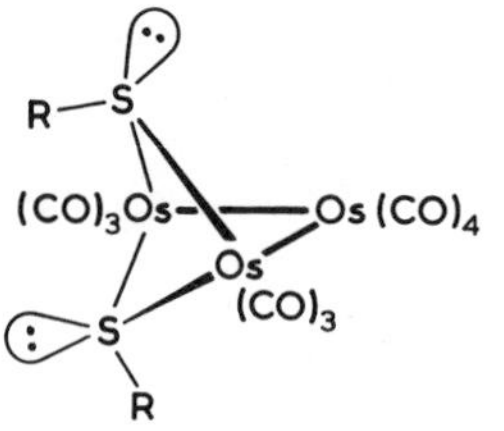

15 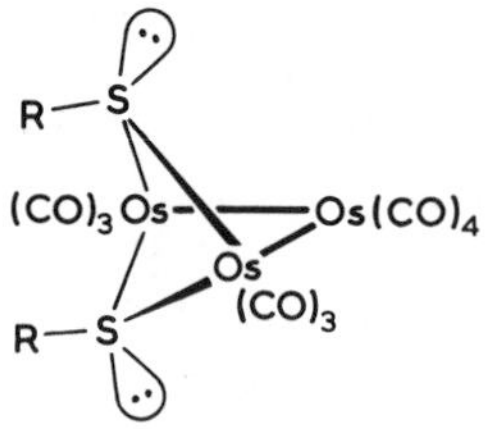   16

Isomers of a different origin are found for $Os_3(\mu\text{-}RCO)_2(CO)_{10}$ (*49*). Acyl ligands are commonly generated by attack of LiR at coordinated CO, and in this case $Os_3(CO)_{12}$ reacts with LiR (R = Me or Et) at 0°C in THF to give $Li_2[Os_3(RCO)_2(CO)_{10}]$, probably containing $\eta^1\text{-}RCO$. Oxidation with

Me_3OSbCl_6 or $CuBr_2$ gives the neutral vicinal diacyl compound (18), which subsequently isomerizes to a mixture of 17 (the geminal diacyl) and 18. An alternative route giving the mixed diacyl compound $Os_3(\mu\text{-MeCO})$-$(\mu\text{-EtCO})(CO)_{10}$ is the hydroformylation reaction of $Os_3H(\mu\text{-MeCO})(CO)_{10}$ with 136 atm $C_2H_4:CO$ (68:1) at 140°C (49).

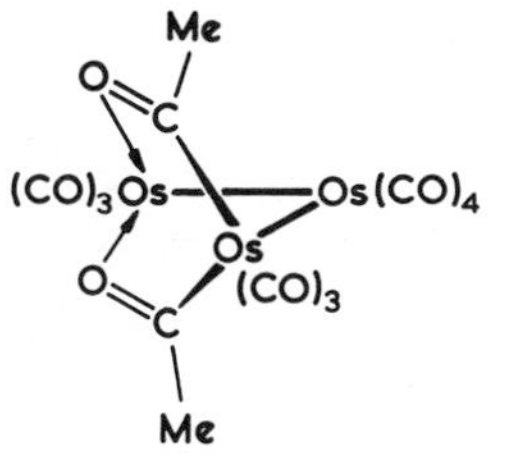

17 (gem) 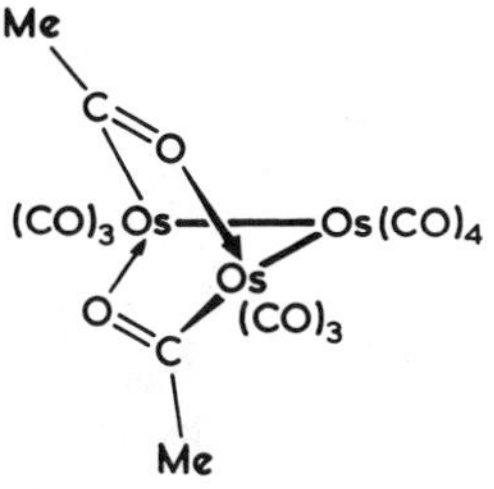   18 (vic)

The third type of system, C (Fig. 3), has both doubly and triply bridging ligands. $Os_3(CO)_{12}$ reacts at 75°C with the acetylenic phosphine $Ph_2PC\equiv CR$ (R = ^iPr or ^tBu) to give $Os_3(CO)_{11}(Ph_2PC\equiv CR)$ which in refluxing decalin undergoes decarbonylation and an internal oxidative addition to yield $Os_3(PPh_2)(C_2R)(CO)_9$ 19 which has a nonbonded Os—Os distance of 3.508(1) Å. The Ru_3 analog readily loses CO reversibly to give an octacarbonyl with three metal–metal bonds, but the Os_3 compound does not (94).

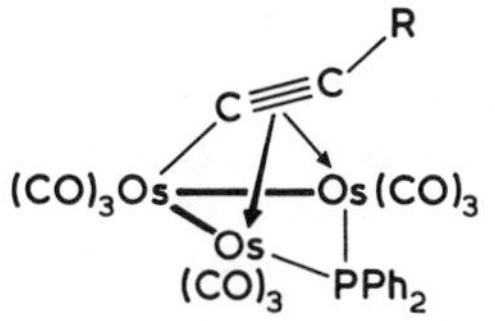

19

(R = ^iPr or ^tBu)

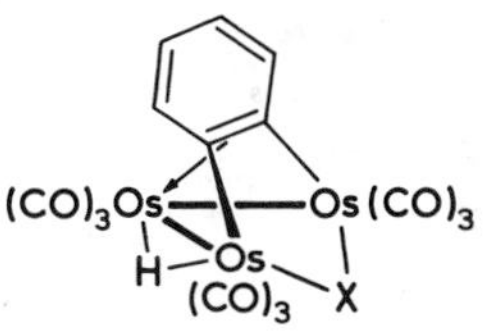

20

(X = SMe or $AsMe_2$)

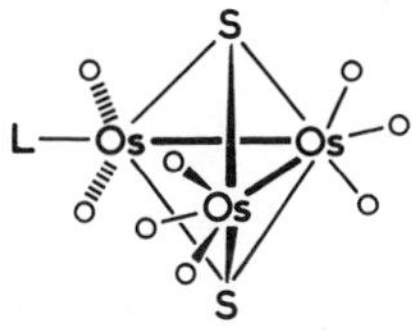

22

Similar cleavages within ligands (S—C and As—C bond fission, respectively) occur on reaction of MeSPh (95) and Me_2AsPh (59) with

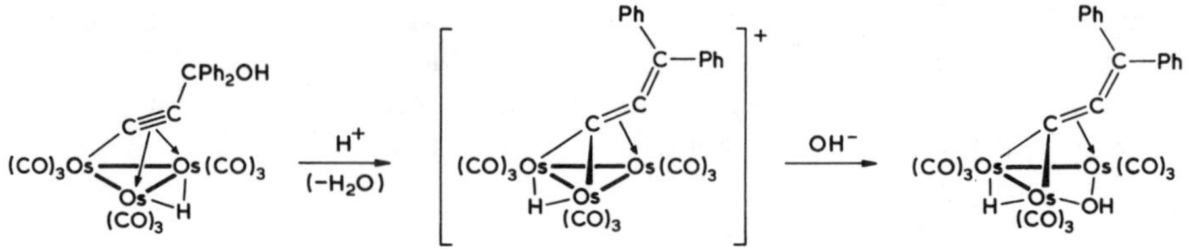

SCHEME 3.

$Os_3(CO)_{12}$, resulting in the related clusters $Os_3H(\mu\text{-}X)(\mu_3\text{-}C_6H_4)(CO)_9$ (**20**) (X = SMe or AsMe$_2$). Two Os—Os bonds are expected for these compounds with 50 valence electrons; the nonbonded Os—Os distances in these compounds are 3.791 (X = SMe) and 3.929 Å (X = AsMe$_2$). The sulfur compound shows two rapidly interconverting isomers in solution which are related by inversion at sulfur. Another example of a type C cluster is derived by a CF_3CO_2H-catalyzed isomerization of $Os_3H(C{\equiv}CCPh_2OH)(CO)_9$ which was formed from the reaction of $Os_3(CO)_{12}$ with the alkyne $HC{\equiv}CCPh_2OH$ at 130°C (Scheme 3). Compound **21** (Scheme 3) is an interesting example of a μ_3-allenylidene complex (*96*). A final example of type C is the cluster Os_3H-$(\mu_3\text{-}S)(\mu\text{-}CH{=}NR)(CO)_9$ (*97*).

The fourth type of cluster, D, has μ_3-capping ligands above and below the Os$_3$ plane. $Os_3(CO)_{12}$ reacts with elemental S, Se, or Te (E) in refluxing octane to give $Os_3H_2(\mu_3\text{-}E)(CO)_9$, $Os_4H_2(\mu_3\text{-}E)_2(CO)_{12}$ and a compound of type D, $Os_3(\mu_3\text{-}S)_2(CO)_9$ (*98,99*). The structures of compounds **22** with L = CO (*60*), PMe$_2$Ph (*60*), or CS (*100*) have been determined. The presence of only two Os—Os bonds is consistent with a valence electron count of 50. The cluster $Os_3(\mu_3\text{-}S)(\mu_3\text{-}CO)(CO)_9$ is superficially very similar, but, having 48 valence electrons, there are three Os—Os bonds and the molecular symmetry is C_{3v} (*101*). The compound **22** (L = CS) was prepared in low yield by uv photolysis of $Os_3H(SPh)(CO)_{10}$ with CS$_2$ (*101*), while **22** (L = CO) is a product of many Os$_3$ reactions involving sulfur ligands. For example, uv treatment of $Os_3H(SPh)(CO)_{10}$ with COS gives $Os_3(S)_2(CO)_9$ (40%) (*101*) and heating $Os_3H(SPh)(CO)_{10}$ under CO pressure gives some of the same compound (*102,103*). However, cluster reforming is common, and Os$_4$, Os$_5$, and Os$_6$ clusters with sulfur ligands are products from various high-temperature treatments of Os$_3$–S clusters. The same skeletal frame as in **22**, however, persists in some cases. Scheme 4 shows that uv expulsion of CO from **22** (L = CO) gives $Os_6(CO)_{16}(\mu_3\text{-}S)_4$; Os—Os bonds are formed between the two Os$_3$ units, but the Os$_3$S$_2$ cages are maintained. This condensation may be reversed, while hydrogenation also gives a cluster with the same type of Os$_3$S$_2$ cage as in **22** (*104*).

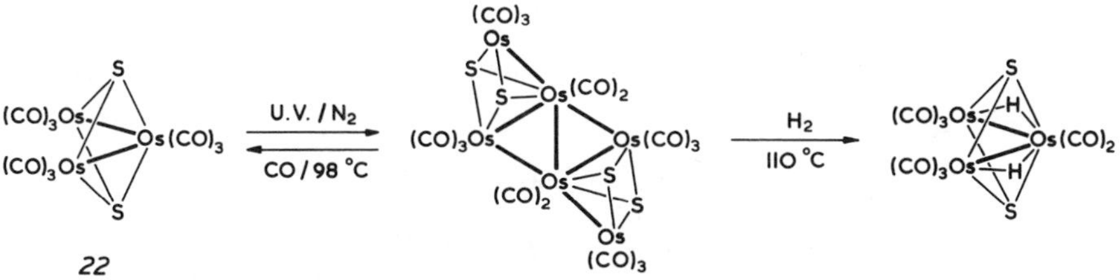

SCHEME 4.

Other clusters of type D are $Os_3(\mu_3\text{-}S)(\mu_3\text{-}NSiMe_3)(CO)_9$ formed from $Os_3(CO)_{12}$ and $S(NSiMe_3)_2$ (105), $Os_3(\mu_3\text{-}PEt)(\mu_3\text{-}C_6H_4)(CO)_9$ obtained by loss of C_6H_6 and CO from $Os_3(CO)_{11}(PEtPh_2)$ (106), and $Os_3(\mu_3\text{-}S)$-$(\mu_3\text{-}CH_2S)(CO)_8(PMe_2Ph)$ from the reaction of $Os_3H_2(CO)_9(PMe_2Ph)$ with CS_2 followed by thermal decarbonylation (64). In all these cases the two capping ligands are four-electron donors.

4. Three Osmium–Osmium Bond with No Bridges

Derivatives of $Os_3(CO)_{12}$ with monodentate ligands such as PR_3, $P(OR)_3$, RNC, RCN, NO, alkenes, and amines such as NH_3, py, and NMe_3 have been made. In most cases the basic structure of $Os_3(CO)_{12}$ is maintained with six axial and six equatorial ligands. Some modifications are found when there is a reduction in coordination number on introducing three-electron donating NO. These compounds will be considered in Section III.

5. Three Osmium–Osmium Bonds with Doubly Bridging Ligands

A major class of Os_3 compounds is of the type $Os_3H(\mu\text{-}X)(CO)_{10}$ where X is formally a three-electron donor. A large number of compounds of this type have been synthesized and over 50 X-ray structures established. There are various methods of synthesis. The main ones are oxidative addition of HX to $Os_3(CO)_{12}$, or better still $Os_3(CO)_{10}(MeCN)_2$ or some similar species with easily displaced ligands, insertion of molecules into Os—H bonds of $Os_3H_2(CO)_{10}$, and nucleophilic attack of Nu^- (amines, RLi, OR^-) at CO of $Os_3(CO)_{12}$ followed by decarbonylation and protonation to give $Os_3H(\mu\text{-}NuCO)(CO)_{10}$. There are many variations. Some of these clusters have a single atom (usually part of a polyatomic ligand) bridging the osmium atoms, but in other compounds there are two or three atom bridges. These clusters have been grouped in Tables III, IV, and V, and an indication is given of the synthetic method.

Insertion reactions of $Os_3H_2(CO)_{10}$ have been used to give the clusters $Os_3H(\mu,\eta^2\text{-}vinyl)(CO)_{10}$ from alkynes (150–153,160,161), and formamido

TABLE III

Clusters of the Type $Os_3H(X)(CO)_{10}$, Where X Is a Single-Atom Bridge[a]

X[b]	Further substituent at metal if any	Bridged Os—Os (Å)	Hydride location	Synthetic method	Ref.
CH		2.910(1)	√	$Os_3H(COMe)(CO)_{10}$ + (i) $[BHEt_3]^-$ (ii) CF_3CO_2H	13
CCH_2CHMe_2		2.870(2)	√	$Os_3H_2(CO)_{10}$ + ◁	107
$C=NH^tBu$		2.812(1)	X	$Os_3H_2(CO)_{10}$ + ^tBuNC	108
$C=CPhPMe_2Ph$		2.802(1)	X	$Os_3H(C{\equiv}CPh)(CO)_{10}$ + PMe_2Ph	108b
COMe	^tBuNC			$[Os_3H(CO)_{11}]^-$ + (i) $MeSO_3F$ (ii) ^tBuNC	109
$CHCH=NEt_2$		2.785(2)	√	$Os_3(CO)_{10}(MeCN)_2$ + Et_3N	110,111
$CHCH_2PMe_2Ph$		2.800(1)	√	$Os_3H(CH=CH_2)(CO)_{10}$ + PMe_2Ph	112,113
$C_6H_4PPhCH_2PPh_2$		2.747(1)	X	$Os_3(CO)_{10}(dppm)$, toluene/110°C	29
$N=CHCF_3$		2.813(4)	√	$Os_3H_2(CO)_{10}$ + CF_3CN	114
$N=CHCF_3$		2.803(5)	X	$Os_3H_2(CO)_{10}$ + CF_3CN	115
$N=CHCF_3$	PMe_2Ph (syn)	2.835(1)	X	$OsH(NCHCF_3)(CO)_{10}$ + PMe_2Ph	116
$N=CHCF_3$	PMe_2Ph (anti)	2.845(1)	X	$Os_3H(NCHCF_3)(CO)_{10}$ + PMe_2Ph	117
$NHCH_2CF_3$		2.792(1)	X	$Os_3H(NCHCF_3)(CO)_{10}$ + H_2	118
$NHSN(SiMe_3)_2$		2.783(1)	X	$Os_3H_2(CO)_{10}$ + $S(NSiMe_3)_2$	119
$NHSO_2C_6H_4Me$		2.814(1)	X	$Os_3H_2(CO)_{10}$ + RSO_2N_3	120
$NHSiMe_3$		2.803(1)	X	$[Os_3H(CO)_{11}]^-$ + Me_3SiN_3	121

14

Ligand		a	d (Å)	Reaction	Ref.
NHCONHNH$_2$		X	2.782(1)	Os$_3$(CO)$_{10}$(MeCN)$_2$ + NH$_2$NH$_2$	*122*
NHCONHNHMe		X	2.788(1)	Os$_3$(CO)$_{10}$(MeCN)$_2$ + NH$_2$NHMe	*122*
N=NC$_6$H$_4$Me		X	2.832(1)	Os$_3$H$_2$(CO)$_{10}$ + [RN$_2$]$^+$	*123*
NHN=CPh$_2$		✓	2.786(1)	Os$_3$H$_2$(CO)$_{10}$ + Ph$_2$CN$_2$	*124,125*
NHN=CMePh		X		Os$_3$H$_2$(CO)$_{10}$ + PhMeCN$_2$	*125*
*O*Me		X	2.812(1)	Os$_3$(CO)$_{12}$ + MeOH	*69*
*S*Et		X	2.863(2)	Os$_3$(CO)$_{12}$ + EtSH	*58*
*S*Me	C$_2$H$_4$	X	2.841(2)	Os$_3$H(SMe)(CO)$_9$ + C$_2$H$_4$	*126,127*
*S*CHPh$_2$		X	2.867(2)	Os$_3$H$_2$(CO)$_{10}$ + Ph$_2$CS	*128*
*S*CHPh$_2$	Ph$_2$C=S	X	2.877(2)	Os$_3$H$_2$(CO)$_9$(NMe$_3$) + Ph$_2$CS	*129*
*S*CH$_2$*S*		X	2.867(1) 2.871(1)	Os$_3$H$_2$(CO)$_{10}$ + CS$_2$	*75,130*
S(CH$_2$)$_3$*S*		X	2.841(3) 2.856(3)	Os$_3$(CO)$_{10}$(MeCN)$_2$ + HS(CH$_2$)$_3$SH	*131a*
$\overline{SCSCH_2CH_2N}$		X	2.842(3)	Os$_3$(CO)$_{10}$(MeCN)$_2$ + $\overline{SCH_2CH_2NHCS}$	*131b*
*S*CH=N*C*$_6$H$_4$F		X	2.870(1)	Os$_3$H$_2$(CO)$_{10}$ + FC$_6$H$_4$NCS	*97*
Cl		X	2.848(1)	Os$_3$H$_2$(CO)$_{10}$ + CH$_2$=CHCH$_2$Cl (55)	*70*
Br			2.851(1) 2.876(1)	Os$_3$H$_2$(CO)$_{10}$ + CH$_2$=CHCH$_2$Br (55)	*132*
Br	CF$_3$CH=CHCF$_3$		2.820(2)	[Os$_3$Br(CO)$_{10}$(CF$_3$CHCHCF$_3$)]$^-$ + MeSO$_3$F	*133*
*P*HPh		✓	2.888(5)	Os$_3$(CO)$_{12}$ + Me$_3$NO + PH$_2$Ph	*134*
*P*HPh		X	2.917(4)	Os$_3$(CO)$_{11}$(PH$_2$Ph), heat	*135,136*

^{a} Including closely related compounds in which there is further substitution at osmium.
^{b} Osmium-bonded atom(s) in italics.

15

TABLE IV
Clusters of the Type $Os_3H(X)(CO)_{10}$[a], Where X Is a Double-Atom Bridge

X[b]	Other substituent at metal if any	Bridged Os—Os (Å)	Hydride location	Synthetic method	Ref.
$PhCH_2CO$		2.884(2)	X	$Os_3(CO)_{10}(MeCN)_2 + PhCH_2CHO$	19
$MeCO$	$CMeOMe$	2.934(1)	✓	$Os_3H(MeCO)(CO)_{10} + $ (i) MeLi (ii) $MeSO_3F$	137,138
^iPrNHCO	^iPrNC	2.915(1)	X	$Os_3H_2(CO)_{10} + (^iPrN)_2C$	139
MeC_6H_4NHCO	PMe_2Ph	2.945(1)	X	$Os_3H_2(CO)_9(PMe_2Ph) + ArNCO$	140
$PhC{=}NMe$		2.918(1)	X	$Os_3(CO)_{12} + PhCH{=}NMe$	141,142
$PhN{=}CH$	$P(OMe)_3$	2.961(1)	✓	$Os_3H(CHNPh)(CO)_9 + P(OMe)_3$	143
$MeNHCO_2C{=}NMe$		2.924(1)	✓	$Os_3H(MeNCOH)(CO)_{10} + MeNCO$	144,145
$CF_3C{=}NH$	PMe_2Ph	2.956(1)	X	$Os_3H_2(CO)_9(PMe_2Ph) + CF_3CN$	117
$^iPrN{=}CNH^iPr$		2.904(1)	X	$Os_3H_2(CO)_{10} + (^iPrN)_2C$	139
$NCCHNHCH$		2.952(1)	X	$Os_3(CO)_{10}(MeCN)_2 + C_3N_2H_4$	146
$NCCHNHCMe$		2.957(1)	X	$Os_3(CO)_{10}(MeCN)_2 + C_3N_2H_3Me$	147
$NCCHCHCHC{-}C_5NH_4$		2.926(4)	X	$Os_3(CO)_{10}(C_8H_{14})_2 + $ bipy	148
$N{=}NPh$		2.895(2)	X	$Os_3H(\eta^1{-}N_2Ph)(CO)_{10}/h\nu$	149
$CH{=}CH_2$		2.845(2)	✓	$Os_3H_2(CO)_{10} + C_2H_2$	150
$CH{=}CHEt$		2.834(1)	X	$Os_3H_2(CO)_{10} + EtC_2H$	151
$CH{=}CH^tBu$		2.814(2)	X	$Os_3H_2(CO)_{10} + {}^tBuC_2H$	152
$CPh{=}CHPh$		2.821(1)	X	$Os_3H_2(CO)_{10} + PhC_2Ph$	153

[a] Including $Os_3H(X)(CO)_9L$, where L is a substituent at osmium.
[b] Osmium-bonded atoms in italics.

16

TABLE V

CLUSTERS OF THE TYPE $Os_3H(X)(CO)_{10}$, WHERE X IS A TRIPLE-ATOM BRIDGE

X^a	Other substituent at metal if any	Bridged Os—Os (Å)	Synthetic method	Ref.
O*CH*O		2.916(1)	$Os_3(CO)_{10}(C_8H_{14})_2$ + HCO_2H	*154*
$[Os_6(CO)_{17}$]-CO_2		2.895(7)	$[Os_3H(CO)_{11}]^-$ + $Os_6(CO)_{18}$	*155,156*
S*CH*S		2.968(1)	$Os_3(CO)_{10}(C_6H_8)$ + (i) $KHCS_2$ (ii) H^+	*157,158*
		2.978(1)		
S*CH*S	PMe_2Ph	2.854(1)	$Os_3H_2(CO)_9(PMe_2Ph)$ + CS_2	*158*
MeC_6H_4*N*CH*O*		2.909(1)	$Os_3H_2(CO)_{10}$ + RNCO	*140*
MeC_6H_4*N*CH*O*	PMe_2Ph	2.940(1)	$Os_3H_2(CO)_9(PMe_2Ph)$ + RNCO	*140*
Ph*NNN*H		2.900(2)	$Os_3H_2(CO)_{10}$ + PhN_3	*159*
H*NNN*H		2.923(2)	$Os_3H_2(CO)_{10}$ + Me_3SiN_3	*121*
Ph_2*P*C_6H_4	PPh_3		$Os_3(CO)_{12}$ + PPh_3	*30*

^{a} Os—bonded atoms in italics.

17

clusters $Os_3H(\mu\text{-}CH{=}NR)(CO)_{10}$ or $Os_3H(\mu\text{-}C{=}NHR)(CO)_{10}$ from iso-cyanides (*108a,143*). The dithioformato cluster $Os_3H(HCS_2)(CO)_9(PMe_2Ph)$ is formed from $Os_3H_2(CO)_9(PMe_2Ph)$ and CS_2, but a cluster with the SCH_2S ligand bridging two $Os_3H(CO)_{10}$ groups is formed from CS_2 and $Os_3H_2(CO)_{10}$ (*15,130,157,158*). CO_2 does not insert but $^iPrN{=}C{=}N^iPr$ gives $Os_3H(^iPrNHC{=}N^iPr)(CO)_{10}$ (*139*). Isocyanates RNCO insert with hydride transfer to carbon or nitrogen (*140,144,145*). Ketene ($CH_2{=}C{=}O$) inserts to give $Os_3H(OCH{=}CH_2)(CO)_{10}$ (*162,163*) while allene reacts readily with $Os_3H_2(CO)_{10}$, but a simple insertion product has not been obtained. Heteroallenes insert to give products with the heteroatom coordinated through a lone pair of electrons; no such electrons are available on allene itself, and further reaction with allene occurs. Sulfur dioxide adds to give $Os_3H_2(SO_2)(CO)_{10}$, but insertion to give a compound of the type considered here does not occur (*164*). The reactions of $Os_3H_2(CO)_{10}$ will be considered further in Section IV.

Oxidative addition reactions of $Os_3(CO)_{10}L_2$ (L = CO, MeCN, or alkene or L_2 = diene) with HX to give $Os_3H(X)(CO)_{10}$ have been applied to simple species such as HCl, HBr, HI (*16,21,22,71*), and RCO_2H (*22,154*), but oxidative additions are also possible when the hydrogen atom transferred to the metal is not acidic. *N*-Heterocycles add with hydrogen transfer from the 2 position, and this method has been applied with pyridine to give the μ-2-pyridyl complex $Os_3H(\mu\text{-}NC_5H_4)(CO)_{10}$ (*21,45,148*) and substituted pyridines, 1,2-, 1,3-, 1,4-diazines, 2,2′-dipyridyl (*148*), imidazole and substituted imidazoles (*146,147*) and quinoline (*45*) have been used in the same way. Alcohols, phenols (*21,69,156–159*), thiols (*21,22,46,58*), amines (*21,22,167,169,170*), formamidines RNHCH${=}$NR (*171–173*), amides (*19*), 2-aminopyridine, 2-pyridone (*19,171,173,174*), imines (*141,142*), aldehydes (*19,163,175,176*), and ketones (*163*) have all been used to add oxidatively to Os_3 clusters to give monohydrides of type $Os_3H(X)(CO)_{10}$. This provides a very general route to Os_3 derivatives since the initially formed hydrido clusters may be further modified by decarbonylation, substitution, or chemical modification of the group X in $Os_3H(X)(CO)_{10}$. For example, HNCO adds to $Os_3(CO)_{10}$-$(MeCN)_2$ to give $Os_3H(\mu\text{-}NCO)(CO)_{10}$, which readily undergoes nucleophilic addition by amines and alcohols to give $Os_3H(\mu\text{-}NHCOR)(CO)_{10}$, where R = R′O or R′NH (*177*). Hydrazines react with the isocyanato compound to give $Os_3H(\mu\text{-}NHCONHNHR)(CO)_{10}(R = H$ or Me), and the same compounds were obtained in low yield from $Os_3(CO)_{12-x}(MeCN)_x$ (x = 0, 1, or 2) and hydrazine or methylhydrazine (*177,122*).

The last main route to $Os_3H(X)(CO)_{10}$ is nucleophilic attack at CO. $Os_3(CO)_{12}$ reacts with $PhCH_2NH_2$ at high temperatures to give Os_3H-($\mu,\eta^2\text{-}PhNHCO)(CO)_{10}$, in which the carbonyl of the amide is bridging

through carbon and oxygen atoms (*169*). High temperatures are not necessary since $Os_3(CO)_{12}$ dissolves in neat Me_2NH at $0°C$ to give
$[Os_3(\eta^1-Me_2NCO)(CO)_{11}]^-$ which decarbonylates much more readily
than $Os_3(CO)_{12}$ itself to give $[Os_3(\mu,\eta^2-Me_2NCO)(CO)_{10}]^-$. Since the
hydrides in $Os_3H(X)(CO)_{10}$ are only very weakly basic, protonation of this
anion occurs even in the presence of amine to give $Os_3H(\mu,\eta^2-Me_2NCO)$-
$(CO)_{10}$, which was isolated after the removal of solvent (*178*). This reaction
applies generally to primary and secondary amines but also to alkoxide
or organolithium reagents (*179,180*). Further nucleophilic addition to Os_3H-
$(\mu,\eta^2-MeCO)(CO)_{10}$ by MeLi followed by alkylation with $MeSO_3CF_3$
gives $Os_3H(\mu,\eta^2-MeCO)(CMeOMe)(CO)_9$ (*137*). $Os_3(CO)_{12}$ reacts with an
excess of MeLi (or other organolithiums) to give $[Os_3(\eta^1-MeCO)_2(CO)_{10}]^{2-}$,
which is oxidized by Me_3OSbCl_6 or $CuBr_2$ to give $Os_3(\mu,\eta^2-MeCO)_2(CO)_{10}$
as vicinal and geminal isomers (*49*).

An overriding feature of the structures of $Os_3H(X)(CO)_{10}$ is that the metal
atoms closely maintain octahedral geometries and so the cluster geometry
depends upon the nature of X and primarily upon the number of atoms in the
bridge. With single atom bridges (Fig. 4, a), a structure is adopted with the
hydride and the bridging atom A positioned equally above and below
the metal plane respectively. As a result the CO ligands trans to H and A are
also positioned equally above and below this plane. The bridged Os—Os distance varies a little depending upon the bridging atom, being slightly less for
first row atoms than for second row ones (Table VI). The nonbridged Os—Os
distances vary very little in any of the compounds in Tables III to V. Figure 4 shows that as the bridge increases from one to two to three bridging
atoms, octahedral geometry is maintained by a twist which moves the
hydride ligand and the CO ligands trans to it into the metal plane. The CO
ligands trans to bridging atoms A and C in the three-atom bridge are now
strictly axial. These observations support the idea that any bonding between
the bridged osmium atoms is not direct even though they are no further
apart, in general, than the unsupported bonds to the $Os(CO)_4$ unit.

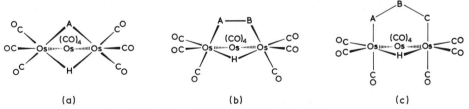

(a)   (b)   (c)

FIG. 4.   Views through the clusters of $Os_3H(X)(CO)_{10}$, where X is a one-, two-, or three-atom
bridge, showing coordination geometries at the bridged metal atoms. Examples are given in
Tables III to V.

TABLE VI

MEAN Os—Os DISTANCES IN CLUSTERS OF THE TYPE
$Os_3H(\mu\text{-}X)(CO)_{10}$[a]

Type	Bridged Os—Os (Å)	Nonbridged Os—Os (Å)
One-atom bridges		
First row atoms	2.810(23)	2.852(16)
Second row atoms	2.862(15)	2.851(14)
Two-atom bridges		
All examples	2.904(49)	2.878(19)
$\mu\text{-}\eta^2$-vinyl	2.829(12)	2.879(27)[b]
All except vinyl	2.929(25)	2.878(15)
Three-atom bridges	2.929(29)	2.892(15)

[a] See Tables III–V.
[b] On side of η^1-bond: 2.858(13) Å; on side of η^2-bond: 2.900(20) Å.

Bridged Os—Os distances with two-atom bridges are longer than those with one-atom bridges but the same as with three-atom bridges if one regards the μ,η^2-vinyl ligand as an exceptional case and so not included (Table VI). The α-carbon atom in the vinyl is bonded to both osmium atoms, shortening the Os—Os distance to that found for first-row single-atom bridges. There are two observed stereochemistries for the μ,η^2-vinyl compounds. Most are of type $Os_3H(\mu,\eta^2\text{-}trans\text{-}CH{=}CHR)(CO)_{10}$ with configuration A (Fig. 5) (150–152), but $Os_3H(\mu,\eta^2\text{-}cis\text{-}CPh{=}CHPh)(CO)_{10}$ (181) has configuration B (153). The Ph substituent at the α-carbon would clash with the axial CO of the $Os(CO)_4$ unit if A were adopted in the $CPh{=}CHPh$ compound. Rapid oscillation of the μ,η^2 ligands between the bridged osmium atoms occurs for both configurations A and B (153,182), but nucleophilic attack at the β-carbon atom by PMe_2Ph, etc. (112,183), occurs only for configuration A and not for B (184). The alkyne $CF_3C{\equiv}CCF_3$ inserts into $Os_3H_2(CO)_{10}$ to give $Os_3H(\mu_3,\eta^2\text{-}CF_3CCHCF_3)(CO)_{10}$, which is structurally unlike other compounds since it adopts structure C (185). This ligand can doubly bridge since it

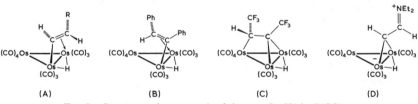

(A) (B) (C) (D)

FIG. 5. Structures of compounds of the type $Os_3H(vinyl)(CO)_{10}$.

does so in the PEt₃ adduct of C, $Os_3H(\mu,\eta^2\text{-}CF_3C=CHCF_3)(CO)_{10}(PEt_3)$ (**6**) (*63,185–187*). Note that the tertiary phosphine adds at osmium and not at the β-carbon atom as for most of the other vinyl compounds.

The compound $Os_3H(\mu\text{-CHCHR})(CO)_{10}$ where R = NEt₂ is another exceptional case. Since the NEt₂ group is a good π-donor, the vinyl converts to the alkylidene bridge as shown in Fig. 5, D, rather than adopting the normal μ,η^2-vinyl structure, A (*110,111*).

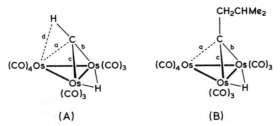

All the compounds in Tables III to V have the ligands H and X bridging the same pair of metal atoms except $Os_3H(CHS_2)(CO)_9(PMe_2Ph)$ (*158*). In most PR₃-substituted derivatives of $Os_3H(X)(CO)_{10}$ the phosphine is coordinated at one of the bridged Os atoms, but in this compound substitution is at the Os(CO)₄ group. As a consequence of preferential bonding of the hydride ligands to PR₃-substituted metal atoms, the hydride bridges as shown in **24** rather than as in the unsubstituted form **23** (*158*). Although the hydride ligands were not located, their positions were inferred from Os—Os lengths and Os—Os—CO angles.

An interesting comparison can be made between the compounds $Os_3H(CR)(CO)_{10}$ [R = H (*13*) or CH₂CHMe₂ (*107*)] (see Fig. 6 and Table III). The distances of the alkylidyne carbon from the third osmium atom are very different. Whereas the distance of 2.640(26) Å when R = CH₂CHMe₂ is

FIG. 6. Comparison of structures of $Os_3H(CR)(CO)_{10}$. Interatomic distances (Å): *molecule A*: a = 2.353(10), b = 2.011(12), c = 2.003(11), d = 2.36(14); *molecule B*: a = 2.640(26), b = 2.020(24), c = 1.966(22).

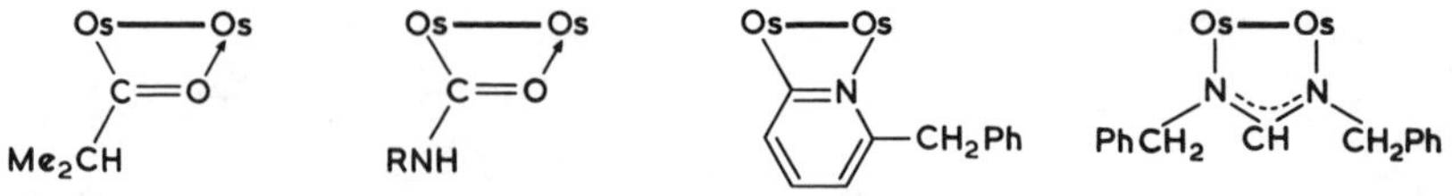

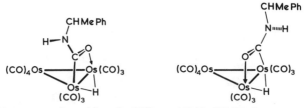

FIG. 7. Bridges containing diastereotopic atoms in Me_2CH or $PhCH_2$ groups in compounds of the type $Os_3H(\mu-X)(CO)_{10}$.

well beyond bonding distance, the CH ligand with a corresponding distance of 2.353(10) Å could be regarded as a semi-μ_3-bridging ligand instead of a μ_2 ligand. The distance seems to depend upon the requirement of the sp^2-hybridized CR atom to be stablized by accepting electrons from the $Os(CO)_4$ group. Although the C—H bond is tilted over toward this metal atom, there does not appear to be donation from the C—H bond as is believed to be the case in $[Fe_4H(CH)(CO)_{12}]^-$ (188), that is of the usual agostic type of bonding (189). In this case the CH is electrophilic and is attacked by 4-methylpyridine or hydride ions. The clusters $Os_3H(CR)(CO)_{10}$ (R = $SiMe_3$ or Ph) are also known, but their structures have not been reported (190).

An interesting aspect of compounds of type B and C in Fig. 4 is that, when the ends of the ligand are different, the cluster is chiral. Chirality is not maintained on an nmr time scale for $Os_3H(\mu,\eta^2\text{-vinyl})(CO)_{10}$ since the vinyl ligand rapidly oscillates to generate a plane of symmetry (153,182). In most cases, however, where bonding is through lone pairs on the heteroatoms the ligand is rigidly fixed. In the nmr spectra of the clusters $Os_3H(\mu-X)(CO)_{10}$, where X is a bridging ligand such as those shown in Fig. 7, separate signals for the diastereotopic atoms within the isopropyl or benzyl groups are observed. The ligands in these cases do not turn about between the bridged metal atoms. The introduction of a chiral ligand can then lead to diastereomers which do not rapidly interconvert. $Os_3(CO)_{12}$ dissolves in (+)-1-phenylethylamine to give the diastereomers shown in Fig. 8, which give separate sets of ^{1}H-nmr signals and can be separated by tlc (SiO_2) (180). Similar mixtures of

FIG. 8. Diastereomers formed from $Os_3(CO)_{12}$ and $PhMeCHNH_2$ that are separable on SiO_2 (180).

diastereomers are formed by reacting (S)-$(-)$-nicotine with $Os_3(CO)_{12}$ to give $Os_3H(C_{10}H_{13}N_2)(CO)_{10}$ as isomers. Metallation occurs at each of the nonequivalent positions ortho to the pyridine nitrogen atom to give positional isomers, each of which exists as diastereomers. In this case the isomers are only partially separable (*192*).

There are other specific compounds that should be included in this section. One product from the reaction of $Os_3(CO)_{12}$ and diphenylacetylene is $Os_3(C_4Ph_4)(CO)_9$ (**25**) in which there is a metallacyclopentadiene ring connecting two of the three Os atoms (*193*).

$(CO)_4Os$

Ph

Ph

$(CO)_2Os$ ——— $Os(CO)_3$

Ph

Ph

25

The final class of compound in this section is that with two-electon donor alkylidene bridges. Compounds of this type such as $Os_3(\mu\text{-}CH_2)(\mu\text{-}CO)(CO)_{10}$ and $Os_3(\mu\text{-}CHSiMe_3)(\mu\text{-}CO)(CO)_{10}$ will be considered in Section VI (*11,12,40,47,48,194–198*).

6. *Three Osmium–Osmium Bonds with Triply Bridging Ligands*

Interest in μ_3 ligands lies in the variety of molecules that may be used, in the variation of the number of electrons donated by the bridge, and in the potential the ligands have to rotate with respect to the three metal atoms. Many compounds may be described by the general formula Os_3H_{6-n}-$(\mu_3\text{-}L)(CO)_9$ where n is the number of electrons (3 to 6) donated by L (Fig. 9). Sometimes there are variants of this which result from replacement of CO or hydride by other ligands or by protonation or deprotonation to give cationic or anionic clusters, but even so many compounds are of the types A to D in Fig. 9. These compounds will be discussed in order of increasing n.

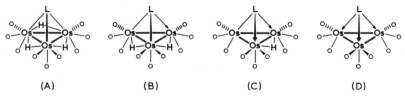

(A)   (B)   (C)   (D)

Fig. 9. Compounds of the type $Os_3H_{6-n}(\mu_3\text{-}L)(CO)_9$, where L is an n-electron donor ($n = 3$ to 6).

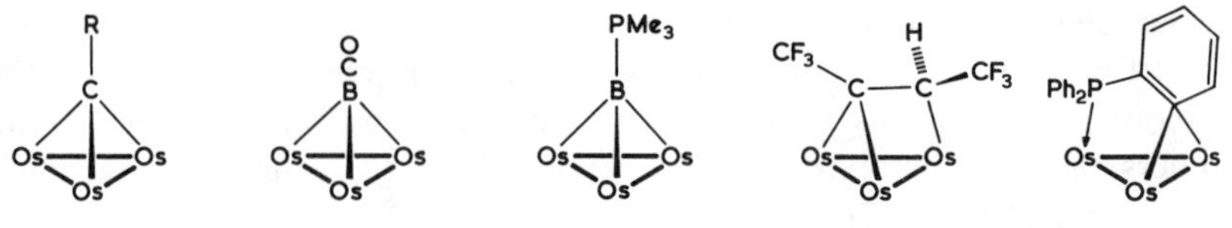

FIG. 10. Triply bridging ligands as three-electron donors in $Os_3H_3(\mu_3$-L)(CO)_9$ or $Os_3H(\mu_3$-L)(CO)_{10}$.

a. *Three-electron donors.* Several examples of μ_3-alkylidyne complexes A (Fig. 10) have been described: $Os_3H_3(\mu_3$-CR)(CO)_9$ where R = H *(41,163,198)*, CH_3 *(34,37)*, Cl, Br *(199,200)*, or OMe *(200–202)*. The area is far less developed than that of $Co_3(\mu_3$-CR)(CO)_9$. The dihydride $Os_3H_2(CO)_{10}$ reacts with CH_2N_2 to give $Os_3H_2(\mu$-CH_2)(CO)_{10}$ and $Os_3H(\mu$-CH_3)(CO)_{10}$ which thermally decarbonylate to give $Os_3H_3(\mu$-CH)(CO)_9$ *(41)*. The same CH compound is formed from the thermolysis of the isomers of Os_3H-$(\mu$-X)(CO)_{10}$, where X = MeCO or CH_2=CHO *(163,198)*, and also in low yield from the high-temperature reaction of $Os_3(CO)_{12}$ with dimethylaniline *(203)*. Presumably in the last case the CH ligand is derived from the *N*-methyl groups since $(4\text{-}Me_2NC_6H_4)_2CH_2$ is formed catalytically by methyl transfer under these conditions.

Ethylene reacts with $Os_3(CO)_{12}$ to give $Os_3H_2(\mu$-C=CH_2)(CO)_9$, which reacts slowly with H_2 (1 atm) at 125°C to give $Os_3H_3(\mu$-CCH_3)(CO)_9$ *(34,37)*. While the Ru analog was established by X-ray and neutron diffraction, that of Os was studied by nematic-phase nmr. Assuming certain interatomic distances (H—H in Me groups, C—C, Os—C, and Os—Os), the hydride positions were determined (Os—H 1.82 Å, Os—H—Os 103°) *(37)*. The COMe compound is formed on hydrogenation (1 atm, 120°C) of $Os_3H(COMe)(CO)_{10}$, itself formed by $MeSO_3F$ treatment of $[Os_3H(CO)_{11}]^-$ *(202)*. Treatment of $Os_3H_3(COMe)(CO)_9$ with BBr_3 gave $Os_3H_3(CBr)(CO)_9$ *(200)*.

Reaction of $Os_3H_2(CO)_{10}$ with B_2H_6 in CH_2Cl_2 in the presence of thf gave $[Os_3H_3(CO)_9(\mu_3$-CO$)\}_3B_3O_3$ (26), which contains a B_3O_3 ring linking the three Os_3 clusters. The hydride ligands were not located crystallographically but were observed in the nmr spectrum, and so the compound relates closely to $Os_3H_3(COMe)(CO)_9$ and, like the methoxymethylidyne compound, reacts with BX_3 to give $Os_3H_3(\mu_3$-CX)(CO)_9$ (X = Cl or Br). With BF_3 in benzene it appears that a transient carbocation is formed since $Os_3H_3(\mu_3$-CPh)(CO)_9$ is obtained *(199)*. The oxygen atoms bonded to boron in 26 are probably derived from thf since butane was shown to be a by-product. In the absence of thf (using B_2H_6 and BH_3NEt_3 in CH_2Cl_2) or on carrying out the reaction in dilute solution a different product was obtained: the carbonyl borylidyne compound $Os_3H_3(\mu_3$-BCO)(CO)_9$, B in Fig. 10 *(14)*. This compound is of C_{3v}

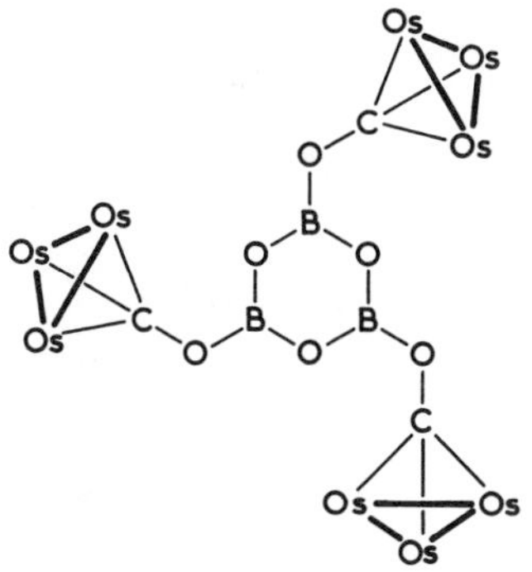

26 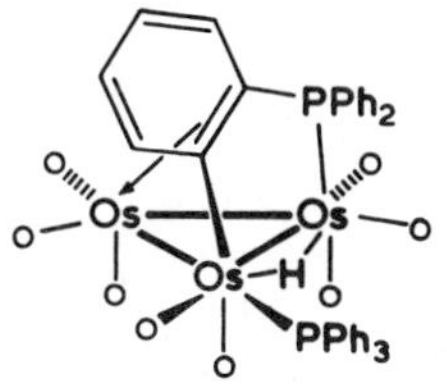   27

symmetry (vertical BCO). There is probably some B—C $p\pi-p\pi$ bonding since the bond is short [1.469(15) Å], but this is not enough to reduce $v(CO)$ for the BCO ligand below 2120 cm^{-1}. The CO is easily replaced by PMe$_3$ to give Os$_3$H$_3(\mu_3$-BPMe$_3)(CO)_9$, C in Fig. 10 (14). The μ_3-CF$_3$CCHCF$_3$ complex Os$_3$H(CF$_3$CCHCF$_3)(CO)_{10}$ contains the three-electron donating triyl ligand, D, but there is no evidence for such a ligand in the Os$_3$H$_3(\mu_3$-L)(CO)$_9$ series (185–187).

The compound Os$_3$H(μ_3-PPh$_2$C$_6$H$_4)(CO)_8$(PPh$_3$) (27) is formally a 46-electron molecule if the μ_3 ligand is a three-electron donor (204). This would imply electron deficiency or multiple metal–metal bonding. However, the aryl bridge is very unsymmetrical (Os—C 2.34 and 2.16 Å) and may actually involve a μ,η^2-aryl, as established in 1984 for the first time in the cluster MoRhPt(CO)$_2(\mu$-PPh$_2)(\mu,\eta^2$-C$_6$H$_5)(PPh$_3)_2$(C$_5H_5$) (205). Full crystallographic details of 27 have not been given, but distortions apparent in the illustration in the preliminary communication support the view that the PPh$_2$C$_6$H$_4$ ligand is a five-electron donor as illustrated for 27 rather than as E in Fig. 10.

b. *Four-Electron Donors.* Examples of four-electron donating μ_3 ligands are common (see Fig. 11). Type A is typified by the molecules Os$_3$H$_2$-(μ_3-X)(CO)$_9$, where X = S (46,206–209), NPh (170,159), NMe (144,212), NCH$_2$CF$_3$ (118,213), or PPh (134,135). The molecule for which X = O has not been made although the precursors Os$_3$H(OH)(CO)$_{10}$ and Os$_3$H(SH)(CO)$_{10}$ are both known (22).

The molecule with X = S is simply and quantitatively made by passing H$_2$S though a refluxing octane solution of Os$_3$(CO)$_{12}$ (46). The X-ray and neutron diffraction structures of Os$_3$H$_2$(S)(CO)$_9$ (206) and the X-ray structure of the anion [Os$_3$H(S)(CO)$_9]^-$ (207) are known. In the neutral compound the hydride ligands occupy octahedral sites at osmium so that Os—H—Os

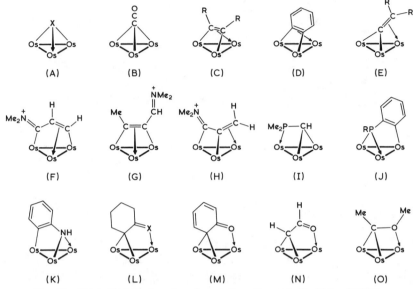

FIG. 11.   Triply bridging ligands as four-electron donors in $Os_3H_2(\mu_3\text{-L})(CO)_9$.

groups are best considered to be bound by open three-center two-electron bonds with no direct Os—Os bond as discussed earlier for compounds of type $Os_3H(\mu\text{-X})(CO)_{10}$. Representation **28** shows the octahedral environment at

$$28$$

each osmium with the longer Os—Os vectors not drawn as bonds. The lack of direct Os—Os bonds is supported by photoelectron data and calculations (*208*), although as described earlier it is usually convenient to consider the Os—H—Os bridges as involving metal–metal bonding and to draw in these bonds. In a vibrational analysis $v(M\text{—}S)$ and $v(M\text{—}M)$ absorptions have been assigned and a comparison made between $\mu_3\text{-}S$ and chemisorbed sulfur at a metal surface (*209*). The ^{13}C-nmr spectra of $M_3H_2(\mu\text{-S})(CO)_9$ (M = Ru or Os) are best interpreted in terms of dynamic hydride migrations and localized axial–equatorial exchange leading to complete CO exchange (*210*).

Hydride migration over all three Os—Os edges is supported by changes with temperature of the ^{187}Os satellites on the hydride nmr signals, which show single satellites with $J_{OsH} = 20.5$ Hz at room temperature but a pair of satellites with $J_{OsH} = 27.8$ and 31.9 Hz at $-60°C$ (211).

The clusters $Os_3H_2(NR)(CO)_9$ are not so straightforwardly synthesized from $Os_3(CO)_{12}$ and RNH_2. Aniline gives $Os_3H(NHPh)(CO)_{10}$ which ortho metallates on decarbonylation at $150°C$ to give $Os_3H_2(NHC_6H_4)(CO)_{10}$ rather than its isomer $Os_3H_2(NPh)(CO)_9$ (167,170). Only at $196°C$ is a little of the NPh compound formed (170). This compound is better made by loss of N_2 and CO from $Os_3H(NHNNPh)(CO)_{10}$ at $85°C$, and its X-ray structure was established (159). Thermolysis of $Os_3H(MeNHCO)(CO)_{10}$ or of $Os_3H(MeNCHO)(CO)_{10}$ at $150°C$ gives $Os_3H_2(NMe)(CO)_9$ (144), and its X-ray structure has also been determined (212). Insertion of CF_3CN into $Os_3H_2(CO)_{10}$ gives $Os_3H(CF_3CH{=}N)(CO)_{10}$ as the major product which may be hydrogenated (49 atm, $140°C$) to give three products: $Os_3H(NHCH_2CF_3)(CO)_{10}$ and two compounds containing μ_3-NCH_2CF_3 bridges, $Os_3H_2(NCH_2CF_3)(CO)_9$ and $Os_3H_4(NCH_2CF_3)(CO)_8$. The crystal structures of all three are known, the octacarbonyl being a rare example of a cluster with a terminal hydride (118,213).

Direct reaction of $Os_3(CO)_{12}$ with PH_2Ph gives simple substitution derivatives, $Os_3(CO)_{10}(PH_2Ph)_2$ and $Os_3(CO)_{11}(PH_2Ph)$, as well as PHPh bridged compounds and $Os_3H_2(PPh)(CO)_9$. Ortho metallation products are largely insignificant, but $Os_3H_2(PHC_6H_4)(CO)_9$ is obtained as a low yield product (134,135) and this presumably contains the bridge (J).

The ketenylidene ligand in the cluster $Os_3H_2(\mu_3\text{-}CCO)(CO)_9$, B in Fig. 11, is formed by C—C bond formation on heating $Os_3(CO)_{11}(CH_2)$ (11,12,196) and is oriented vertically (12) as originally proposed but not established for $[Co_3(\mu_3\text{-}CCO)(CO)_9]^+$ (214). In contrast, however, $[Fe_3(\mu_3\text{-}CCO)(CO)_9]^{2-}$ has an essentially linear CCO group (C—C—O 173°) tilted by 33.5° out of the direction perpendicular to the metal plane (215). In $[CoFe_2(\mu_3\text{-}CCO)(CO)_9]^-$ the ligand is also tilted but only by 24° from the vertical (216). Within experimental error there are no differences between the C—C and C—O distances for the ketenylidene ligand in these different orientations. So far no adequate account has been given of the reasons for these different orientations, but the neutral CCO ligand is a four-electron donor in each case and presumably only small energies are required for it to reorientate.

The vinylidene ligand in $Os_3H_2(C{=}CR_2)(CO)_9$ (E) adopts a tilted orientation in established cases. The 1,1 compound $Os_3H_2(\mu_3\text{-}C{=}CH_2)(CO)_9$ is formed by direct reaction of ethylene (1 atm) with $Os_3(CO)_{12}$ at $125°C$ in preference to the 1,2 isomer $Os_3H_2(\mu_3CH{=}CH)(CO)_9$ (34–36,46). Substituted vinylidene compounds are also obtained by inserting terminal alkynes RC_2H into $Os_3H_2(CO)_{10}$ to give $Os_3H(\mu,\eta^2\text{-}CH{=}CHR)(CO)_{10}$, which decarbonylates thermally to $Os_3H_2(\mu_3,\eta^2\text{-C}{=}CHR)(CO)_9$ [R = H (161), Me, Et, ^tBu, Ph (218), or ^iPr (107)]. The direct oxidative addition of the appropriate alkene to $Os_3(CO)_{12}$, as with ethylene, is rarely a good route but is successful when the 1,2 isomer cannot be formed. Thus methylenecyclobutane reacts with $Os_3(CO)_{12}$ at 145°C to give $Os_3H_2(C{=}\overline{CCH_2CH_2CH_2})(CO)_9$ (219). The dynamic behavior of these compounds will be described in Section VI.

Vinylidene ligands are also formed from $AsMe_2CH{=}CH_2$ (220). The ligand breaks down to give $Os_3H(AsMe_2)(C_2H_2)(CO)_9$, which exists as two isomers containing $C{=}CH_2$ and $CH{=}CH$, respectively. The related compound $Os_3H(PEt_2)(C_2H_2)(CO)_9$ has only been observed in the $CH{=}CH$ form on thermal breakdown of $Os_3H(Et_2PCH{=}CH_2)(CO)_9$ (221). These compounds probably contain only two Os—Os bonds and should not really be in this section. This is also true for $Os_3H(OH)$ (μ_3,η^2-C$=$C$=$CPh$_2$)(CO)$_9$, which contains an allenylidene ligand closely related to vinylidene (96).

29   30   31

μ_3-Alkyne compounds of type C may be prepared from the alkyne or the alkene. Oxidative addition of 1,2-disubstituted alkenes at $Os_3(CO)_{12}$ can give compounds of type 29 directly [cyclopentene (35), cyclooctene (217), but-2-ene (217)]. Reactions of $Os_3(CO)_{10}(MeCN)_2$ (or related compounds) with alkyne (C_2H_2, RC_2H or RC_2R') gives $Os_3(CO)_{10}(\mu_3,\eta^2$-alkyne), and this seems to be generally applicable even with functionalized alkynes such as $HOCH_2C{\equiv}CCH_2OH$ (223). A less general route is by reaction of $Os_3H_2(CO)_{10}$ with the alkyne, which sometimes gives significant amounts of $Os_3(CO)_{10}(\mu_3,\eta^2$-alkyne) as well as the insertion product (161,181). Examples of clusters of this type contain the alkynes C_2H_2 (22,161), MeC_2Me (161), MeC_2H (161,218), PhC_2Ph (181,218,224), $^tBuC_2H$ (152), PhC_2H (22), EtC_2Et (224), and $HOCH_2C_2CH_2OH$ (223). Most compounds of this type appear to have structure 30 with a μ-CO (~1850 cm^{-1}), but the one with a known structure is $Os_3(CO)_{10}(PhC_2Ph)$, which appears anomalous in having only semibridging CO ligands as shown, compound 31 (181,223). The attachment of the alkyne is normal, however, with two σ-Os—C bonds and one η^2-alkene bond. Vacuum sublimation at 250°C gives the coordinatively unsaturated red compound $Os_3(CO)_9(PhC_2Ph)$ (224). The structure 32 is based on those of $Fe_3(CO)_9(PhC_2Ph)$ (225) and $Fe_2Ru(CO)_9(PhC_2Ph)$ (226). A closo trigonal

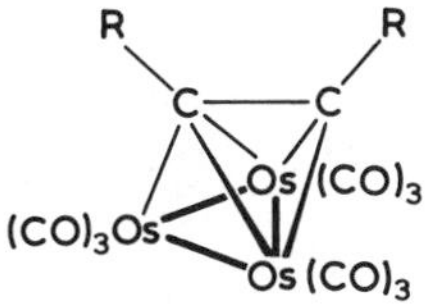

32

bipyramidal rather than a nido square pyramidal M_3C_2 geometry is found as expected when there are two less valence electrons. Compound **32** reacts with CH_2N_2 to give $Os_3(PhC_2Ph)(CO)_9(CH_2N_2)$ and $Os_3(PhC_2Ph)(CO)_9(CH_2)$ which is like **30** except that μ-CO is replaced by μ-CH_2 (*224*). The compound $Os_3(\mu_3,\eta^2\text{-}PhC_2Ph)(CO)_7(Ph_4C_4)$ also contains a normal μ_3-alkyne (*225*).

Other compounds containing μ_3-alkynes are $Os_3H_2(\mu_3\text{-indyne})(CO)_9$ (**34**) formed by reacting $Os_3(CO)_{12}$ with indene (vinylic rather than more acidic allylic C—H bonds being cleaved) (*228*) and $Os_3H_2(\mu_3\text{-}C_4H_2NR)(CO)_9$ (**35**)

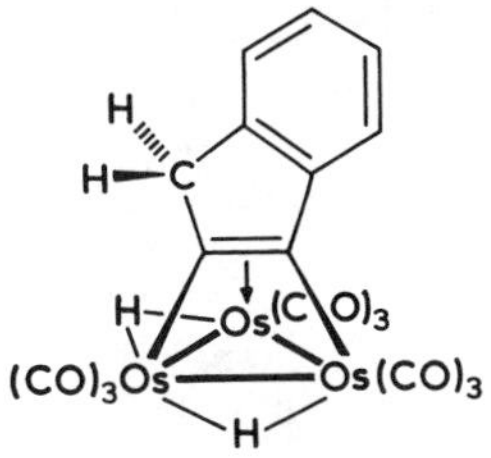

34 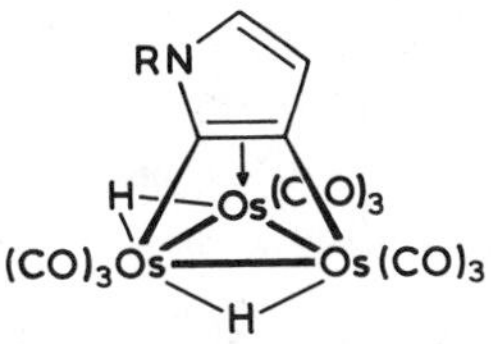   *35*

formed by oxidative addition of pyrrole or N-methylpyrrole (C_4H_4NR, R = H or Me) at $Os_3(CO)_{12}$ (*229*). The formation of other compounds of this type depends upon the electrophilicity of certain cluster-bound carbon atoms. Scheme 5 shows that thermal decarbonylation of $Os_3(HC_2CH_2CH_2OH)(CO)_{10}$ leads to a transient μ_3-alkynyl compound which undergoes an intramolecular nucleophilic attack at the α-carbon atom to give the dihydride (**33**) (*230*). Nucleophilic attack by PMe_2Ph at the corresponding carbon atom in $Os_3H(\mu_3\text{-}C\equiv CH)(CO)_9$ gives compound **36**, while addition at $Os_3H(\mu_3\text{-}MeC=C=CH_2)(CO)_9$ gives compound **37**.

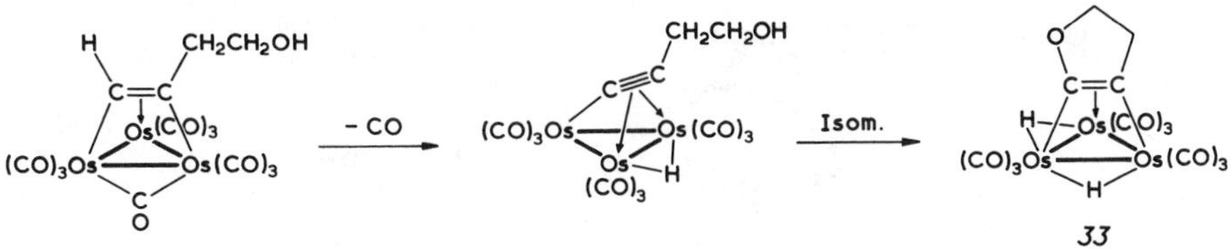

SCHEME 5.

Clusters **36** and **37** are closely related zwitterions with negative charge formally located at the metal atoms. This is apparent from the lower values for $v(CO)$, and the monohydride form of these compounds is also consistent with this (*108b*).

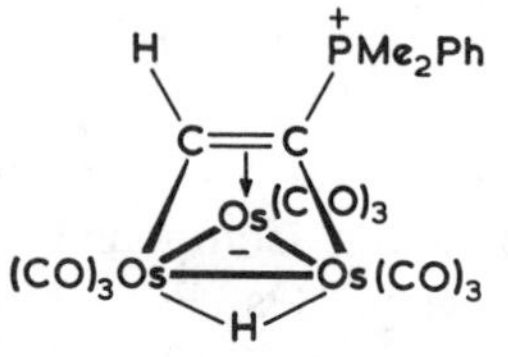

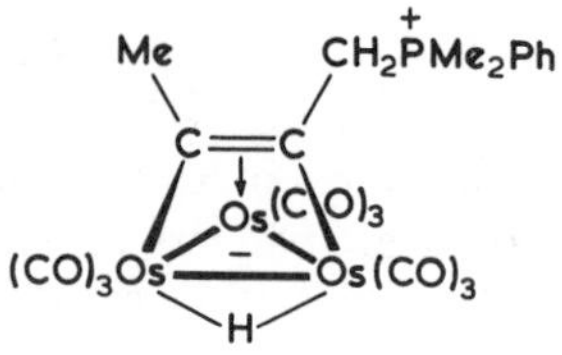

Triply bridging benzyne (*ortho*-phenylene) clusters are also common. Direct reaction of benzene with $Os_3(CO)_{12}$ at around 190°C (*35,46*) or with $Os_3(CO)_{10}(MeCN)_2$ at much lower temperatures (*20*) gives $Os_3H_2(\mu_3,\eta^2\text{-}C_6H_4)(CO)_9$ (X-ray structure determined). This compound is also formed on thermal decarbonylation of $Os_3H(\mu_2\text{-}PhCO)(CO)_{10}$ (*163*) or of $Os_3H(OCH_2Ph)(CO)_{10}$ (*168,169*). C_6H_4 ligands commonly appear as products of thermolysis of various aryl compounds. Thermolysis of $Os_3(CO)_{11}(PMe_2Ph)$ gives $Os_3H(PMe_2)(C_6H_4)(CO)_9$, while 1,2-$Os_3(CO)_{10}(PMe_2Ph)_2$ gives $Os_3(PMe_2)_2(C_6H_4)(CO)_7$ as the major products (*231,232*). Related $\mu_3\text{-}C_6H_4$ compounds have been synthesized from PPh_3 (*30,204,233,234*), $AsMe_2Ph$ (*59,235*), diars [1,2-$(Me_2As)_2C_6H_4$] (*232*), $PEtPh_2$ (*106*), or $MeSPh$ (*95*). Some of these compounds have three Os—Os bonds, e.g., $Os_3(PR_2)_2(C_6H_4)(CO)_7$ (*30,231,232,234*), while others have two, e.g., $Os_3H(X)(C_6H_4)(CO)_9$ [X = PMe_2 (*232*), $AsMe_2$ (*59*), or SMe (*95*)] (see Fig. 2) and $Os_3(\mu_3\text{-}PEt_3)(\mu_3,\eta^2\text{-}C_6H_4)(CO)_9$ (*106*). Structures like D in Fig. 11 are found for all known C_6H_4—Os_3 clusters except for the last named compound, structure **38**, which has one carbon of the *ortho*-phenylene bonded to only one osmium atom, and $Os_3H_3(\mu,\eta^2\text{-}C_6H_4)(\mu,\eta^2\text{-}CH{=}NPh)(CO)_8$ formed in low yield from thermolysis of $Os_3H_2(CO)_{10}(PhNC)$. This has structure **39**; the three hydride ligands are not shown since they were not located (*236*).

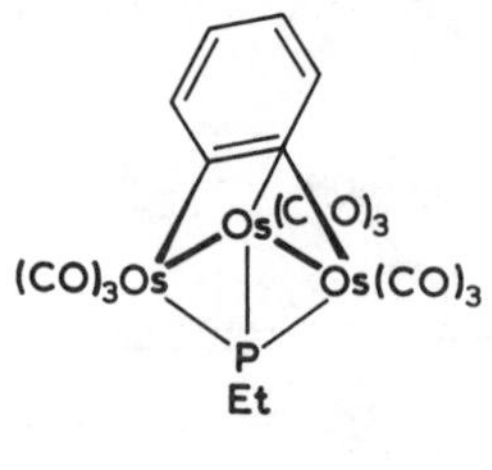

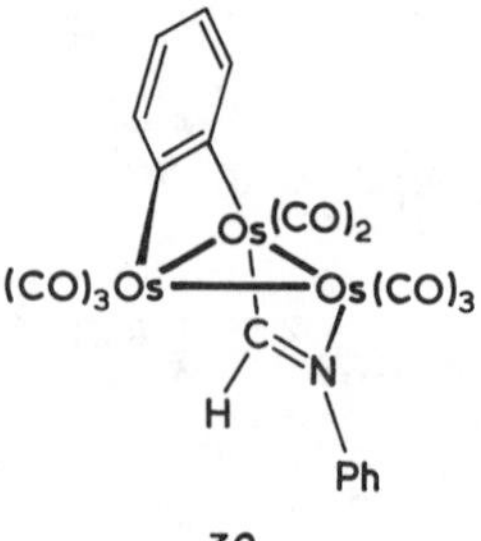

Triply bridging *ortho*-phenylene complexes sometimes arise in curious ways. In the reaction of $Os_3(CO)_{10}(MeCN)_2$ with salicylaldehyde benzylimine one would naively expect that coordination would be through N and O atoms as with other metal systems. The first formed compound (**40**, Scheme 6) results from ortho metallation while, of the three isomeric decarbonylation products (**41** to **43**), compound **43** contains only Os—C bonds to the ligand and is a ring-substituted version of $Os_3H_2(C_6H_4)(CO)_9$ (Scheme 6) (*237*).

The ligands F to H in Fig. 11 are found in derivatives of $Ru_3(CO)_{12}$ or $Os_3(CO)_{12}$ with the alkynes $HC\equiv CCH_2NMe_2$ or $MeC\equiv CCH_2NMe_2$. The clusters $M_3H(Me_2N=CC=CH_2)(CO)_9$ (X-ray structure determined for M = Ru) (*238,239*) adopt the ligand geometry shown rather than that found without the NMe_2 substituent: $M_3H(RC=C=CH_2)(CO)_9$ (Fig. 13, C, see later). π Donation from the NMe_2 group leads to conversion from a five- to a four-electron donor and the zwitterionic form shown. Similar behavior is found for $M_3H(Me_2N=CCH=CH)(CO)_9$ (type F) and $M_3H(Me_2N=CHC=CMe)(CO)_9$ (type G) (structures established for M = Ru) (*240*).

The ligand system I (Fig. 11) has been found on decarbonylation of $Os_3(CO)_{11}(PMe_3)$ at 150°C. This gives $Os_3H_2(Me_2PCH)(CO)_9$ as the main product, and $Os_3H_2(Et_2PCCH_3)(CO)_9$ is formed similarly from the PEt_3 compound (*221,241*). The cluster $Os_3H_2(PHC_6H_4)(CO)_9$ (type J) (*135*) has a direct analog in the aniline derivative $Os_3H_2(NHC_6H_4)(CO)_9$ (*170*), and a

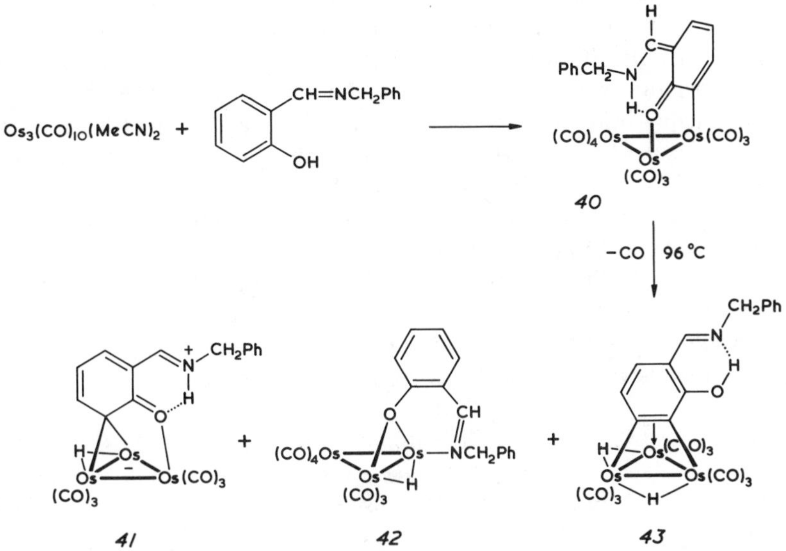

SCHEME 6.

comparable μ_3 ligand is found in $Os_3(\mu\text{-Ph})(\mu\text{-PPh}_2)(\mu_3\text{-PPhC}_6H_4)(CO)_8$ (30). The X-ray structure of $Os_3H_2(NHC_6H_3F)(CO)_9$ formed from 4-fluoroaniline shows the ligand arrangement K (167). It was initially thought that the dihydride $Os_3H_2(OC_6H_4)(CO)_9$ derived similarly from phenol had a structure directly related to K (165), but it was later showed that structure M is adopted (166). The ligand has rotated by 60° from the orientation K so that a carbon rather than an oxygen atom is bridging and a nonaromatic cyclohexadienone ligand is formed with an alkylidene bridge. Figure 12 shows details confirming this interpretation.

Unlike the phenol and aniline derivatives which have different structures, the compounds formed from cyclohexanone or its corresponding imine are isostructural. The X-ray structure of $Os_3H_2(NHC_6H_8)(CO)_9$ is known (242a) while the structure of $Os_3H_2(OC_6H_8)(CO)_9$ may be established spectroscopically by comparison with the phenol derivative (163); both are described by L (X = O or NH) (Fig. 11). Compound **40** in Scheme 6 is believed to have the same ligand arrangement (237).

The simplest possible μ_3 ligand of type N is CHCHO, and both Os_3H_2-$(\mu_3\text{-CHCHO})(CO)_9$ (163,175) and $Os_3(CHCHO)(CO)_{10}$ (196,198) are known. The dihydride is a low yield product from the thermolysis of the vinyloxy compound $Os_3H(OCH{=}CH_2)(CO)_{10}$, or it may be obtained by hydrogenation of $Os_3(CHCHO)(CO)_{10}$ which is obtained by decarboxylation of vinylene carbonate, $\overline{OCH{=}CHOCO}$, on reaction with $Os_3(CO)_{10}(MeCN)_2$. Structures with formylmethylidene rather than the enolate valence tautomer are adopted. The X-ray structure of $Os_3(CHCHO)(CO)_{10}$ (242b) is shown in Fig. 12 to demonstrate the similarity of this μ_3 ligand to the cyclohexadienone ligand. $Os_3(CHCHO)(CO)_{10}$ is isomeric with $Os_3(CH_2)(CO)_{11}$ and has a similar structure with μ-CO and μ-alkylidene bridges. Since the CHCHO ligand is triply bridging, the OsCOs plane of the alkylidene bridge is more nearly vertical while the μ-CO is closer to being in the Os_3 plane than in $Os_3(CH_2)(CO)_{11}$ (195) or $Os_3(CHSiMe_3)(CO)_{11}$ (194,197).

Finally $CH_2{=}CHOMe$ inserts into $Os_3H_2(CO)_{10}$ to give Os_3H-$(\mu\text{-MeCHOMe})(CO)_{10}$, which thermally decarbonylates to give various compounds including one thought to be $Os_3H_2(\mu_3\text{-MeCOMe})(CO)_9$ with framework O in Fig. 11. An alternative formulation is $Os_3H_2(\mu\text{-OMe})$-$(\mu_3\text{-CMe})(CO)_9$, which would contain two Os—Os bonds, but since the hydride ligands are different this is unlikely (243).

c. Five-Electron Donors. Deprotonation of $Os_3H_2(\mu_3\text{-S})(CO)_9$ leads to an anion isolated as $[PPN][Os_3H(\mu_3\text{-S})(CO)_9]$. Alkylation of this with $[R_3O]^+$ (R = Me or Et) gave $Os_3H(\mu_3\text{-SR})(CO)_9$, representation A in Fig. 13. This complex cannot be made by decarbonylation of $Os_3H(SR)(CO)_{10}$ but

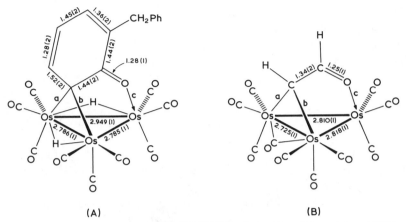

FIG. 12. Structures of $Os_3H_2(OC_6H_3CH_2Ph)(CO)_9$ derived from 2-benzylphenol (166) and of $Os_3(CHCHO)(CO)_{10}$ from $Os_3(CO)_{10}(MeCN)_2$ and vinylene carbonate (242b) (hydrogen atoms are not located). Bond lengths (Å): *molecule A*: a = 2.21(1), b = 2.24(1), c = 2.09(1); *molecule B*: a = 2.26(1), b = 2.22(1), c = 2.18(1).

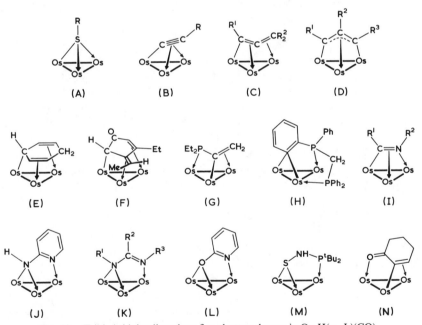

FIG. 13. Triply bridging ligands as five-electron donors in $Os_3H(\mu_3\text{-L})(CO)_9$.

readily picks up two-electron donors, L (CO or C_2H_4), to give $Os_3H(\mu\text{-SR})$-$(CO)_9L$. Thus the five-electron donor SR bridge is readily converted to a three-electron donor. The Fe and Ru analog with μ_3-SR bridges are more stable. Iodide also forms a five-electron donating bridge in $Os_3H(\mu_3\text{-I})(CO)_9$ (126).

The hydrocarbon bridges B to D in Fig. 13 are common. There are various routes to the alkynyl ligand B in clusters $Os_3H(\mu_3C{\equiv}CR)(CO)_9$. Direct reaction of $Os_3(CO)_{12}$ with the alkyne $RC{\equiv}CH$ is rarely a good route. $Os_3H(\mu_3\text{-}C{\equiv}C^tBu)(CO)_9$ is formed from $^tBuC{\equiv}CH$ in 20% yield (152,244), but with use of $PhC{\equiv}CH$ or $MeC{\equiv}CH$ many products are obtained and the yields of the $\mu_3\text{-}C{\equiv}CR$ compounds are low (245). It is usually better to go via $Os_3(\mu_3\text{-}CH{\equiv}CR)(CO)_{10}$, which is easy to decarbonylate (161). Carbon–carbon bond cleavage can lead to $\mu_3\text{-}C{\equiv}CR$ compounds. Attempts to alkylate the anion $[Os_3H(trans\text{-}CF_3CH{=}CHCF_3)(CO)_{10}]^-$ with $[Me_3O][BF_4]$ gave four neutral products, none obviously resulting from methylation, but one resulted from $C{-}CF_3$ bond cleavage: Os_3H-$(\mu_3\text{-}C{\equiv}CCF_3)(CO)_9$ (X-ray structure 44) (246).

44   **45**

Reaction of $Os_3(CO)_{12}$ with $HOCR_2C{\equiv}CCR_2OH$ (R = Me or Ph) gave $Os_3H(\mu_3\text{-}C{\equiv}CCR_2OH)(CO)_9$ and R_2CO (247). The group R cannot be hydrogen because different chemistry involving dehydration follows (222). Theoretical and uv photoelectron work on $\mu_3\text{-}C{\equiv}CR$ compounds of Os and Ru indicate, as with other systems, that there is no direct M—M bonding associated with the hydride bridge and that the bonding of the μ_3 ligand naively illustrated by formula 44 is a fair representation (248). Calculations indicating that the α-carbon is associated with some positive charge are borne out by observed nucleophilic attack at this atom (108b,230).

The isomeric ligands C and D (Fig. 13) are encountered when Os_3-$(\mu_3\text{-}MeC{\equiv}CMe)(CO)_{10}$ is thermolyzed to give $Os_3H(\mu_3\text{-}MeC{=}C{=}CH_2)$-$(CO)_9$ (type C) which subsequently isomerizes to $Os_3H(\mu_3\text{-}MeCCHCH)(CO)_9$ (type D) (161). The parent compound $Os_3H(\mu_3\text{-}CH{=}C{=}CH_2)(CO)_9$ containing C ($R^1 = R^2 = H$) (249) was made by the uv treatment of the allene cluster $Os_3(C_3H_4)(CO)_{11}$, compound 13. Subsequent thermal isomerization gave $Os_3H(\mu_3\text{-}CHCHCH)(CO)_9$ (249). Photolytic conditions are necessary because thermolysis of 13 gives the alkyne cluster $Os_4(CHCMe)(CO)_{12}$ (87). Heating $Os_3(HOCH_2C{\equiv}CCH_2OH)(CO)_{10}$ at 96°C induces dehydration and decarbonylation to give $Os_3H(\mu_3\text{-}CH_2\!=\!C\!=\!CCHO)(CO)_9$ (type C) (R^1 = CHO, R^2 = H), which isomerizes to $Cs_3H(\mu_3\text{-}CHCHCCHO)(CO)_9$ (type D) (R^1 = R^2 = H, R^3 = CHO) (X-ray structure determined) (222). Reaction of $Os_3(CO)_{10}(MeCN)_2$ with $MeC\!\equiv\!CCH_2NMe_2$ at room temperature gives $Os_3H(\mu_3\text{-}MeC\!=\!C\!=\!CH_2)(CO)_9$, but the mechanism is unknown (239). Acetylene reacts with $Os_3(CO)_{11}(MeCN)$ in dichloromethane with a trace of water to give $Os_3H(\mu_3\text{-}CHCHCOH)(CO)_9$, containing a hydroxy-substituted allyl ligand of type D (X-ray structure) (250). Coupling of an alkyne with CO is also observed in the treatment of $Os_3(\mu_3\text{-}RC\!\equiv\!CH)(CO)_{10}$ with $BHEt_3^-$ Following acidification various products including $Os_3H_2(RC\!\equiv\!CH)(CO)_9$, $Os_3H(\mu_3\text{-}RC\!=\!C\!=\!CH_2)(CO)_9$, and $Os_3H(\mu_3\text{-}RCCHCH)(CO)_9$ are obtained. CO coupling and reduction occur (251). $Os_3(PhC\!\equiv\!CPh)(CO)_9$ reacts with CH_2N_2 to give $Os_3(PhC\!\equiv\!CPh)(CO)_9(CH_2)$ which thermally isomerizes to $Os_3H(\mu_3\text{-}PhCCPhCH)(CO)_9$, which has also been made by reaction of $Os_3(CO)_{10}(MeCN)_2$ with diphenylcyclopropene (224).

Cyclic organic systems have been studied. The reaction of $Os_3H_2(CO)_{10}$ with the allene cyclonona-1,2-diene (C_9H_{14}) gave $Os_3H(\mu_3\text{-}C_9H_{13})(CO)_9$, which may have the structure **45** related to D, but there are other possibilities (252). The fully methylated form of C, $Os_3H(\mu_3\text{-}MeC\!=\!C\!=\!CMe_2)(CO)_9$ cannot isomerize to an isomer of type D, which is normally more stable (253). Ultraviolet photoelectron spectra of osmium species with ligands of types C and D have been described (254).

Cyclohexadiene (C_6H_8) reacts with $Os_3H_2(CO)_{10}$ at 25°C to give mainly $Os_3(CO)_{10}(C_6H_8)$ and a small amount of another compound which becomes the major product if the reaction is carried out at 125°C. This is the oxidative addition compound $Os_3H(C_6H_7)(CO)_9$ containing the cyclohexadienyl ligand (type E). The X-ray structure indicates that the double bonds are more localized than when the ligand is bonded to a single metal atom (255). One product from the reaction of $EtC\!\equiv\!CH$ with $Os_3H_2(CO)_{10}$ is $Os_3(C_9H_{12}O)(CO)_9$, containing the ligand shown in Figure 14, C with two butyne molecules and a CO coupled. On thermolysis this isomerizes to $Os_3H(C_9H_{11}O)(CO)_9$, which contains the five-electron donating ligand of type F (256).

The systems G and H are generated by C—H cleavage of coordinated tertiary phosphines. Thermolysis of $Os_3(CO)_{11}(PEt_3)$ gives mainly $Os_3H_2(Et_2PCMe)(CO)_9$ and a trace of $Os_3H(Et_2PC\!=\!CH_2)(CO)_9$ containing G. The dehydration may be achieved by successively removing H^- then H^+ as shown in Scheme 7 (221). Thermolysis of $1,2\text{-}Os_3(CO)_{10}$-(dppm) (dppm = $Ph_2PCH_2PPh_2$) in refluxing toluene gives Os_3H-$(Ph_2PCH_2PPhC_6H_4)(CO)_8$, which contains an essentially symmetrical phenyl bridge as shown in H (see Table III) (29). This could be regarded as a substituted derivative of the unknown compound $Os_3H(\mu\text{-}Ph)(CO)_{10}$.

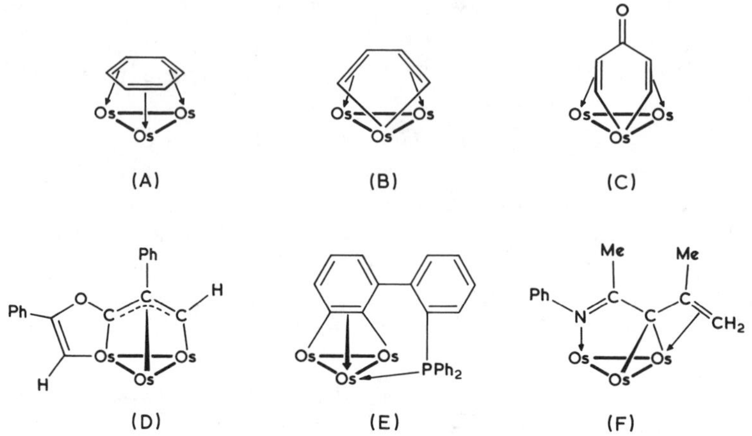

FIG. 14. Triply bridging ligands as six-electron donors.

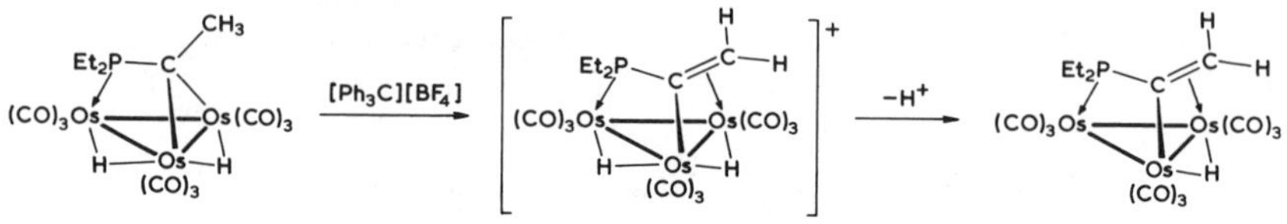

SCHEME 7.

A symmetrical phenyl (i.e., not μ,η^2) is a one-electron donor and the cluster a formally unsaturated 46-electron system. Clusters of the type Os_3H-$(\mu_3$-RC=$NR')(CO)_9$ (type I) have been synthesized with the μ_3 ligands: CH=NMe (142), CH=NPh (X-ray structure determined) (257), CF_3C=NH (117) (X-ray structure determined) (258), and ^iPrNHC=N^iPr (139). These are made by the insertion of PhNC, CF_3CN, or ^iPrN=C=N^iPr into Os_3H_2-$(CO)_{10}$ followed by heating to decarbonylate, except for the CH=NMe compound which was formed by the high-temperature reaction of Me_2NCH_2Ph and $Os_3(CO)_{12}$.

The ligands J to M are all of a type, being formed from 2-aminopyridine (159,174), ^iPrNHCH=N^iPr or $PhCH_2NHCH$=NCH_2Ph (171), PhNHCPh=NH or NH_2CMe=NH (172), 2-pyridone (173), or $^tBu_2PN$=S=NP^tBu_2 (174). The crystal structures of the compounds $Os_3H(\mu_3$-L)(CO)_9$ have been determined, where L is the ligand shown in J (173), K ($R^1 = R^2 = Ph$; $R^3 = H$) (172), and M (174). Decarbonylation of $Os_3H(\mu_2$-NHCHO)(CO)_{10}$ in refluxing nonane was reported to give the closely related compound $Os_3H(\mu_3$-NHCHO)(CO)_9$ (19), but this product is more likely to be $Os_3H(\mu_3$-NH)(CO)_9$.

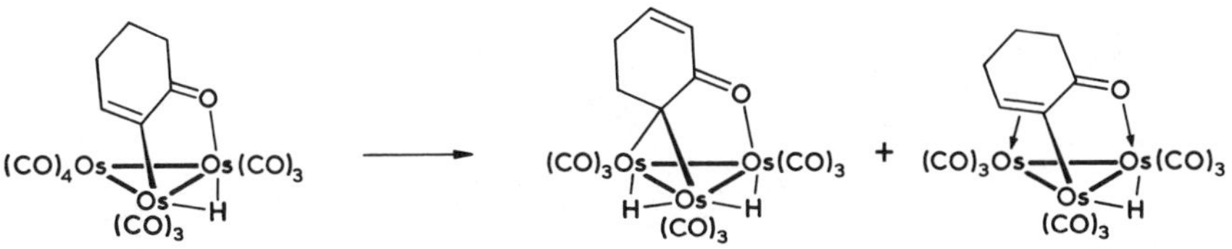

SCHEME 8.

Oxidative addition of cyclohexenone at the vinylic site to $Os_3(CO)_{10}$-$(MeCN)_2$ gives compound **46** (Scheme 8). Carbon monoxide is lost on heating to give two isomers. Compound **47** is directly comparable to the phenol derivative in Fig. 12 and the cyclohexanone-derived species $Os_3H_2(\mu_3\text{-}C_6H_8O)$-$(CO)_9$, so the μ_3 ligand in **47** is a four-electron donor. The other product (**48**) retains the ligand in **46** intact but must now be a five-electron donor probably as shown in Scheme 8 and N in Fig. 13 (*163*). Indeed **48** is the major product on uv photolysis of **46** and converts to **47** on heating (*259*). Similar decarbonylation of $Os_3H(CH=CHCOMe)(CO)_{10}$ (**8**) occurs on heating, and the product contains $\mu_3\text{-}CH=CHCOMe$ as a five-electron donor (*83*).

d. Six-Electron Donors. Some of the ligands in Fig. 14 are made up of components of simpler ligands. Thus the cluster $Os_3H(PPh_2)$-$(PPh_2C_6H_4C_6H_3)(CO)_7$, formed thermally from $Os_3(CO)_{12}$ and PPh_3, contains ligand D with a four-electron donor *ortho*-phenylene linked to a two-electron donor tertiary phosphine (*204*). The compound with the apparent formula $Os_3(CO)_{10}(PhC\equiv CH)_2$ has two alkynes and a CO ligand coupled to give the μ_3 ligand C. This can be considered as a five-electron donor allyl-type ligand as in D (Fig. 13) with a σ-Os—C bond at a side chain (*260*). A compound with apparent formula $Ru_3(CO)_7(^tBuC\equiv CH)_4$ has been shown (X-ray structure) to be $Ru_3(\mu_3\text{-}^tBuC=CC^tBuCH_2)(\mu_3\text{-}CHC^tBuCOC^t\text{-}BuCH)(CO)_7$, in which the second named ligand is like C (Fig. 14) but with Ph replaced by ^tBu. The compound $Os_3(CO)_7(^tBuC\equiv CH)_4$ may have a similar or related structure (*261*).

Reaction of $PhN=CMeCH=CMe_2$ with $Os_3H_2(CO)_{10}$, possibly by insertion followed by a double oxidative addition and CO dissociation, gives $Os_3H_2(\mu_3\text{-}PhN=CMeC^iPr)(CO)_9$ containing a four-electron donor related to L in Fig. 11. Dehydrogenation and decarbonylation at 125°C give $Os_3H_2(PhN=CMeCCMe=CH_2)(CO)_9$ (see F in Fig. 14).

The coupled-alkyne ligand in $Os_3(CO)_9(C_4R_4)$ forms a μ_2 bridge rather than the μ_3 bridge (type B) in Fig. 14 where R = Ph (*193*), whereas when CO is coupled into the ligand the μ_3 bridge (type C) is found as in Os_3-$(C_4Me_4CO)(CO)_9$ formed from $MeC\equiv CMe$ (*263*). Coupling of $EtC\equiv CH$

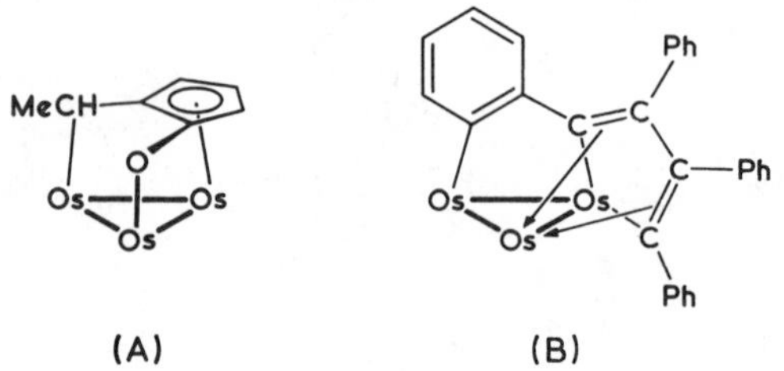

(A)   (B)

FIG. 15.   Triply bridging ligands as seven-electron donors.

with CO gives two isomers of $Os_3\{(EtC_2H)_2CO\}(CO)_9$ with different orientations of coupling. Thermolysis of these isomers gives different products. The isomer with Et groups distant from the ketonic carbonyl isomerizes as already described to a compound containing ligand F in Fig. 13 (256), while the one with Et groups adjacent to the carbonyl decarbonylates to give $Os_3H(C_9H_{11}O)(CO)_8$ containing the μ_3 ligand A in Fig. 15.

The most important μ_3 ligand is benzene. The cluster $Os_3(\mu_3\text{-}C_6H_6)(CO)_9$ is formed by successive removal of H^- and H^+ from the cyclohexadienyl cluster $Os_3H(\mu_3\text{-}C_6H_7)(CO)_9$ (E in Fig. 13). Nonnucleophilic base must be used for the deprotonation. The X-ray crystal structure establishes C_{3v} symmetry and points to a partial localization of single and double bonds as implied by drawing A (Fig. 14), but the X-ray data are not good enough to be confident about this (10). The remarkable simple and stable μ_3 arrangement of benzene has only recently been discovered and is not formed in the reactions of benzene with $Os_3(CO)_{12}$ (34,46) or $Os_3(CO)_{10}(MeCN)_2$ (20) which give the isomer $Os_3H_2(\mu_3\text{-}C_6H_4)(CO)_9$. $Os_3(C_6H_6)(CO)_9$ does not isomerize to the dihydride on heating. It is probable that oxidative addition of benzene to give $Os_3H(Ph)(CO)_{10}$, prior to the loss of the necessary extra CO ligand, prevents the $\mu_3\text{-}C_6H_6$ compound from being formed directly. A simple precursor to $Os_3(CO)_9$ clusters would be valuable in this and other examples of triosmium chemistry.

e. Seven-Electron Donors.   Clusters of type $Os_3H(\mu_3\text{-}L)(CO)_8$ where L is a seven-electron donor are rare; Fig. 15 shows two such examples. Ligand A is ultimately derived from $EtC{\equiv}CH$ (264,265) and B from $PhC{\equiv}CPh$ (266,267).

7. Four Osmium–Osmium Bonds

Although compounds of 46 valence electrons require multiple metal–metal bonds to fit the 18-electron rule, there is no evidence for these in Os_3 clusters such as $Os_3H_2(CO)_{10}$, $Os_3H(AuPR_3)(CO)_{10}$, $Os_3(AuPR_3)_2(CO)_{10}$, and

$Os_3H(Ph_2PCH_2PPhC_6H_4)(CO)_8$. This was discussed in Section II,A, while the syntheses and properties of the gold compounds will be described in Section VIII,B. There has been no consideration of Os_3 clusters with other metal atoms in this section since these will be considered in Sections VII and VIII.

III

DODECACARBONYLTRIOSMIUM AND SIMPLE SUBSTITUTED COMPOUNDS

A. *Dodecacarbonyltriosmium*

1. *Synthesis and Structure*

$Os_3(CO)_{12}$ was first reported by Hieber as $Os_2(CO)_9$ in 1943 (*268*). The reaction of OsO_4 with HI gave an oxyiodide which was treated with Ag powder in benzene at 150°C under CO (200 atm) to give $Os_3(CO)_{12}$ in poor yield. Good synthetic routes were reported in 1967 (*269*) and 1968 (*270,271*), and these mark the beginning of its study. Our method of synthesis is treatment of a solution of OsO_4 (10 g) in ethanol (500 cm^3) with CO (80 atm at 20°C) at 175°C for 4 hours in a 2-liter rocking autoclave. Essentially pure air-stable yellow crystals of the carbonyl (>90% yield) separate on cooling, but the use of lower CO pressures leads to contamination with a dark impurity. Purification can be recrystallization from refluxing toluene or sublimation *in vacuo* at 403 K, although this is rarely necessary.

The solid-state molecular symmetry is approximately D_{3h}, and this structure apparently persists in other phases (*272,273*). The six axial and six equatorial CO ligands define a cubo-octahedron (Fig. 16). In the crystal there is a very small distortion to C_3 symmetry by a slight rotation of each $Os(CO)_4$ unit. This results in equatorial ligands being alternately above and below the metal plane and in the CO(axial)—Os—CO(axial) vectors being slightly out of parallel (see Fig. 16). Such distortions are sometimes more pronounced in substituted derivatives. Some structural parameters are given in Table VII. Competition between axial ligands for metal π-electron density results in these being associated with the longer M—C and the shorter C—O lengths. The average distances between axial carbon atoms is 2.869(10) Å, and this crowding means that only simple rod-like ligands and hydride will coordinate axially in substituted compounds.

EXAFS spectroscopy of $Os_3(CO)_{12}$ was conducted as a preliminary to a

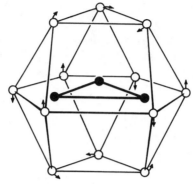

FIG. 16. The CO polyhedron in $Os_3(CO)_{12}$ indicating small distortions from D_{3h} symmetry in the crystal.

TABLE VII

AVERAGE BOND DISTANCES (Å) AND ANGLES (°) IN $Os_3(CO)_{12}$[a]

Os—Os	2.8771(27)	Os—Os—C(ax)	89.9(9)
Os—C(ax)	1.946(6)	Os—Os—C(eq)	98.2(17)[b]
Os—C(eq)	1.912(7)	C(eq)—Os—C(eq)	103.5(7)
C—O(ax)	1.134(8)		
C—O(eq)	1.145(5)		

[a] Data from ref. 273.
[b] Smaller angles.

study of high nuclearity clusters (274) and surface-supported clusters (275). Distances obtained were 1.94 (Os—C), 2.79 (Os—Os), and 2.93 Å (Os—O).

Infrared (276,277) and Raman (277–279) spectra have been reported. Four ir-active (2004.7, 2015.6, 2036.5, and 2069.0 cm^{-1}) and three Raman-active (2004.8, 2037.3, and 2128.2 cm^{-1}) $\nu(CO)$ absorptions and two Raman absorptions associated with metal triangle motions (158.9 and 113 cm^{-1}) have been assigned. Unlike large clusters like $Os_{10}H_2C(CO)_{24}$, smaller clusters like $Os_6(CO)_{18}$ and $Os_3(CO)_{12}$ are nonparamagnetic even below 70 K (280). The electronic spectra down to 77 K and the electronic structure of $Os_3(CO)_{12}$ and the Fe and Ru analogs have been discussed (281). The mass spectrum shows the parent molecular ion, $[Os_3(CO)_{12}]^+$, and the successive loss of the ligands as well as quite high abundances of $[Os_3(CO)_n]^{2+}$ ions but no mono- and dinuclear ions (282). More often than not, mass spectra of Os_3 clusters show the parent molecular ions, and these are recorded routinely by many research groups. Negative-ion mass spectra have been recorded for $Os_3(CO)_{12}X_2$ and $Os_3(CO)_{10}X_2$ (X = Cl or Br). Parent molecular ions are not observed (283).

The ^{13}C-nmr spectrum of $Os_3(CO)_{12}$ shows signals for axial and equatorial carbonyls which coalesce around 70°C (284,285). Corresponding spectra of $^{187}Os_3(^{13}CO)_{12}$ (99.8% ^{187}Os, up to 60% ^{13}C) have shown that the axial CO doublet (δ 182.3, J_{OsC} 90 $\pm$ 2 Hz) coalesces with the equatorial CO doublet (δ 170.4, J_{OsH} 115 $\pm$ 2 Hz) to give a quartet (δ 176.4, J_{OsC} 33 $\pm$ 1 Hz) at 150°C. Axial CO signals are downfield of equatorial ones and have smaller J_{OsC} values, consistent with their longer Os—C bond lengths. Total intramolecular CO scrambling over all sites within the molecule is occurring (286). Although localized (turnstile) rotations within each $Os(CO)_4$ unit do not explain these results, different delocalized mechanisms might operate. Exchange between the D_{3h} structure and a C_{2v} one with two bridging CO like in $Fe_3(CO)_{12}$ is the most likely process, but another involving a concerted motion of the total CO polyhedron about the metal triangle must be considered (287).

Table VIII shows how metal–metal and metal–CO bond enthalpy contributions for $Os_3(CO)_{12}$ compare with those for adjacent metal carbonyls and with Fe and Ru analogs. These values indicate that CO dissociation and M—M bond cleavage for $Os_3(CO)_{12}$ are more endothermic than for $Ru_3(CO)_{12}$ (288). Substitution reactions of $Ru_3(CO)_{12}$ have associative as well as dissociative rate terms (289), whereas $Os_3(CO)_{12}$ has substitution that is independent of the entering ligand. The first-order rate coefficient for reaction in decane at 92°C with PPh_3 is 1.12 $\pm$ 0.14 second^{-1}; $\Delta H^\ddagger$ is 161 $\pm$ 2 kJ mol^{-1} and $\Delta S^\ddagger$ is 101 $\pm$ 6 J K^{-1} mol^{-1}, and a simple dissociative mechanism seems likely (290).

2. Reactivity

The nuclearity of $Os_3(CO)_{12}$ is decreased on gaining CO ligands and increased on losing them. $Os_3(CO)_{12}$ reacts with CO (200 atm) at 280°C to give $Os(CO)_5$ (60%) which only reverts to $Os_3(CO)_{12}$ in normal laboratory

TABLE VIII

BOND ENTHALPIES OF METAL CARBONYLS

	$\Delta H_{f,g}^0$	ΔH_D	ΔH_D per M atom	Bond enthalpy contributions (kJ mol^{-1})	
				M—M	M—CO
$Re_2(CO)_{10}$	-1559 ± 21	2029	1015	80	195
$Os_3(CO)_{12}$	-1644 ± 28	2690	897	94	201
$Ir_4(CO)_{12}$	-1715 ± 26	3051	763	117	196
$Fe_3(CO)_{12}$	-1753 ± 28	1676	559	52	126
$Ru_3(CO)_{12}$	-1820 ± 28	2414	805	78	182

lighting at a significant rate at 80°C (24). Ultraviolet irradiation of $Os(CO)_5$ in heptane at $-40°C$ gives $Os_2(CO)_9$, which decomposes to $Os_3(CO)_{12}$ at room temperature (23). Breakdown of $Os_3(CO)_{12}$ to the mononuclear species also occurs with dienes under uv irradiation; compounds of the type $Os(CO)_3$(diene), where diene is butadiene, 1,3-pentadiene, or $1,4,5,6-\eta^4$-cyclooctatetraene (291), are formed in this way. Irradiation of $Os_3(CO)_{12}$ in the presence of methyl acrylate gives $Os(CO)_4(CH_2{=}CHCO_2Me)$ and $Os_2(CO)_8(\mu,\eta^2-CH_2CHCO_2Me)$ (292).

At high temperatures $Os_3(CO)_{12}$ loses CO. At 210°C $Os_6(CO)_{18}$ is the major product (80%), but small amounts of $Os_5(CO)_{16}$, $Os_7(CO)_{21}$, and $Os_8(CO)_{23}$ are also obtained. At even higher temperatures (250°C) carbides are also obtained: $Os_8C(CO)_{21}$ and $Os_5C(CO)_{15}$ (293). Anions also are formed in low yields from the vacuum pyrolysis of $Os_3(CO)_{12}$. The carbido species $[Os_{10}C(CO)_{24}]^{2-}$ (294) and $[Os_{11}C(CO)_{27}]^{2-}$ (295) have been structurally characterized. In high-boiling polar solvents (e.g., isobutanol) $Os_3(CO)_{12}$ reacts to give non-carbido anions such as $[Os_{10}H_4(CO)_{24}]^{2-}$ among other products (296).

49

Electrophilic addition at $Os_3(CO)_{12}$ occurs at metal atoms while nucleophilic addition is at carbon atoms. Protonation in concentrated H_2SO_4 as the solvent gives $[Os_3H(CO)_{12}]^+$ (297,298), which was shown by the ^{13}C-nmr spectrum of the 99.8% ^{187}Os-enriched sample to have structure 49(286). ^{187}Osmium coupling to carbon was observed to be 84 and 85 Hz for the axial ligands A and D and 120, 116, and 121 Hz for the equatorial ligands B, C, and E, respectively. Hydride coupling to B (3.7 Hz) and to C (6.6 Hz) was also identified. Increasing the basicity of the cluster by PR_3 substitution allows double protonation in concentrated H_2SO_4 solution (297). There is spectroscopic evidence for the electrophilic addition of $Ag(O_2CCF_3)$ to $Os_3(CO)_{12}$, the adduct probably having an osmium-bridging Ag atom as in $Ru_3(CO)_8(dppm)_2(AgO_2CCF_3)$ (299).

Halogenation reactions of $Os_3(CO)_{12}$ have already been described (78–81). Initially trinuclear compounds are formed, but easy breakdown to mono- and dinuclear compounds occurs with an excess of halogen (81,300). Hydrogenation of $Os_3(CO)_{12}$ at atmospheric pressure gives good yields of

$Os_3H_2(CO)_{10}$ (15), but if high pressures of hydrogen are used (120 atm) then $Os_4H_4(CO)_{12}$ is the major product (301).

Nucleophilic addition at CO as a route to compounds of type Os_3H-$(\mu$-X$)(CO)_{10}$ was considered in Section II,B,5. Thus amines (RNH_2) give $[Os_3(\eta^1$-CONHR$)(CO)_{11}]^-$ as the primary product, but these readily decarbonylate (178,179). The extensive use of Me_3NO to substitute CO at metal carbonyls depends upon its nucleophilicity (302,303). Attack at CO gives M—CO_2NMe_3 which eliminates CO_2, and if better ligands than Me_3N are present these will coordinate. For example, treatment of $Os_3(CO)_{12}$ with Me_3NO in the presence of 1,2-diazine $(C_4H_4N_2)$ gives a high yield of $Os_3(C_4H_4N_2)(CO)_{10}$ (304). Reaction of $Os_3(CO)_{12}$ with Me_3NO in the presence of MeCN is a good route to $Os_3(CO)_{11}(MeCN)$ (16,17,305) and $Os_3(CO)_{10}(MeCN)_2$ (16,18) which are useful materials for the synthesis of many Os_3 clusters.

The addition of hydride ions at $Os_3(CO)_{12}$ using $K[BH(O^iPr)_3]$ gives $K[Os_3(\eta^1$-CHO$)(CO)_{11}]$ (306), which may be protonated with H_3PO_4 to give $Os_3(CH_2)(CO)_{11}$ (48). Presumably the formyl anion thermally decarbonylates because treatment of $Os_3(CO)_{12}$ with $NaBH_4$ in refluxing thf gives $Na[Os_3H(CO)_{11}]$ (56%), which may be separated from the tetranuclear products $[Os_4H_3(CO)_{12}]^-$ and $[Os_4H_2(CO)_{12}]^{2-}$ by fractional crystallization of the PPN salts (307). Nucleophilic addition of $[PPN][NO_2]$ at $Os_3(CO)_{12}$ gives an intermediate which is probably $[Os_3(\eta^1$-NO$)(CO)_{10}]^-$ but this isomerises to $[Os_3(\mu$-NO$)(CO)_{10}]^-$ which protonates to give $Os_3H(\mu$-NO$)(CO)_{10}$ (308–310). The same species is formed from $[NO][PF_6]$ and $[Os_3H(CO)_{11}]^-$ (309). Presumably the nitrite attacks coordinated CO to give $[M$—$CO_2NO]^-$ which eliminates CO_2 to give the nitrosyl ligand. Substitution of CO by PR_3 leads to reduced reactivity of the cluster toward the nucleophile. Whereas $Os_3(CO)_{12}$ dissolves in $^iPrNH_2$ to give Os_3H-$(^iPrNHCO)(CO)_{10}$, $Os_3(CO)_{11}(PMe_2Ph)$ gives only $[Os_3(\eta^1$-$^iPrNHCO)$-$(CO)_{10}(PMe_2Ph)]^-$ reversibly, while $Os_3(CO)_9(PMe_2Ph)_3$ dissolves in the neat amine without any reaction (191).

The cluster $Ru_3(CO)_{12}$ is very reactive toward ions of the form $[PPN][X]$ in dichloromethane, reacting readily with $[PPN][Cl]$ to give $[PPN][Ru_3$-$(\mu$-Cl$)(CO)_{10}]$ (311). Furthermore, $[PPN][CH_3CO_2]$ and $[PPN][CN]$ catalyze CO substitution by PPh_3 (312). As yet no such reactions have been reported for $Os_3(CO)_{12}$.

The cluster $Os_3(CO)_{12}$ is more difficult to reduce than the Fe or Ru analogs but shows a reduction at -1.16 V (versus Ag/AgCl) which is irreversible even at 195 K (50,313,314). A second ill-defined reduction was also described. Chemical reduction with $K[Ph_2CO]$, however, gave $[Os_3(CO)_{11}]^{2-}$, which was isolated, and ^{13}C-nmr and IR data (μ-CO at 1625 cm^{-1}) agree with structure 50 (315). The protonated form of this is well-known (307).

50

In general reactions of $Os_3(CO)_{12}$ with most common reagents (PR_3, RNC, alkenes, etc.) require temperatures greater than $100°C$; reactions are presumably limited by the rate of unimolecular CO dissociation. Apart from the use of Me_3NO to overcome this inertness, substitution has been catalyzed by the radical anion $[Ph_2CO]^-$ (*316*).

B. *Simple Substitution Compounds*

1. *Synthesis*

The direct reaction of tertiary phosphines or arsines (L) with $Os_3(CO)_{12}$ in refluxing toluene leads to a mixture of $Os_3(CO)_{11}L$, $Os_3(CO)_{10}L_2$, and $Os_3(CO)_9L_3$ which is easily separated by chromatography. This means that the rates of the second and third substitutions are not very different from that of the first, unlike the substitution of $Ru_3(CO)_{12}$ where the first substitution is the slowest. Substitution of MeCN from $Os_3(CO)_{11}(MeCN)$ or $Os_3(CO)_{10}(MeCN)_2$ or $[Ph_2CO]^-$-catalyzed substitutions have been used. All three derivatives are reported for PPh_3 (*233,316*), PMe_3 (*241*), PEt_3 (*221,241,317*), PMe_2Ph and $AsMe_2Ph$ (*21,232,318*), PEt_2Ph (*221*), $AsMe_2(CH{=}CH_2)$ (*220*), and various mono- and disubstituted compounds for $AsMe_2Ar$ (Ar = various substituted phenyl groups) (*59,235*), PPh_2Me and PPh_2Et (*21,106*), $P^tBu_2NH_2$ (*174*), $P(OMe)_3$ (*21,319*), and PH_2Ph (*134,135*).

Reaction of $Os_3(CO)_{11}(MeCN)$ with dppm ($Ph_2PCH_2PPh_2$) gives Os_3-$(CO)_{11}(\eta^1$-dppm). The rate of decarbonylation to give $Os_3(CO)_{10}(\mu$-dppm) is independent of [dppm], [CO], or $[O_2]$. Absence of retardation by CO, indicating that reversible CO dissociation is not rate-determining, and the negative $\Delta S^‡$ value indicate that bridge closure is associative. In contrast, the positive $\Delta S^‡$ value for the $[PBu_3]$-independent term for the reaction of PBu_3 with $Os_3(CO)_{11}(PBu_3)$ supports a normal dissociative pathway (*319*). The compound $Os_3(CO)_{10}(dppm)$ is also formed directly from $Os_3(CO)_{12}$, dppm, and Me_3NO (*29*) or by substitution of $Os_3(CO)_{10}(MeCN)_2$ or 1,1- or 1,2-$Os_3(CO)_{10}(butadiene)$ (*320*). The use of the 1,1 and 1,2 isomers of the butadiene complex has allowed 1,1 and 1,2 isomers of $Os_3(CO)_{10}(diphos)$ to be synthesized, where diphos = $Ph_2P(CH_2)_nPPh_2$ ($n = 2$, 3, or 4). This interestingly complete series of chelating (1,1) and bridging (1,2) diphosphine isomers lacks only $1,1\text{-}Os_3(CO)_{10}(dppm)$ (320).

Direct reaction of isocyanides with $Os_3(CO)_{12}$ in refluxing toluene (110°C) gives $Os_3(CO)_{11}(RNC)$ and $Os_3(CO)_{10}(RNC)_2$ (R = Me, "Bu, 'Bu, Ph, etc.) and the tri- and tetrasubstituted compounds for R = "Bu, 'Bu, or Ph (321,322). Higher temperatures should not be used since at 125°C hexanuclear species $Os_6(CO)_{18-n}(RNC)_n$ are formed (321,323). $[Ph_2CO]^-$ catalyzes the substitution (316), and the MeCN ligands in $Os_3(CO)_{10}(MeCN)_2$ may be replaced by RNC (21).

The MeCN ligand in the compound $Os_3(CO)_{11}(MeCN)$ may be replaced by NH_3 (305) or pyridine (16,224), but the bis(acetonitrile) compound gives $Os_3H(NH_2)(CO)_{10}$ and $Os_3H(NC_5H_4)(CO)_{10}$, respectively. Treatment of $Os_3(CO)_{12}$ with Me_3NO and 1,2-diazine gives $Os_3(CO)_{10}(1,2\text{-diazine})$ (304).

51   52

Simple alkene Os_3 compounds are rare. $Os_3(CO)_{11}(C_2H_4)$ can only be isolated by evaporating solvent from a solution of $Os_3(CO)_{11}(MeCN)$ by passing C_2H_4 through it (16). Diene compounds are more robust. Reaction of butadiene with $Os_3H_2(CO)_{10}$ gives 1,1- and $1,2\text{-}Os_3(CO)_{10}(C_4H_6)$ (51 and 52) (325–326). The corresponding norbornadiene (327) and cyclohexadiene (328) compounds exist only in the 1,1 form. Vinyl acetate reacts with $Os_3\text{-}(CO)_{10}(MeCN)_2$ or $Os_3H_2(CO)_{10}$ to give $Os_3(CO)_{10}(CH_2{=}CHOCOMe)$ in which the ester group is coordinated axially (329). The cluster $Os_3H_2(CO)_9\text{-}(PPh_3)$ reacts with cyclohexadiene to give $Os_3(C_6H_8)(CO)_9(PPh_3)$, which relates directly to the unsubstituted cyclohexadiene compound (330).

2. Structure and Dynamic Behavior

Tertiary phosphines have been found only in equatorial sites, as in the crystal structures of $Os_3(CO)_{11}L$ where L = $P(OMe)_3$ (331) or $P^tBu_2NH_2$ (174). Isomers are possible within this constraint (21,318). Reactions of $Os_3(CO)_{10}(MeCN)_2$ or the 1,1- and 1,2-butadiene compounds with PMe_2Ph give two separable isomers: 1,1- and $1,2\text{-}Os_3(CO)_{10}(PMe_2Ph)_2$. Variable temperature ^{31}P-, ^{13}C- and ^1H-nmr spectra show that there are two rapidly interconverting 1,2-isomers. On the basis of these data, the isomers 53, 54, and

55 have been identified (L = PMe_2Ph). Similar isomers exist for other tertiary phosphines. Although **54** and **55** rapidly interconvert, they do not interchange with **53** up to its decomposition at $110°C$ to $Os_3H(\mu\text{-}Me_2PC_6H_4)(CO)_9$-($PMe_2Ph$); which still contains the phosphorus atoms at the same osmium (*318*). The 1,1,2 and 1,2,3 isomers (**56** and **57**) of $Os_3(CO)_9(PMe_2Ph)_3$ have been separated (*318*). Likewise $Os_3(CO)_{10}\{Ph_2P(CH_2)_nPPh_2\}$ (n = 2, 3, or 4) exists in bridging and chelating forms corresponding to **53** and **54** (except that the P nuclei must be adjacent, of course) (*320*).

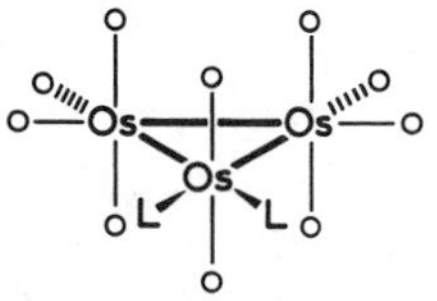

53

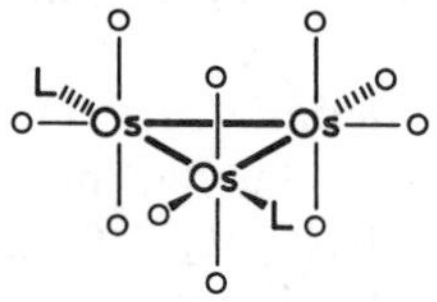

54

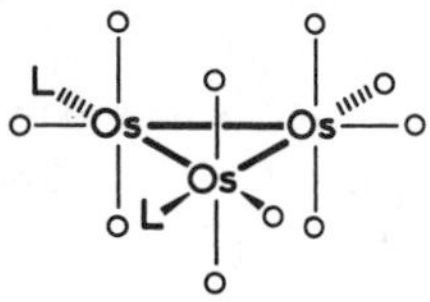

55

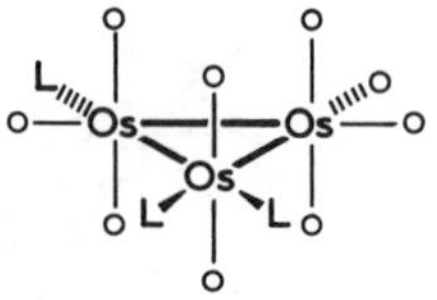

56

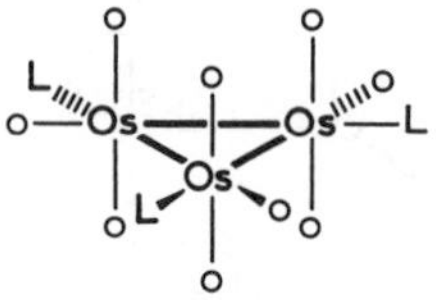

57

In contrast, organic cyanides and isocyanides predominantly coordinate axially. Compounds $Os_3(CO)_{11}(MeCN)$ and $Os_3(CO)_{10}(MeCN)_2$ have structures **58** and **59**, and whereas axial CO ligands in $Os_3(CO)_{12}$ and the $Os(CO)_4$ groups of **58** and **59** are the more weakly bound, the axial CO ligands trans to MeCN have the shortest Os—C bonds (*332*). The overall structure is maintained in $[Os_3H(CO)_{10}(MeCN)_2]^+$ (*18*). Compounds with RNC ligands adopt structures analogous to **58** and **59** except for ${}^tBuNC$. Equilibrium mixtures of axial and equatorial forms of $Os_3(CO)_{11}({}^tBuNC)$ and of diaxial (trans) and diequatorial forms of $Os_3(CO)_{10}({}^tBuNC)_2$ exist in solution (*322*). The crystal structures of the corresponding Ru compounds are axial and trans diaxial, respectively (*333*). Whereas 1,2-$Os_3(CO)_{10}$(diphosphine) is diequatorial, 1,2-$Os_3(CO)_{10}$(1,2-diazine) is diaxial (*304*). Various diene ligands

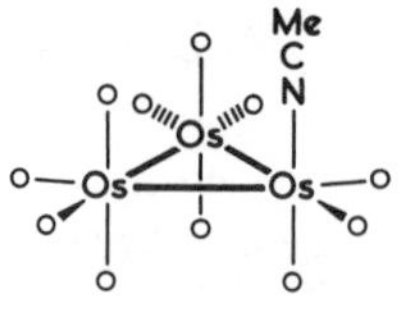

58

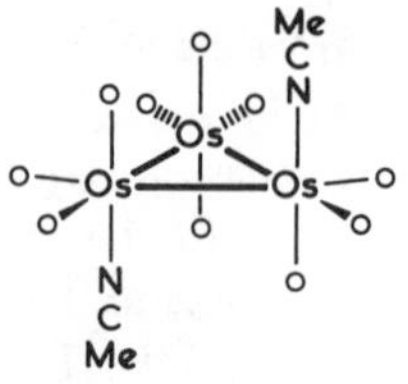

59

in compounds of the type $1,1\text{-}Os_3(CO)_{10}(diene)$ occupy one axial and one equatorial site; the structure is known where diene $= C_4H_6$ $(325,326)$, but the others are based on nmr data $(325-328)$. Substitution of CO by NO gives $Os_3(CO)_9(NO)_2$ (three Os—Os bonds) which takes up CO to give $Os_3(CO)_{10}(NO)_2$ (two Os—Os bonds). The cluster $Os_3(CO)_8(NO)_2\text{-}\{P(OMe)_3\}$ is a derivative of the nonacarbonyl compound with structure 60 $(91,92)$.

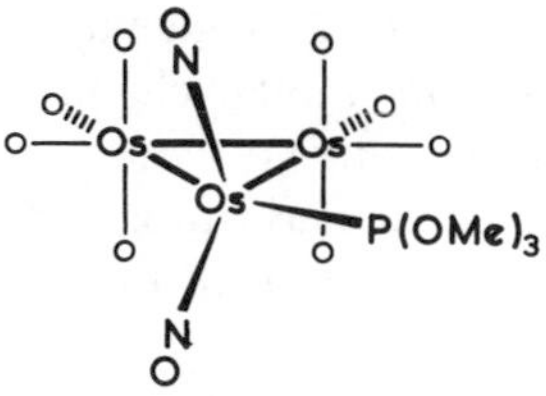

60

The intramolecular behavior of $Os_3(CO)_{12}$ and substituted derivatives has been discussed in terms of different mechanisms:

a. *One-center* exchange leading to axial–equatorial exchange by a turnstile pathway

b. *Two-center* merry-go-round exchange involving two equatorial and four axial CO ligands in a vertical plane *via* di-μ-CO $Fe_3(CO)_{12}$-type structures

c. *Three-center* merry-go-round exchange involving six equatorial CO ligands in a horizontal plane via tri-μ-CO intermediates

d. *Three-center* exchange of three axial CO ligands in a conical surface on the same side of the metal plane *via* tri-μ-CO intermediates

e. *Overall* motion of a CO polyhedron about a metal triangle (287)

Recent reviews on the intramolecular processes in clusters have been published $(334,335)$.

In PR_3-substituted compounds of Os_3 and Ru_3 the fastest process is b (317), and in general Ru_3 compounds give more rapid exchange especially where μ-CO intermediates are involved. For example, the ^{13}C-nmr spectrum of $Ru_3(CO)_{11}\{P(OMe)_3\}$ 61 at $-80°C$ contains CO signals in the ratio 10:1

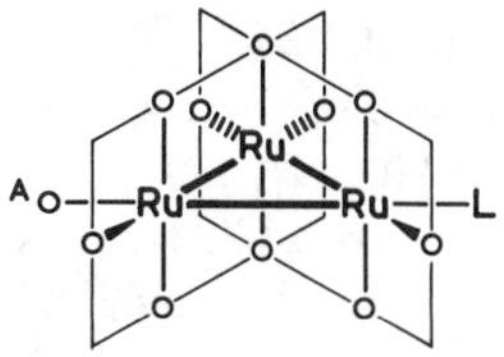

61

since the unique CO ligand A directly in line with the phosphite ligand is protected from exchange by process b operating rapidly in the two planes shown (336). At higher temperatures the carbonyl A exchanges with the others. In general, process b operates as the fastest process if there is an appropriate plane of six CO ligands for it to operate (317,318,320). In a report on the dynamic behavior of PEt_3 derivatives it was not recognized that 1,2-$Os_3(CO)_{10}(PEt_3)_2$ exists in solution as isomers 54 and 55 (317). In compound 54 [L = $P(OMe)_3$] all CO ligands exchange at the same rate by process b whereas in 55 only six CO ligands are exchangeable by this process (336). Slower processes lead to the exchange of nonequivalent ligands L in 55 at a somewhat faster rate than the isomerization of 54 to 55. The complex 53 (L = PMe_2Ph) undergoes exchange of all the CO ligands except the axial ones in the $Os(CO)_2L_2$ group which give a ^{13}C-nmr triplet up to over 90°C. These axial ligands could only exchange by a process such as d which must therefore have a high activation energy in this system (336).

62

The diphosphine complexes 1,2-$Os_3(CO)_{10}$(diphos) (62) where diphos = $Ph_2P(CH_2)_nPPh_2$ ($n = 1$ to 4), have b as the fastest process exchanging carbonyls A with B with increasing rates in the order n = 4 > 3 > 2 > 1 (320). $Ru_3(CO)_{10}$(dppm) behaves likewise (336). The next fastest process for this compound is the single-center exchange of C with D. Provided the dppm ligand remains diequatorial, the highest activation energy process leading to total CO exchange must involve process d. The crystal structure of $Ru_3(CO)_{10}$(dppm) shows a marked twist of the axail groups in the direction required by d (338). However, the suggestion (338) that this three-center exchange could be the fastest one observed is invalid because the observed process involves exchange of four axial and two equatorial CO ligands and not six axial ones as required by d.

There is evidence for process c in some systems where axial substitution blocks process d: $Os_3(CO)_{10}(MeCN)_2$ (334), $Os_3(CO)_{10}$(1,2-diazine) (304), and the Ru analog which has the tri-μ-CO form as the ground-state structure (339). Process c can also be identified for isomer 55 in the Ru_3 series (336) and for the compounds 1,1-$Os_3(CO)_{10}$(diene) (327,328). The fastest process in $Os_3(CO)_{10}(C_7H_8)$ (C_7H_8 = norbornadiene) involves pairwise exchange of all

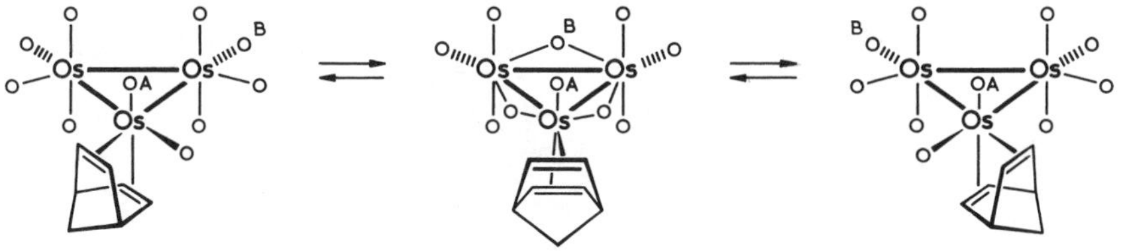

Scheme 9.

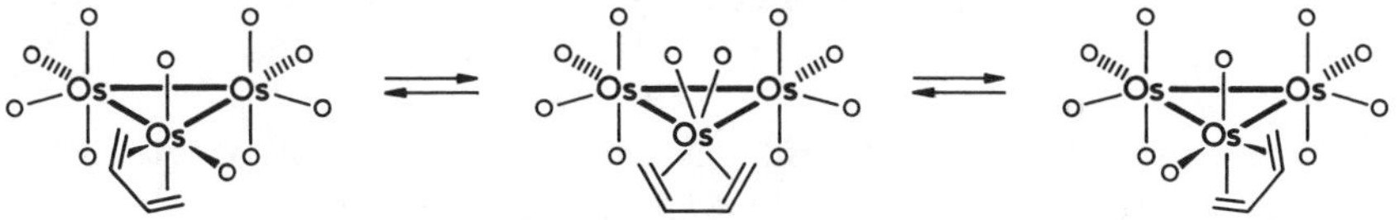

Scheme 10.

the CO ligands except two (A and B in Scheme 9). Consistent with the mechanism in Scheme 9 is the exchange of the bridgehead protons without the exchange of the double bonds (327). At higher temperatures total exchange of CO occurs. In the cyclohexadiene compound $Os_3(CO)_{10}(C_6H_8)$, however, the fastest process (of four) has been identified as that in Scheme 10. Pairwise exchange of all 10 CO carbonyls is occurring. One of the higher energy processes involves a tri-μ-CO intermediate (328).

IV

DIHYDRIDODECACARBONYLTRIOSMIUM

A. Structure and Simple Properties

The synthesis of $Os_3H_2(CO)_{10}$ from H_2 and $Os_3(CO)_{12}$ (15) and the X-ray and neutron diffraction structures (63) have been described (Table II) (58,66–68). The hydrides are not easily lost as protons but easily exchange with D_2O on silica (340). Molecule 63 shows only slow one-center CO exchange between carbonyls A and B (341,342). The exchange rate of 0.36 second^{-1} at 300 K was measured by saturation-transfer methods; coalescence occurs only at high temperatures. The carbonyls A and B couple to hydride with $^2J_{CH}$ 10.7 Hz (trans) and 0.8 Hz (cis), respectively and the coupling $^2J_{CC}$ between carbonyls C and D is 3.2 Hz, an order of magnitude less than trans couplings (343,344). ^{13}C-Chemical shift anisotropy and the ^{17}O quadrupolar coupling constant have been measured (342,345) and proton chemical shift tensors determined

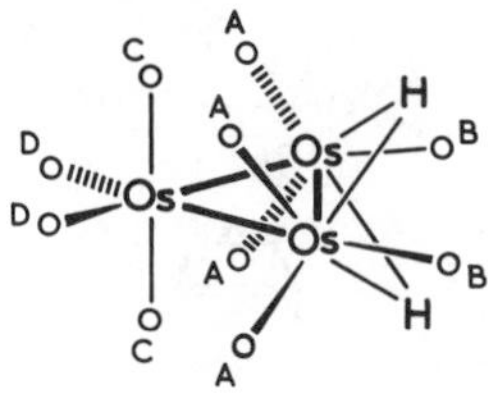

63

from multiple-pulse nmr techniques, using polycrystalline samples. The proton chemical shift anisotropy for the hydrides is less than 30 ppm (*346*).

Analysis of the vibrational spectra of **63** has given v(Os—H) values of 1228 (asym) and 1177 (sym), the intensity and sharpness of these absorptions being very temperature dependent so that low temperatures are needed for clear assignment. There is a good correlation between the M—H—M angle in a range of hydrido-bridged compounds (92.6° in this case) and the ratio of these two frequencies (*347*). The "plastic metal atom cluster" model has been applied to the prediction of v(M—M) frequencies which are experimentally observed at 187, 142, and 97 cm^{-1}. The estimated force constant associated with the bridged osmium atom pair is low, supporting the "no direct Os—Os bond" picture of the OsH$_2$Os bridge (*349*). PES data and MO calculations have been reported for Os$_3$(CO)$_{12}$, Os$_3$H$_2$(CO)$_{10}$, and other species of the type Os$_3$H(μ-X)(CO)$_{10}$ (*74,350–352*).

B. Addition of Nucleophiles

Whatever the electronic structure of Os$_3$H$_2$(CO)$_{10}$, it certainly shows unsaturation by its reactions with donor molecules. If CO is bubbled through a purple solution of Os$_3$H$_2$(CO)$_{10}$, a yellow solution of Os$_3$H$_2$(CO)$_{11}$ is formed, and, if N$_2$ is bubbled through this solution, the decacarbonyl is regenerated (*53–55*). The addition of neutral ligands such as PR$_3$, AsR$_3$, P(OR)$_3$, PhCN, MeNC, and ^{t}BuNC gives stable adducts which may be isolated. The structures of Os$_3$H$_2$(CO)$_{11}$ (**64**, L = CO) (*53,273*),

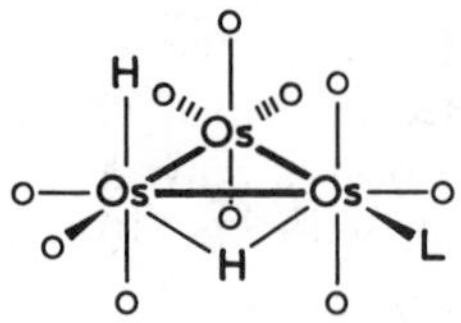

64 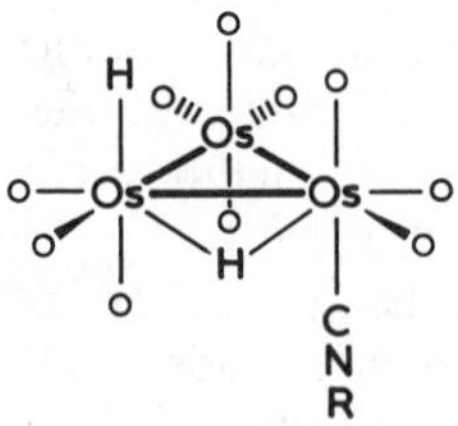   **65** 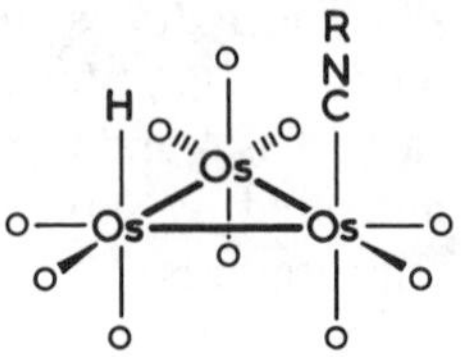   **66**

$Os_3H_2(CO)_{10}(PPh_3)$ (**64**, L = PPh_3) (*353*), and $Os_3H_2(CO)_{10}$ (^tBuNC) (**65**, R = ^tBu) (*354*) in the crystal have been determined. The structures relate closely to that of $Os_3(CO)_{12}$ with an axial terminal hydride replacing one CO ligand. As with substitution derivatives of $Os_3(CO)_{12}$, bulky ligands like tertiary phosphines are equatorial while hydride and isocyanides are axial. Two axial isomers are possible, **65** and **66**. Both are present in solution but the trans isomer (**65**) is crystallized.

Several papers have been published on the mechanism of interchange of the terminal hydride ($\delta = \sim -10$) and the bridging hydride ($\delta = \sim -20$) ligands (*341,354–356*). The process involves the generation of a time-averaged symmetry plane through the metal atoms. Specific CO exchanges are observed, notably carbonyl ligands A and B exchange at the same rate as H^a with H^b. Scheme 11 shows the accepted mechanism. A kinetic isotope effect means that carbonyl A exchanges with B more slowly in the dideuteride ($k_H/k_D = 1.5 \pm 0.1$) (*356*). There is no kinetic effect for the exchange of carbonyls A with B in compound **63**, comparing rates for the dihydride with those of the dideuteride (*356*). As expected for the mechanism in Scheme 11, the isomers **65** and **66** are interconverted by this process. The nonequivalent axial ligands on the $Os(CO)_4$ unit are not expected to interconvert, and this was confirmed. Furthermore, it was shown by magnetization-transfer experiments below coalescence temperatures that the axial hydride in one isomer exchanges with the bridging hydride of the other and vice versa (*355*).

The clusters $Os_3H_2(CO)_{10}(PR_3)$ decarbonylate on heating, but loss of CO is also induced photochemically. An examination of the photolysis at 77 K indicates that the primary photoproduct is a decacarbonyl which only subsequently loses CO (*358*). The products $Os_3H_2(CO)_9(PR_3)$ are purple and show addition of nucleophiles like the parent decacarbonyl. The reactions of these compounds [$PR_3 = PPh_3$, PPh_2Et, or $P(OMe)_3$] with nucleophiles and their insertion reactions have been examined. Whereas the ethylene adduct of $Os_3H_2(CO)_{10}$ is unobservable, there is spectroscopic evidence for $Os_3H_2(CO)_9(PR_3)(C_2H_4)$ (*357*). The decacarbonyl must, however, form this adduct because this leads to insertion, ethane elimination, and oxidative addition of C_2H_4 to give $Os_3H(CH{=}CH_2)(CO)_{10}$ (*161,162*).

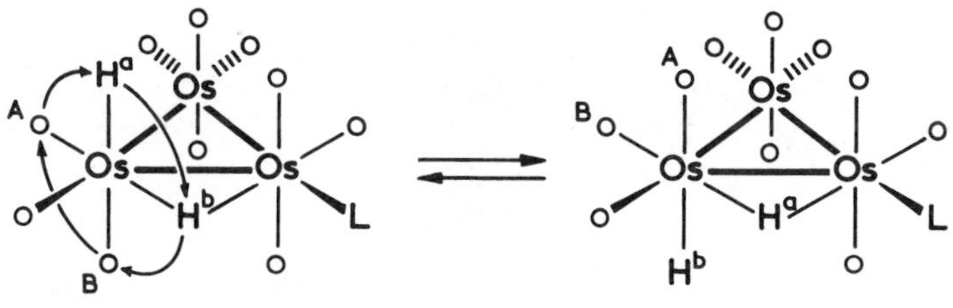

SCHEME 11.

Anion addition [halide and hydride (*359*) and cyanide (*360*)] at $Os_3H_2(CO)_{10}$ readily occurs to give the species $[Os_3H_2(CO)_{10}X]^-$. The cluster anion $[Os_3H_2(CO)_{10}(CN)]^-$ exists as three isomers, and protonation leads, via CNH complexes, to the aminomethylidyne cluster Os_3H-$(\mu\text{-}CNH_2)(CO)_{10}$ (*360*). The anion $K[Os_3H_3(CO)_{10}]$ is formed initially on adding KH, but a second KH will react to liberate H_2 and form $K_2[Os_3H_2(CO)_{10}]$, the same species obtained by reducing $Os_3H_2(CO)_{10}$ with $K[Ph_2CO]$ (*359*).

C. Addition of Electrophiles

A solution of $Os_3H_2(CO)_{10}$ in CF_3CO_2H gives ^1H-nmr hydride signals in intensity ratio 2:1, and the species formed is believed to be $[Os_3H_3(CO)_{10}]^+$ (**67**), although the trifluoroacetate ion may be coordinated and the unsatu-

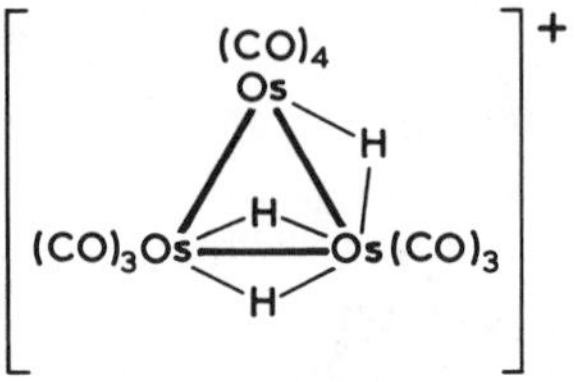

67

ration removed (*362*). Most Os_3 clusters are protonated at osmium by strong acids, and the cluster cations $[Os_3H_3(CO)_9(PEt_3)]^+$, $[Os_3H_4(CMe)(CO)_9]^+$, $[Os_3H_2(CH{=}CHPh)(CO)_{10}]^+$, and $[Os_3H(HC_2Me)(CO)_{10}]^+$ have been reported (*362*). Addition of electrophiles E^+ to $Os_3H_2(CO)_{10}$ can remove unsaturation if the radical E can behave as a three-electron donor. Thus $Os_3H_2(CO)_{10}$ reacts with $[PhN_2][BF_4]$ in refluxing dichloromethane to give a cation which deprotonates to give cluster **68**, which on photolysis gives cluster **69** (Scheme 12). The structures of both have been established (Tables III and IV) (*123,149*). Sulfur dioxide reacts with $Os_3H_2(CO)_{10}$ in various

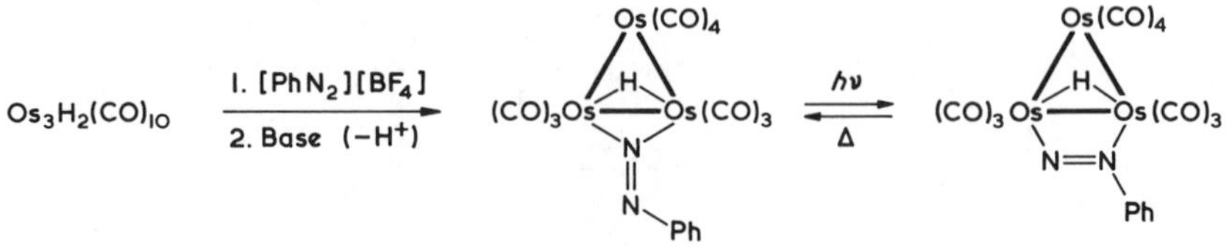

68   *69*

SCHEME 12.

solvents to give the adduct **70** (*164*) which is closely related to Os_3H_2-$(CH_2)(CO)_{10}$ (*43,44*), assuming SO_2 is a two-electron donor. There is no ambiguity in an electrophile adding to a metal system and formally behaving as a donor in the adduct.

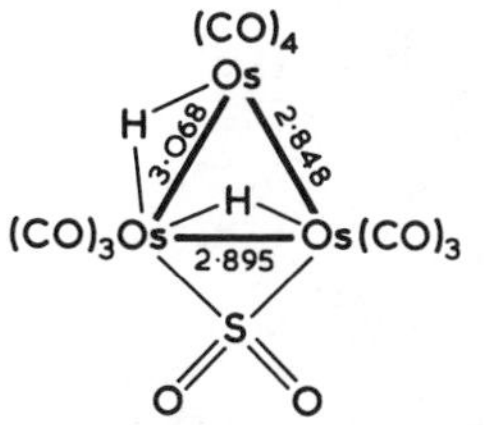

70

D. *Insertion Reactions*

Many insertion reactions have been described in Section II,B,5 on the synthesis of clusters of the type $Os_3H(\mu\text{-}X)(CO)_{10}$. It is usually assumed but without much evidence that an adduct is formed prior to the hydrogen transfer from Os to the new ligand. Scheme 13 represents the course of many reactions of $Os_3H_2(CO)_{10}$ with molecules L. The insertion compound may be the final product, but often L reacts further to displace the hydrogenated compound LH_2, allowing L to add oxidatively to give $Os_3H(L - H)(CO)_{10}$. Decarbonylation of $Os_3H(LH)(CO)_{10}$ or $Os_3H(L - H)(CO)_{10}$ can occur especially if above ambient temperatures are used and then double oxidative additions and other reactions can follow. Simple insertions stopping at $Os_3H(LH)(CO)_{10}$ are found for alkynes (*153,160,161,181,185–187,218,263–264*), but the alkene (LH_2) is often lost to give other products by secondary reactions of the alkyne.

Insertion of ketenes (*162*), CS_2 (*64,75,130,157–158*), isocyanates RNCO (*140,144,145*), $^iPrN{=}C{=}N^iPr$ (*139*), azides (RN_3) (*121,159*), diazo compounds (R_2CN_2) (*124,125*), $Ph_2C{=}S$ (*128*), isonitriles (RNC) (*108,143,257*), and nitriles (RCN) (*116–118,213,258*) give simple insertion compounds. The

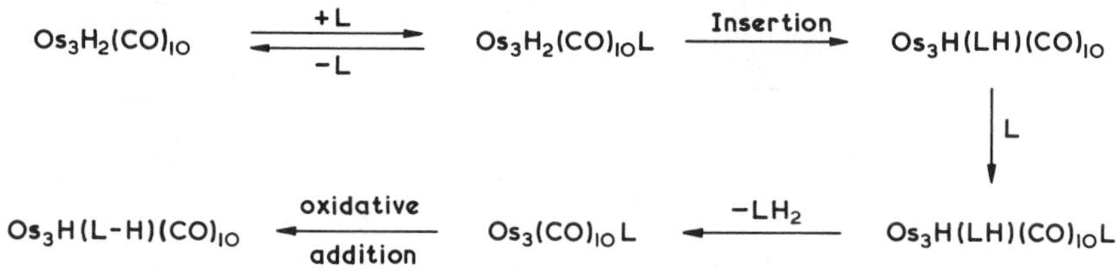

Scheme 13.

imine $PhCH=NMe$ undergoes insertion, decarbonylation, and oxidative addition to give $Os_3H_2(MeNCH_2C_6H_4)(CO)_9$ and elimination and oxidative addition to give $Os_3H(PhC=NMe)(CO)_{10}$ (*365*). The imine $Me_2C=CHCMe=NMe$ undergoes decarbonylation and oxidative addition reactions following the insertion into $Os_3H_2(CO)_{10}$ to give Os_3H_2-($^iPrCCMe=NMe)(CO)_9$ and $Os_3H_2(CH_2CMeCCMe=NMe)(CO)_8$ (*262*). The compound $Me_3SiN=S=NSiMe_3$ inserts, but rearrangement also occurs to give $Os_3H\{NHSN(SiMe_3)_2\}(CO)_{10}$ (*119*). Sometimes insertions occur with a fragment of the adding molecule being lost as when $\overline{OCH=CHOCO}$ reacts with $Os_3H_2(CO)_{10}$ to give $Os_3H(OCH=CH_2)(CO)_{10}$ (CO_2 lost) (*196*), when $^tBu_2PN=S=NP^tBu_2$ gives $Os_3H(\mu_3-SNHP^tBu_2)(CO)_9$ ($^tBu_2PN$ and CO lost) (*174*), when $CH_3C_6H_4SO_2N_3$ gives $Os_3H(NHSO_2C_6H_4Me)$-$(CO)_{10}$ (N_2 lost) (*120*), or when CH_2N_2 gives $Os_3H(CH_3)(CO)_{10}$ and $Os_3H_2(CH_2)(CO)_{10}$ (N_2 lost) (*41–44*).

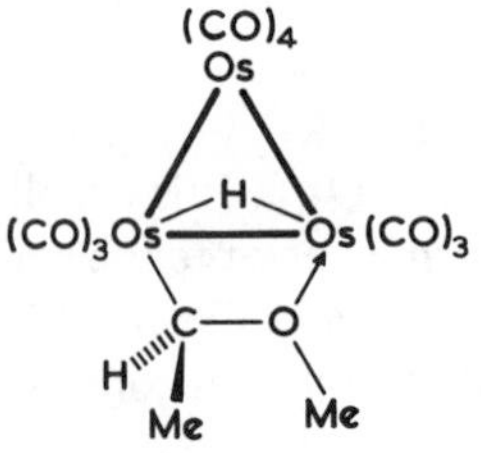

71 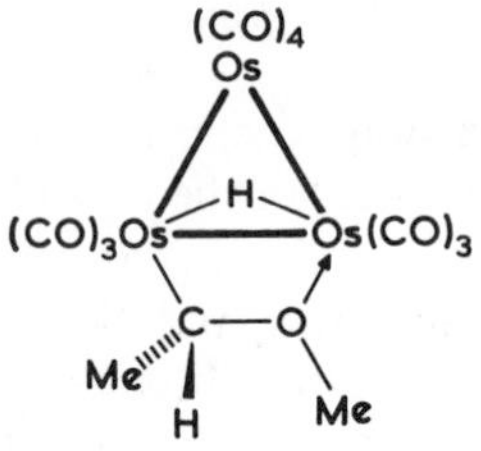   72

Insertion of alkenes into $Os_3H_2(CO)_{10}$ gives compounds of the type $Os_3H(alkyl)(CO)_{10}$ which are usually reactive toward β-elimination to regenerate alkene (see Section V) or reductive elimination of alkane to allow oxidative addition of the alkene (*161,162*). Sometimes, however, if the alkene is bi- or polyfunctional, stable insertion products are formed. For example, $CH_2=CHOMe$ inserts to give a mixture of diastereomers **71** and **72**. Ether coordination reduces the rate of β-elimination (*243*). Similar stabilization occurs on inserting α,β-unsaturated esters (*51*), although α,β-unsaturated ketones $RCH=CHCOMe$ (R = H, Me, or Ph) insert, then eliminate the

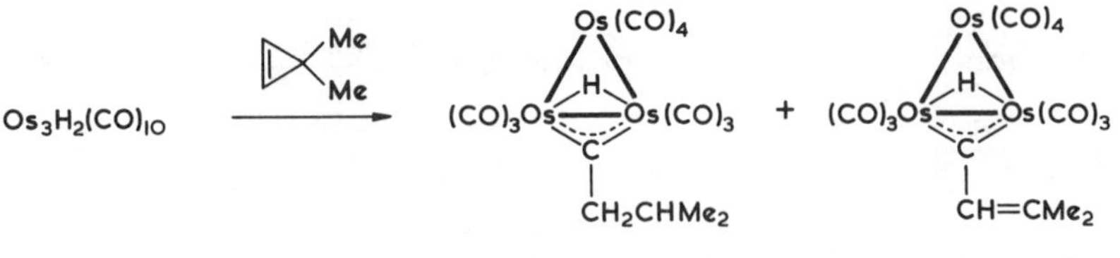

73   74

SCHEME 14.

saturated ketones, and oxidatively add to give $Os_3H(RC=CHCOMe)$-$(CO)_{10}$ (*83*). Likewise cyclohexadiene gives the cyclohexadienyl cluster $Os_3H(\mu_3\text{-}C_6H_7)(CO)_9$ (*255*). Dimethylcyclopropene reacts according to Scheme 14. Compound **73** isomerizes above 110°C to give Os_3H-$(\mu,\eta^2\text{-}CH=CHCHMe_2)(CO)_{10}$, which decarbonylates to $Os_3H_2\text{-}(\mu_3\text{-}C=CHCHMe_2)(CO)_9$ (*107*).

V

SOME ASPECTS OF HYDROCARBON LIGAND CHEMISTRY

A. *One-Carbon Ligands*

Schemes 15 and 16 summarize several papers on the synthesis and reactions of the ligands CH, CH_2, and CH_3 in triosmium clusters, and their formation from or transformation into C_2 ligands (see the schemes for references). The CH_2 ligand in $Os_3(CH_2)(CO)_{11}$ has been formed in three ways: from a CO ligand of $Os_3(CO)_{12}$, from CH_2N_2, and from CH_2CO (Scheme 15). The reduction of CO to CH_2 relates interestingly to Fischer–Tropsch chemistry. There is good evidence that $[BH(O^iPr)_3]^-$ or $[BHEt_3]^-$ generate

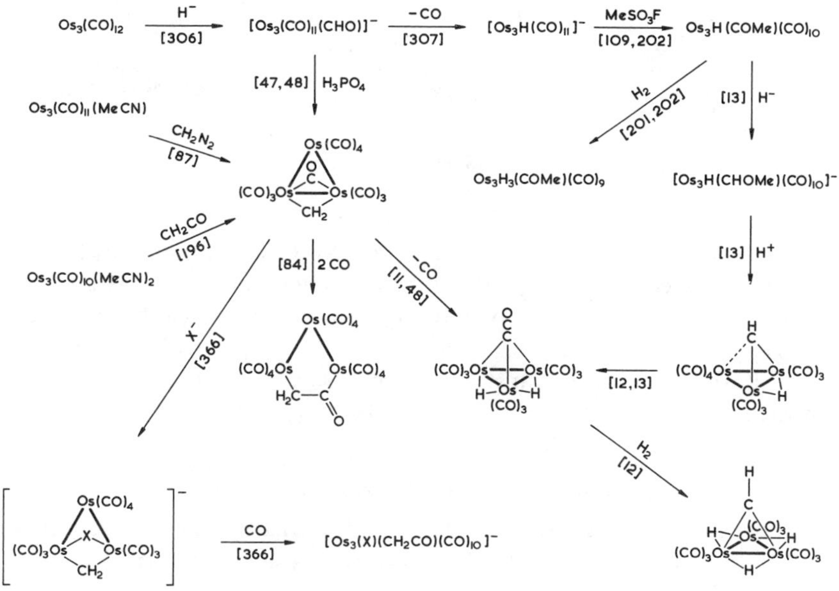

Scheme 15.

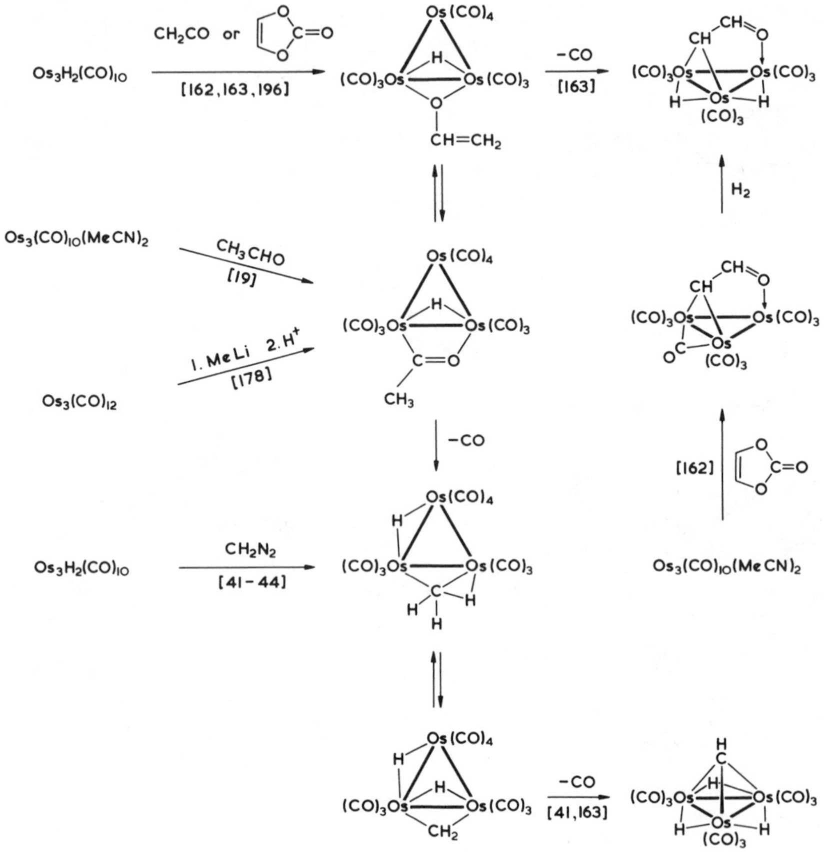

SCHEME 16.

$[Os_3(CHO)(CO)_{11}]^-$ from $Os_3(CO)_{12}$ (306). No CH_2 compound is formed if acid is added only after the formyl compound has converted to $[Os_3H(CO)_{11}]^-$, and the use of $[BDEt_3]^-$ followed by acidification with H_3PO_4 gave $Os_3(CD_2)(CO)_{11}$ which was free from hydrogen atoms. $[Me_3O][BF_4]$ may be used instead of protons as the acid. The mechanism consistent with these observations is given below (47,48):

$$Os_3(CO)_{12} + [BHR_3]^- \longrightarrow [Os_3(CHO)(CO)_{11}]^- + BR_3$$
$$[Os_3(CHO)(CO)_{11}]^- + H^+ \longrightarrow Os_3(CHOH)(CO)_{11}$$
$$Os_3(CHOH)(CO)_{11} + [Os_3(CHO)(CO)_{11}]^- \longrightarrow [Os_3(CH_2OH)(CO)_{11}]^- + Os_3(CO)_{12}$$
$$[Os_3(CH_2OH)(CO)_{11}]^- + H^+ \longrightarrow Os_3(CH_2)(CO)_{11} + H_2O$$

The cluster $Os_3(CH_2)(CO)_{11}$ undergoes coupling of the CH_2 carbon atom with CO under conditions of *both* loss and gain of CO. Addition of CO induces coupling of CH_2 to a prebound CO ligand, while thermal decarbonylation leads to the ketenylidene compound $Os_3H_2(C_2O)(CO)_9$. When CO is lost from $Os_3(CH_2)(CO)_{11}$ the bonding C—H electrons must be used and presumably $Os_3H(CH)(CO)_{10}$ is formed first; this CH compound readily isomerizes to the ketenylidene cluster. Unfortunately it is not known whether hydrogen transfer from CH to Os leads to the transient carbide $Os_3H_2(C)(CO)_{10}$ prior to C—C coupling, but more likely a ketenyl ligand is involved (see Scheme 17) (*12*).

The cluster $Os_3(CH_2)(CO)_{11}$ is substitutionally labile, giving $[Os_3(\mu\text{-}X)\text{-}(\mu\text{-}CH_2)(CO)_{11}]^-$ very readily on reaction with [PPN][X]; X-ray structures are reported for X = Cl, I, or NCO (*366*). These anions have 50 valence electrons and should therefore have two Os—Os bonds. The bridged Os—Os distance (3.112 Å) is fairly long and leads to an Os—C—Os angle of 92°, again quite large. The molecule has tight bridges which hold the nonbonded Os atoms relatively close. CO insertion gives μ-ketene complexes more rapidly than for $Os_3(CH_2)(CO)_{11}$. Scheme 15 shows that the C—C bond of CH_2=C=O may be broken, reformed, and broken again in the transformations of ketene to $Os_3(CH_2)(CO)_{11}$, hence to the ketenylidene compound which is hydrogenated to the cluster $Os_3H_3(CH)(CO)_9$.

Scheme 16 shows how clusters with C_2 ligands may be synthesized and broken down into clusters with C_1 ligands. The isomers $Os_3H(\mu\text{-}X)(CO)_{10}$ with X = CH_3CO and OCH=CH_2 do not interconvert except at a temperature (125°C) at which CO is displaced to give $Os_3H_3(CH)(CO)_9$ and $Os_3H_2(CHCHO)(CO)_9$. The third isomer $Os_3H(COMe)(CO)_{10}$ is given in Scheme 15. Thermal decarbonylation of $Os_3H(CH_3CO)(CO)_{10}$ presumably occurs via the CH_3 and CH_2 clusters, but since these readily decarbonylate to give $Os_3H_3(CH)(CO)_9$ they are not observed in this reaction. Synthesis of these intermediates from $Os_3H_2(CO)_{10}$ and CH_2N_2 is possible below room temperature. Their interconversion is sufficiently rapid to permit observation of magnetization transfer effects in their ^1H-nmr spectra. Crystallization of the $Os_3H(CH_3)(CO)_{10}$ compound was not possible, so its structure involving agostic bonding is based upon ^1H-nmr spectra. By analogy with the structure established for $Os_3H(SnHR_2)(CO)_{10}$ (see Section VII) (*367*), the hydride ligand possibly does not bridge the same metal atoms as the CH_3 group. The

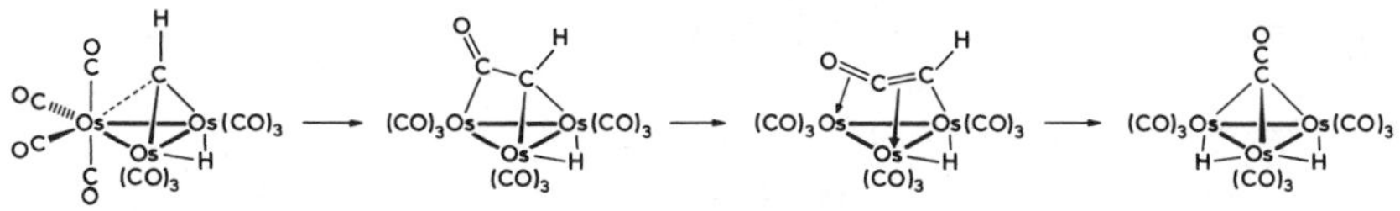

SCHEME 17.

bridging and terminal hydrogen atoms of the CH_3 group average on the nmr time-scale to give a singlet in the range $\delta -3.5$ to -4.5 which is an average of a high-field signal for the agostically bonded hydrogen and normal signals for the other hydrogens. Since there is a significant equilibrium isotope effect favoring H rather than D atoms at the agostic site, the CH_3, CH_2D, and CHD_2 compounds give very different (and temperature-dependent) chemical shifts in the above range. Similar effects have been observed for other agostically bonded μ_2-CH_3 groups.

Decarbonylations of other acyl complexes of the type $Os_3H(RCO)(CO)_{10}$ do not give simple alkyl or aryl clusters. When R = Ph the cluster Os_3-$H_2(C_6H_4)(CO)_9$ is obtained (163). Thermolysis of $Os_3H(OCH{=}CPh_2)$-$(CO)_{10}$, formed by inserting Ph_2CCO into $Os_3H_2(CO)_{10}$, causes isomerization to the acyl cluster $Os_3H(Ph_2CHCO)(CO)_{10}$ as a transient species which loses CO and H_2 to give $Os_3H(PhCC_6H_4)(CO)_9$ (163), a compound directly analogous to the Ru species formed earlier from $Ru_3(CO)_{12}$ and PhLi (368).

B. Mobility of Bridging Hydrocarbon Ligands

Carbonyl mobility (Section III) is a feature of most Os_3 clusters but will not be considered here. Hydride ligand mobility is also often high, but hydrocarbon ligands are only sometimes mobile. The vinyl ligand in Os_3H-$(\mu,\eta^2$-$CH{=}CH_2)(CO)_{10}$ rapidly oscillates in solution between the two metal atoms it bridges, as seen by the pairwise coalescence of ^{13}C- and ^{17}O-nmr signals for the two $Os(CO)_3$ groups (153,182,369). This oscillation occurs without hydrogen atom exchange within the $CH{=}CH_2$ ligand. Since the two isomers of $Os_3H(CH{=}CH_2)(CO)_9L$ [L = PPh$_3$, PPh$_2$Et, or P(OMe)$_3$] are noninterconverting, the likelihood is that they are not simply related by interchange of the η^1 and η^2 components of the vinyl bridge (357). Related vinyl oscillations occur when the vinyl is part of a μ_3 ligand such as $Et_2PC{=}CH_2$ (221) and $Me_2N{=}CC{=}CH_2$ (238).

Insertion of C_2D_2 into $Os_3H_2(CO)_{10}$ stereospecifically gives $Os_3H(cis$-$CD{=}CHD)(CO)_{10}$, which on addition of nucleophiles such as pyridine converts to an equilibrium mixture of the cis and trans $CD{=}CHD$ isomers. Since other better nucleophiles (CN^- and PMe_2Ph, for example) add to give isolable species (112,183,184), the mechanism in Scheme 18 was proposed (364). Decarbonylation of the vinyl compound gives Os_3H_2-$(\mu_3$-$C{=}CH_2)(CO)_9$ which undergoes exchange of the nonequivalent C—CH_2 protons much more rapidly than hydride ligand exchange, which could be explained by hydride migration or rotation about the $C{=}CH_2$ bond. Since the rate of H^a–H^b exchange is slower for $Os_3HD(C{=}CH_2)$-$(CO)_9$ (T_c raised from 80 ± 2 to $88 \pm 2°C$ on deuteration), a hydride

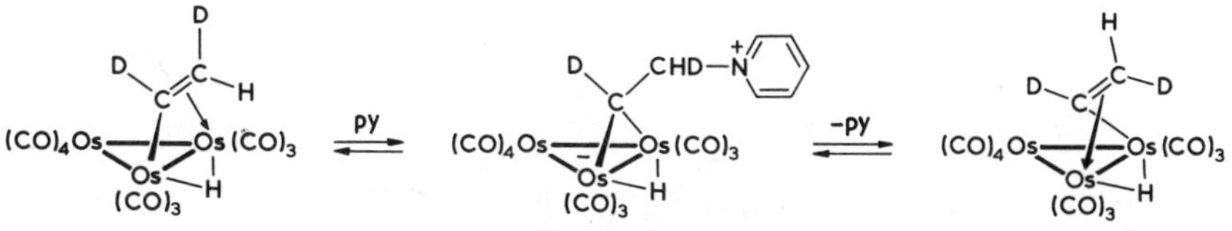

SCHEME 18.

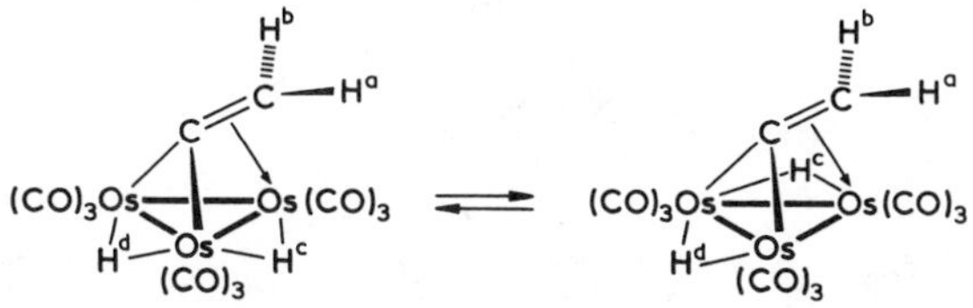

SCHEME 19.

migration is favored (Scheme 19). This is confirmed for $Os_3H_2\{C=CMe-(CH_2CHMe_2)\}(CO)_9$ for which the diastereomers, which are related to the two forms in Scheme 19, interconvert without exchange of the diastereotopic isopropyl Me groups. Chirality of the ligand is retained and so rotation about the $C=C$ bond is not the mechanism for isomer interconversion (370).

This contrasts with the results on $[Co_3(\mu_3-C=CHR)(CO)_9]^+$. When $R = CHMe_2$, the exchange of the diastereotopic Me groups confirms that there is a dynamic enantiomerization, probably a disrotatory rotation of the $C=CHCHMe_2$ ligand correlated with a rotation of the organic ligand from one site at the metal atoms to another (317). The theory behind this has been tackled (372). For the protonated compound $[Os_3H_3(C=CH_2)(CO)_9]^+$, the exchange of the two equivalent hydride ligands with the third could be explained by hydride or ligand motions (36). Hydride rigidity was claimed for the methylenecyclobutane derivative $[Os_3H_3(C=CCH_2CH_2CH_2)(CO)_9]^+$ (75), but this was based on a misinterpretation of the ^{187}Os satellites of the hydride ^1H-nmr signal. Even so there was evidence for a rotation about the $C=C$ bond to give a time-averaged plane of symmetry through the organic ring (219). It is reasonable that μ_3 ligand motion would be faster in the protonated form.

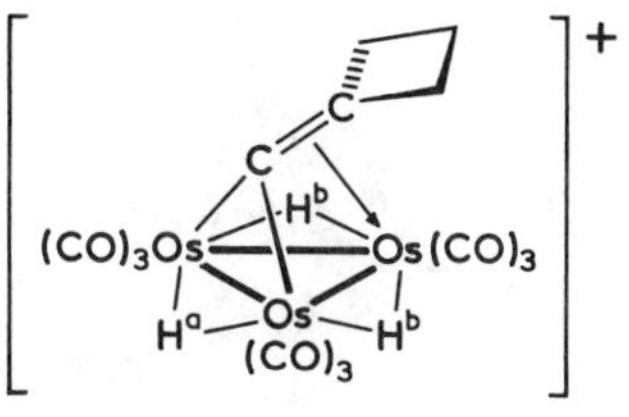

75

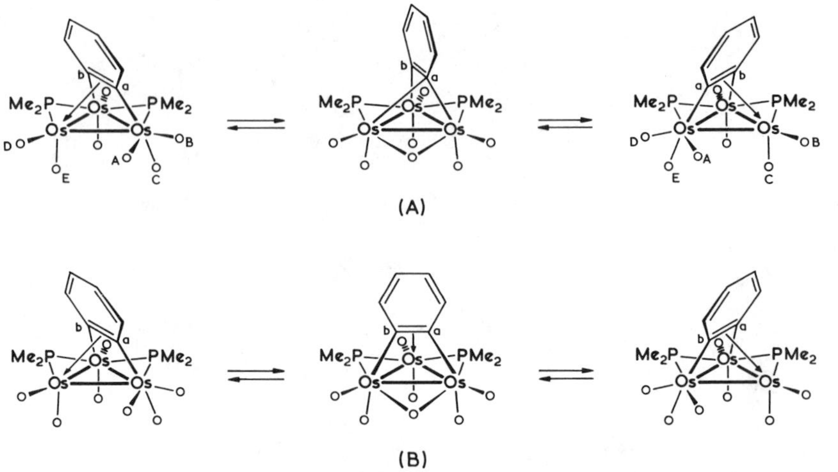

(A)

(B)

SCHEME 20.

Fluxionality is a common feature of μ_3-R_2C_2 and μ_3-C_6H_4 ligands. As an example, the cluster $Os_3(PMe_2)_2(C_6H_4)(CO)_7$ undergoes two separate processes both involving 60° rotations of the C_6H_4 ligand about the metal triangle. Process A (Scheme 20) exchanges the PMe_2 ligands, generates a plane of symmetry, and equilibrates carbonyl ligands A to E. Migration of carbonyl A is required as is also some one-center exchange. Process B is slower and involves two 60° rotations to symmetrize the C_6H_4 ligand (59,231,232). Studies on $Os_3H(AsMe_2)(\mu_3$-$C_6H_3{}^iPr)(CO)_9$ have established that the rotation process involves exchange of the diastereotopic isopropyl methyl groups and hence a time-averaged symmetry plane through the organic ring (59). Observations on $Os_3H(S^iPr)(\mu_3$-$C_6H_4)(CO)_9$ have been interpreted in terms of inversion at sulfur and hydride migration, but μ_3-C_6H_4 mobility was not ruled out (95). C_6H_4 motions are implicated in other studies (20,106,221,235). In $Os_3(\mu_3$-$C_6H_4)(\mu_3$-$PEt)(CO)_9$ with two Os—Os bonds like in $Os_3H(X)$-$(C_6H_4)(CO)_9$ [X = $AsMe_2$ (59) or SMe (95)], the μ_3-C_6H_4 ligand adopts a different orientation (76) but still undergoes rotation (106). Since C_6H_4 rota-

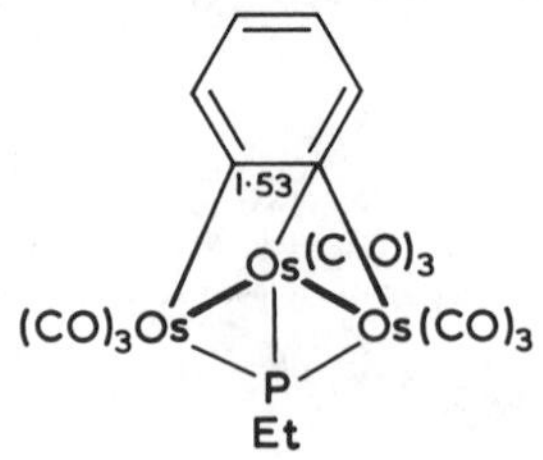

76

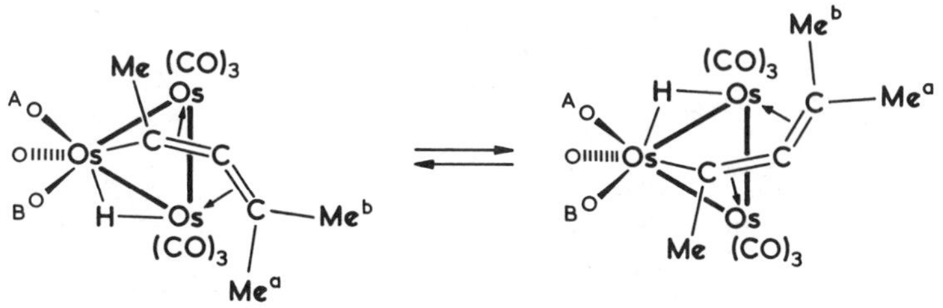

SCHEME 21.

tions are so fast, different orientations in the ground state are expected, and that in **76** could correspond to an intermediate in processes like those in Scheme 20. The long C—C distance is, however, unlike that in normal μ_3-arynes.

Alkyne rotation with a simultaneous flip from one face to another is detected for the clusters $Os_3(CO)_{11}(alkyne)$, where the alkyne is EtC_2Et, HC_2CMe_2OH, or $HOCH_2C_2CH_2OH$, containing diastereotopic groups which exchange in this process (222,223,373). The same applies to the clusters $Os_3H_2(alkyne)(CO)_9$ [alkyne = indyne (228), $\overline{OCH_2CH_2C\!=\!C}$ (230)], $Os_3H(MeC_2CH_2PMe_2Ph)(CO)_9$, and $Os_3H(HC_2PMe_2Ph)(CO)_9$ (108b). The dynamic behavior of other μ_3-alkyne clusters has been studied (220,217); often CO exchange and hydride migrations are also involved.

The μ_3 ligands in the clusters $M_3H(\mu_3\text{-}MeC\!=\!C\!=\!CMe_2)(CO)_9$ (M = Ru, Os) are mobile (mobility for Ru > Os) leading to the exchange of Me^a with Me^b and of CO^A with CO^B by a process shown in Scheme 21 (253). Ligands of the type $\mu_3\text{-}C\!\equiv\!CR$ appear to be rigid (244).

It has been shown that [187]Os satellites on [1]H-nmr hydride signals may be used to identify otherwise unobservable dynamic processes. So far, this has not been applied extensively and not to hydrocarbon ligands (374).

C. Dimerization and Trimerization of Alkynes

$Os_3(CO)_{12}$ catalyzes the formation of hexaphenylbenzene from diphenyl-acetylene, and Scheme 22 summarizes the known reactions of Ph_2C_2 with Os_3 clusters. Direct reaction of $Os_3(CO)_{12}$ leads to products containing two or three alkynes, mainly $Os_3(C_4Ph_4)(CO)_9$ and $Os_3H(C_6H_4C_4Ph_4)(CO)_8$ (193,227,245,266,267,375–377). The compound $Os_3(C_4Ph_4)(CO)_9$, containing the metallacyclopentadiene ring, reacts further with PhC_2Ph to introduce a μ_3-alkyne. Unfortunately the structure of $Os_3(CO)_6(C_6Ph_6)$ is unknown, but it clearly does not correspond with that of $Os_3(CO)_9(C_6H_6)$ (10) containing a μ_3-benzene ligand. Probably the C_6 ring has not yet formed in $Os_3(CO)_6(C_6Ph_6)$. Compounds with only one PhC_2Ph ligand are available by

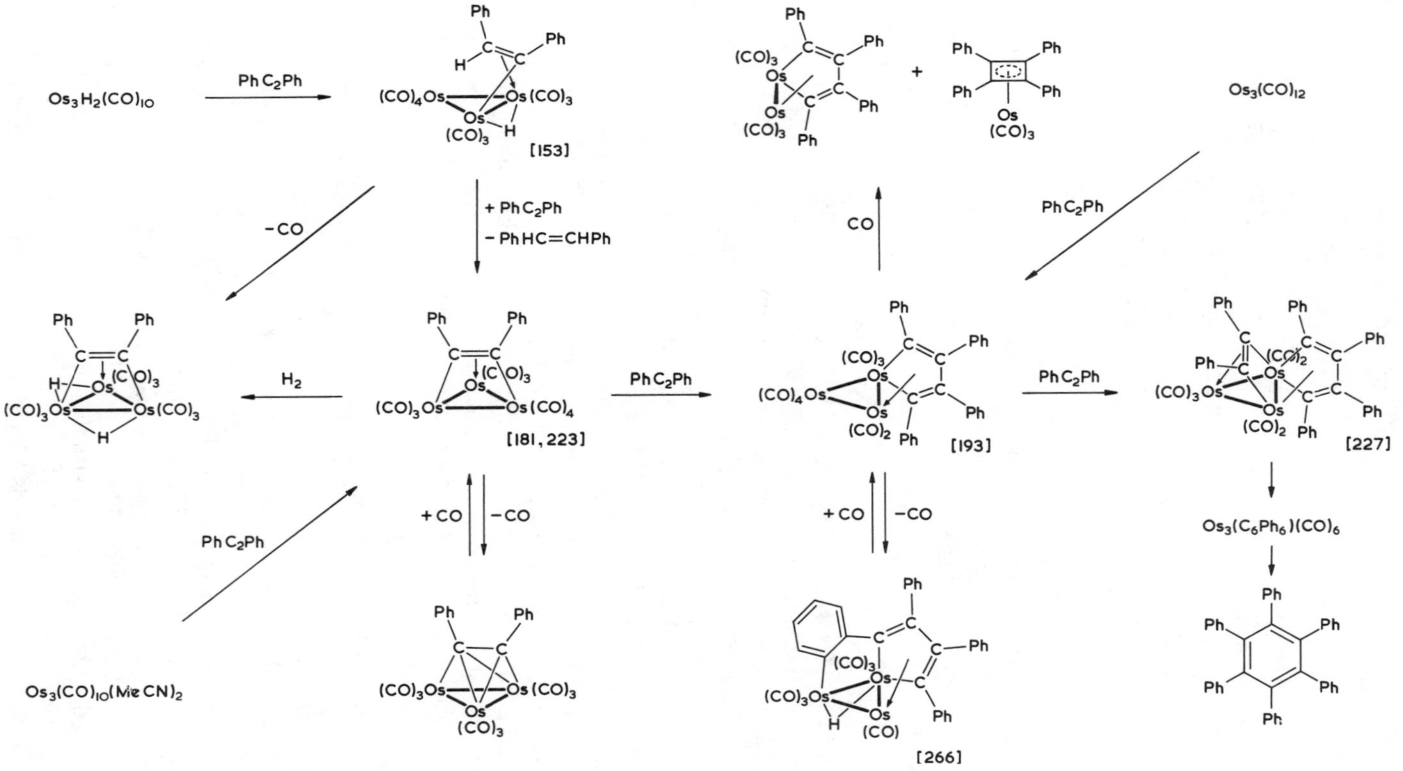

SCHEME 22.

reaction of $Os_3H_2(CO)_{10}$ ($161,181,218$) or $Os_3(CO)_{10}(MeCN)_2$ (21) with the alkyne. The whole sequence of reactions leading to C_6Ph_6 is established, but this may, of course, not correspond to the catalytic route. References to X-ray structures are given next to the compounds in Scheme 22. The structure of $Os_3(PhC_2Ph)(CO)_9$ is unknown, but may be like those of the Fe_3 (225) or Fe_2Ru (226) analogs.

D. *Catalysis and Related Topics*

1. *Supported Os$_3$ Clusters*

Linking a cluster to a surface might be an advantage in catalysis by allowing homogeneous-type chemistry without problems of separating catalyst from reagents and products and also by suppressing bimolecular reactions between clusters (especially in unsaturated or catalytically active forms) that could lead to cluster reformation and/or deactivation. This is the philosophy behind attempts to bind clusters to oxide supports by direct reaction with surface OH groups, by functionalizing the surface with ligands such as tertiary phosphines and isocyanides before binding the cluster, or by forming clusters with groups such as —$Si(OEt)_3$ which are known to couple irreversibly with oxide surfaces (378). Chromatography on SiO_2 or Al_2O_3 is one of the main synthetic tools in cluster synthesis, so binding of clusters to oxide surfaces does not usually occur spontaneously on contact.

Physisorbed $Os_3(CO)_{12}$ on silica, alumina, etc. (usually at least partially dehydroxylated) requires higher temperatures, e.g., 150°C, to decarbonylate and form bonds to the oxide ($379-382$). Lower temperatures are needed for $Os_3(CO)_{10}(MeCN)_2$ or $Os_3(CO)_{10}(diene)$ to give similar (or the same) chemisorbed cluster ($330,379$). This species is generally believed to be $Os_3H(CO)_{10}(O$—$M)$, where O—M represents a terminal oxide at the surface, formed by oxidative addition of HO—M as in the formation of $Os_3H(OSiPh_3)(CO)_{10}$ from $HOSiPh_3$ ($330,379-382$). At various temperatures between 100 and 400°C cluster breakdown gives ill-defined species exhibiting new $v(CO)$ absorptions. Infrared spectra, gas phase analysis, and EXAFS (274) are the main techniques for identifying supported clusters.

By adding groups such as $HS(CH_2)_3Si(OMe)_3$ ($383,384$) or Ph_2P-$(CH_2)_2Si(OEt)_3$ ($385,386$) to the oxide surface, substitution at the $Si(OR)_3$ group leads to attachment, and the sulfur and phosphorus atoms can subsequently be used to link a cluster. Cluster-supported species such as Os_3H_2-$(CO)_x(PPh_2CH_2CH_2SIL)$ ($x = 9$ or 10), $Os_3(CO)_{11}(PPh_2CH_2CH_2SIL)$ ($385,386$), and $Os_3H(CO)_{10}(SCH_2CH_2CH_2SIL)$ ($383,384$) have been made on silica (SIL) and related compounds on other oxides. Polystyrene-linked clusters have been made ($387-389$). Poly(4-vinylpyridine) has allowed

Os_3 clusters to be supported which have infrared spectra like that of $Os_3H(NC_5H_4)(CO)_9(py)$ *(45,388)*. Polystyrene functionalized with PPh_2 groups has been used to bind $Os_3H_2(CO)_9(PPh_2\text{-polystyrene})$ and $HAuOs_3\text{-}(CO)_{10}(PPh_2\text{-polystyrene})$ *(387)*.

The alternative approach is to prepare Os_3 clusters substituted so that they will readily bind to oxide surfaces *(385,386,390,391)*. The isonitrile $CN(CH_2)_3Si(OEt)_3$ reacts with $Os_3H_2(CO)_{10}$ to give $Os_3H_2(CO)_{10}\text{-}[CN(CH_2)_3Si(OEt)_3]$, which isomerizes by insertion to give $Os_3H(CO)_{10}\text{-}[CNH(CH_2)_3Si(OEt)_3]$. Both isomers bind to an oxide surface to give the same species $Os_3H(CO)_{10}(CNHCH_2CH_2CH_2SIL)$, where SIL is silica linked to the cluster through the substituted $Si(OEt)_3$ group *(391)*. Likewise $CH_2{=}CHSi(OEt)_3$ reacts with $Os_3(CO)_{10}(MeCN)_2$ to give $Os_3H(CO)_9\text{-}[C{\equiv}CSi(OEt)_3]$, which can then be bound to SiO_2 *(386)*. $Ph_2PC_2H_4Si\text{-}(OEt)_3$-substituted Os_3 clusters can be prepared prior to supporting on an oxide surface *(385,386)*. The use of $HS(CH_2)_3Si(OMe)Me_2$ has allowed Os_3 clusters to be linked to silica; both approaches were used *(390)*.

2. Catalysis by Supported or Free Os₃ Clusters

Catalysis by clusters has been a popular topic for discussion in spite of (or because of) our limited knowledge. Analogies with metal surfaces have been developed by some researchers while others have particularly considered supported clusters (see, for example, refs. *378* and *395–399*).

Directly supported clusters of type $Os_3H(CO)_{10}(O\text{—metal oxide})$ break down at quite low temperatures to give species which have a high selectivity to methane from CO and H_2 *(381,400)*. Similar behavior has been reported for $Os_3(CO)_{12}$ itself *(401)*, but it is difficult to rule out metal as the catalyst. $Os_3(CO)_{12}$ also leads to methanol, methyl and ethyl formate, and acetone by reaction with CO and H_2 (190°C, 180 atm) in glyme solvents *(402)*. The water–gas-shift reaction is catalyzed by $Os_3(CO)_{12}$, using KOH or even sodium sulfide in methanol as the base *(403)*, although ruthenium catalysts are better *(404)*.

The cluster $Os_3H_2(CO)_{10}$ catalyzes the isomerization of 1-alkenes to internal alkenes at room temperature or above *(51,55,394,405)*. $Os_3(CO)_{12}$ is also a catalyst at over 100°C *(406)*, and the supported clusters $Os_3H_2\text{-}(CO)_9(PPh_2\text{—surface})$ on polystyrene or silica *(394)*, $HAuOs_3(CO)_{10}\text{-}(PPh_2\text{—surface})$ on silica *(387)*, and $Os_3H(CO)_{10}(O\text{—Al})$ directly attached to alumina and promoted by H_2 *(393)* will also catalyze alkene isomerization. The catalyst $Os_3H_2(CO)_{10}$ is purple but solutions eventually become yellow and deactivated as species such as $Os_3H(vinyl)(CO)_{10}$ are formed. Simple alkenes are known to insert reversibly, and $Os_3H_2(CO)_{10}$ induces scrambling to give $C_2H_{4-x}D_x$ ($x = 0$ to 4) starting with *trans-*

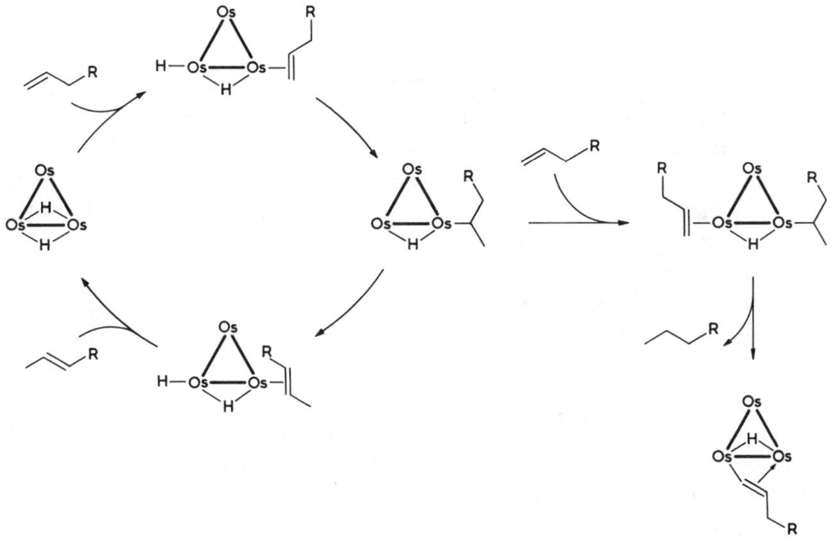

SCHEME 23.

$C_2H_2D_2$ (405). Scheme 23 shows the most probable mechanism and deactivation route. The structure of the intermediate $Os_3H(alkyl)(CO)_{10}$ is unknown but is likely to involve agostic bonding through the alkyl rather than as shown. The only evidence for the nature of such species comes from $Os_3H(CH_3)$-$(CO)_{10}$ (41,42) or where there are donor atoms to satisfy unsaturation such as in $Os_3H(MeCHOMe)(CO)_{10}$ (243). If anything the supported cluster is more active than the unsupported one (394).

Other reactions that have been found to be catalyzed by Os_3 clusters are alkene hydrogenation (51,385,392,407) and hydroformylation (408). Other researchers have considered pathways for alkyne hydrogenation at Os_3 clusters (186) and for isonitrile (409) and nitrile (118,213) reduction. Os_3-$(CO)_{12}$ also catalyzes the conversion of dimethylaniline to $(4\text{-}Me_2NC_6H_4)_2$-CH_2 (203) and exchange of D_2O with trialkylamines (410,411).

VI

CHANGES IN NUCLEARITY OF TRIOSMIUM CLUSTERS

A. Reduction in Nuclearity

Ultraviolet photolysis of $Os_3(CO)_{12}$ in the presence of dienes such as butadiene (291), cyclooctatetraene (291), or cyclooctadienes (412) leads to mononuclear species of type $Os(CO)_3(diene)$. Irradiation of methylacrylate with $Os_3(CO)_{12}$ gives $Os(CO)_4(CH_2CHCO_2Me)$ and $Os_2(CO)_8$-(CH_2CHCO_2Me) (292); the latter has a μ,η^1,η^1 alkene bridge. The parent species $Os_2(CO)_8(C_2H_4)$ was synthesized from dinuclear precursors (413).

Dinuclear osmium(I) compounds of general formula $Os_2(\mu\text{-}X)_2(CO)_6$ have been synthesized by oxidation of $Os_3(CO)_{12}$ with X_2 (X = I) (300), but the high-temperature treatment of $Os_3(CO)_{12}$ with various reagents leads initially to trinuclear clusters which at the very highest temperatures (e.g., above 180°C) break down to these dinuclear compounds. Acetic acid reacts with $Os_3(CO)_{12}$ to give the compound with X = CH_3CO_2 (414); pyridine at 184°C gives the μ-2-pyridyl species $Os_2(NC_5H_4)_2(CO)_6$ as two isomers (cis and trans) with the nitrogen atoms bonded to the same or different osmium atoms, respectively (45). Related compounds are formed similarly: $Os_2(RNCHNR)_2$-$(CO)_6$ (R = ^iPr or $PhCH_2$) from the appropriate formamidines (171) and cis- and trans-$Os_2(NC_5H_4X)_2(CO)_6$ (X = O or NH) from 2-pyridone or 2-aminopyridine (173). Dinuclear compounds are formed in very low yield along with trinuclear species from the thermolysis of $Os_3(CO)_{10}L_2$ (L = PMe_2Ph or $AsMe_2Ph$). The species $Os_2(PMe_2)$ $(PMe_2C_6H_4)(CO)_6$ and $Os_2(C_6H_4)$ $(AsMe_2)_2(CO)_6$ were identified spectroscopically (232). The dihydride $Os_3H_2(CO)_{10}$ reacts with cyclonona-1,2-diene to give tri- and dinuclear compounds (252); $Os_2(C_9H_{14})_2(CO)_6$ has two $Os(CO)_3$ groups bridged by μ,η^1,η^3-allene ligands (415). Reactions leading to mono- or dinuclear species are much rarer for $Os_3(CO)_{12}$ than for $Ru_3(CO)_{12}$, but probably many reactions reported in this review give trace quantities of dinuclear compounds that have never been characterized in a search for clusters.

B. *Increase in Nuclearity*

Exhaustive reduction of $Os_3(CO)_{12}$ with Na sand leads to $Na[OsH(CO)_4]$ (416), while reduction in a similar way but with 2, 2'-bipyridyl as catalyst gives $Na_2[Os(CO)_4]$, which reacts readily with traces of moisture to give the monohydride (417). On the other hand partial reduction of $Os_3(CO)_{12}$ using only 1 mol Na/mol Os_3 in diglyme at 162°C gives $[Os_6(CO)_{18}]^{2-}$ (65%), which is probably formed by condensation of $[Os_3(CO)_{11}]^{2-}$ (315) with $Os_3(CO)_{12}$ (418). Oxidation of $[Os_6(CO)_{18}]^{2-}$ with iron(III) gives $Os_6(CO)_{18}$, which is formed as the major product from vacuum thermal decarbonylation of $Os_3(CO)_{12}$ at 150°C (293). Various other low-yield products are formed from this and similar vacuum pyrolyses: neutral binary carbonyls like $Os_5(CO)_{16}$, $Os_7(CO)_{21}$, and $Os_8(CO)_{23}$ (293), carbides such as $Os_8C(CO)_{21}$ and $Os_5C(CO)_{15}$ (293), and anions such as $[Os_{11}C(CO)_{27}]^{2-}$ (295) and $[Os_{10}C(CO)_{24}]^{2-}$ (294). The Os_{10} dianion is formed from $[Os_6(CO)_{18}]^{2-}$ in triglyme at 216°C (418).

Other methods have been applied to obtaining higher clusters. Vacuum thermolysis of $Os_3(CO)_{11}(py)$ gives, after the addition of [PPN]Cl in solvent, $[PPN][Os_5H(CO)_{15}]$, $Os_3H(2\text{-pyridyl})(CO)_{10}$, and $[PPN]_2[Os_{10}C(CO)_{24}]$. Increasing the duration of the heat treatment gives more of the Os_{10} cluster and also $Os_5C(CO)_{14}H(2\text{-pyridyl})$, which contains the NC_5H_4 ligand bridging across a nonbonded pair of osmium atoms (324). Thermolysis of $Os_3(CO)_{12}$ at 230°C in the presence of a little water (0.02 g/0.5 g cluster) gives a range of hydrides each in low yield: $Os_3H(OH)(CO)_{10}$, $Os_4H_2(CO)_{13}$, $Os_4H_4(CO)_{12}$, $Os_5H_2(CO)_{10}$, $Os_5H_2(CO)_{15}$, $Os_6H_2(CO)_{18}$, and $Os_7H_2C\text{-}$ $(CO)_{19}$ (419). Heating $Os_3(CO)_{12}$ in ^iBuOH gave $[Os_{10}H_4(CO)_{24}]^{2-}$, $[Os_8H(CO)_{22}]^-$, $[Os_{10}C(CO)_{24}]^{2-}$, and another species presumably derived from the solvent, $[Os_9(CO)_{21}(CHCMeCH)]^-$ (420). The same reaction under basic conditions (KOH added) gave $[Os_3H(CO)_{11}]^-$, $Os_3H(O^iBu)(CO)_{10}$, and $[Os_4H_3(CO)_{12}]^-$ over a short reaction time but also $Os_6(CO)_{18}$, $[Os_6H(CO)_{18}]^-$, $[Os_8H(CO)_{22}]^-$, and $[Os_{10}HC(CO)_{24}]^-$ over longer periods (421). Clearly, complex chemistry is occurring in these solutions, and a wide range of nuclearities and compound types are obtainable directly from $Os_3(CO)_{12}$ (7,8).

Heating $Os_3(CO)_{12}$ in hexane with H_2 under pressure gives $Os_4H_4(CO)_{12}$ quantitatively (301), while a similar treatment with C_2H_4 (20 atm, 160°C) gives $Os_3H_2(C{=}CH_2)(CO)_9$ as the major product (422), which is also formed under milder conditions (35). Other products isolated from the pressure treatment were $Os_4(CO)_{12}(CH{\equiv}CH)$ and $Os_4(CO)_{12}(CH{\equiv}CEt)$, the X-ray structures of which involve Os_4C_2 octahedra (422).

The clusters $Os_3(CO)_{12-n}(CNR)_n$ ($n = 1$ to 4) readily form the higher clusters $Os_6(CO)_{18-n}(CNR)_n$ ($n = 1$ to 5) in refluxing octane (321,323). This process occurs much more readily than for $Os_3(CO)_{12}$, which is indefinitely stable in refluxing octane and only gives $Os_6(CO)_{18}$ at a significant rate at 150°C or above. The structure of $Os_6(CO)_{16}(CN^tBu)_2$ relates directly to that of $Os_6(CO)_{18}$ (423).

The building of higher clusters is often facilitated by the use of strongly bridging ligands such as sulfur, which has been examined extensively. Direct reaction of $Os_3(CO)_{12}$ with elemental sulfur at 200 to 260°C gives $Os_3(S)_2\text{-}$ $(CO)_9$, $Os_5(S)(CO)_{15}$, $Os_6(CO)_{18}$, $Os_7(S)_2(CO)_{20}$, and $Os_{10}(S)_2(CO)_{23}$ (99), while in refluxing octane hydrides such as $Os_4H_2(S)(CO)_{12}$ are formed (98). Ultraviolet photolysis of $Os_3(S)_2(CO)_9$ gives $Os_6(S)_4(CO)_{16}$ (see Scheme 4) (104). Thermal decomposition of $Os_3H(SPh)(CO)_{10}$ is quite unlike that of $Os_3H(OPh)(CO)_{10}$ which gives the cyclohexadienone derivative $Os_3H_2\text{-}$ $(C_6H_4O)(CO)_9$ (see Fig. 12) (166). Sulphur-carbon bond cleavage occurs to give a range of Os_3 to Os_6 clusters with μ_3-S and μ_4-S bridges; the X-ray structures of several have been determined, including $Os_4(\mu_3\text{-}S)_2(CO)_{12}$, $Os_3(\mu_3\text{-}S)(\mu_3\text{-}CO)(CO)_9$, and $Os_6(\mu_4\text{-}S)(\mu_3\text{-}S)(CO)_{16}$ (101–103). A similar treatment of $Os_3H(SCH_2Ph)(CO)_{10}$ gives an even wider range of S-bridged clusters (*424*). A more rational synthesis based on the excess donor capacity of μ_3-S bridges and the ease of displacement of MeCN from $Os_3(CO)_{10}(MeCN)_2$ is reaction of the bis(acetonitrile) compound with $Os_3(\mu_3$-S)(CO)_{10}$ to give $Os_6(\mu_3$-S)(CO)_{19}$ and $Os_5(\mu_4$-S)(CO)_{15}$. The Os_6 compound thermolyzes to $Os_6(\mu_4$-S)(CO)_{17}$. The structures of many of these species are known (*425,426*). A similar method was used to react $Os_3(CO)_{10}(MeCN)_2$ with $Os_4(\mu_3$-S)(CO)_{12}$ to give $Os_7(\mu_4$-S)(CO)_{19}$ (*427*). Thermolysis of $Os_3H(\mu_3$-S)-$($\mu$-CH$=$NR)(CO)_9$ (R = aryl) gives the clusters $Os_6(\mu_4$-S)(CH$=$NR)_2$-$(CO)_{18}$, $Os_6H_2(\mu_4$-S)(\mu_3$-S)(CH$=$NR)_2(CO)_{17}$ (two isomers), $Os_6H_2(\mu_4$-S)-$(\mu_3$-S)(CH$=$NR)_2(CO)_{16}$ (three isomers), and $Os_6(\mu_4$-S)_2(CH$=$NR)_2(CO)_{15}$ (*428–430*), while hydrogenation of the same compound gives $Os_6H_4(\mu_4$-S)-$(\mu_3$-S)(CH$=$NR)_2(CO)_{15}$ (two isomers) and $Os_6H_6(\mu_4$-S)(\mu_3$-S)(CH$=$NR)_2$-$(CO)_{14}$ (*431*). These clusters are held together by μ_3-S, μ_4-S, and μ-CH$=$NR bridges.

The possibilities of PR or PR_2 bridges in holding clusters together are suggested by the thermolysis products of $Os_3(CO)_{11}[P(OMe)_3]$ (*331*) which are hexanuclear, unlike the thermolysis products of similar clusters with PPh_3 (*30,204*), PMe_2Ph (*231*), $PEtPh_2$ (*106*), PMe_3 (*221,241*), PEt_3 (*221*), etc., which are all trinuclear. Two Os_5 products have been structurally characterized: $Os_5(\mu_4$-POMe)(CO)_{15}$ and $Os_5H(C)(CO)_{13}[OP(OMe)OP(OMe)_2]$ (*432*). The cluster $Os_3(S)_2(CO)_8(PPh_2H)$ condenses through PPh_2 bridges giving $Os_6(S)_4(\mu$-PPh$_2)_2(CO)_{14}$ on heating at 125°C (*433*).

The reaction of $Os_3H_2(CO)_{10}$ with azobenzene PhN$=$NPh does not occur under mild conditions, but at 125°C cluster reformation leads to Os_5H-$(CO)_{13}(PhNC_6H_4N)$ (*434*). This compound forms simple adducts with PEt_3, CN^{t}Bu, or CO which surprisingly have essentially the same cluster skeleton as their parent (*435*).

This survey of higher Os clusters covers their direct formation from Os_3 clusters only. Studies of the structures, properties, and reactions of high nuclearity osmium clusters encompass a large area, too big to be considered in this article.

VII

TRIOSMIUM CLUSTERS WITH GROUP IV ELEMENTS OTHER THAN CARBON

A natural extension of oxidative addition reactions of Os_3 clusters with C—H bond cleavage is that with Si—H, Ge—H, or Sn—H bond cleavage. $Os_3(CO)_{12}$ reacts at high temperatures (e.g., 140°C) or under uv photolysis with Me_3SiH to give mono- and dinuclear compounds, exemplified by $Os(SiMe_3)_2(CO)_4$, $OsH(SiMe_3)(CO)_4$, and $Os_2(SiMe_3)_2(CO)_8$ (436). Some related compounds of Ge and Sn are also formed. No simple adducts of type $Os_3H(SiMe_3)(CO)_{12}$ were obtained that are analogous to the oxidative addition product $Os_3Cl(SnCl_3)(CO)_{12}$ from $Os_3(CO)_{12}$ and $SnCl_4$, which presumably has a chain of osmium atoms and terminal Cl and $SnCl_3$ groups (437). Trinuclear clusters can, however, be regenerated by heating $Os(GeMe_3)_2(CO)_4$, which gives $Os_3(GeMe_2)_3(CO)_9$ (77), having

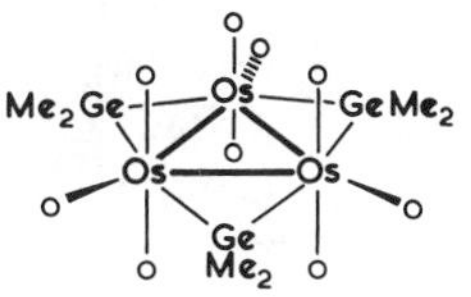

77 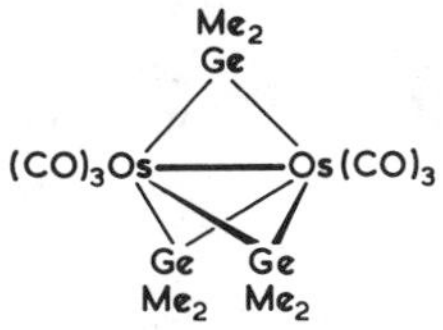   78 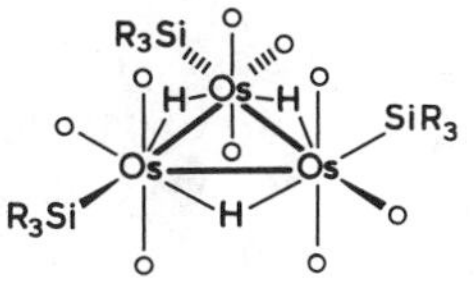   79

the same structure as that established for the Ru_3 analog (X-ray structure) (438). The dinuclear compound $Os_2(GeMe_2)_3(CO)_6$ (78) is also formed in this pyrolysis and is structurally like the iron compound for which the structure is established (439). These species suggest the possibility of forming μ-alkylidene compounds such as $Os_2(\mu\text{-}CH_2)_3(CO)_6$ and $Os_3(\mu\text{-}CH_2)_3(CO)_9$ that are at present unknown. In fact, it has been speculated that, in most of these cases at least, the trinuclear clusters are formed from mononuclear compounds. Thus $Os_3(CO)_{12}$ reacts with $SiHCl_3$ to give cis-$OsH(SiCl_3)(CO)_4$ initially, but this spontaneously decarbonylates to give $Os_3H_3(SiCl_3)_3(CO)_9$ (79) (440). Ultraviolet irradiation of the trimer under CO (2 atm) leads back to the monomer. Thermally $Os_3(CO)_{12}$ reacts with $SiHCl_3$ under an atmosphere of CO to given trans-$Os(SiCl_3)_2(CO)_4$ and $Os_3(SiCl_3)_2(CO)_{12}$ (5) (81). X-Ray structures of 79, where $SiR_3 = SiCl_3$ (81) or $SiMeCl_2$ (441), and of 5 have been determined. Unlike $Os_3I_2(CO)_{12}$, in which the terminal I ligands are cis to the Os—Os bonds in a linear Os_3 unit (78), the five metal atoms in 5 are close to being linear.

An alternative approach is to add oxidatively to $Os_3H_2(CO)_{10}$. $SiHPh_3$ adds to give $Os_3H_3(CO)_9(SiPh_3)$ (80), which is formally unsaturated like the parent dihydride. The hydride ligands A and B rapidly exchange intramolecularly, but the exchange of these with C is slow and only detected by spin saturation transfer experiments. Os—Os distances are related to those in

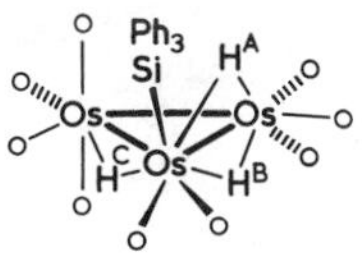

80 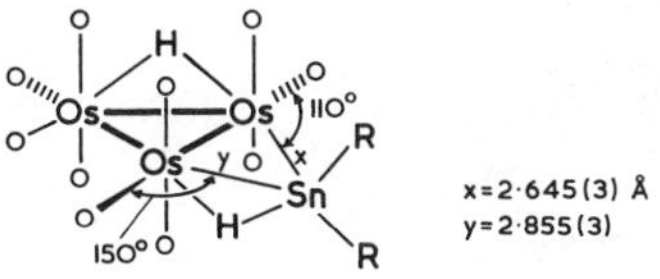   81

$Os_3H_2(CO)_{10}$ in that the shortest distance [2.7079(4) Å] is associated with the two metal atoms spanned by the double hydride bridge. As expected the single hydride bridge leads to a long Os—Os distance (442). It is not clear yet whether **80** undergoes simple ligand addition as does $Os_3H_2(CO)_{10}$.

Perhaps the most interesting group IV system is that obtained by insertion of SnR_2 [R = $CH(SiMe_3)_2$] into $Os_3H_2(CO)_{10}$ (367). The product $Os_3H(SnHR_2)(CO)_{10}$ (**81**) relates interestingly to $Os_3H(CH_3)(CO)_{10}$, formed by CH_2 insertion using CH_2N_2. By a combination of ^1H nmr with Sn coupling and an X-ray structure (hydrogen atoms not located) the structure was shown to have an Os—H—Sn bridge related to the agostic bridge found in the methyl compound. Only one of the hydride ligands shows coupling to ^{117}Sn and ^{119}Sn nuclei [258.7 and 298.0 Hz, respectively (367)]. The methyl compound might not have the CH_3 and H bridges between the same Os atoms as originally proposed (41,42), but rather as in **81**.

VIII

TRIOSMIUM CLUSTERS WITH INTRODUCED TRANSITION METAL ATOMS

A. *General Considerations*

A metal atom may be added to an Os_3 cluster along with its associated set of ligands or, in the case of Ag, Au, and Hg, without ligand. For example, these metal atoms may bridge Os_3 units as in $\{Os_3(CO)_{11}\}_3Hg_3$ (443), $[\{Os_3H(CO)_{10}\}_2Ag]^-$ (444), and the corresponding gold compound (445). The Os_3 clusters with added metal atoms are, of course, heterometallic clusters and usually discussed in terms of the overall metal geometry. $Os_3NiH_3(CO)_9(C_5H_5)$ (**82**) is a 60-electron cluster, and six metal–metal bonds are predicted; it adopts a tetrahedral arrangement (446,447). $Os_3Re_2H_2$-$(CO)_{20}$ (**83**), on the other hand, should have five metal–metal bonds since it is an 80-electron cluster; it was shown to have three Os–Os and two Os–Re

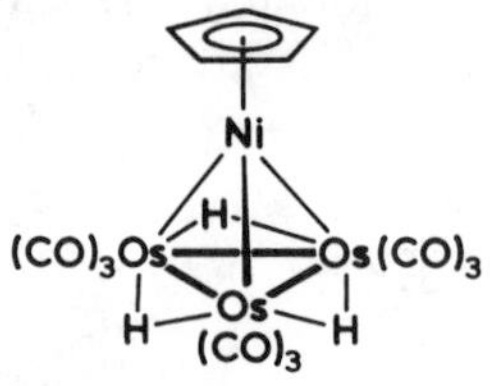

82 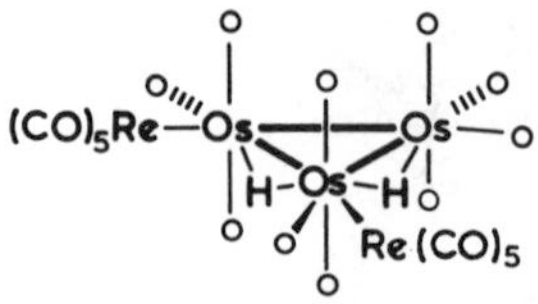   83

bonds (*448*). One could, alternatively and appropriately when emphasizing Os_3 chemistry, consider these as Os_3 clusters with μ_3-Ni(C_5H_5) or μ_1-Re(CO)$_5$ as ligands. The relationship between the 15-electron Ni(C_5H_5) group and the 5-electron CR one is apparent, and both behave as 3-electron donors X in molecules of type $Os_3H_3(\mu_3$-X)(CO)$_9$. Likewise the 17-electron group Re(CO)$_5$ will act as a 1-electron donor; there are two such terminal ligands in $Os_3Re_2H_2(CO)_{20}$. Considering Os_3 clusters in this way, we can identify three types of clusters in which the added metal atom is terminally (μ_1), doubly (μ_2), or triply (μ_3) bridging, and these will be dealt with in turn.

B. *Added d-Block Metals as Terminal Ligands*

The metal terminal ligand systems have been prepared in various ways. Three main methods are briefly described here.

1. *Nucleophilic Addition at $Os_3H_2(CO)_{10}$*

While $Os_3H_2(CO)_{10}$ undergoes nucleophilic addition of neutral nucleophiles (PR$_3$, CO, RNC, etc.) to give simple adducts (Section IV), it also adds anions (halide, H$^-$, or metal carbonyl anions) (*359*). The addition of K[Mn(CO)$_5$] gave K[Os$_3$H$_2$(CO)$_{10}${Mn(CO)$_5$}], isolated as a bright yellow solid, which was stable at room temperature and which contains one terminal hydride (δ −10.4, d) and one bridging (δ −20.4, d) like better characterized adducts with neutral donors. Adducts with Na$_2$[Fe(CO)$_4$], K[Mo(CO)$_3$-(C$_5$H$_5$)], K[Fe(CO)$_2$(C$_5$H$_5$)], or K$_2$[V(CO)$_3$(C$_5$H$_5$)] are also formed but decompose at room temperature. Presumably the adducts have structures like that shown for K[Os$_3$H$_2$(CO)$_{10}${Mn(CO)$_5$}] (**84**) (*359*).

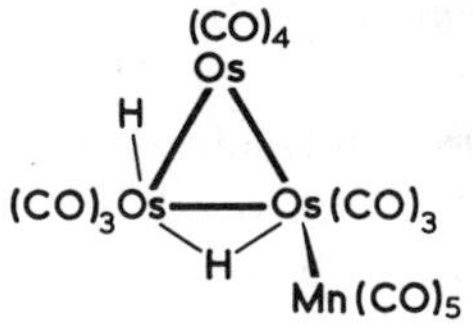

84 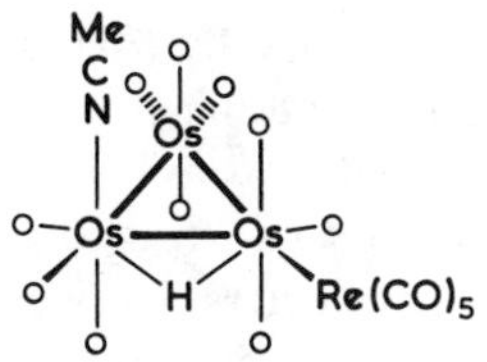   **85**

2. *Oxidative Addition of Metal Hydrides*

This route has been applied to oxidative addition of the mononuclear species ReH(CO)$_5$ or OsH$_2$(CO)$_4$ to Os$_3$(CO)$_{11}$(MeCN) or to Os$_3$(CO)$_{10}$L$_2$ (L = MeCN or cyclooctene). The compounds Os$_3$H(CO)$_{11}${Re(CO)$_5$} (*449*) and $Os_3H(CO)_{11}\{OsH(CO)_4\}$ (*450*), formed by addition to $Os_3(CO)_{11}$-(MeCN), are probably structurally related to $Os_3H_2(CO)_{11}$ (*53*), with the terminal hydride replaced by a terminal metal group. This was confirmed by the X-ray structure of the derivative $Os_3H(CO)_{10}(MeCN)\{Re(CO)_5\}$ (**85**), except that the Re group was found to be equatorial (*451,452*). This acetonitrile derivative was made by treating $Os_3H_2(CO)_{10}\{Re(CO)_5\}_2$ with refluxing acetonitrile; an acetonitrile molecule replaces a molecule of $ReH(CO)_5$. The double oxidative addition product $Os_3H_2(CO)_{10}\{Re(CO)_5\}_2$ **83** was prepared from $ReH(CO)_5$ and $Os_3(CO)_{10}(C_8H_{14})_2$ (*449*).

An alternative route to Os_3Re clusters is to add $[Re(CO)_5]^-$ and H^+ successively to $Os_3(CO)_{12}$, rather than by oxidative addition of $ReH(CO)_5$ (*453*), but a wider range of compounds in lower yield was obtained: $Os_3H(CO)_{11}\{Re(CO)_5\}$ but also $Os_3ReH(CO)_{15}$ (**86**) and $Os_3ReH_3(CO)_{13}$

86   87

(*453*). Another compound to add to this series is $Os_3ReH_5(CO)_{12}$ (**87**) which is formed by treating **86** with Me_3NO and H_2 in acetonitrile solution (*452*). As far as one can tell from the structures determined these Os_3Re derivatives have the number of metal–metal bonds predicted by electron counting. Thus there are four metal–metal bonds in the 64-electron compound $Os_3H(CO)_{10}$-(MeCN)$\{Re(CO)_5\}$ (**85**), five metal–metal bonds in the 62-electron compound $Os_3ReH(CO)_{15}$ (**86**) (*454*), and six metal–metal bonds in the 60-electron compound $Os_3ReH_5(CO)_{12}$ (**87**) (*452*). This system provides a clear picture of how a single metal atom can be added to a trinuclear cluster initially into a terminal position but then, as ligands are lost and metal–metal bonds created, the system closes up to a tetrahedron. The oxidative addition route may also be applied to larger clusters. Thus $Os_6(CO)_{17}(MeCN)$ reacts with $OsH_2(CO)_4$ to give $Os_7H_2(CO)_{20}$ (*450*). In this case CO loss occurs spontaneously and a spiked structure is not obtained.

3. Reaction of $Os_3H_2(CO)_{10}$ with an Unsaturated Ligand in a Mononuclear Compound

There is only one clear example of the reaction of $Os_3H_2(CO)_{10}$ with an unsaturated ligand leading to a terminal metal group. The vinylidene complex $(C_5H_5)(CO)_2Mn=C=CHPh$ reacts with $Os_3H_2(CO)_{10}$ to give $Os_3H(CO)_{10}\{Mn(CH=CHPh)(CO)_2(C_5H_5)\}$ (**88**); hydride addition at the vinylidene ligand has given a vinyl bridge as shown (*455*). The approach of using the bridging potential of terminal ligands has been applied in other cases, such as the reaction of $(C_5H_5)(CO)_2W\equiv CC_6H_4Me$ with $Os_3H_2(CO)_{10}$ to give various mixed osmium–tungsten clusters (*456*), but the manganese case is the only one to give a spiked product of a type which is probably involved as intermediates in many reactions of mononuclear with trinuclear compounds. The $Mn(CO)_2(C_5H_5)$ unit is readily lost from **88** to give the known organoclusters $Os_3H(CH=CHPh)(CO)_{10}$ and $Os_3H_2(\mu\text{-}X)(CO)_9$ (X = $C=CHPh$ or $CH=CPh$) (*161*).

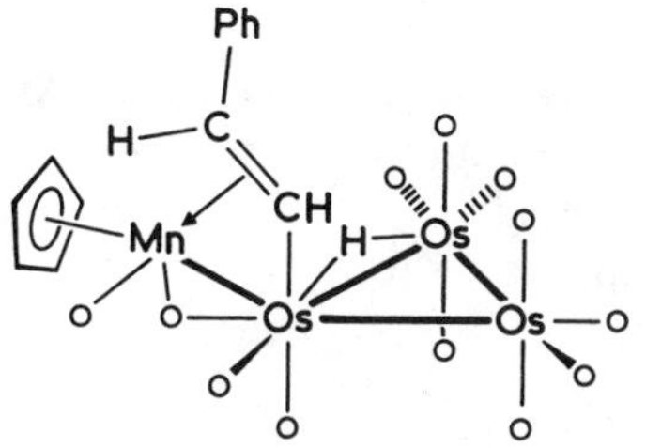

88

C. Added d-Block Metals as Doubly Bridging Ligands

Metal atoms as doubly bridging ligands have been found only in the later part of the d block, notably with Hg, Au, Ag, Cu, and Pt. There are now many examples of the $Au(PR_3)$ group being used as a pseudohydride, the two being isolobal (*457*). Its use is not limited by any means to osmium chemistry, but in Os_3 systems each of the following clusters was synthesized directly or indirectly from the related hydride: $Os_3H\{\mu\text{-}Au(PR_3)\}(CO)_{10}$ (**89**) (*52*), Os_3-$\{\mu\text{-}Au(PR_3)\}_2(CO)_{10}$ (**90**) (*458*), $Os_3(\mu\text{-}X)\{\mu\text{-}Au(PR_3)\}(CO)_{10}[X = SCN$ (*52*), NCO (*459,460*), Cl (*461*), OH (*462*), or $NHSO_2\text{-}p\text{-tolyl}$ (*462*)], Os_3-

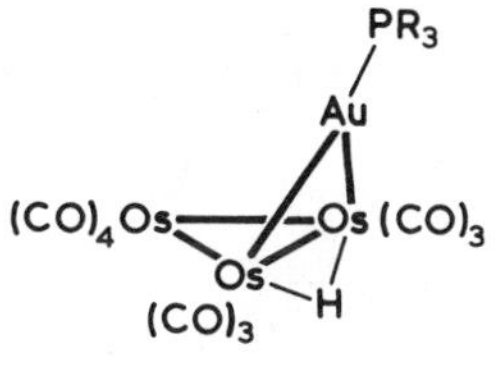

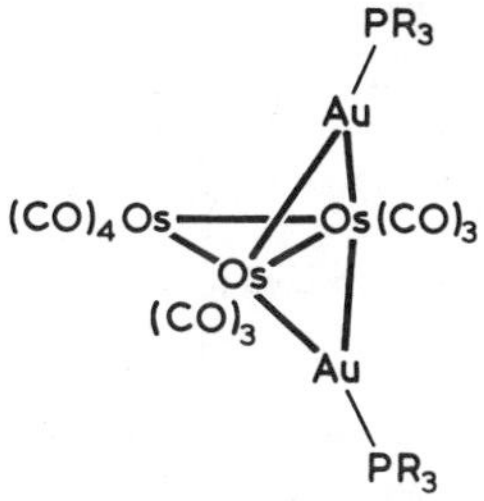

89   **90**

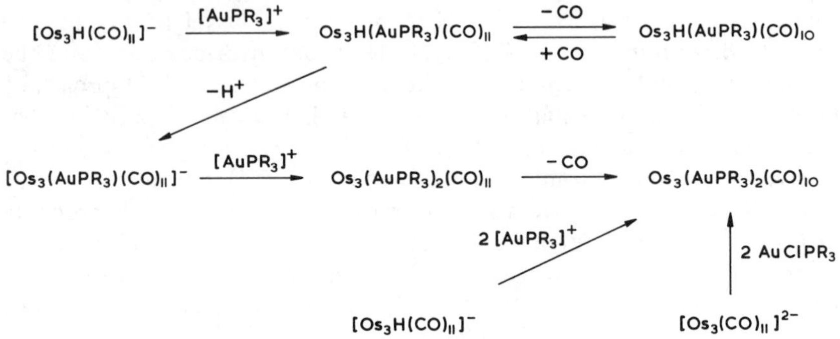

$[Os_3H(CO)_{11}]^- \xrightarrow{[AuPR_3]^+} Os_3H(AuPR_3)(CO)_{11} \underset{+CO}{\overset{-CO}{\rightleftarrows}} Os_3H(AuPR_3)(CO)_{10}$

$\xrightarrow{-H^+}$

$[Os_3(AuPR_3)(CO)_{11}]^- \xrightarrow{[AuPR_3]^+} Os_3(AuPR_3)_2(CO)_{11} \xrightarrow{-CO} Os_3(AuPR_3)_2(CO)_{10}$

$\xrightarrow{2\,[AuPR_3]^+}$ $\uparrow 2\,AuClPR_3$

$[Os_3H(CO)_{11}]^-$ $[Os_3(CO)_{11}]^{2-}$

SCHEME 24.

$(NCO)\{Au(PEt_3)\}(CO)_{11}$ (459,460), and $Os_3H_2\{\mu\text{-}AuPPh_3\}\{\mu_3\text{-}Ni(C_5H_5)\}$-$(CO)_9$ (463). There are two main methods of synthesis: by the reaction of $AuX(PR_3)$ (X = halide, PF_6, or pseudohalide) with the appropriate Os_3 hydride or by reaction of $AuX(PR_3)$ with the anion formed by deprotonation of the hydride. Scheme 24 shows reactions that have been used for the synthesis of $Os_3H\{Au(PR_3)\}(CO)_{10\ or\ 11}$ and $Os_3\{Au(PR_3)\}_2(CO)_{10\ or\ 11}$.

A third method of synthesis is oxidative addition. $Os_3(CO)_{12}$ reacts with $AuX(PPh_3)$ (X = Cl, Br, I, or SCN) in refluxing xylene to give the clusters $Os_3(X)\{Au(PPh_3)\}(CO)_{10}$; the X-ray structures for the compounds with X = Cl or Br have been determined but were not reported in detail (461). Under milder conditions $Au(NCO)(PEt_3)$ adds to $Os_3(CO)_{11}(MeCN)$ to give a mixture of $Os_3(NCO)\{\mu\text{-}Au(PEt_3)\}(CO)_{10}$ and $Os_3(NCO)\{\mu\text{-}Au(PEt_3)\}$-$(CO)_{11}$, while $Os_3(CO)_{10}(MeCN)_2$ gives only the decacarbonyl (460).

Perhaps the most interesting aspect of these studies is the relation between $Os_3H_2(CO)_{10}$ and clusters **89** and **90**. The data in Table IX show that as

TABLE IX

RELATION BETWEEN $AuPR_3$ AND H IN THE CLUSTERS OF THE TYPE $Os_3(\mu\text{-}X)(\mu\text{-}Y)(CO)_{10}$, WHERE X AND Y ARE H OR $AuPR_3$

X	Y	Bridged OS—Os (Å)	Nonbridged Os—Os (Å)	Os—X	Os—X—Os	Ref.
H	H	2.683(1)	2.815(1)	1.845(av)	95.5(1) 93.1(1)	68[a]
H	AuPPh_3	2.699(1)	2.834(1)	2.755(2)[b]	58.7(1)	52
AuPEt_3	AuPEt_3	2.684(1)	2.830(1)	2.761(av)	58.2(1)	458

[a] Neutron diffraction results.
[b] Hydride not located.

bridging hydrides are successively replaced by $AuPR_3$ the Os—Os distance in the bridge remains short and varies only slightly. To maintain this short distance the Os—Au—Os angles are very acute ($\sim 58°$). This is very much in contrast to the increase to 2.848(1) Å in $Os_3HCl(CO)_{10}$ (70) and to 3.233(2) Å in $Os_3Cl_2(CO)_{10}$ (71). This increase is always observed where the introduced bridging ligand (Cl in this case) is a three-electron donor. The bridged Os—Os distance does not simply depend upon the size of the atom in the bridge but rather on its electronic characteristics. $AuPR_3$ is a one-electron donor while Cl is a three-electron donor. It has been proposed (458) that Au is sp hybridized, with a hybrid orbital directed between the osmium atoms corresponding to the $1s$ orbital of the hydride in $Os_3H_2(CO)_{10}$. The Os_2H_2 ring in the dihydride and the Os_2Au_2 ring in 90 would involve four-center four-electron bonds (68). The chemistry indicated in Scheme 24 shows some parallel to that of $Os_3H_2(CO)_{10}$, which adds CO to give $Os_3H_2(CO)_{11}$. However, $Os_3H_2(CO)_{11}$ contains a bridging and a terminal hydride ligand and does not relate directly to $Os_3\{Au(PPh_3)\}_2(CO)_{11}$, which contains a Au—Au bond of length 2.845(1) Å (459).

91

(M = Ag or Au)

92

Although $AuPR_3$-bridged systems are now common, there are interesting examples of naked metal atom bridges. The cluster $Os_3H\{Au(PR_3)\}(CO)_{10}$ (R = Ph or Et) reacts with [PPN]Cl with loss of the $[Au(PR_3)_2]$Cl set of atoms to give $[PPN][\{Os_3H(CO)_{10}\}_2Au]$ (91), in which each $Os_3H(CO)_{10}$ unit is structurally close to that in the parent $AuPR_3$ complex (445). The short bridged Os—Os distances of 2.698 and 2.689 Å should be compared with data in Table IX. The corresponding Ag compound was synthesized by the reaction of $[PPN][Os_3H(CO)_{11}]$ with $Ag[PF_6]$, and its structure was also determined (444). The neutral cluster $\{Ru_3(\mu_3\text{-}C_2{}^t Bu)(CO)_9\}_2Hg$ (465) is closely related structurally to 91 except that the silver and gold atoms are strictly planar, but there is a severe twist from a square-planar to a tetrahedral geometry in the Ru_6Hg compound that creates chirality. These clusters could be regarded as containing two tetranuclear butterfly clusters fused at a wing tip atom. It is argued that, as in the $Au(PR_3)$ bridged compounds, the gold or silver atoms are *sp* hydridized and behave as one-electron donors to each Os_3 group of atoms *(444,445)*.

The dihydride cluster $Os_3H_2(CO)_{10}$ reacts with $\{CuH(PPh_3)\}_6$ to give $Os_3H\{CuH_2(PPh_3)\}(CO)_{10}$ **(92)**, which is analogous to the gold compound except for the extra two hydride ligands that were not located in the X-ray structure but were deduced to be as shown. The presence of extra hydrogen atoms removes any analogies with $Os_3H_2(CO)_{10}$, and the bridged Os—Os distance is long, 3.026(3) Å *(466)*. There is a dynamic exchange of the Os*H*Os atom (δ −20.47) with the Os*H*Cu atoms (δ −11.80).

In the same way that **91** (M = Ag) was prepared by reaction of Ag^+ with $[Os_3H(CO)_{11}]^-$, the reaction of this Os_3 anion with mercury(II) or mercury(I) salts gives the neutral cluster $\{Os_3(CO)_{11}\}_3Hg_3$ **93** *(443)*. The 12 metal atoms in this cluster form a planar triangulated array, and one could regard the Hg_3 triangle as a six-electron donor, two electrons being donated to each $Os_3(CO)_{11}$ unit. Unlike the Ag- or Au-bridged systems, short Os—Os bonds are not found. Related copper compounds are also formed, $[Os_3(CO)_{11}Cu]$ from copper(II) salts and $[Os_3(CO)_{11}Cu_2]$ from copper(I) salts, but their structures and states of aggregation are so far unknown *(443)*. There is an intriguing possibility of generating two-dimensional arrays related to **93** if clusters with lower CO content could be made.

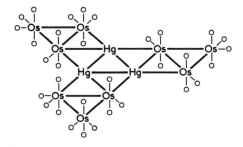

93

Doubly bridging metal atoms in Os_3 clusters can be synthesized in some cases by opening up closo tetrahedra of metal atoms by the addition of ligands. $Os_3H_2(CO)_{10}$ reacts with $Pt(C_2H_4)_2(PR_3)$ (R = Ph or Cy, where Cy = cyclohexyl) to give the closo clusters $Os_3PtH_2(CO)_{10}(PR_3)$ **(94)** *(467,468)*. Scheme 25 shows how the reversible addition of CO to **94** (R = Cy) *(469,470)* or the addition of PPh_3 to **94** (R = Ph) *(471)* leads to the opening of the tetrahedron to give the butterfly structures **95** and **96**. Compounds **95** and **96** might be considered to be of the general type $Os_3H(\mu\text{-}X)(CO)_{10}$, where X, $PtH(CO)(PR_3)$ in these cases, is doubly bridging. On forming **95** and **96** the electron count changes from 58 to 60, and assuming that the Pt atoms are associated with 16 rather than 18 electrons (they have essentially

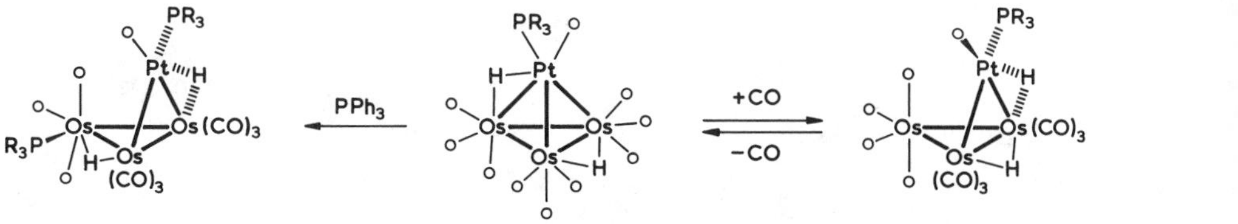

96 (R = Ph) 94 (R = Ph or Cy) 95 (R = Cy)

SCHEME 25.

square-planar configurations) then the correct number of metal–metal bonds is found. Interestingly, **95** and **96** have somewhat different geometries: the Os*H*Os atoms have different positions, and the angles between the metal triangular planes are 100.0° for **95** and 88.4° for **96**.

The addition of two-electron donors to **94** does not always lead to an opening of the closo structure. For example, in the additions of H_2 to give $Os_3PtH_4(CO)_{10}(PCy_3)$ (*269,270*) or of CH_2 (using CH_2N_2) to give two isomers of $Os_3PtH_2(CH_2)(CO)_{10}(PCy_3)$ (*469,472*) the final products maintain the closo tetrahedral geometries.

The addition of $Rh(acac)(CO)_2$ (acac = MeCOCHCOMe) to $Os_3H_2(CO)_{10}$ gives another rather more complex example of a doubly bridging metal system: $Os_3H_2\{Rh(acac)(CO)\}(CO)_9$ (**97**), in which the wing tip atoms (Rh and Os) are bridged by the five-electron donor acac ligand. The distance

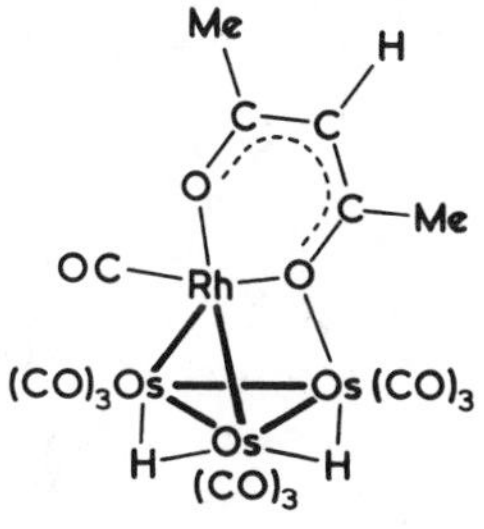

97

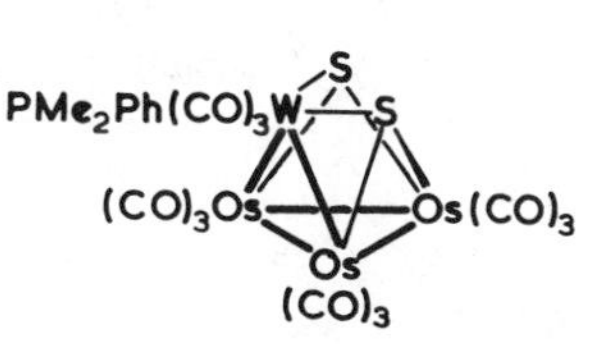

98

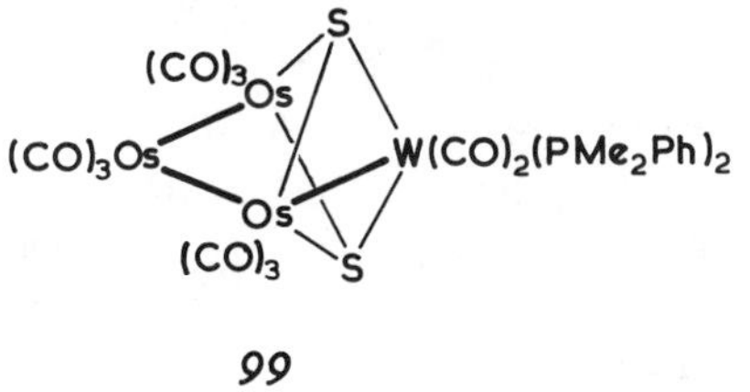

99

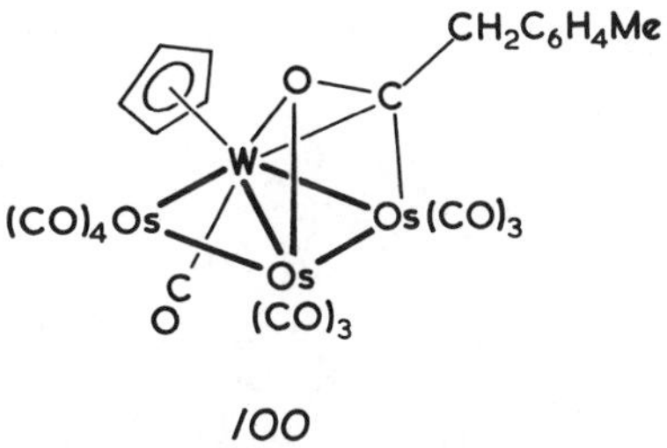

100

between the wing tips (3.292 Å) seems to rule out a Rh—Os bond between them even though the electron count of the cluster is 60 (473). Another cluster with five metal–metal bonds is $Os_3W(CO)_{12}(PMe_2Ph)(\mu_3\text{-}S)_2$ (98), which is formed as the major product (28%) along with several other clusters from the reaction of $Os_3(CO)_9(\mu_3\text{-}S)_2$ with $W(CO)_5(PMe_2Ph)$. The donor properties of the μ_3-S are used to attach the tungsten atom in the initial stages of the reaction (474,475). The cluster 99 is a low yield product from the same reaction and only contains one W—Os bond, and so perhaps should have been included in the previous section (VIII,B).

The clusters $Os_3ReH(CO)_{15}$ (453) and $Os_3W(CO)_{11}(\mu_3\text{-}MeC_6H_4CH_2CO)$-$(C_5H_5)$ (456) adopt structures with five metal–metal bonds, but in these cases tungsten and rhenium atoms do not bridge edges of Os_3 triangles but rather osmium bridges the edge of MOs_2 (M = Re or W) triangles (see 86 and 100).

D. Added d-Block Metals as Triply Bridging Ligands

The rapidly growing area of triply bridging d-block metal ligands is represented by the compounds in Tables X and XI. This section, however, will do little more than indicate the extent and variety of the chemistry without attempting to treat each system separately.

Table X lists examples of clusters in which a μ_3 group such as $Ni(C_5H_5)$, $Co(C_5H_5)$, $Rh(C_6H_5Me)$, $Co(CO)_3$, $Ir(CO)_3$, $Pt(PR_3)_2$, $Pt(CO)(PR_3)$, or $W(C_5H_5)(CO)_2$ caps a triangle of osmium atoms to which it is attached only by metal–metal bonds. Many are made by direct reaction of $Os_3H_2(CO)_{10}$ with an appropriate mononuclear precursor, but less systematic routes are clearly also available. Two of the examples are paramagnetic.

The hexanuclear Os_3Ni_3 cluster, which is a by-product of the reaction of $Os_3H_2(CO)_{10}$ or $Os_3(CO)_{12}$ with $[Ni(C_5H_5)(CO)]_2$, has 87 valence electrons. In spite of their different electron counts, the clusters $Os_3H_3(CO)_9\{\mu_3\text{-}M(C_5H_5)\}$ where M = Ni (60 electrons) or Co (59 electrons) are structurally extremely similar except that the cyclopentadienyl ligand is rotated by 180° from one to the other. Such a minor structural variation could not be explained convincingly. There is, however, a very big difference in reactivity. The 59-electron Co compound is extremely reactive and air-sensitive while the 60-electron Ni compound is much easier to handle and has been used as an alkene hydrogenation catalyst (486).

Another example of a small structural variation is found for the orange cluster capped by μ_3-Pt(CO)L in Table X which isomerizes over several days to give an equilibrium mixture with a red species in which one of the hydrides has moved from bridging two osmiums to bridging a Pt–Os edge. In the isomerization the Pt(CO)L group rotates by 180°.

TABLE X

TETRAHEDRAL Os_3M CLUSTERS WITH NO LIGANDS BRIDGING M AND Os

Cluster	Valence electrons	Synthesis	Ref.
$Os_3H_3(CO)_9\{\mu_3\text{-}Ni(C_5H_5)\}$	60	$Os_3(CO)_{12} + H_2 + [Ni(C_5H_5)(CO)]_2$	446,447,476
$[Os_3H_2(CO)_9\{\mu_3\text{-}Ni(C_5H_5)\}]^-$	60	By deprotonation of trihydride	464
$Os_3H_2(AuPPh_3)(CO)_9\{\mu_3\text{-}Ni(C_5H_5)\}$	60	Anion + $AuCl(PPh_3)$	464
$Os_3(CO)_9\{Ni_3(C_5H_5)_3\}$	87	$Os_3(CO)_{12} + [Ni(C_5H_5)(CO)]_2$	477,478
$Os_3H_3(CO)_9\{\mu_3\text{-}Co(C_5H_5)\}$	59	$Os_3(CO)_{12} + H_2 + Co(C_5H_5)(CO)_2$	479
$Os_3H_4(CO)_9\{\mu_3\text{-}Co(C_5H_5)\}$	60	$Os_3(CO)_{12} + H_2 + Co(C_5H_5)(CO)_2$	479
$Os_3H_3(CO)_9\{\mu_3\text{-}Rh(C_6H_5Me)\}$	60	$Os_3H_2(CO)_{10} + H_2 + Rh(C_5H_5)(CO)_2$/toluene	480,481
$Os_3H_3(CO)_9\{\mu_3\text{-}Co(CO)_3\}$	60	$Os_3H_2(CO)_{10} + Co_4(CO)_{12}$	482
$Os_3H_2Cl(CO)_9\{\mu_3\text{-}Ir(CO)_3\}$	62	$Os_3H_2(CO)_{10} + IrCl(CO)_2(NH_2R)$	483
$Os_3H_2(CH_2)(CO)_9\{\mu_3\text{-}Pt(CO)L\}$ (orange)	60	$Os_3PtH_2(CO)_{10}L + CH_2N_2$ (L = tricyclohexylphosphine)	472
$Os_2H(CO)_8(C_2R_2)\{\mu_3\text{-}W(C_5H_5)(CO)_2\}$	60	$Os_3WH(CO)_{12}(C_5H_5) + Me_3NO + C_2R_2$	484
$Os_3(\mu_3\text{-}S)(CO)_8L\{\mu_3\text{-}PtL_2\}$	60	$Os_3(S)(CO)_{10} + PtL_4(L = PMe_2Ph)$	485

TABLE XI

Os$_3$M Tetrahedral Clusters of 60 Valence
Electrons with Bridges between Os and M

Cluster	Ref.
With CO bridges	
Os$_3$H$_2$(CO)$_9$(μ-CO)$_2${μ_3-Fe(CO)$_2$}^{a}	487
[PPN][Os$_3$H(CO)$_9$(μ-CO)$_2${μ_3-Fe(CO)$_2$}]^a	488
Os$_3$Cl(CO)$_8$(μ-CO){μ_3-Ir(CO)(PPh$_3$)}	489
Os$_3$H$_2$(CO)$_9$(μ-CO){μ_3-Co(C$_5$H$_5$)}	490
Os$_3$H$_2$(CO)$_9$(μ-CO){μ_3-Rh(C$_5$H$_5$)}	491
Os$_3$H(CO)$_9$(μ-CO){μ_3-W(C$_5$H$_5$)(CO)$_2$}	492,493
With H bridges	
Os$_3$H$_2$(CO)$_9$(μ-H)$_3${μ_3-Re(CO)$_3$}	452
Os$_3$H$_2$(CO)$_9$(μ-H){μ_3-Ir(CO)$_2$(PPh$_3$)}	466
Os$_3$H(CH$_2$)(CO)$_9$(μ-H){μ_3-Pt(CO)(PCy$_3$)}	469
Os$_3$H$_2$(CO)$_9$(μ-H)$_2${μ_3-Pt(CO)(PCy$_3$)}	469,470
Os$_3$H(CO)$_9$(μ-H){μ_3-Pt(CO)(PCy$_3$)}^{b}	467,468
Os$_3$H(CO)$_9$(μ-H)$_2${μ_3-W(C$_5$H$_5$)(CO)$_2$}	494
With H and CO bridges	
Os$_3$H(CO)$_8$(PPh$_3$)(μ-H)(μ-CO){μ_3-NiPPh$_3$}	473
With Other bridges	
Os$_3$H(CO)$_9$(μ-C=CH^{t}Bu){Ni(C$_5$H$_5$)}	496
Os$_3$H(CO)$_9$(μ-S){W(C$_5$H$_5$)(CO)$_2$}	497
Os$_3$(CO)$_9$(μ-CR){W(C$_5$H$_5$)(CO)$_2$}	498–500
Os$_3$(CO)$_9$(μ-CCH$_2$C$_6$H$_4$Me)(μ-O){W(C$_5$H$_5$)}	501
Os$_3$H(CO)$_9$(μ-C$_2$H$_2$)(μ-O){W(C$_5$H$_5$)}	502

^{a} The μ-CO are semibridging.
^{b} Total of 58 valence electrons.

The Os$_3$W cluster in Table X is shown in Scheme 26. The alkyne ligand when first introduced into the Os$_3$W cluster behaves as a μ_3 four-electron donor in a normal manner. Decarbonylation leads to major reorganization. By cleaving to give two μ_3-alkylidene ligands the alkyne allows the cluster to remain electron precise. Such is not always the case because decarbonylation of Os$_3$(μ_3-C$_2$Ph$_2$)(CO)$_{10}$ gives Os$_3$(μ_3-C$_2$Ph$_2$)(CO)$_9$ without C—C bond fission but instead a reorganization of the alkyne–triosmium geometry (see Scheme 22). It may be that having different metal atoms in the cluster facilitates C—C cleavage. The product cluster Os$_3$WH(CO)$_9$(CAr)(C$_5$H$_5$) is quite unusual in having a terminal hydride ligand at tungsten (δ 3.75; J_{WH} 89 Hz).

Table XI presents a series of compounds mostly with the Os$_3$(CO)$_9$ framework to which is attached a transition metal atom with various bridging

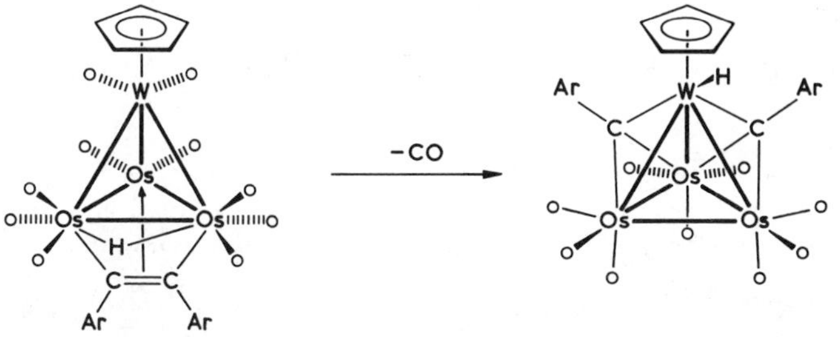

101 102

SCHEME 26.

ligands (hydride, CO, vinylidene, oxide, sulfide, alkylidyne, or alkyne) bridging it to the osmium atoms. Just to demonstrate such systems a few Os_3W clusters have been illustrated. In addition to compounds **98** to **100** in the last section and compounds **101** and **102** in Scheme 26, we also illustrate compounds **103** (*494*), **104** (*493,495*), **105** (*498,499*), **106** (*502*), and **107** (*497*). In addition to these species, OsW_2 clusters, such as $OsW_2(C_2R_2)(CO)_7(C_5H_5)$, are also formed from reactions of mononuclear W compounds with Os_3 clusters (*498–500,503*).

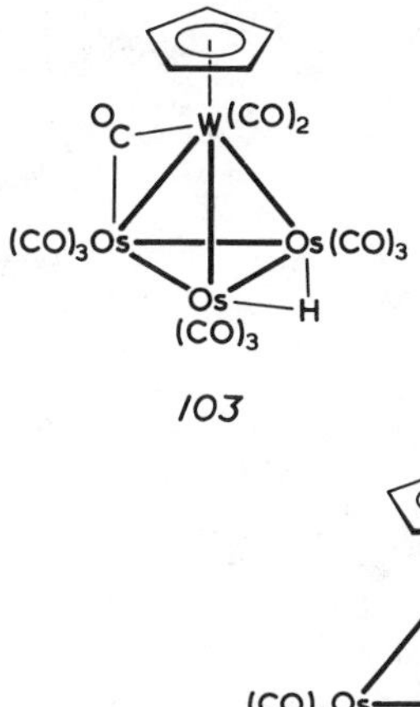

103

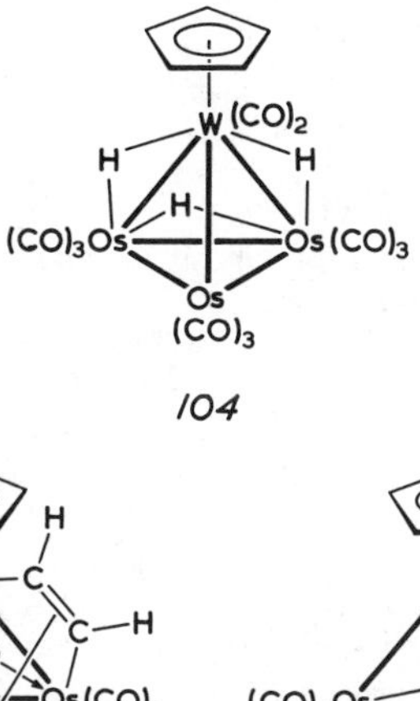

104

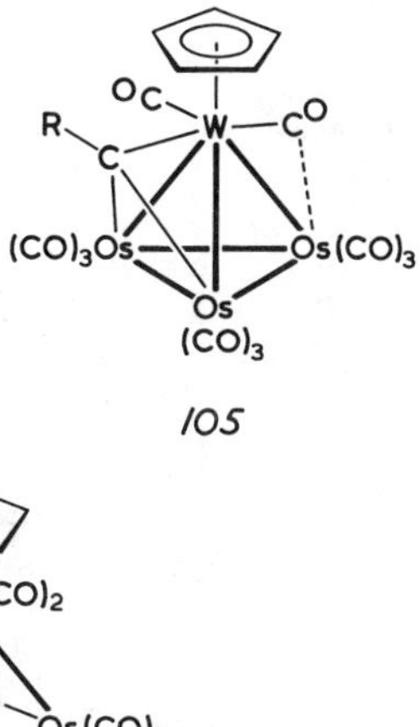

105

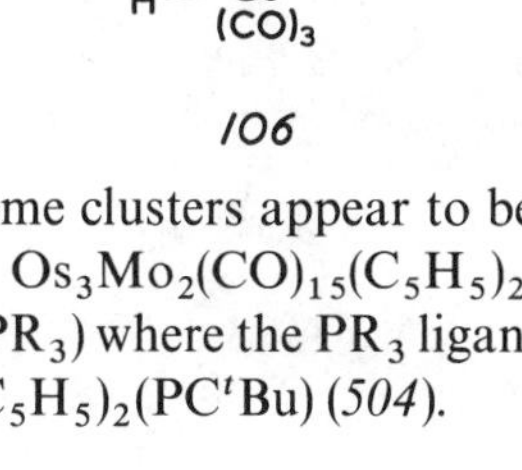

106

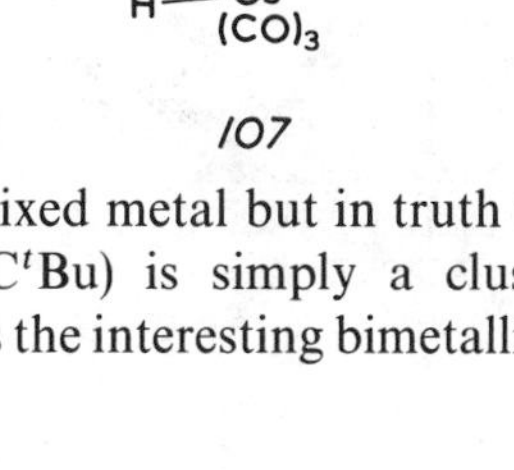

107

Finally, some clusters appear to be mixed metal but in truth they are not. The cluster $Os_3Mo_2(CO)_{15}(C_5H_5)_2(PC^tBu)$ is simply a cluster of type $Os_3(CO)_{11}(PR_3)$ where the PR_3 ligand is the interesting bimetallic compound $Mo_2(CO)_4(C_5H_5)_2(PC^tBu)$ (*504*).

References

1. P. O. Nubel, S. R. Wilson, and T. L. Brown, *Organometallics* **2**, 515 (1983).
2. D. R. Gard and T. L. Brown, *Organometallics* **1**, 1143 (1982); P. O. Nubel and T. L. Brown, *Organometallics* **3**, 29 (1984); K.-H. Franzreb and C. G. Kreiter, *Z. Naturforsch.* **39b**, 81 (1984).
3. K.-H. Franzreb and C. G. Kreiter, *J. Organomet. Chem.* **246**, 189 (1983); P. O. Nubel and T. L. Brown, *J. Am. Chem. Soc.* **104**, 4955 (1982); *idem, ibid.*, **106**, 644 (1984); *idem, ibid.*, **106**, 3474 (1984).
4. C. G. Kreiter, K.-H. Franzreb, and W. S. Sheldrick, *J. Organomet. Chem.* **270**, 71 (1984).
5. S. C. Tripathi, S. C. Srivastava, R. P. Mani, and A. K. Shrimal, *Inorg. Chim. Acta* **15**, 249 (1975).
6. R. D. Adams and J. P. Selegue, in "Comprehensive Organometallic Chemistry" (G. Wilkinson, F. G. A. Stone, and E. W. Abel, eds.), Chap. 33, p. 968. Pergamon, Oxford, 1982.
7. B. F. G. Johnson (ed.), "Transition Metal Clusters." Wiley (Interscience), New York, 1980.
8. B. F. G. Johnson and J. Lewis, *Adv. Inorg. Chem. Radiochem.* **24**, 225 (1981); *idem, Pure Appl. Chem.* **54**, 97 (1982).
9. B. F. G. Johnson (ed), *Polyhedron* **3**, 1277 (1984)
10. M. P. Gomez-Sal, B. F. G. Johnson, J. Lewis, P. R. Raithby, and A. H. Wright, *J. Chem. Soc., Chem. Commun.*, 1682 (1985).
11. A. C. Sievert, D. S. Strickland, J. R. Shapley, G. R. Steinmetz, and G. L. Geoffroy, *Organometallics* **1**, 214 (1982).
12. J. R. Shapley, D. S. Strickland, G. M. St. George, M. R. Churchill, and C. Bueno, *Organometallics* **2**, 185 (1983).
13. J. R. Shapley, M. E. Cree-Uchiyama, G. M. St. George, M. R. Churchill, and C. Bueno, *J. Am. Chem. Soc.* **105**, 140 (1983).
14. S. G. Shore, D.-Y. Jan, L.-Y. Hsu, and W.-L. Hsu, *J. Am. Chem. Soc.* **105**, 5923 (1983).
15. S. A. R. Knox, J. W. Koepke, M.A. Andrews, and H.D. Kaesz, *J. Am. Chem. Soc.* **97**, 3942 (1975).
16. B. F. G. Johnson, J. Lewis, and D. A. Pippard, *J. Organomet. Chem.* **145**, C4 (1978).
17. B. F. G. Johnson, J. Lewis, and D. A. Pippard, *J. Chem. Soc., Dalton Trans.*, 407 (1981).
18. C. E. Anson, E. J. Ditzel, M. Fajardo, H. D. Holden, B. F. G. Johnson, J. Lewis, J. Puga, and P. R. Raithby, *J. Chem. Soc., Dalton Trans.*, 2723 (1984).
19. B. F. G. Johnson, J. Lewis, T. I. Odiaka, and P. R. Raithby, *J. Organomet. Chem.* **216**, C56 (1981).
20. R. J. Goudsmit, B. F. G. Johnson, J. Lewis, P. R. Raithby, and M. J. Rosales, *J. Chem. Soc., Dalton Trans.*, 2257 (1983).
21. M. Tachikawa and J. R. Shapley, *J. Organomet. Chem.* **124**, C19 (1977).
22. E. G. Bryan, B. F. G. Johnson, and J. Lewis, *J. Chem. Soc., Dalton Trans.*, 1328 (1977).
23. J. R. Moss and W. A. G. Graham, *J. Chem. Soc., Dalton Trans.*, 95, (1977).
24. P. Rushman, G. N. van Buuren, M. Shiralian, and R. K. Pomeroy, *Organometallics* **2**, 693 (1983).
25. J. C. Bricker, N. Bhattacharyya, and S. G. Shore, *Organometallics* **3**, 201 (1984).
26. D. H. Farrar, B. F. G. Johnson, J. Lewis, J. N. Nicholls, P. R. Raithby, and M. J. Rosales, *J. Chem. Soc., Chem. Commun.*, 273 (1981).
27. J. N. Nicholls, D. H. Farrar, P. F. Jackson, B. F. G. Johnson, and J. Lewis, *J. Chem. Soc., Dalton Trans.*, 1395 (1982).
28. P. V. Broadhurst, B. F. G. Johnson, J. Lewis, and P. R. Raithby, *J. Chem. Soc., Dalton Trans.*, 1641 (1982).

29. J. A. Clucas, D. F. Forster, M. M. Harding, and A. K. Smith, *J. Chem. Soc., Chem. Commun.,* 949 (1984).
30. C. W. Bradford, R. S. Nyholm, G. J. Gainsford, J. M. Guss, P. R. Ireland, and R. Mason, *J. Chem. Soc., Chem. Commun.,* 87 (1972).
31. K. Wade, *J. Chem. Soc., Chem. Commun.,* 792 (1971); *idem, Inorg. Nucl. Chem. Lett.* **8,** 559 (1972); *idem,* "Electron Deficient Compounds." Nelson, London, 1971.
32. D. M. P. Mingos, *Nature, Phys. Sci.* **236,** 99 (1972); *idem, Acc. Chem. Res.* **17,** 311 (1984).
33. B. K. Teo, *Inorg. Chem.* **23,** 1251 (1984); B. K. Teo, G. Longoni, and F. R. K. Chung, *Inorg. Chem* **23,** 1257 (1984).
34. A. J. Deeming and M. Underhill, *J. Chem. Soc., Chem. Commun.,* 277 (1973).
35. A. J. Deeming and M. Underhill, *J. Chem. Soc., Dalton Trans.,* 1415 (1974).
36. A. J. Deeming, S. Hasso, M. Underhill, A. J. Canty, B. F. G. Johnson, W. G. Jackson, J. Lewis, and T. W. Matheson, *J. Chem. Soc., Chem. Commun.,* 807 (1974).
37. J. P. Yesinowski and D. Bailey, *J. Organomet. Chem.* **65,** C27 (1974).
38. G. A. Somerjai, *Chem. Soc. Rev.* **13,** 321 (1984).
39. C. E. Anson, B. T. Keiller, I. A. Oxton, D. B. Powell, and N. Sheppard, *J. Chem. Soc., Chem Commun.,* 470 (1983).
40. I. A. Oxton, D. B. Powell, N. Sheppard, K. Burgess, B. F. G. Johnson, and J. Lewis, *J. Chem. Soc., Chem. Commun.,* 719 (1982).
41. R. B. Calvert and J. R. Shapley, *J. Am. Chem. Soc.* **99,** 5225 (1977).
42. R. B. Calvert and J. R. Shapley, *J. Am. Chem. Soc.* **100,** 7726, (1978).
43. R. B. Calvert, J. R. Shapley, A. J. Shultz, J. M. Williams, S. L. Suib, and G. D. Stucky, *J. Am. Chem. Soc.* **100,** 6240 (1978).
44. A. J. Shultz, J. M. Williams, R. B. Calvert, J. R. Shapley, and G. D. Stucky, *Inorg. Chem.* **18,** 319 (1979).
45. C. Choo Yin and A. J. Deeming, *J. Chem. Soc., Dalton Trans.,* 2091 (1975).
46. A. J. Deeming and M. Underhill, *J. Organomet. Chem* **42,** C60 (1972).
47. G. R. Steinmetz and G. L. Geoffroy, *J. Am. Chem. Soc.* **103,** 1278 (1981).
48. G. R. Steinmetz, E. D. Morrison, and G. L. Geoffroy, *J. Am. Chem. Soc.* **106,** 2559 (1984).
49. C. M. Jensen, Y. J. Chen, and H. D. Kaesz, *J. Am. Chem. Soc.* **106,** 4046 (1984).
50. B. M. Peake, B. H. Robinson, J. Simpson, and D. J. Watson, *J. Chem. Soc., Chem. Commun.,* 945 (1974).
51. J. B. Keister and J. R. Shapley, *J. Am. Chem. Soc.* **98,** 1056 (1976).
52. B. F. G. Johnson, D. A. Kaner, J. Lewis, and P. R. Raithby, *J. Organomet. Chem.* **215,** C33 (1981).
53. J. R. Shapley, J. B. Keister, M. R. Churchill, and B. G. De Boer, *J. Am. Chem. Soc.* **97,** 4145 (1975).
54. A. J. Deeming and S. Hasso, *J. Organomet. Chem.* **88,** C21 (1975).
55. A. J. Deeming and S. Hasso, *J. Organomet. Chem.* **114,** 313 (1976).
56. M. Gochin and J. R. Moss, *J. Organomet. Chem.* **192,** 409 (1980).
57. J. R. Moss and W. A. G. Graham, *J. Chem. Soc., Dalton Trans.,* 89 (1977).
58. V. F. Allen, R. Mason, and P. B. Hitchcock, *J. Organomet. Chem.* **140,** 297 (1977).
59. A. J. Deeming, I. P. Rothwell, M. B. Hursthouse, and J. D. J. Backer-Dirks, *J. Chem. Soc. Dalton Trans.,* 1879 (1981).
60. R. D. Adams, I. T. Horváth, B. E. Segmüller, and L.-W. Yang, *Organometallics* **2,** 144 (1983).
61. S. Bellard and P. R. Raithby, *Acta Cryst.* **B36,** 705 (1980).
62. K. Burgess, H. D. Holden, B. F. G. Johnson, J. Lewis, M. B. Hursthouse, N. P. C. Walker, A. J. Deeming, P. J. Manning, and R. Peters, *J. Chem. Soc., Dalton Trans.,* 85 (1985).
63. Z. Dawoodi, M. J. Mays, and P. R. Raithby, *J. Chem. Soc., Chem. Commun.,* 721 (1979).
64. R. D. Adams, N. M. Golembeski, and J. P. Selegue, *J. Am. Chem. Soc.* **101,** 5862 (1979).

65. R. D. Adams, N. M. Golembeski, and J. P. Selegue, *Organometallics* **1**, 240 (1982).
66. M. R. Churchill, F. J. Hollander, and J. P. Hutchinson, *Inorg. Chem.* **16**, 2697 (1977).
67. A. G. Orpen, A. V. Rivera, D. Pippard, G. M. Sheldrick, and K. D. Rouse, *J. Chem. Soc., Chem. Commun.*, 723 (1978).
68. R. W. Broach and J. M. Williams, *Inorg. Chem.* **18**, 314 (1979).
69. M. R. Churchill and H. J. Wasserman, *Inorg. Chem.* **19**, 2391 (1980).
70. M. R. Churchill and R. A. Lashewycz, *Inorg. Chem.* **18**, 1926 (1979).
71. F. W. B. Einstein, T. Jones, and K. G. Tyers, *Acta Cryst.* **B38**, 1272 (1982).
72. R. Mason and D. M. P. Mingos, *J. Organomet. Chem.* **50**, 53 (1973); R. Mason, *Pure Appl. Chem.* **33**, 513 (1973).
73. B. K. Teo, M. B. Hall, R. F. Fenske, and L. F. Dahl, *J. Organomet. Chem.* **70**, 413 (1974).
74. D. E. Sherwood and M. B. Hall, *Inorg. Chem.* **21**, 3458 (1982).
75. R. D. Adams and N. M. Golembeski, *J. Am. Chem. Soc.* **101**, 1306 (1979).
76. R. D. Adams, N. M. Golembeski, and J. P. Selegue, *J. Am. Chem. Soc.* **103**, 546 (1981).
77. R. D. Adams and D. A. Katahira, *Organometallics* **1**, 460 (1982).
78. B. F. G. Johnson, J. Lewis, and P. A. Kilty, *J. Chem. Soc. A*, 2859 (1968).
79. N. Cook, L. Smart, and P. Woodward, *J. Chem. Soc., Dalton Trans.*, 1744 (1977).
80. J. P. Candlin and J. Cooper, *J. Organomet. Chem.* **15**, 230 (1968).
81. A. J. Deeming, B. F. G. Johnson, and J. Lewis, *J. Chem. Soc. A*, 897 (1970).
82. A. C. Willis, G. N. van Buuren, R. K. Pomeroy, and F. W. B. Einstein, *Inorg. Chem.* **22**, 1162 (1983).
83. A. J. Deeming, P. J. Manning, I. P. Rothwell, M. B. Hursthouse, and N. P. C. Walker, *J. Chem. Soc., Dalton Trans.*, 2039 (1984).
84. E. D. Morrison, G. R. Steinmetz, G. L. Geoffroy, W. C. Fultz, and A. L. Rheingold, *J. Am. Chem. Soc.* **105**, 4104 (1983).
85. F. W. B. Einstein, S. Nussbaum, D. Sutton, and A. C. Willis, *Organometallics* **2**, 1259 (1983).
86. F. W. B. Einstein, S. Nussbaum, D. Sutton, and A. C. Willis, *Organometallics* **3**, 568 (1984).
87. B. F. G. Johnson, J. Lewis, P. R. Raithby, and S. W. Sankey, *J. Organomet. Chem.* **231**, C65 (1982).
88. K. Burgess, B. F. G. Johnson, J. Lewis, and P. R. Raithby, *J. Organomet. Chem.* **224**, C40 (1982).
89. K. Burgess, B. F. G. Johnson, J. Lewis, and P. R. Raithby, *J. Chem. Soc., Dalton Trans.*, 2119 (1982).
90. B. F. G. Johnson, J. Lewis, P. R. Raithby, and C. Zuccaro, *J. Chem. Soc., Chem. Commun.*, 916 (1979).
91. S. Bhaduri, B. F. G. Johnson, J. Lewis, D. J. Watson, and C. Zuccaro, *J. Chem. Soc., Chem. Commun.*, 477 (1977); idem, *J. Chem. Soc., Dalton Trans.*, 557 (1979).
92. A. V. Rivera and G. M. Sheldrick, *Acta Cryst.* **B34**, 3372 (1978).
93. P. V. Broadhurst, B. F. G. Johnson, and J. Lewis, *J. Chem. Soc., Dalton Trans.*, 1881 (1982).
94. A. J. Carty, S. A. Maclaughlin, and N. J. Taylor, *J. Organomet. Chem.* **204**, C27 (1981).
95. R. D. Adams, D. A. Katahira, and L.-W. Yang, *Organometallics* **1**, 235 (1982).
96. S. Aime, A. J. Deeming, M. B. Hursthouse, and J. D. J. Backer-Dirks, *J. Chem. Soc., Dalton Trans.*, 1625 (1982).
97. R. D. Adams and Z. Dawoodi, *J. Am. Chem. Soc.* **103**, 6510 (1981); R. D. Adams, Z. Dawoodi, D. F. Foust, and B. E. Segmüller, *Organometallics* **2**, 315 (1983).
98. B. F. G. Johnson, J. Lewis, P. G. Lodge, P. R. Raithby, K. Henrick, and M. McPartlin, *J. Chem. Soc., Chem. Commun.*, 719 (1979).
99. J. P. Attard, B. F. G. Johnson, J. Lewis, J. M. Mace, M. McPartlin, and A. Sironi, *J. Chem. Soc., Chem. Commun.*, 595 (1984).
100. P. V. Broadhurst, B. F. G. Johnson, J. Lewis, and P. R. Raithby, *J. Organomet. Chem.* **194**, C35 (1980).

101. R. D. Adams, I. T. Horváth, and H.-S. Kim, *Organometallics* **3**, 548 (1984).
102. R. D. Adams, I. T. Horváth, B. E. Segmüller, and L.-W. Yang, *Organometallics* **2**, 1301 (1983).
103. R. D. Adams and L.-W. Yang, *J. Am. Chem. Soc.* **104**, 4115 (1982).
104. R. D. Adams and I. T. Horváth, *J. Am. Chem. Soc.* **106**, 1869 (1984).
105. G. Süss-Fink, U. Thewalt, and H.-P. Klein, *J. Organomet. Chem.* **224**, 59 (1982).
106. S. C. Brown, J. Evans, and L. E. Smart, *J. Chem. Soc., Chem. Commun.*, 1021 (1980).
107. M. Green, A. G. Orpen, and C. J. Schaverien, *J. Chem. Soc., Chem. Commun.*, 37 (1984)
108a. R. D. Adams and N. M. Golembeski, *Inorg. Chem.* **18**, 2255 (1979).
108b. K. Henrick, M. McPartlin, A. J. Deeming, S. Hasso, and P. J. Manning, *J. Chem. Soc., Dalton Trans.*, 899 (1982).
109. P. D. Gavens and M. J. Mays, *J. Organomet. Chem.* **162**, 389 (1978).
110. J. R. Shapley, M. Tachikawa, M. R. Churchill, and R. A. Lashewycz, *J. Organomet. Chem.* **162**, C39 (1978).
111. M. R. Churchill and R. A. Lashewycz, *Inorg. Chem.* **18**, 848 (1979).
112. M. R. Churchill, B. G. De Boer, J. R. Shapley, and J. B. Keister, *J. Am. Chem. Soc.* **98**, 2357 (1976).
113. M. R. Churchill and B. G. De Boer, *Inorg. Chem.* **16**, 1141 (1977).
114. Z. Dawoodi, M. J. Mays, and A. G. Orpen, *J. Organomet. Chem.* **219**, 251 (1981).
115. M. Laing, P. Sommerville, P. J. Wheatley, M. J. Mays, and Z. Dawoodi, *Acta Cryst.* **B37**, 2230 (1981).
116. R. D. Adams, D. A. Katahira, and L.-W. Yang, *J. Organomet. Chem.* **219**, 241 (1981).
117. R. D. Adams, D. A. Katahira, and L.-W. Yang, *J. Organomet. Chem.* **219**, 85 (1981).
118. Z. Dawoodi, M. J. Mays, and K. Henrick, *J. Chem. Soc., Dalton Trans.*, 433 (1984).
119. G. Süss-Fink, W. Bühlmeyer, M. Herberhold, A. Gieren, and T. Hübner, *J. Organomet. Chem.* **280**, 129 (1985).
120. M. R. Churchill, F. J. Hollander, J. R. Shapley, and J. P. Keister, *Inorg. Chem.* **19**, 1272 (1980).
121. B. F. G. Johnson, J. Lewis, P. R. Raithby, and S. W. Sankey, *J. Organomet. Chem.* **228**, 135 (1982).
122. A. J. Deeming, I. G. Ghatak, D. W. Owen, M. B. Hursthouse, N. P. C. Walker, and J. D. J. Backer-Dirks, unpublished results.
123. M. R. Churchill and H. J. Wasserman, *Inorg. Chem.* **20**, 1580 (1981).
124. M. R. Churchill and H. J. Wasserman, *Inorg. Chem.* **20**, 2905 (1981).
125. K. Burgess, B. F. G. Johnson, J. Lewis, and P. R. Raithby, *J. Chem. Soc., Dalton Trans.*, 263 (1982).
126. B. F. G. Johnson, J. Lewis, D. Pippard, and P. R. Raithby, *J. Chem. Soc., Chem. Commun.*, 551 (1978).
127. B. F. G. Johnson, J. Lewis, D. Pippard, and P. R. Raithby *Acta Cryst.* **B36**, 703 (1980).
128. H. D. Holden, B. F. G. Johnson, J. Lewis, P. R. Raithby, and G. Uden, *Acta Cryst.* **C39**, 1200 (1983).
129. H. D. Holden, B. F. G. Johnson, J. Lewis, P. R. Raithby, and G. Uden, *Acta Cryst.* **C39**, 1197 (1983).
130. R. D. Adams, N. M. Golembeski, and J. P. Selegue, *J. Am. Chem. Soc.* **103**, 546 (1981).
131a. H. D. Holden, B. F. G. Johnson, J. Lewis, P. R. Raithby, and G. Uden, *Acta Cryst.* **C39**, 1203 (1983).
131b. A. M. Brodie, H. D. Holden, J. Lewis, and M. J. Taylor, *J. Organomet. Chem.* **253**, C1 (1983).
132. M. R. Churchill and R. A. Lashewycz, *Inorg. Chem.* **18**, 3261 (1979).
133. Z. Dawoodi, M. J. Mays, P. R. Raithby, K. Henrick, W. Clegg, and G. Weber, *J. Organomet. Chem.* **249**, 149 (1983).
134. K. Natarajan, L. Zsolnai, and G. Huttner, *J. Organomet. Chem.* **220**, 365 (1981).

135. F. Iwasaki, M. J. Mays, P. R. Raithby, P. L. Taylor, and P. J. Wheatley, *J. Organomet. Chem.* **213**, 185 (1981).
136. M. J. Mays, F. Pavelčik, P. R. Raithby, P. J. Taylor, and P. J. Wheatley, *Acta Cryst.* **B37**, 2228 (1981).
137. C. M. Jensen, T. J. Lynch, C. B. Knobler, and H. D. Kaesz, *J. Am. Chem. Soc.* **104**, 4679 (1982).
138. C. M. Jensen, C. B. Knobler, and H. D. Kaesz, *J. Am. Chem. Soc.* **106**, 5926 (1984).
139. R. D. Adams, D. A. Katahira, and J. P. Selegue, *J. Organomet. Chem.* **213**, 259 (1981).
140. R. D. Adams, N. M. Golembeski, and J. P. Selegue, *Inorg. Chem.* **20**, 1242 (1981); R. D. Adams and N. M. Golembeski, *J. Organomet. Chem.* **171**, C21 (1979).
141. R. D. Adams and N. M. Golembeski, *Inorg. Chem.* **17**, 1969 (1978).
142. C. Choo Yin and A. J. Deeming, *J. Organomet. Chem.* **133**, 123 (1977).
143. R. D. Adams and N. M. Golembeski, *J. Am. Chem. Soc.* **101**, 2579 (1979).
144. Y.-C. Lin, C.B. Knobler, and H. D. Kaesz, *J. Am. Chem. Soc.* **103**, 1216 (1981).
145. Y.-C. Lin, A. Mayr, C. B. Knobler, and H. D. Kaesz, *J. Organomet. Chem.* **272**, 207 (1984).
146. J. R. Shapley, D. E. Samkoff, C. Bueno, and M. R. Churchill, *Inorg. Chem.* **21**, 634 (1982).
147. M. R. Churchill and J. R. Missert, *J. Organomet. Chem.* **256**, 349 (1983).
148. A. J. Deeming, R. Peters, M. B. Hursthouse, and J. D. J. Backer-Dirks, *J. Chem. Soc., Dalton Trans.*, 787 (1982).
149. D. E. Samkoff, J. R. Shapley, M. R. Churchill, and H. J. Wasserman, *Inorg. Chem.* **23**, 397 (1984).
150. A. G. Orpen, D. Pippard, G. M. Sheldrick, and K. D. Rouse, *Acta Cryst.* **B34**, 2466 (1978).
151. J. J. Guy, B. E. Reichert, and G. M. Sheldrick, *Acta Cryst.* **B32**, 3319 (1976).
152. E. Sappa, A. Tiripicchio, and A. M. Manotti Lanfredi, *J. Organomet. Chem.* **249**, 391 (1983).
153. A. D. Clauss, M. Tachikawa, J. R. Shapley, and C. G. Pierpont, *Inorg. Chem.* **20**, 1528 (1981).
154. J. R. Shapley, G. M. St. George, M. R. Churchill, and F. J. Hollander, *Inorg. Chem.* **21**, 3295 (1982).
155. C. R. Eady, B. F. G. Johnson, J. Lewis, and M. C. Malatesta, *J. Chem. Soc., Dalton Trans.*, 1358 (1978).
156. J. J. Guy and G. M. Sheldrick, *Acta Cryst.* **B34**, 1718 (1978).
157. R. D. Adams, N. M. Golembeski, and J. P. Selegue, *J. Am. Chem. Soc.* **103**, 546 (1981).
158. R. D. Adams and J. P. Selegue, *J. Organomet. Chem.* **195**, 223 (1980).
159. K. Burgess, B. F. G. Johnson, J. Lewis, and P. R. Raithby, *J. Chem. Soc., Dalton Trans.*, 2085 (1982).
160. A. J. Deeming, S. Hasso, and M. Underhill, *J. Organomet. Chem.* **80**, C53 (1974).
161. A. J. Deeming, S. Hasso, and M. Underhill, *J. Chem. Soc., Dalton Trans.*, 1614 (1975).
162. K. A. Azam and A. J. Deeming, *J. Chem. Soc., Chem. Commun.*, 472 (1977).
163. K. A. Azam, A. J. Deeming, and I. P. Rothwell, *J. Chem. Soc., Dalton Trans.*, 91 (1981).
164. G. D. Jarvinen and R. R. Ryan, *Organometallics* **3**, 1434 (1984).
165. K. A. Azam, A. J. Deeming, R. E. Kimber, and P. R. Shukla, *J. Chem. Soc., Dalton Trans.*, 1853 (1976).
166. K. A. Azam, A. J. Deeming, I. P. Rothwell, M. B. Hursthouse, and L. New, *J. Chem. Soc., Chem. Commun.*, 1086 (1978).
167. K. A. Azam, A. J. Deeming, I. P. Rothwell, M. B. Hursthouse, and J. D. J. Backer-Dirks, *J. Chem. Soc., Dalton Trans.*, 2039 (1981).
168. K. A. Azam and A. J. Deeming, *J. Chem. Soc., Chem. Commun.*, 852 (1976).
169. C. Choo Yin, K. A. Azam, and A. J. Deeming, *J. Chem. Soc., Dalton Trans.*, 1201 (1978).
170. C. Choo Yin and A. J. Deeming, *J. Chem. Soc., Dalton Trans.*, 1013 (1974).
171. A. J. Deeming and R. Peters, *J. Organomet Chem.* **202**, C39 (1980).

172. K. Burgess, H. D. Holden, B. F. G. Johnson, and J. Lewis, *J. Chem. Soc., Dalton Trans.*, 1199 (1983).

173. A. J. Deeming, R. Peters, M. B. Hursthouse and J. D. J. Basker-Dirks, *J. Chem. Soc., Dalton Trans.*, 1205 (1982).

174. W. Ehrenreich, M. Herberhold, G. Süss-fink, H.-P. Klein, and V. Thewalt, *J. Organomet. Chem.* **248**, 171 (1983).

175. K. A. Azam, A. J. Deeming, I. P. Rothwell, *J. Organomet. Chem.* **178**, C20 (1979); K. A. Azam and A. J. Deeming, *J. Mol. Cat.* **3**, 207 (1977).

176. K. Burgess, B. F. G. Johnson, and J. Lewis, *J. Organomet. Chem.* **233**, C55 (1982).

177. A. J. Deeming, I. Ghatak, D. W. Owen, and R. Peters, *J. Chem. Soc., Chem. Commun.*, 392 (1982).

178. A. Mayr, Y. C. Lin, N. M. Boag and H. D. Kaesz, *Inorg. Chem.* **21**, 1704 (1982).

179. H. D. Kaesz, C. B. Knobler, M. A. Andrews, G. van Buskirk, R. Szostak, C. E. Strouse, Y. C. Lin, and A. Mayr, *Pure Appl. Chem.* **54**, 131 (1982).

180. A. J. Arce and A. J. Deeming, *J. Chem. Soc., Chem. Commun.*, 1102 (1980).

181. M. Tachikawa, J. R. Shapley, and C. G. Pierpont, *J. Am. Chem. Soc.* **97**, 7172 (1975).

182. J. R. Shapley, S. I. Richter, M. Tachikawa, and J. B. Keister, *J. Organomet. Chem.* **94**, C43 (1975).

183. A. J. Deeming and S. Hasso, *J. Organomet. Chem.* **112**, C39 (1976)

184. A. J. Deeming and P. J. Manning, *J. Organomet. Chem.* **265**, 87 (1984).

185. M. Laing, P. Sommerville, Z. Dawoodi, M. J. Mays, and P. J. Wheatley, *J. Chem. Soc., Chem. Commun.*, 1035 (1978).

186. Z. Dawoodi, K. Henrick, and M. J. Mays, *J. Chem. Soc., Chem. Commun.*, 696 (1982).

187. Z. Dawoodi and M. J. Mays, *J. Chem. Soc., Dalton Trans.*, 1931 (1984).

188. M. A. Beno, J. M. Williams, M. Tachikawa, and E. L. Muetterties, *J. Am. Chem. Soc.* **102**, 4542 (1980); M. Tachikawa and E. L. Muetterties, *J. Am. Chem. Soc.* **102**, 4541 (1980).

189. M. Brookhart and M. L. H. Green, *J. Organomet. Chem.* **250**, 395 (1983).

190. J. S. Holmgren and J. R. Shapley, *Organometallics* **3**, 1322 (1984).

191. A. J. Deeming and N. P. Randle, unpublished results.

192. A. J. Arce, A. J. Deeming, M. B. Hursthouse, L. New, and N. P. C. Walker, *J. Chem. Soc., Dalton Trans.*, to be submitted.

193. G. Ferraris and G. Gervasio, *J. Chem. Soc., Dalton Trans.*, 1813 (1974).

194. J. R. Shapley, A. C. Sievert, M. R. Churchill, and H. J. Wasserman, *J. Am. Chem. Soc.* **103**, 6975 (1981).

195. M. R. Churchill and H. J. Wasserman, *Inorg. Chem.* **21**, 825 (1982).

196. A. J. Arce and A. J. Deeming, *J. Chem. Soc., Chem. Commun.*, 364 (1982).

197. M. R. Churchill and H. J. Wasserman, *J. Organomet. Chem.* **248**, 365 (1983).

198. A. J. Deeming, *J. Mol. Cat.* **21**, 25 (1983).

199. S. G. Shore, D.-Y. Jan. W.-L. Hsu, L.-Y. Hsu, S. Kennedy, J. C. Huffman, T.-C. Lin Wang and A. G. Marshall, *J. Chem. Soc., Chem. Commun.*, 392 (1984).

200. J. B. Keister and T. L. Horling, *Inorg. Chem.* **19**, 2304 (1980).

201. J. B. Keister, *J. Chem. Soc., Chem. Commun.*, 214 (1979).

202. J. B. Keister, M. W. Payne, and M. J. Muscatella, *Organometallics* **2**, 219 (1983).

203. C. Choo Yin and A. J. Deeming, *J. Organomet. Chem.* **144**, 351 (1978).

204. G. J. Gainsford, J. M. Guss, P. R. Ireland, R. Mason, C. W. Bradford, and R. S. Nyholm, *J. Organomet. Chem.* **40**, C70 (1972).

205. L. J. Farrugia, A. D. Miles, and F. G. A. Stone, *J. Chem. Soc., Dalton Trans.*, 2415 (1984).

206. B. F. G. Johnson, J. Lewis, D. Pippard, P. R. Raithby, G. M. Sheldrick, and K. D. Rouse, *J. Chem. Soc., Dalton Trans.*, 616 (1979).

207. B. F. G. Johnson, J. Lewis, D. Pippard, and P. R. Raithby, *Acta Cryst.* **B34**, 3767 (1978).

208. G. Granozzi, R. Benoni, M. Acampora, S. Aime, and D. Osella, *Inorg. Chim. Acta* **84,** 95 (1984).
209. A. J. Hempleman, I. A. Oxton, D. B. Powell, P. Skinner, A. J. Deeming, and L. Marko, *J. Chem. Soc., Faraday Trans.,* 2 **77,** 1669 (1981).
210. A. Forster, B. F. G. Johnson, J. Lewis, and T. W. Matheson, *J. Organomet. Chem.* **104,** 225 (1976).
211. M. J. Taylor, personal communication; A. J. Deeming and D. Markham, unpublished results.
212. Y. C. Lin, C. B. Knobler, and H. D. Kaesz, *J. Organomet. Chem.* **213,** C41 (1981).
213. J. Banford, Z. Dawoodi, K. Henrick, and M. J. Mays, *J. Chem. Soc., Chem. Commun.,* 554 (1982).
214. D. Seyferth, J. E. Hallgren, and C. S. Eschbach, *J. Am. Chem. Soc.* **96,** 1730 (1974); D. Seyferth, G. H. Williams, C. L. Nivert, *Inorg. Chem.* **16,** 758 (1977).
215. J. W. Kolis, E. M. Holt, and D. F. Shriver, *J. Am. Chem. Soc.* **105,** 7307 (1983); J. W. Kolis, E. M. Holt, M. Drezdzon, K. H. Whitmire, and D. F. Shriver, *J. Am. Chem. Soc.* **104,** 6134 (1982).
216. J. W. Kolis, E. M. Holt, J. A. Hriljac, and D. F. Shriver, *Organometallics* **3,** 496 (1984).
217. J. Evans, B. F. G. Johnson, J. Lewis, and T. W. Matheson, *J. Organomet. Chem.* **97,** C16 (1975).
218. W. G. Jackson, B. F. G. Johnson, J. W. Kelland, and J. Lewis, *J. Organomet. Chem.* **87,** C27 (1975).
219. A. A. Koridze, O. A. Kizas, N. E. Kolobova, P. V. Petrovskii, and E. I. Fedin, *J. Organomet. Chem.* **265,** C33 (1984); *idem, ibid.* **272,** C31 (1984).
220. C. J. Cooksey, A. J. Deeming, and I. P. Rothwell, *J. Chem. Soc., Dalton Trans.,* 1719 (1981).
221. A. J. Deeming, *J. Organomet. Chem.* **128,** 63 (1977).
222. S. Aime, A. Tiripicchio, M. Tiripicchio Camellini, and A. J. Deeming, *Inorg. Chem.* **20,** 2027 (1981).
223. C. G. Pierpont, *Inorg, Chem.* **16,** 636 (1977).
224. A. D. Clauss, J. R. Shapley, and S. R. Wilson, *J. Am. Chem. Soc.* **103,** 7387 (1981).
225. J. F. Blount, L. F. Dahl, C. Hoogzand, and W. Hübel, *J. Am. Chem. Soc.* **88,** 292 (1966); G. Granozzi, E. Tondello, M. Casarin, S. Aime, and D. Osella, *Organometallics* **2,** 430 (1983).
226. V. Busetti, G. Granozzi, S. Aime, R. Gobetto, and D. Osella, *Organometallics* **3,** 1510 (1984).
227. G. Ferraris and G. Gervasio, *J. Chem. Soc., Dalton Trans.,* 1933 (1973).
228. A. J. Deeming, *J. Organomet. Chem.* **150,** 123 (1978).
229. C. Choo Yin and A. J. Deeming, *J. Chem. Soc., Dalton Trans.,* 2563 (1982).
230. S. Aime and A. J. Deeming, *J. Chem. Soc., Dalton Trans.,* 1807 (1983).
231. A. J. Deeming, R. S. Nyholm, and M. Underhill, *J. Chem. Soc., Chem. Commun.,* 224 (1972).
232. A. J. Deeming, R. E. Kimber, and M. Underhill, *J. Chem. Soc., Dalton Trans.,* 2589 (1973).
233. C. W. Bradford and R. S. Nyholm, *J. Chem. Soc., Dalton Trans.,* 529 (1973).
234. M. I. Bruce, J. M. Guss, R. Mason, B. W. Skelton, and A. H. White, *J. Organomet. Chem.* **251,** 261 (1983).
235. A. J. Arce and A. J. Deeming, *J. Chem. Soc., Dalton Trans.,* 1155 (1982).
236. R. D. Adams and N. M. Golembeski, *J. Organomet. Chem.* **172,** 239 (1979).
237. A. J. Arce, A. J. Deeming, and R. Shaunak, *J. Chem. Soc., Dalton Trans.,* 1023 (1983).
238. S. Aime, D. Osella, A. J. Arce, A. J. Deeming, M. B. Hursthouse, and A. M. R. Galas, *J. Chem. Soc., Dalton Trans.,* 1981 (1984)
239. S. Aime, G. Jannon, D. Osella, A. J. Arce, and A. J. Deeming *J. Chem. Soc., Dalton Trans.,* 1987 (1984).
240. S. Aime, D. Osella, A. J. Deeming, A. J. Arce, M. B. Hursthouse, and H. M. Dawes, *J. Chem. Soc., Dalton Trans.,* in press.

241. A. J. Deeming and M. Underhill, *J. Chem. Soc., Dalton Trans.*, 2727 (1973).
242a. G. Süss-fink and P. R. Raithby, *Inorg. Chim. Acta* **71**, 109 (1983).
242b. A. J. Arce, A. J. Deeming, M. B. Hursthouse, and N. P. C. Walker, unpublished results.
243. E. Boyar, A. J. Deeming, A. J. Arce, and Y. De Sanctis, *J. Organomet. Chem.* **276**, C45 (1984).
244. S. Aime, O. Gambino, L. Milone, E. Sappa, and E. Rosenberg, *Inorg. Chim. Acta* **15**, 53 (1975).
245. O. Gambino, R. P. Ferrari, M. Chinone, and G. A. Vaglio, *Inorg. Chim. Acta* **12**, 155 (1975).
246. Z. Dawoodi, M. J. Mays, and K. Henrick, *J. Chem. Soc., Dalton Trans.*, 1769 (1984)
247. S. Aime, L. Milone, and A. J. Deeming, *J. Chem. Soc., Chem. Commun.*, 1168 (1980).
248. G. Granozzi, E. Tondello, R. Bertoncello, S. Aime, and D. Osella, *Inorg. Chem.* **22**, 744 (1983).
249. A. J. Arce and A. J. Deeming, unpublished results.
250. B. E. Hanson, B. F. G. Johnson, J. Lewis, and P. R. Raithby, *J. Chem. Soc., Dalton Trans.*, 1852 (1980).
251. A. J. Deeming and P. J. Manning, *Philos. Trans. R. Soc. London A* **308**, 59 (1982).
252. E. G. Bryan, B. F. G. Johnson, and J. Lewis, *J. Organomet. Chem.* **122**, 249 (1976).
253. S. Aime, R. Gobetto, D. Osella, L. Milone, and E. Rosenberg, *Organometallics* **1**, 640 (1982).
254. G. Granozzi, R. Bertoncello, S. Aime, and D. Osella, *J. Organomet. Chem.* **229**, C27 (1982).
255. E. G. Bryan, B. F. G. Johnson, J. W. Kelland, J. Lewis, and M. McPartlin, *J. Chem. Soc., Chem. Commun.*, 254 (1976).
256. M. R. Churchill and R. A. Lashewycz, *Inorg. Chem.* **18**, 156 (1979).
257. R. D. Adams and N. M. Golembeski, *J. Am. Chem. Soc.* **100**, 4622 (1978)
258. Z. Dawoodi, M. J. Mays, and P. R. Raithby, *J. Organomet. Chem.* **219**, 103 (1981).
259. E. Boyar and A. J. Deeming, unpublished results.
260. G. Gervasio, *J. Chem. Soc., Chem. Commun.*, 25 (1976).
261. G. Gervasio, S. Aime, L. Milone, E. Sappa, and M. Franchini-Angela, *Transition Met. Chem.* **1**, 96 (1976).
262. R. D. Adams and J. P. Selegue, *Inorg. Chem.* **19**, 1795 (1980).
263. W. G. Jackson, B. F. G. Johnson, J. W. Kelland, J. Lewis, and K. T. Schorpp, *J. Organomet. Chem.* **88**, C17 (1975).
264. M. R. Churchill, R. A. Lashewycz, M. Tachikawa, and J. R. Shapley, *J. Chem. Soc., Chem. Commun.*, 699 (1977).
265. M. R. Churchill and R. A. Lashewycz, *Inorg. Chem.* **17**, 1291 (1978).
266. G. Ferraris and G. Gervasio, *J. Chem. Soc., Dalton Trans.*, 1057 (1972).
267. R. P. Ferrari and G. A. Vaglio, *Transition Met. Chem.* **8**, 155 (1983).
268. W. Hieber and H. Stallman, *Z. Electrochem.* **49**, 288 (1943).
269. C. W. Bradford and R. S. Nyholm, *Chem. Commun.*, 384 (1967); C. W. Bradford, *Platinum Met. Rev.* **11**, 104 (1967).
270. B. F. G. Johnson, J. Lewis, and P. A. Kilty, *J. Chem. Soc. A*, 2859 (1968).
271. B. F. G. Johnson and J. Lewis, *Inorg. Synth.* **13**, 92 (1971).
272. E. R. Corey and L. F. Dahl, *Inorg. Chem.* **1**, 521 (1962).
273. M. R. Churchill and B. G. DeBoer, *Inorg. Chem.* **16**, 878 (1977).
274. S. L. Cook, J. Evans, G. N. Greaves, B. F. G. Johnson, J. Lewis, P. R. Raithby, P. B. Wells, and P. Worthington, *J. Chem. Soc., Chem. Commun.*, 777 (1983).
275. S. J. Cook, J. Evans, and G. N. Greaves, *J. Chem. Soc., Chem. Commun.*, 1287 (1983).
276. C. O. Quicksall and T. G. Spiro, *Inorg. Chem.* **7**, 2365, (1968).
277. G. A. Battiston, G. Bor, U. K. Dietler, S. F. A. Kettle, R. Rosetti, G. Sbrignadello, and P. L. Stanghellini, *Inorg. Chem.* **19**, 1961 (1980).
278. B. Delley, M. C. Manning, D. E. Ellis, and J. Berkowitz, *Inorg. Chem.* **21**, 2247 (1982).
279. T. R. Gilson, *J. Chem. Soc., Dalton Trans.*, 149 (1984).
280. R. E. Benfield, P. P. Edwards, and A. M. Stacy, *J. Chem. Soc., Chem. Commun.*, 525 (1982).

281. D. R. Tyler, R. A. Levenson, and H. B. Gray, *J. Am. Chem. Soc.* **100**, 7888 (1978).
282. J. Lewis and B. F. G. Johnson, *Acc. Chem. Res* **1**, 245 (1968).
283. G. A. Vaglio, *J. Organomet. Chem.* **169**, 83 (1979).
284. A. Forster, B. F. G. Johnson, J. Lewis, T. W. Matheson, B. H. Robinson, and W. G. Jackson, *J. Chem. Soc., Chem. Commun.*, 1042 (1974).
285. S. Aime, O. Gambino, L. Milone, E. Sappa, and E. Rosenberg, *Inorg. Chim. Acta* **15**, 53 (1975).
286. A. A. Koridze, O. A. Kizas, N. M. Astakhova, P. V. Petrovskii, and Yu. K. Grishin, *J. Chem. Soc., Chem. Commun.*, 853 (1981).
287. R. E. Benfield and B. F. G. Johnson, *Transition Met. Chem.* **6**, 131 (1981).
288. J. A. Connor, "Transition Metal Clusters" (B. F. G. Johnson, ed.), pp. 345–389. John Wiley, New York, 1980.
289. J. P. Candlin and A. C. Shortland, *J. Organomet. Chem.* **16**, 289 (1969); A. J. Poë and M. V. Twigg, *J. Organomet. Chem.* **50**, C39 (1973); *idem, J. Chem. Soc., Dalton Trans.*, 1860 (1974); S. K. Malik and A. J. Poë, *Inorg. Chem.* **17**, 1484 (1978); *idem, ibid.* **18**, 1241 (1979); D. P. Keeton, S. K. Malik, and A. J. Poë, *J. Chem. Soc., Dalton Trans.*, 233 (1977).
290. A. Shojaie and J. D. Atwood, *Organometallics* **4**, 187 (1985).
291. S. Zobl-Ruh and W. von Philipsborn, *Helv. Chim. Acta* **63**, 733 (1980); M. I. Bruce, M. Cooke, M. Green, and D. J. Westlake, *J. Chem. Soc. A*, 987 (1969).
292. M. R. Burke, J. Takats, F.-W. Grevels, and J. G. A. Reuvers, *J. Am. Chem. Soc.* **105**, 4092 (1983).
293. C. R. Eady, B. F. G. Johnson, and J. Lewis, *J. Chem. Soc., Dalton Trans.*, 2606 (1975).
294. P. F. Jackson, B. F. G. Johnson, J. Lewis, M. McPartlin, and W. J. H. Nelson, *J. Chem. Soc., Chem. Commun.*, 224 (1980); *idem, J. Chem. Soc., Dalton Trans.*, 2099 (1982).
295. D. Braga, K. Henrick, B. F. G. Johnson, J. Lewis, M. McPartlin, W. J. H. Nelson, A. Sironi, and M. D. Vargas, *J. Chem. Soc., Chem. Commun.*, 1131 (1983).
296. D. Braga, J. Lewis, B. F. G. Johnson, M. McPartlin, W. J. H. Nelson, and M. D. Vargas, *J. Chem. Soc., Chem. Commun.*, 241 (1983).
297. A. J. Deeming, B. F. G. Johnson, and J. Lewis, *J. Chem. Soc. A*, 2967 (1970).
298. J. Knight and M. J. Mays, *J. Chem. Soc. A*, 711 (1970).
299. J. A. Ladd, H. Hope, and A. L. Balch, *Organometallics* **3**, 1838 (1984).
300. E. E. Sutton, M. L. Niven, and J. R. Moss, *Inorg. Chim. Acta* **70**, 207 (1983).
301. B. F. G. Johnson, J. Lewis P. R. Raithby, G. M. Sheldrick, K. Wong, and M. McPartlin, *J. Chem. Soc., Dalton Trans.*, 673 (1978).
302. T.-Y. Luh, *Coord. Chem. Rev.* **60**, 255 (1984).
303. M. O. Albers and N. J. Coville, *Coord. Chem. Rev.* **53**, 227 (1984).
304. F. A. Cotton and B. E. Hanson, *Inorg. Chem.* **16**, 2820 (1977).
305. G. Süss-Fink, *Z. Naturforsch.* **35B**, 454 (1980).
306. B. F. G. Johnson, R. L. Kelly, J. Lewis, and J. R. Thornback, *J. Organomet. Chem.* **190**, C91 (1980); R. L. Pruett, R. C. Schoening, J. L. Vidal, and R. A. Fiato, *J. Organomet Chem.* **182**, C57 (1979); R. C. Schoening, J. L. Vidal, and R. A. Fiato, *J. Organomet. Chem.* **206**, C43 (1981).
307. B. F. G. Johnson, J. Lewis, P. R. Raithby, G. M. Sheldrick, and G. Süss-Fink, *J. Organomet. Chem.* **162**, 179 (1978).
308. R. E. Stevens and W. L. Gladfelter, *Inorg. Chem.* **22**, 2034 (1983).
309. B. F. G. Johnson, P. R. Raithby, and C. Zuccaro, *J. Chem. Soc., Dalton Trans.*, 99 (1980).
310. R. E. Stevens, T. J. Yanta, and W. L. Gladfelter, *J. Am. Chem. Soc.* **103**, 4981 (1981).
311. C. E. Kampe and H. D. Kaesz, *Inorg. Chem.* **23**, 1390 (1984).
312. G. Lavigne and H. D. Kaesz, *J. Am. Chem. Soc.* **106**, 4647 (1984).
313. A. M. Bond, P. A. Dawson, B. M. Peake, B. H. Robinson, and J. Simpson, *Inorg. Chem.* **16**, 2199 (1977).

314. P. Lemoine, *Coord. Chem. Rev.* **47**, 55 (1982).
315. C. C. Nagel, J. C. Bricker, D. G. Alway, and S. G. Shore, *J. Organomet. Chem.* **219**, C9 (1981).
316. M. I. Bruce, D. C. Kehoe, J. G. Matisons, B. K. Nicholson, P. H. Rieger, and M. L. Williams, *J. Chem. Soc., Chem. Commun.*, 442 (1982).
317. B. F. G. Johnson, J. Lewis, B. E. Reichart, and K. T. Schorpp, *J. Chem. Soc., Dalton Trans.*, 1403 (1976).
318. A. J. Deeming, S. Donovan-Mtunzi, S. E. Kabir, and P. J. Manning, *J. Chem. Soc., Dalton Trans.*, 1037 (1985).
319. A. Poë and V. C. Sekhar, *J. Am. Chem. Soc.* **106**, 5034 (1984).
320. A. J. Deeming, S. Donovan-Mtunzi, and S. E. Kabir, *J. Organomet. Chem.* **276**, C65 (1984).
321. M. J. Mays and P. D. Gavens, *J. Organomet. Chem.* **124**, C37 (1977).
322. M. J. Mays and P. D. Gavens, *J. Chem. Soc., Dalton Trans.*, 911 (1980).
323. M. J. Mays and P. D. Gavens, *J. Organomet. Chem.* **177**, 433 (1979).
324. P. F. Jackson, B. F. G. Johnson, J. Lewis, W. J. H. Nelson, and M. McPartlin, *J. Chem. Soc., Dalton Trans.*, 2099 (1982).
325. M. Tachikawa, J. R. Shapley, R. C. Haltiwanger, and C. G. Pierpont, *J. Am. Chem. Soc.* **98**, 4651 (1976).
326. C. G. Pierpont, *Inorg. Chem.* **17**, 1976 (1978).
327. M. Tachikawa, S. I. Richter, and J. R. Shapley, *J. Organomet. Chem.* **128**, C9 (1977).
328. E. G. Bryan, B. F. G. Johnson, and J. Lewis, *J. Chem. Soc., Dalton Trans.*, 144 (1977).
329. E. Boyar, A. J. Deeming, I. P. Rothwell, K. Henrick, and M. McPartlin, *J. Chem. Soc., Dalton Trans.*, in press.
330. P. L. Watson and G. L. Schrader, *J. Mol. Cat* **9**, 129 (1980).
331. R. E. Benfield, B. F. G. Johnson, P. R. Raithby, and G. M. Sheldrick, *Acta Cryst.* **B34**, 666 (1978).
332. P. A. Dawson, B. F. G. Johnson, J. Lewis, J. Puga, P. R. Raithby, and M. J. Rosales, *J. Chem. Soc., Dalton Trans.*, 233 (1982).
333. M. I. Bruce, J. G. Matisons, R. C. Wallis, J. M. Patrick, B. W. Skelton, and A. H. White, *J. Chem. Soc., Dalton Trans.*, 2365 (1983).
334. E. Band and E. L. Muetterties, *Chem. Rev.* **78**, 39 (1978).
335. B. F. G. Johnson and R. E. Benfield, *in* "Transition Metal Clusters" (B. F.G. Johnson, ed.), p. 471. Wiley (Interscience), New York, 1980.
336. S. Aime, A. J. Deeming, S. Donovan-Mtunzi, and S. E. Kabir, unpublished results.
337. F. A. Cotton, and B. E. Hanson, *Inorg. Chem.* **16**, 3369 (1977).
338. A. W. Coleman, D. F. Jones, P. H. Dixneuf, C. Brisson, J.-J. Bonnet, and G. Lavigne, *Inorg. Chem.* **23**, 952 (1984).
339. F. A. Cotton, B. E. Hanson, and J. D. Jamerson, *J. Am. Chem. Soc.* **99**, 6588 (1977).
340. M. A. Andrews, S. W. Kirtley, and H. D. Kaesz, *Inorg. Chem.* **16**, 1556 (1977).
341. S. Aime, D. Osella, L. Milone, and E. Rosenberg, *J. Organomet. Chem.* **213**, 207 (1981).
342. G. E. Hawkes, E. W. Randall, S. Aime, D. Osella, and J. E. Elliot, *J. Chem. Soc., Dalton Trans.*, 279 (1984)
343. S. Aime and D. Osella, *J. Chem. Soc., Chem. Commun.*, 300 (1981).
344. S. Aime, *Inorg. Chim. Acta* **62**, 51 (1982).
345. S. Aime, R. Gobetto, D. Osella, L. Milone, G. E. Hawkes, and E. W. Randall, *J. Chem. Soc., Chem. Commun.*, 794 (1983).
346. A. T. Nicol and R. W. Vaughan, *J. Am. Chem. Soc.* **101**, 583 (1979).
347. M. W. Howard, U. A. Jayasooriya, S. F. A. Kettle, D. B. Powell, and N. Sheppard, *J. Chem. Soc., Chem. Commun.*, 18 (1979).
348. J. R. Andrews, S. F. A. Kettle, D. B. Powell, and N. Sheppard, *Inorg. Chem.* **21**, 2874 (1982) and references therein.
349. S. F. A. Kettle and P. L. Stanghellini, *Inorg. Chem.* **21**, 1447 (1982).

350. J. C. Green, D. M. P. Mingos, and E. A. Seddon, *J. Organomet. Chem.* **185,** C20 (1980); *idem, Inorg. Chem.* **20,** 2595 (1981).
351. C. F. Brucker, T. N. Rhodin, S. Wijeyesekara, G. George, and J. R. Shapley, *J. Chim. Phys. Phys.-Chim. Biol.* **78,** 897 (1981).
352. G. Granozzi, R. Benoni, E. Tondello, M. Casarin, S. Aime, and D. Osella, *Inorg. Chem.* **22,** 3899 (1983).
353. M. R. Churchill and B. G. DeBoer, *Inorg. Chem.* **16,** 2397 (1977).
354. R. D. Adams and N. M. Golembeski, *Inorg. Chem.* **18,** 1909 (1979).
355. J. B. Keister and J. R. Shapley, *Inorg. Chem.* **21,** 3304 (1982).
356. E. Rosenberg, E. V. Anslyn, C. Barner-Thorsen, S. Aime, D. Osella, R. Gobetto, and L. Milone, *Organometallics* **3,** 1790 (1984).
357. S. C. Brown and J. Evans, *J. Chem. Soc., Dalton Trans.,* 1049 (1982).
358. J. G. Bentsen and M. S. Wrighton, *Inorg. Chem.* **23,** 512 (1984).
359. S. Kennedy, J. J. Alexander, and S. G. Shore, *J. Organomet. Chem.* **219,** 385 (1981).
360. A. J. Arce, A. J. Deeming, S. Donovan-Mtunzi, and S. E. Kabir, *J. Chem. Soc., Dalton Trans.,* submitted.
361. J. B. Keister and J. R. Shapley, *J. Organomet. Chem.* **85,** C29 (1975).
362. E. G. Bryan, W. G. Jackson, B. F. G. Johnson, J. W. Kelland, J. Lewis, and K. T. Schorpp, *J. Organomet. Chem.* **108,** 385 (1976).
363. J. Evans and G. S. McNulty, *J. Chem. Soc., Dalton Trans.,* 639 (1983).
364. J. Liu, E. Boyar, A. J. Deeming, and S. Donovan-Mtunzi, *J. Chem. Soc., Chem. Commun.,* 1182 (1984).
365. R. D. Adams and J. P. Selegue, *Inorg. Chem.* **19,** 1791 (1980).
366. E. D. Morrison and G. L. Geoffroy, *J. Am. Chem. Soc.* **107,** 254 (1985).
367. C. J. Cardin, D. J. Cardin, H. E. Parge, and J. M. Power, *J. Chem. Soc., Chem. Commun.,* 609 (1984).
368. A. W. Parkins, E. O. Fischer, G. Huttner, and D. Regler, *Angew. Chem. Int. Ed. Eng.* **9,** 633 (1970).
369. S. Aime, R. Gobetto, D. Osella, G. E. Hawkes, and E. W. Randall, *J. Chem. Soc., Dalton Trans.,* 1863 (1984).
370. W. G. Jackson, B. F. G. Johnson, and J. Lewis, *J. Organomet. Chem.* **90,** C13 (1975); *idem, ibid.,* **139,** 125 (1977).
371. R. T. Edidin, J. R. Norton, and K. Mislow, *Organometallics* **1,** 561 (1982).
372. B. E. R. Schilling and R. Hoffmann, *J. Am. Chem. Soc.* **101,** 3456 (1979).
373. S. Aime and A. J. Deeming, *J. Chem. Soc., Dalton Trans.,* 828 (1981).
374. E. C. Constable, B. F. G. Johnson, J. Lewis, G. N. Pain, and M. J. Taylor, *J. Chem. Soc., Chem. Commun.,* 754 (1982).
375. O. Gambino, G. A. Vaglio, R. P. Ferrari, and G. Cetini, *J. Organomet. Chem.* **30,** 381 (1971).
376. R. P. Ferrari, G. A. Vaglio, O. Gambino, M. Valle, and G. Cetini, *J. Chem. Soc., Dalton Trans.,* 1998 (1972).
377. G. A. Vaglio, O. Gambino, R. P. Ferrari, and G. Cetini, *Inorg. Chim. Acta* **7,** 193 (1973).
378. J. Evans, *Chem. Soc. Rev.* **10,** 159 (1981).
379. B. Besson, B. Moraweck, A. K. Smith, J. M. Basset, R. Psaro, A. Fusi, and R. Ugo, *J. Chem. Soc., Chem. Commun.,* 569 (1980).
380. M. Deeba and B. C. Gates, *J. Catal.* **67,** 303 (1981).
381. R. Psaro, R. Ugo, G. M. Zanderighi, B. Besson, A. K. Smith, and J. M. Basset, *J. Organomet. Chem.* **213,** 215 (1981).
382. G. Collier, D. J. Hunt, S. D. Jackson, R. B. Moyes, I. A. Pickering, P. B. Wells, A. K. Simpson, and R. Whyman, *J. Catal.* **80,** 154 (1983).
383. J. Evans and B. P. Gracey, *J. Chem. Soc., Chem. Commun.,* 852 (1980).

384. J. Evans and B. P. Gracey, *J. Chem. Soc., Dalton Trans.*, 1123 (1982).
385. S. C. Brown and J. Evans, *J. Organomet. Chem.* **194,** C53 (1980).
386. S. C. Brown and J. Evans, *J. Chem. Soc., Chem. Commun.,* 1063 (1978).
387. R. Pierantozzi, K. J. McQuade, B. C. Gates, M. Wolf, H. Knözinger, and W. Ruhmann, *J. Am. Chem. Soc.* **101,** 5436 (1979).
388. S. Bhaduri, H. Khwaja, and B. A. Narayanan, *J. Chem. Soc., Dalton Trans.*, 2327 (1984).
389. S. Bhaduri, H. Khwaja, and K. R. Sharma, *Indian J. Chem.* **21,** 155 (1982).
390. T. Catrillo, K. Knözinger, and M. Wolf, *Inorg. Chim. Acta* **45,** L235 (1980).
391. J. Evans and B. P. Gracey, *J. Organomet. Chem.* **228,** C4 (1982).
392. B. Besson, A. Choplin, L. D'Ornelas, and J.-M Basset, *J. Chem. Soc., Chem. Commun.,* 843 (1982).
393. X. J. Li, J. H. Onuferko, and B. C. Gates, *J. Catal.* **85,** 176 (1984).
394. M. B. Freeman, M. A. Patrick, and B. C. Gates, *J. Catal.* **73,** 82 (1982).
395. R. D. Adams, Z. Dawoodi, D. F. Foust, and L. W. Yang, *Ann. N.Y. Acad. Sci.* **415,** 47 (1983).
396. E. L. Muetterties and M. J. Krause, *Angew. Chem. Int. Ed. Eng.* **22,** 135 (1983).
397. E. L. Muetterties, *Chem. Soc. Rev.* **11,** 283 (1982).
398. R. D. Adams, *Acc. Chem. Res.* **16,** 67 (1983).
399. R. Whyman, *in* "Transition Metal Clusters" (B. F. G. Johnson, ed.), p. 545. John Wiley, New York, 1980.
400. A. K. Smith, A. Theolier, J.-M. Basset, R. Ugo, D. Commerceuc, and Y. Chauvin, *J. Am. Chem. Soc.* **100,** 2590 (1978).
401. M. G. Thomas, B. F. Beier, and E. L. Muetterties, *J. Am. Chem. Soc.* **98,** 1296 (1976).
402. R. D. Daroda, J. R. Blackborrow, and G. Wilkinson, *J. Chem. Soc., Chem. Commun.,* 1101 (1980).
403. A. D. King, R. B. King, and D. B. Young, *J. Chem. Soc., Chem. Commun.,* 529 (1980).
404. P. C. Ford, *Acc. Chem. Res.* **14,** 31 (1981).
405. R. P. Ferrari, G. A. Vaglio, and M. Valle, *Inorg. Chim. Acta* **31,** 177 (1978).
406. R. P. Ferrari and G. A. Vaglio, *Inorg. Chim. Acta.* **20,** 141 (1976).
407. R. A. Sánchez-Delgado, J. Puga, and M. Rosales, *J. Mol. Cat.* **24,** 221 (1984).
408. H. Kang, C. H. Mauldin, T. Cole, W. Slegeir, K. Cann, and R. Pettit, *J. Am. Chem. Soc.* **99,** 8323 (1977).
409. R. D. Adams and N. M. Golembeski, *J. Am. Chem. Soc.* **101,** 2579 (1979).
410. Y. Shvo and R. M. Laine, *J. Chem. Soc., Chem. Commun.,* 753 (1980)
411. Y. Shvo, D. W. Thomas, and R. M. Laine, *J. Am. Chem. Soc.* **103,** 2461 (1981).
412. A. J. Deeming, S. S. Ullah, A. J. P. Domingos, B. F. G. Johnson, and J. Lewis, *J. Chem. Soc., Dalton Trans.*, 2093 (1974).
413. K. M. Motyl, J. R. Norton, C. K. Shauer, and O. P. Anderson, *J. Am. Chem. Soc.* **104,** 7325 (1982).
414. G. R. Crooks, B. F. G. Johnson, J. Lewis, I. G. Williams, and G. Gamlen, *J. Chem. Soc. A*, 2761 (1969); J. G. Bullitt and F. A. Cotton, *Inorg. Chim. Acta* **5,** 406 (1971).
415. B. E. Reichert and G. M. Sheldrick, *Acta Cryst.* **B33,** 175 (1977).
416. F. L'Eplattenier, *Inorg. Chem.* **8,** 965 (1969).
417. F. L'Eplattenier and C. Pelichet, *Helv. Chim. Acta* **53,** 1091 (1970).
418. C.-M. T. Hayward and J. R. Shapley, *Inorg. Chem.* **21,** 3816 (1982).
419. C. R. Eady, B. F. G. Johnson, and J. Lewis, *J. Chem. Soc., Dalton Trans.*, 838 (1977).
420. B. F. G. Johnson, J. Lewis, M. McPartlin, W. J. H. Nelson, P. R. Raithby, A. Sironi, and M. D. Vargas, *J. Chem. Soc., Chem. Commun.,* 1476 (1983).
421. B. F. G. Johnson, J. Lewis, W. J. H. Nelson, M. D. Vargas, D. Braga, K. Henrick, and M. McPartlin, *J. Chem. Soc., Dalton Trans.*, 2151 (1984).

422. R. Jackson, B. F. G. Johnson, J. Lewis, P. R. Raithby, and S. W. Sankey, *J. Organomet. Chem.* **193**, Cl (1980).
423. A. G. Orpen and G. M. Sheldrick, *Acta Cryst.* **B34**, 1989 (1978).
424. R. D. Adams, I. T. Horváth, P. Mathur, and B. E. Segmüller, *Organometallics* **2**, 996 (1983).
425. R. D. Adams, I. T. Horváth, and L.-W. Yang, *J. Am. Chem. Soc.* **105**, 1533 (1983).
426. R. D. Adams, I. T. Horváth, and P. Mathur, *Organometallics* **3**, 623 (1984).
427. R. D. Adams, D. F. Foust, and P. Mathur, *Organometallics* **2**, 990 (1983).
428. R. D. Adams, Z. Dawoodi, and D. F. Foust, *Organometallics* **1**, 411 (1982).
429. R. D. Adams and D. F. Foust, *Organometallics* **2**, 323 (1983).
430. R. D. Adams, Z. Dawoodi, D. F. Foust, and B. E. Segmüller, *J. Am. Chem. Soc.* **105**, 831 (1983).
431. R. D. Adams, D. F. Foust, and B. E. Segmüller, *Organometallics* **2**, 308 (1983).
432. J. M. Fernandez, B. F. G. Johnson, J. Lewis, and P. R. Raithby, *J. Chem. Soc., Chem. Commun.*, 1015 (1978); *idem, Acta Cryst.* **B35**, 1711 (1979); A. G. Orpen and G. M. Sheldrick, *Acta Cryst.* **B34**, 1992 (1978).
433. R. D. Adams, I. T. Horváth, and B. E. Segmüller, *J. Organomet. Chem.* **262**, 243 (1984).
434. Z. Dawoodi, M. J. Mays, and P. R. Raithby, *J. Chem. Soc., Chem. Commun.*, 712 (1980).
435. Z. Dawoodi, M. J. Mays, and P. R. Raithby, *J. Chem. Soc., Chem. Commun.*, 801 (1981).
436. S. A. R. Knox and F. G. A. Stone, *J. Chem. Soc. A*, 3147 (1970).
437. J. R. Moss and W. A. G. Graham, *J. Organomet. Chem.* **18**, P24 (1969).
438. J. A. K. Howard, S. A. R. Knox, F. G. A. Stone, and P. Woodward, *J. Chem. Soc., Chem. Commun.*, 1477 (1970).
439. M. Elder and D. Hall, *Inorg. Chem.* **8**, 1424 (1969).
440. R. K. Pomeroy, *J. Organomet. Chem.* **221**, 323 (1981).
441. G. N. van Buuren, A. C. Wills, F. W. B. Einstein, L. K. Peterson, R. K. Pomeroy, and D. Sutton, *Inorg. Chem.* **20**, 4361 (1981).
442. A. C. Wills, F. W. B. Einstein, R. M. Ramadan, and R. K. Pomeroy, *Organometallics* **2**, 935 (1983).
443. M. Fajardo, H. D. Holden, B. F. G. Johnson, J. Lewis, and P. R. Raithby, *J. Chem. Soc., Chem. Commun.*, 24 (1984).
444. M. Fajardo, M. P. Gómez-Sal, H. D. Holden, B. F. G. Johnson, J. Lewis, R. C. S. McQueen, and P. R. Raithby, *J. Organomet. Chem.* **267**, C25 (1984).
445. B. F. G. Johnson, D. A. Kaner, J. Lewis, and P. R. Raithby, *J. Chem. Soc., Chem. Commun.*, 753 (1981).
446. M. R. Churchill and C. Bueno, *Inorg. Chem.* **22**, 1510 (1983).
447. G. Lavigne, F. Papageorgiou, C. Bergounhou, and J.-J. Bonnett, *Inorg. Chem.* **22**, 2485 (1983).
448. M. R. Churchill and F. J. Hollander, *Inorg. Chem.* **17**, 3546 (1978).
449. J. R. Shapley, G. A. Pearson, M. Tachikawa, G. E. Schmidt, M. R. Churchill, and F. J. Hollander, *J. Am. Chem. Soc.* **99**, 8064 (1977).
450. E. J. Ditzel, H. D. Holden, B. F. G. Johnson, J. Lewis, A. Sanders, and M. J. Taylor, *J. Chem. Soc., Chem. Commun.*, 1373 (1982).
451. M. R. Churchill and F. J. Hollander, *Inorg. Chem.* **20**, 4124 (1981).
452. M. R. Churchill, F. J. Hollander, R. A. Lashewycz, G. A. Pearson, and J. R. Shapley, *J. Am. Chem. Soc.* **103**, 2430 (1981).
453. J. Knight and M. J. Mays, *J. Chem. Soc., Dalton Trans.*, 1022 (1972).
454. M. R. Churchill and F. J. Hollander, *Inorg. Chem.* **16**, 2493 (1977).
455. A. B. Antonova, S. V. Kovalenko, E. D. Korniyets, A. A. Johansson, Yu. T. Struchkov, and A. I. Yanovsky, *J. Organomet. Chem.* **267**, 299 (1984).
456. J. T. Park, J. R. Shapley, M. R. Churchill, and C. Bueno, *Inorg. Chem.* **22**, 1579 (1983).

457. D. G. Evans and D. M. P. Mingos, *J. Organomet. Chem.* **232**, 171 (1982).
458. K. Burgess, B. F. G. Johnson, D. A. Kaner, J. Lewis, P. R. Raithby, S. N. Azman, and B. Syed-Mustaffa, *J. Chem. Soc., Chem. Commun.*, 455 (1983).
459. J. A. K. Howard, L. Farrugia, C. Foster, F. G. A. Stone, and P. Woodward, *Eur. Cryst. Meeting* **6**, 73 (1980).
460. K. Burgess, B. F. G. Johnson, and J. Lewis, *J. Chem. Soc., Dalton Trans.*, 1179 (1983).
461. K. Burgess, B. F. G. Johnson, and J. Lewis, *J. Organomet. Chem.* **247**, C42 (1983).
462. C. W. Bradford, W. van Bronswijk, R. J. H. Clark, and R. S. Nyholm, *J. Chem. Soc. A*, 2889 (1970).
463. K. Burgess, B. F. G. Johnson, J. Lewis, and P. R. Raithby, *J. Chem. Soc., Dalton Trans.*, 1661 (1983).
464. P. Braunstein, J. Rosé, A. M. Manotti Lanfredi, A. Tiripicchio, and E. Sappa, *J. Chem. Soc., Dalton Trans.*, 1843 (1984).
465. S. Ermer, K. King, K. I. Hardcastle, E. Rosenberg, A. M. Manotti Lanfredi, A. Tiripicchio, and M. Tiripicchio Camellini, *Inorg. Chem.* **22**, 1339 (1983).
466. B. F. G. Johnson, J. Lewis, P. R. Raithby, S. N. Azman, B. Syed-Mustaffa, M. J. Taylor, K. H. Whitmire, and W. Clegg, *J. Chem. Soc., Dalton Trans.*, 2111 (1984).
467. L. J. Farrugia, J. A. K. Howard, P. Mitrprachachon, J. L. Spencer, F. G. A. Stone, and P. Woodward, *J. Chem. Soc., Chem. Commun.*, 260 (1978).
468. L. J. Farrugia, J. A. K. Howard, P. Mitrprachachon, F. G. A. Stone, and P. Woodward, *J. Chem. Soc., Dalton Trans.*, 155 (1981).
469. L. J. Farrugia, M. Green, D. R. Hankey, M. Murray, A. G. Orpen, and F. G. A. Stone, *J. Chem. Soc., Dalton Trans.*, 177 (1985).
470. L. J. Farrugia, M. Green, D. R. Hankey, A. G. Orpen, and F. G. A. Stone, *J. Chem. Soc., Chem. Commun.*, 310 (1983).
471. L. J. Farrugia, J. A. K. Howard, P. Mitrprachachon, F. G. A. Stone, and P. Woodward, *J. Chem. Soc., Dalton Trans.*, 162 (1981).
472. M. Green, D. R. Hankey, M. Murray, A. G. Orpen, and F. G. A. Stone, *J. Chem. Soc., Chem. Commun.*, 689 (1981).
473. L. J. Farrugia, J. A. K. Howard, P. Mitrprachachon, F. G. A. Stone, and P. Woodward, *J. Chem. Soc., Dalton Trans.*, 171 (1981).
474. R. D. Adams, I. T. Horváth, and P. Mathur, *J. Am. Chem. Soc.* **105**, 7202 (1983).
475. R. D. Adams, I. T. Horváth, and P. Mathur, *J. Am. Chem. Soc.* **106**, 6296 (1984).
476. E. Sappa, M. Valle, G. Predieri, and A. Tiripicchio, *Inorg. Chim. Acta* **88**, L23 (1984).
477. E. Sappa, M. Lanfranchi, A. Tiripicchio, and M. Tiripicchio Camellini, *J. Chem. Soc., Chem. Commun.*, 995 (1981).
478. M. Castiglioni, E. Sappa, M. Valle, M. Lanfranchi, and A. Tiripicchio, *J. Organomet. Chem.* **241**, 99 (1983).
479. S. G. Shore, W.-L. Hsu, C. R. Weisenberger, M. L. Castle, M. R. Churchill, and C. Bueno, *Organometallics* **1**, 1405 (1982).
480. S. G. Shore and W.-L. Hsu, *J. Am. Chem. Soc.* **105**, 655 (1983).
481. M. R. Churchill and C. Bueno, *J. Organomet. Chem.* **256**, 357 (1983).
482. S. Bhaduri, B. F. G. Johnson, J. Lewis, P. R. Raithby, and D. J. Watson, *J. Chem. Soc., Chem. Commun.*, 343 (1978).
483. L. J. Farrugia, A. G. Orpen, and F. G. A. Stone, *Polyhedron* **2**, 171 (1983).
484. J. T. Park, J. R. Shapley, M. R. Churchill, and C. Bueno, *J. Am. Chem. Soc.* **105**, 6182 (1983).
485. R. D. Adams, T. S. A. Hor, and P. Mathur, *Organometallics* **3**, 634 (1984).
486. M. Castiglioni, R. Giordano, E. Sappa, G. Predieri, and A. Tiripicchio, *J. Organomet. Chem.* **270**, C7 (1984).

487. M. R. Churchill, C. Bueno, W.-L. Hsu, J. S. Plotkin, and S. G. Shore, *Inorg. Chem.* **21,** 1958 (1982).
488. D. T. Eadie, H. D. Holden, B. F. G. Johnson, and J. Lewis, *J. Chem. Soc., Dalton Trans.*, 301 (1984).
489. R.D. Adams, I. T. Horváth, and B. E. Segmüller, *Organometallics* **1,** 1537 (1982).
490. M. R. Churchill, C. Bueno, S. Kennedy, J. C. Bricker, J. S. Plotkin, and S. G. Shore, *Inorg. Chem.* **21,** 627 (1982).
491. J. S. Plotkin, D. G. Alway, C. R. Weisenberger, and S. G. Shore, *J. Am. Chem. Soc.* **102,** 6156 (1980).
492. L.-Y. Hsu, W.-L. Hsu, D.-Y. Jan, A. G. Marshall, and S. G. Shore, *Organometallics* **3,** 591 (1984).
493. M. R. Churchill, F. J. Hollander, J. R. Shapley, and D. S. Foose, *J. Chem. Soc., Chem. Commun.*, 534 (1978).
494. M. R. Churchill and F. J. Hollander, *Inorg. Chem.* **18,** 843 (1979).
495. M. R. Churchill and F. J. Hollander, *Inorg. Chem.* **18,** 161 (1979).
496. E. Sappa, A. Tiripicchio, and M. Tiripicchio Camellini, *J. Organomet. Chem.* **246,** 287 (1983); M. Castiglioni, R. Giordano, and E. Sappa, *J. Organomet. Chem.* **258,** 217 (1983).
497. G. Süss-Fink, U. Thewalt, and H.-P. Klein, *J. Organomet. Chem.* **262,** 315 (1984).
498. L. Busetto, M. Green, J. A. K. Howard, B. Hessner, J. C. Jeffery, R. M. Mills, F. G. A. Stone, and P. Woodward, *J. Chem. Soc., Chem. Commun.*, 1101 (1981).
499. L. Busetto, M. Green, B. Hessner, J. A. K. Howard, J. C. Jeffery, and F. G. A. Stone, *J. Chem. Soc., Dalton Trans.*, 519 (1983).
500. J. R. Shapley, J. T. Park, M. R. Churchill, C. Bueno, and H. J. Wasserman, *J. Am. Chem. Soc.* **103,** 7385 (1981).
501. J. R. Shapley, J. T. Park, M. R. Churchill, J. W. Ziller, and L. R. Beanan, *J. Am. Chem. Soc.* **106,** 1144 (1984).
502. M. R. Churchill, C. Bueno, J. T. Park, and J. R. Shapley, *Inorg. Chem.* **23,** 1017 (1984).
503. M. R. Churchill, C. Bueno, and H. J. Wasserman, *Inorg. Chem.* **21,** 640 (1982).
504. R. Bartsch, P. B. Hitchcock, M. F. Meidine, and J. F. Nixon, *J. Organomet. Chem.* **266,** C41 (1984).

Chemistry of
1,3-Ditungstacyclobutadienes

MALCOLM H. CHISHOLM and
JOSEPH A. HEPPERT[1]

Department of Chemistry
Indiana University
Bloomington, Indiana 47405

I

INTRODUCTION: METALLACYCLES

Metallacycles have enjoyed a pivotal position in recent advances in organometallic chemistry (*1*). The investigations by Whitesides and co-workers (*2*) into the thermal decomposition of $L_2Pt(CH_2)_n$ compounds, where $n = 4$, 5 and 6, were among the first to place emphasis on the importance of steric accessiblity for the β-hydrogen atom abstraction reaction. The $Cp_2Ti(CH_2)_4$ molecule, under thermoloysis, was found to undergo a competitive C—C bond cleavage to give ethylene and 1-butene (*3,4*), while for nickellacyclopentanes, $L_nNi(CH_2)_4$, Grubbs and co-workers (*5,6*) showed that reductive elimination may lead to cyclobutane, 1-butene, or ethylene (2 equiv) depending upon the number of appending phosphine ligands, $n = 3$, 2, or 1 (*7*). The metallacyclobutanes are worthy of special note in the "saturated" metallacycles since these are now generally accepted as the reactive intermediates in the olefin metathesis reaction (*8,9*), while for platinum the elegant work of Puddephatt (*10*) has elucidated the mechanistic details of the rearrangements within the PtC_3 ring.

Unsaturated metallacycles have also had a fascinating history. Often these are readily formed by the coupling of alkyne units and may be involved in the catalytic cyclization of alkynes. There are numerous examples of structurally characterized metallacyclopentadienes, and even some examples of metallacycloheptatrienes and η^4-benzene complexes (*11*). More recently metallacyclobutadienes have been implicated as the active intermediates in alkyne

---

[1] Present address: Department of Chemistry, University of Kansas, Lawrence, Kansas 66046.

metathesis reactions [Eq. (1)] (*12,13*).

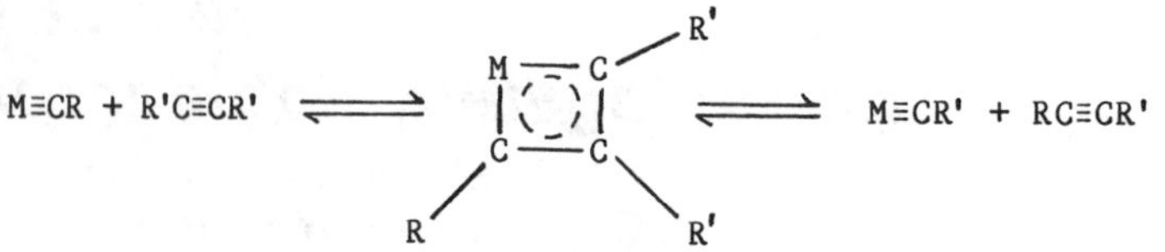

$$M\equiv CR + R'C\equiv CR' \quad \rightleftharpoons \quad \begin{matrix} M \text{---} C \diagup^{R'} \\ | \; \overset{\frown}{(\;\;)} \; | \\ C \text{---} C \\ \diagup_R \qquad ^{R'}\diagdown \end{matrix} \quad \rightleftharpoons \quad M\equiv CR' + RC\equiv CR' \qquad (1)$$

Schrock and co-workers, in their pioneering studies of alkyne metathesis reactions, have actually isolated and structurally characterized one such compound (**I**) (*14,15*) and shown that addition of a nitrogen base can convert the metallacyclobutadiene to **II**, which is formally a metallatetrahedrane or a

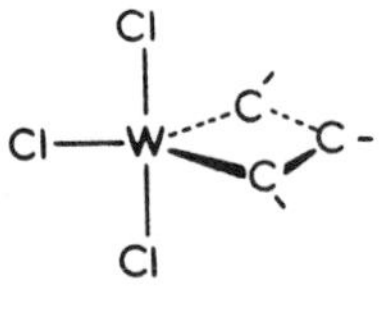

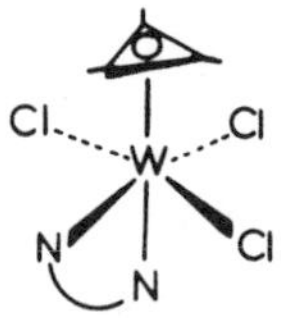

cyclopropenium $(C_3R_3^+)$ metal complex (*16*). When the metallacyclo-butadiene, such as **I**, is stable, then alkyne metathesis is slow and other reactions compete (*4,17*), but with OBu^t ligands, alkyne metathesis is fast and the proposed intermediate metallacyclobutadienes are not detected.

Our interest in dimetallacyclobutadienes was triggered when Schrock reported the synthesis of alkylidyne tungsten compounds in a metathesis-like reaction involving W≡W bonds [Eq. (2)] (*18*).

$$W_2(OBu^t)_6 + RC\equiv CR \;\rightarrow\; 2(t\text{-BuO})_3W\equiv CR \qquad (2)$$

where R = Me, Et, Pr

Subsequently, the MIT workers showed that this reaction could be extended to a wide variety of R groups, even those containing other functionalities such as C=O and C=C (*19*), and also that the reaction was not restricted to Bu^{t}O ligands but also occurred for groups such as OPr^i and OC(CF$_3$)Me$_2$ (*20*).

II

DIMETALLATETRAHEDRANES AND DIMETALLACYCLOBUTADIENES

In our laboratories, we had prepared a variety of so-called perpendicular alkyne adducts (*21*) from the reactions between Mo$_2$(OR)$_6$ compounds and simple alkynes such as C$_2$H$_2$, MeC≡CH, and MeC≡CMe (*22*) and in extending these studies to W$_2$(OR)$_6$ compounds discovered evidence for a facile equilibrium involving the μ-perpendicular alkyne adducts and the alkylidyne complexes. This was first discovered for the compound $W_2(OBu^t)_6(py)(\mu\text{-}C_2H_2)$ [Eq. (3)] on the basis of labeling experiments involving $\mu\text{-*}C_2H_2$ and $\mu\text{-}C_2D_2$, where *C represents 92.5 g atom % ^{13}C (23). Subsequently we have found that a variety of alkyne adducts (MeCCMe, EtCCEt, and PhCCPh) of $W_2(OCH_2Bu^t)_6$ behave similarly but that the equilibrium favors the alkyne adduct (24,25).

$$W_2(OBu^t)_6(py)(\mu\text{-}C_2H_2) \ \rightleftharpoons \ 2\,(t\text{-BuO})_3W\equiv CH + py \qquad (3)$$

It should be noted that we view the μ-perpendicular alkyne adducts of $W_2(OR)_6$ compounds as dimetallatetrahedranes (III) on the basis of W—W, W—C, and C—C bond distances together with the remarkably small $^1J_{^{13}C-^{13}C}$ values of approximately 10–15 Hz for the W_2C_2 moiety (23,24). The latter may be compared to $^1J_{^{13}C-^{13}C} = 9$ Hz found for the central C_4 core in the tetrahedrane $C_4(Bu^t)_4$ (26). Thus addition of the alkyne across the $(W\equiv W)^{6+}$ moiety leads to a formal oxidation to $(W-W)^{10+}$ in the dimetallatetrahedranes or to two W^{6+} units if complete cleavage of the W—W and C—C bonds is achieved.

III

The question which thus arises is how can a dimetallatetrahedrane (III) break apart reversibly to give two $W\equiv CR$ groups. The two most plausible pathways involve a twisting to give a 1,2-ditungstacyclobutadiene (IV) or a flattening to a 1,3-dimetallacyclobutadiene (V). At first glance the 1,2-ditungstacyclobutadiene appears like a parallel alkyne adduct (21), several of which are known [e.g., $M_2Cl_2(dpm)_2\{\mu\text{-}C_2(CF_3)_2\}$ where $M = Rh$ and Pd] but all in reality are better viewed as 1,2-dimetallacyclobutenes, having a 10-electron M_2C_2 core with M—M and M—C single bonds and a C—C double bond. 1,3-Dimetallacyclobutadienes of type V have been known for some

IV   V

years for Nb (27), Ta (27), W $(28,29)$, and Re (30), each having a structure based on two fused distorted tetraheda (**VI**).

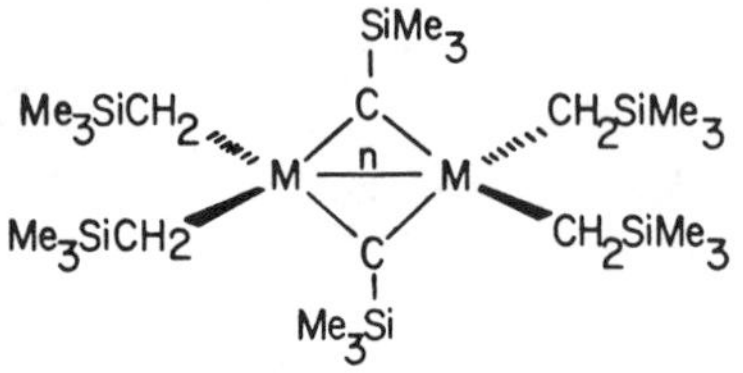

VI

The formal oxidation state of the metal atoms in each of these compounds is $+5$, and it seemed a simple and obvious task merely to oxidize the tungsten compound to W^{6+} and affect cleavage of the $W_2(\mu\text{-CR})_2$ unit to give two $W\equiv CR$ units. Alternatively, by an oxidative-induced internal redox reaction $(31,32)$, coupling of the two $\mu\text{-CSiMe}_3$ fragments gives a $W_2\{\mu\text{-C}_2\text{-}(SiMe_3)_2\}$ containing compound or, by release of $TMSC\equiv CTMS$, effects the reverse of the reaction shown in Eq. (2).

In practice none of these objectives have been realized and the detailed pathway interconverting the ditungstatetrahedranes and the tungsten alkylidyne compounds remains to be discovered. However, what we describe here are the results of our first 2 years' work dealing with the comparative chemistry of 1,3-dimetallacyclobutadienes as a function of both d^n configuration ($n = 1$ or 0) and substituent ligands (X = Me_3SiCH_2 or OR).

III

ELECTRONIC STRUCTURE OF THE $[X_2M(\mu\text{-CSiMe}_3)]_2$ MOLECULES (where M = Nb, Ta, W, or Re and X = CH_2SiMe_3, or in one instance O-i-Pr)

X-Ray crystallographic studies have determined M—M bond distances of 2.90, 2.54, and 2.55 Å for the Nb (27), W (29), and Re (30) derivatives, respectively. M—M bond orders of 0, 1, and 2 have been qualitatively assigned to the complexes in spite of the slight lengthening of the M—M bond which accompanies substitution of Re for W. The results of EHMO and Fenske–Hall calculations, summarized in Fig. 1, are more consistent with the bond distance trend, showing that the electronic configurations of the M—M bonds in the Ta, W, and Re complexes are $\sigma^0\delta^{*0}$, $\sigma^2\delta^{*0}$, and $\sigma^2\delta^{*2}$, respectively (33). The PE spectrum of $[(Me_3SiCH_2)_2W(\mu\text{-CSiMe}_3)]_2$ shows a

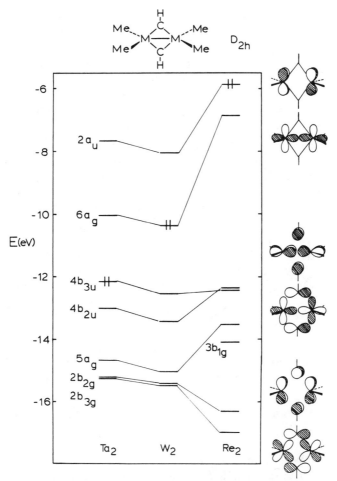

FIG. 1. Molecular orbital energy level diagram of the $Me_4M_2(\mu\text{-CH})_2$ molecules, modeling the $(Me_3SiCH_2)_4M_2(\mu\text{-CSiMe}_3)_2$ molecules, employing the method of Fenske and Hall, and showing the orbital population of the HOMO: M—C and M—M π, $4b_{3u}$ for M = Ta; M—M σ, $6a_g$ for M = W; and M—M δ^*, $2a_u$ for M = Re.

broad isolated ionization at 6.4 eV which originates from the $6a_g$ M—M σ-bonding orbital (33).

Bonding in the M_2C_2 core unit is essentially the same in the Ta, W, and Re complexes. An envelope of four orbitals at significantly lower energy than the M—M σ and δ^* orbitals accounts for the two M—μ—C π bonds and two of the four M—μ—C σ bonds. All of these orbitals are primarily carbon p_x or p_y in character, in constrast to the M—M σ and δ^* orbitals, which have at least

90% metal character. The envelope of four M—C bonding orbitals is represented in the PE spectra of the W and Ta complexes, either in whole or in part, by a broad ionization at about 8.0 eV.

Substitution of the terminal CH_2SiMe_3 ligands by π-donor alkoxide ligands has several effects on bonding in the W_2C_2 core. The most easily noted change is a lengthening of the M—M bond to 2.62 Å in both [(i-PrO)$_2$W-(μ-CSiMe$_3$)]$_2$ (34) and Cotton's compound [(t-BuO)$_2$W(μ-CPh)]$_2$ (25). The M—M σ bond ionization in the former compound appears at 6.1 eV, consistent with the prediction by Fenske–Hall calculations of a slight destabilization of the 6a$_g$ orbital bond through the σ influence of the alkoxide ligands. The 2b$_{2g}$ orbital, which is contained in the M—μ—C bonding envelope and has a δ-bonding metal component, exhibits a stronger destabilization than the M—M σ bond. The destabilization causes a splitting of the PE ionization associated with the M—μ—C envelope into two broad envelopes at 7.2 and 7.9 eV (see Fig. 2). The final predicted effect is a nearly twofold increase in the HOMO–LUMO gap. This stems from a strong destabilization of the δ^* LUMO by mixing with a low-lying π orbital located on the O atoms of the alkoxide ligands.

IV

ADDITION OF X_2 (where X = Cl, Br, I, and OPr^i) TO THE W_2 (d^1—d^1) COMPOUNDS

Oxidative addition of X_2 (X = Cl, Br, I, and OPr^i) to the X_4W_2-(μ-CSiMe$_3$)$_2$ (d^1—d^1) alkylidyne complexes was expected to lead to either cleavage of the $W_2(\mu$-CSiMe$_3$)$_2$ moiety with monomer formation X-(Me$_3$SiCH$_2$)W≡CSiMe$_3$ or dimer formation through the agency of X bridging groups (36) (VII). Alternatively, by induction of internal redox chemistry (31,32), formation of $W_2(\mu$-C$_2$(SiMe$_3$)$_2$)(CH$_2$SiMe$_3$)$_4$X$_2$ or, with release of alkyne, $W_2X_2(CH_2SiMe_3)_4$ (M≡M) (VIII), where R = CH$_2$SiMe$_3$, occurs.

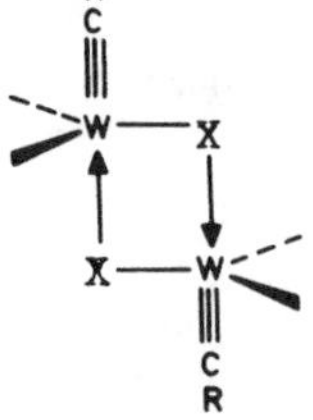

VII 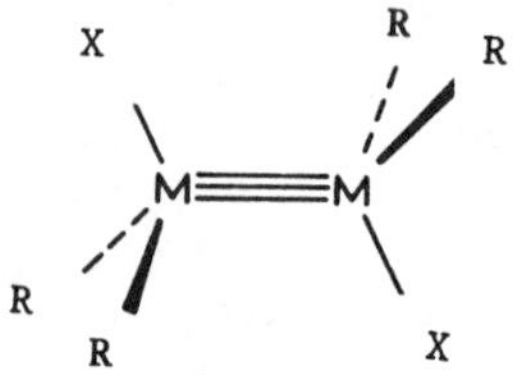   VIII

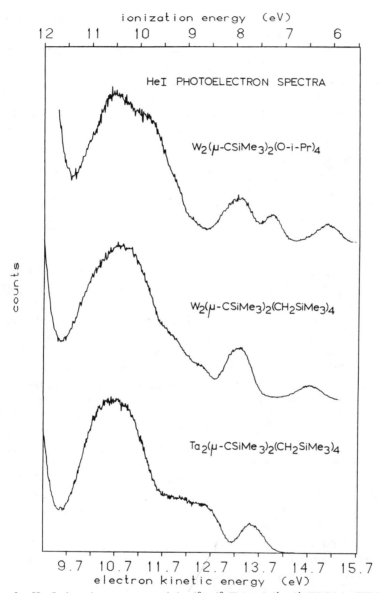

FIG. 2.   He–I photoelectron spectra of the d^0—d^0 (Ta) and d^1—d^1 (W) M$_2$(μ-CSiMe$_3$)$_2$-containing compounds showing the low energy ionizations from M—M σ^2 and the effect of the supporting ligands, Me$_3$SiCH$_2$ versus OPr^i, on ionizations from the M$_2$C$_2$ π molecular orbitals.

It should be noted that several compounds of these types are well known, for example, $(Me_3SiCH_2)_3W \equiv CSiMe_3$ (*37*), $(t\text{-}BuO)_3W \equiv CSiMe_3$ (*18*), and $Mo_2X_2(CH_2SiMe_3)_4$ ($M \equiv M$) where $X = Br$ or OPr^i (*38*). In practice, addition of 1 equiv of the halogens to hydrocarbon solutions of $[(Me_3SiCH_2)_2\text{-}W(\mu\text{-}CSiMe)]_2$ leads to $[X(Me_3SiCH_2)_2W(\mu\text{-}CSiMe_3)]_2$ compounds which, in the solid state, have a molecular structure based on two fused trigonal bipyramids sharing a common equatorial–axial edge with terminal (axial)

IX

W—X bonds as depicted in **IX** (*39*). The W—C distances within the W_2C_2 ring are essentially identical (2.0 Å), and W-to-W distances are 2.75(1) Å in the structurally characterized compounds ($X = Cl$ and Br).

Cryoscopic molecular weight determinations in benzene indicate that the dimer (**IX**) is maintained in solution. The most amazing property of these compounds is that they are paramagnetic, both in solution and in the solid state. Oxidative addition involving Pr^iOOPr^i has not yielded a crystalline product, but the species formed also appears paramagnetic. In the case of compounds **IX** where $X = Cl$ and Br, determination of magnetic susceptibility by Evan's method (toluene solutions) yields approximately one unpaired electron per dinuclear unit at 22°C. Variable-temperature magnetic susceptibility measurements employing a Faraday balance (solid samples) show a decrease in μ_{eff} with a decrease in temperature such that at 91 K (the lowest temperature yet employed in this study) μ_{eff} is only 0.63 BM per ditungsten center.

Considering that the formal oxidation state of tungsten is $+6$ in **IX**, the paramagnetism would seem to be associated with the central W_2C_2 ring, wherein a spin singlet ground state is in equilibrium with a spin triplet. In contrast to the perfectly square cyclobutadiene molecule the degeneracy of the π orbitals of e symmetry is lifted in the $1,3\text{-}M_2C_2$-containing molecules. Indeed an extended Hückel calculation, using weighted H_{ij}'s and previously published tungsten parameters, indicates that the HOMO is an orbital of b_u symmetry having an M—M δ-type interaction with contributions from the carbon p orbitals (see Fig. 3). At a distance of 2.75 Å, overlap of δ-type orbitals of the metal atoms will be very small and thus could result in the spin singlet–triplet equilibrium (*40*). According to the calculations the LUMO is M—M

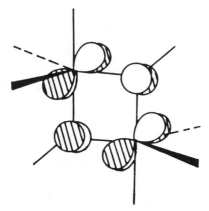

FIG. 3.   Pictorial representation of the M—M δ interaction together with carbon p mixing of the HOMO predicted for $[X(Me_3SiCH_2)_2W(\mu\text{-}CSiMe_3)]_2$ compounds, where X = Cl, Br, or I, on the basis of EHMO calculations, using weighted H_{ij}'s (R. Hoffmann, *J. Chem. Phys.* **39**, 1397, 1963) and previously published parameters for W (R. H. Sommerville and R. Hoffmann, *J. Am. Chem. Soc.* **101**, 3821, 1979).

δ^*, being about 1 eV above the HOMO. It is possible that in reality this energy gap is much smaller—but for the time being we must merely note that the compounds **IX** are most unusual in being paramagnetic and that the initial simple attempt to disrupt the planar $M_2(\mu\text{-}C)_2$ unit failed to give any of the predicted products.

V

REACTIONS WITH ALKYNES

A compound having the central skeleton depicted by **X** is known for $(Bu^tO)_4W_2(\mu\text{-}C_2Ph_2)_2$ (*35*). Addition of alkynes to the tungsten d^1—d^1

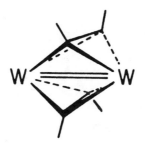

X

dimers (VI) might then yield, by a coupling of the μ-CSiMe$_3$ groups, a compound of type **X**. Alternatively, a coupling of the alkyne moieties might lead to W$_2$(μ-C$_4$R$_4$)-containing compounds (41,42).

Once again the chemistry proved quite different but fascinating inasmuch as the added alkyne was found to insert into one of the μ-CSiMe$_3$ ligands with the formation of W$_2$(μ-CSiMe$_3$)(μ-CRCR'SiMe$_3$)-containing compounds (34,43). Two compounds have been fully characterized by single-crystal X-ray studies, namely, (Me$_3$SiCH$_2$)$_2$W(μ-CPhCPhCSiMe$_3$)-(μ-CSiMe$_3$)W(CH$_2$SiMe$_3$)$_2$ and (Pr^{i}O)$_2$W(μ-CHCHCSiMe$_3$)(μ-CSiMe$_3$)-W(OPr^i)$_2$, and a ball and stick drawing of the latter molecule is shown in Fig. 4.

The essential features seen in both structural determinations are as follows: (1) There is an essentially planar X$_4$W$_2$ unit (X = C or O) and perpendicular to this is the M$_2$(μ-C) plane associated with the μ-CSiMe$_3$ ligand. (2) One tungsten atom lies in a plane of the μ-C$_3$ ring, that is, forms part of a metallacyclobutadiene, while the other tungsten atom is π-bonded to this ring.

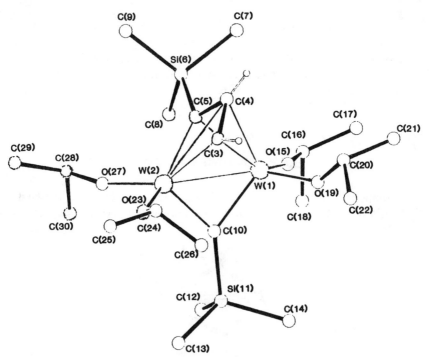

FIG. 4. Ball and stick drawing of the (Pr^{i}O)$_4$W$_2$(μ-CSiMe$_3$)(μ-CHCHCSiMe$_3$) molecule showing how W-2 is contained within the C$_3$ plane (C-3, C-4, and C-5) which in turn is π-bonded to W-1.

A simple counting of valence electrons leads one to predict 12 and 14 valence electrons for the two tungsten atoms, and it is possible to envisage the formation of a W—W single bond which in turn is consistent with the W—W distances [2.548(1) and 2.658(1) Å for X = CH_2SiMe_3 and Pr^iO, respectively]. A valence bond (VB) description of the molecules can thus be drawn as in **XI**.

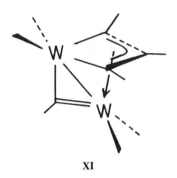

XI

Careful studies of the reactions between the compounds $X_4W_2(\mu\text{-}CSiMe_3)_2$ (where X = CH_2SiMe_3 or OPr^i) and alkynes provide a fairly comprehensive view of the overall reaction sequence leading to insertion (*44*):

1. There is an inital reaction leading to a 1:1 alkyne adduct. In the case of reactions involving C_2H_2 and RC_2H (R = Ph, Bu^t, Me, or Me_3Si), these 1:1 adducts can only be detected by low-temperature NMR studies, but for MeC_2Me and PhC_2Me the 1:1 alkyne adducts are sufficiently kinetically stable to be isolated as solids and studied by NMR spectroscopy at and even above room temperature. The molecular structure of the compound $(Me_3SiCH_2)_2(PhCCMe)W(\mu\text{-}CSiMe_3)_2W(CH_2SiMe_3)_2$ is shown in Fig. 5.

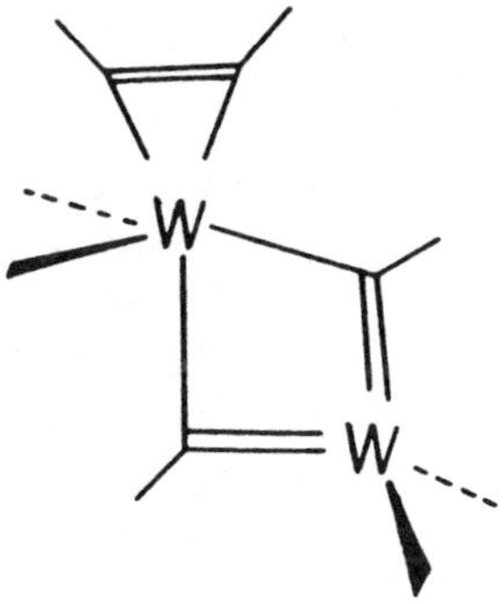

XII

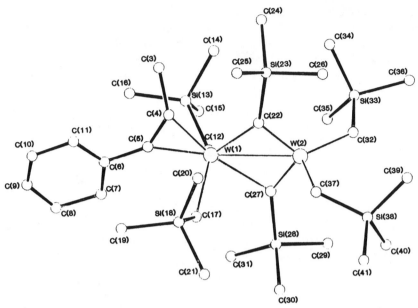

FIG. 5.   Ball and stick drawing of the $W_2(\mu\text{-CSiMe}_3)_2(CH_2SiMe_3)_4\{\eta^2\text{-C}_2(Ph)Me\}$ molecule. The carbon atoms of the alkyne C-4 and C-5 and the μ-CSiMe$_3$ ligands, C-22 and C-27, and the two tungsten atoms lie in a plane.

This is tantamount to the VB description shown in **XII**. The alkyne is formally viewed as a -2 ligand, a metallacyclopropene, and the two electrons of the W—W bond in **VI** have been cleaved in a heterolytic manner. Both tungsten atoms are in the $+6$ oxidation state, with the coordination environment about one being pseudotetrahedral and the other a distorted trigonal–bipyramid. Two further points are worthy of note here. First, no similar reactions occur for the d^0—d^0 tantalum analog, that is, we observe no insertion into the μ-CSiMe$_3$ group and no 1:1 alkyne adduct formation. Second, the alkyne has a significantly higher binding affinity when the supporting ligands are Me$_3$SiCH$_2$, relative to OPr^i. From studies employing the labeled acetylene *C$_2$H$_2$, where *C represents 92.5 g atom % ^{13}C, the latter point is revealed in the magnitude of $J_{^{183}W-^{13}C}$, (33 and 26 Hz), $J_{^{13}C-^{13}C}$ (42 and 49 Hz), and $\delta(C_2H_2)$ (216.8 and 187.3 ppm) for the Me$_3$SiCH$_2$ and OPr^i ligands, respectively. The weaker binding in the presence of alkoxide ligands can be traced to their π-donating properties. Formally the alkyne can be counted as a -2 ligand, and a four-electron π-donor utilizing both sets of filled carbon $p\,\pi$ orbitals. In the case of the OPr^i-substituted compounds there is competition between the oxygen $p\pi$ and the acetylene $\pi_{\perp}$ orbitals for donation to the vacant metal d orbitals.

2. Studies of reaction rates show that the rate of insertion is first order in the 1:1 alkyne adduct and greatly dependent on the substituents on the alkyne: $C_2H_2 \simeq RC_2H > R_2C_2$ where R = alkyl or Ph. In all but reactions involving PhCCPh, the initial rate of adduct formation is faster than insertion. The reactions involving the alkynes Me_3SiC_2H and Me_2C_2 and $(Me_3SiCH_2)_4W_2(\mu\text{-}CSiMe_3)_2$ revealed the following activation parameters: for Me_3SiC_2H, $\Delta H^\dagger = 15.2 \pm 1.2$ kcal mol^{-1} and $\Delta S^\dagger = -15 \pm 5$ e.u.; for C_2Me_2, $\Delta H^\dagger = 20.5 \pm 0.7$ kcal mol^{-1} and $\Delta S^\dagger = -13 \pm 3$ e.u. From the sign and magnitude of $\Delta S^\dagger$ we infer that the transition state is highly ordered.

3. In most cases, and in particular in all reactions employing the Me_3SiCH_2 supporting ligand, the observed insertion/addition product is fluxional and two processes are observed. (1) The $\mu\text{-}C_3R_2SiMe_3$ ligand flip-flops between the two tungsten atoms at a rate which is fast on the NMR time scale even at low temperatures. This process interconverts the π- and σ-bonded tungsten atoms with respect to the $\mu\text{-}C_3R_2SiMe_3$ ligand. (2) The substituent positions among the $\mu\text{-}CR^1CR^2CSiMe_3$ ligands are labile, sometimes on the NMR time scale at or above room temperature. On the chemical reaction time scale, the initially observed $\mu\text{-}CR^1CR^2CSiMe_3$ ligand has often undergone $R^1, R^2, SiMe_3$

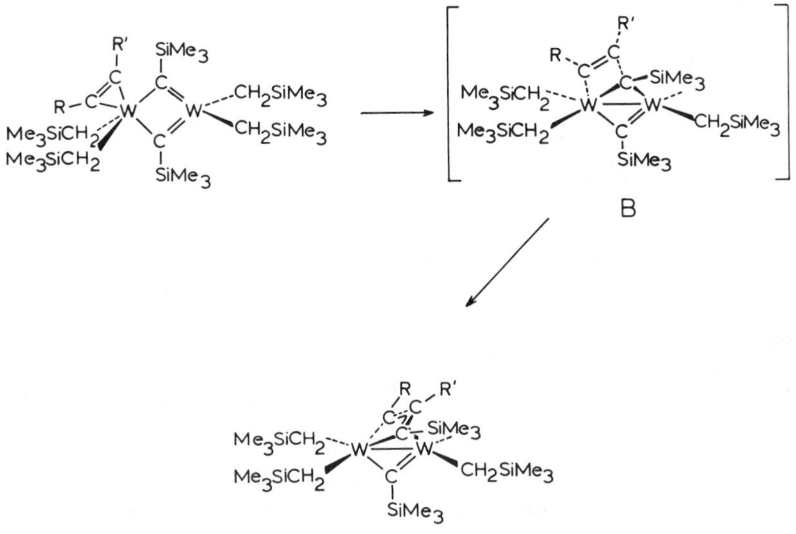

SCHEME 1. Proposed mechanism of migratory insertion/addition of the coordinated alkyne ligand to the $\mu\text{-}CSiMe_3$ ligand generating the $W_2(\mu\text{-}CRCR'SiMe_3)$ moiety. The steric preference for $R' = H$ relative to alkyl or aryl is implied in the proposed transition state, B.

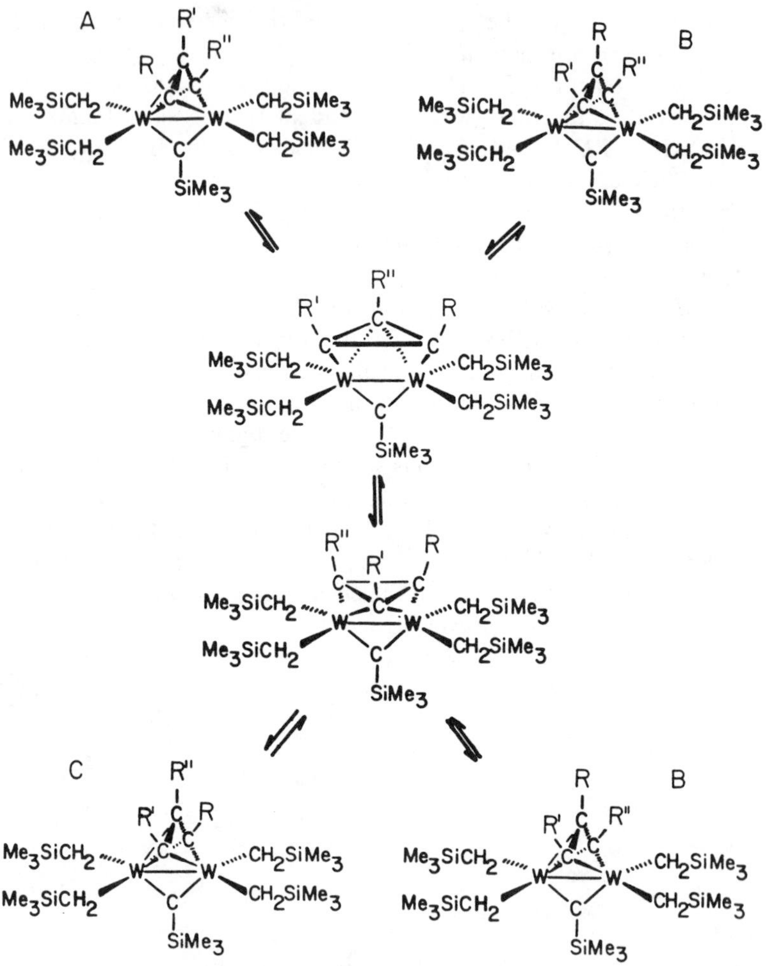

SCHEME 2. General scheme for the rearrangement of the μ-CRCR'CR'' ligand in $W_2(\mu$-CRCR'CR'')-containing compounds through the formation of a μ-CRCR'CR'' cyclopropenyl intermediate. The $\alpha-\beta$ exchange route is shown by the A and B interconversion at the top. The hypothetical $\alpha-\alpha'$ site exchange is illustrated by the A and C interconversion. All site exchanges of CR, CR', and CR'' are accessible by successive 60° rotations of the μ-cyclopropenyl intermediates.

group scrambling at a rate which is faster than the insertion/addition step. However, from low-temperature studies and particularly from studies of the reaction between $(Pr^iO)_4W_2(\mu$-CSiMe$_3)_2$ and Me$_3$SiC$_2$H, we can infer that the alkyne adds intact, that is, formation of the μ-CR^{1}CR^{2}CSiMe$_3$ involves only C—C bond formation, and that the kinetically favored insertion/addition product involves C—C bond formation with the carbon bearing the least sterically encumbered R group. For example, reactions between $(Pr^iO)_4W_2$-$(\mu\text{-}CSiMe_3)_2$ and C_2H_2 give initially $(Pr^iO)_4W_2$-$(\mu\text{-}CHCHCSiMe_3)(\mu\text{-}CSiMe_3)$ (Fig. 5) which then rearranges to the isomer having the $\mu\text{-}CHCSiMe_3CH$ ligand. The reaction between $(Pr^iO)_4W_2$-$(\mu\text{-}CSiMe_3)_2$ and Me_3SiC_2H gives initially $(Pr^iO)_4W_2(\mu\text{-}CSiMe_3CHCSiMe_3)$-$(\mu\text{-}CSiMe_3)$, which then rearranges to give the isomer containing the $\mu\text{-}CHCSiMe_3CSiMe_3$ ligand.

It is thus possible to put together a reasonable, generalized scheme for the insertion/addition reaction and for the subsequent rearrangement of the substituents in the $\mu\text{-}CR^1CR^2CSiMe_3$ ligand. These two processes are shown in Schemes 1 and 2, respectively.

Having studied the reactions involving alkynes in some detail we turned our attention to other unsaturated molecules, and in particular simple hydrocarbons. Alkenes appear unreactive, though allenes, as shown in the following section, do yield novel products.

VI

REACTIONS INVOLVING ALLENES

Allenes have been found to react with $W_2(\mu\text{-}CSiMe_3)_2$-containing compounds to give, by C—C bond formation between the tertiary carbon of the allene and the bridging carbon of one of the $\mu\text{-}CSiMe_3$ ligands, a bridging ligand which is formally a π-allyl to one tungsten atom and a terminal alkylidene to the other tungsten (45,46). The central skeleton of the compounds can be represented by the VB pictorial description shown in **XIII**.

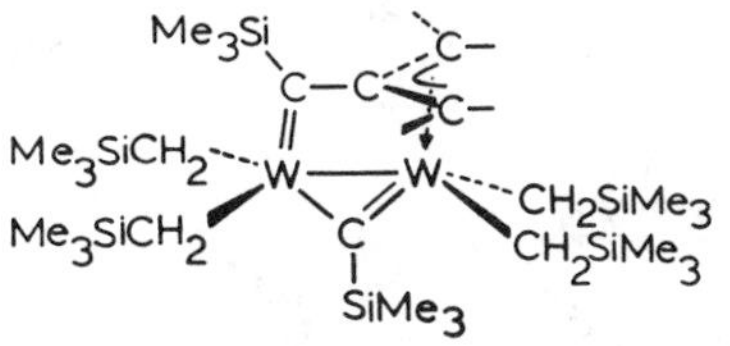

XIII

Most of these compounds are extremely hydrocarbon soluble, yielding waxy solids upon crystallization from pentane at low temperatures. This has not endeared them as a class toward structural studies by single-crystal X-ray crystallography, though the molecular structure has been determined by this means for $(Me_3SiCH_2)_4W_2(\mu\text{-}CSiMe_3)(\mu\text{-}MeCHC(CH_2)(CSiMe_3))$. A ball and stick drawing of the molecule is given in Fig. 6 (45).

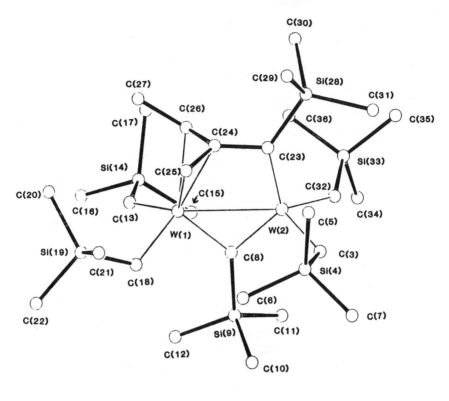

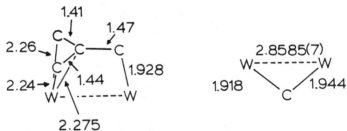

FIG. 6. Ball and stick drawing of the $(Me_3SiCH_2)_4W_2(\mu\text{-}CSiMe_3)\{\mu\text{-}C(SiMe_3)C\text{-}(CH_2)CHMe\}$ molecule. Selected bond distances associated with the central $W_2(\mu\text{-}CSiMe_3)$-$\{\mu\text{-}C(SiMe_3)C(CH_2)CHMe\}$ moiety are shown.

The ^1H- and ^{13}C-NMR data for the other compounds leave no doubt that they are related similarly in structure and bonding. In particular there is one alkylidyne $\mu\text{-}CSiMe_3$ resonance at $\delta \sim 350$ ppm with coupling to two different ^{183}W nuclei ($J_{183_W-13_C} \sim 110$ Hz and $J_{183_W-13_C} \sim 90$ Hz) and one alkylidene carbon resonance at $\delta \sim 240$ ppm which shows coupling to only one ^{183}W

atom. In these and other respects the NMR data conform to expectations based on the template shown in **XIII**.

The relative rates of reactions involving these $X_4W_2(\mu\text{-CSiMe}_3)_2$-containing compounds are dependent upon the substituents on the allene ($CH_2=C=CH_2 \geq CH_2=C=CHR > RCH=C=CHR$) and also upon the nature of the substituents X on the W atoms ($Me_3SiCH_2 > Pr^iO$). The mixed alkoxy trimethylsilylmethyl ligated compound $(Me_3SiCH_2)(Bu^tO)W\text{-}(\mu\text{-CSiMe}_3)_2W(OBu^t)(CH_2SiMe_3)$ which exists in either anti or syn isomers (**XIV** or **XV**, respectively) reacts at an intermediate rate.

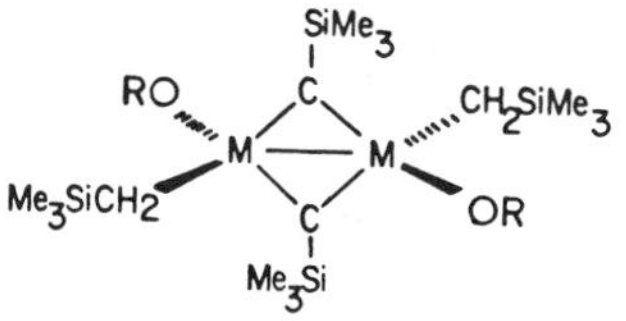

XIV 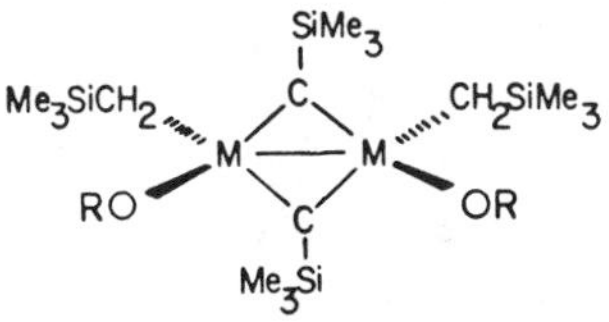   XV

From low-temperature studies of the reaction between $MeCH=C=CHMe$ and $(Me_3SiCH_2)_4W_2(\mu\text{-CSiMe}_3)_2$ utilizing NMR spectroscopy, it is clear that the initial step of the reaction is 1:1 π complex formation. The spectroscopic properties of the 1:1 adduct allow the formulation of the 1:1 π complex in terms of the VB description shown in **XVI**, and in this regard it is

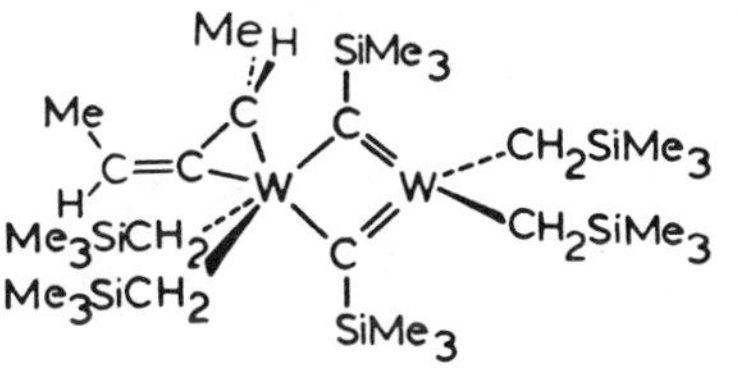

XVI

analogous to the 1:1 alkyne adducts discussed previously. The addition of allene in **XVI** is again seen to represent an asymmetric oxidative addition to the W—W center, and it is interesting to note that no analogous reactions have been found for the tantalum d^0–d^0, $Ta_2(\mu\text{-CSiMe}_3)_2$-containing species.

The attack of the coordinated allene ligand on one of the bridging alkylidyne ligands is found to be stereoselective which together with the relative rates of attack $RCH=C=CH_2 > RCH=C=CHR'$ can be understood in terms of a 1:1 intermediate of the type depicted by **XVI**. (1) Reactions employing $MeCH=C=CH_2$ and $(Me_3SiCH_2)_4W_2(\mu\text{-CSiMe}_3)_2$ yield initially the anti isomer of the allyl ligand, i.e., that found is the molecular structure shown in Fig. 6. This can reasonably be understood in terms of the reaction pathway shown in Scheme 3. However, with time an equilibrium reaction mixture is formed in solution involving both the syn and anti forms of the allyl ligand. The H—H couplings involving the allyl ligands in compounds of type **XIII** conform to those commonly found in mononuclear allyl-containing compounds (47). A reasonable pathway for the interconversion of syn and anti allyl isomers at the dinuclear center is shown in Scheme 4. (2) Reactions employing either anti or syn $(Me_3SiCH_2)(Bu^tO)W(\mu\text{-}CSiMe_3)_2W(OBu^t)$- (CH_2SiMe_3) proceed to give different products in their reactions with allenes. Presumably these are either anti or syn isomers with respect to the disposition of the Bu^tO/CH_2SiMe_3 ligands at the dinuclear center, but at this time the absolute stereochemistry is unknown. The simplest interpretation of the latter finding is that allene addition to the $\mu\text{-}CSiMe_3$ ligand in the 1:1 π complex occurs faster than a pseudorotation about the pseudo-trigonal–bipyramidal W atom. [In the case of the 1:1 alkyne adduct (Me_2C_2)-

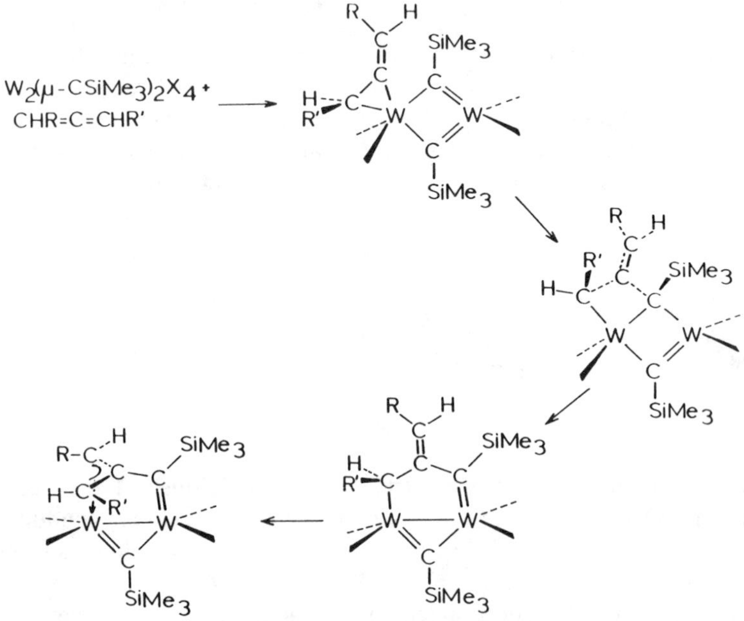

SCHEME 3.   Proposed reaction pathway for the formation of the W_1–π-allyl, σ-alkylidene–W_2 ligand. During the C—C bond forming step there will be a steric preference for the groups R and R′ because R experiences the least steric pressure from neighboring groups on the metal atoms $(CH_2SiMe_3$ or $OPr^i)$. Thus, for insertion involving $MeCH{=}C{=}CH_2$, steric factors will dictate R = Me and R′ = H leading, as shown, to the kinetic preference for the formation of the anti-Me isomer.

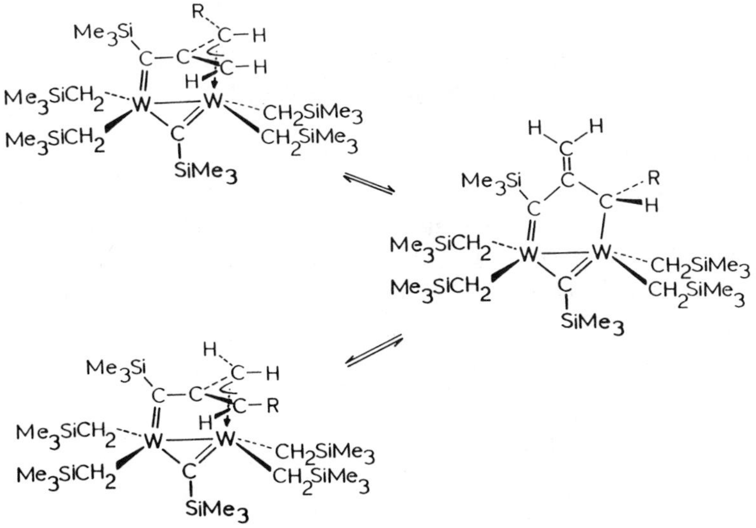

SCHEME 4. Proposed mechanism for syn–anti isomerization of the W_1–π-allyl, σ-alkylidene–W_2 ligand.

$(Me_3SiCH_2)_2W(\mu\text{-}CSiMe_3)_2W(CH_2SiMe_3)_2$ a pseudorotation about the π-alkyne complexed tungsten atom is observed to be rapid on the NMR time scale (43).]

Reactions employing $(Pr^iO)_4W_2(\mu\text{-}CSiMe_3)_2$ and allene (2 equiv) in hydrocarbon solutions proceed via an intermediate of the type **XIII** to give a product lacking any element of symmetry. Although the compound was not structurally characterized by a single-crystal X-ray study, the NMR data clearly indicate that a further insertion/addition of allene to the initially formed π-allyl ligand has occurred. The $W_2(\mu\text{-}CSiMe_3)$ moiety has been retained with formation of a new alkylidene "wrap-around" ligand of the type shown pictorially in **XVII** (45). This type of reaction has some analogy with coupling of alkyne ligands at dimetal centers (48,49).

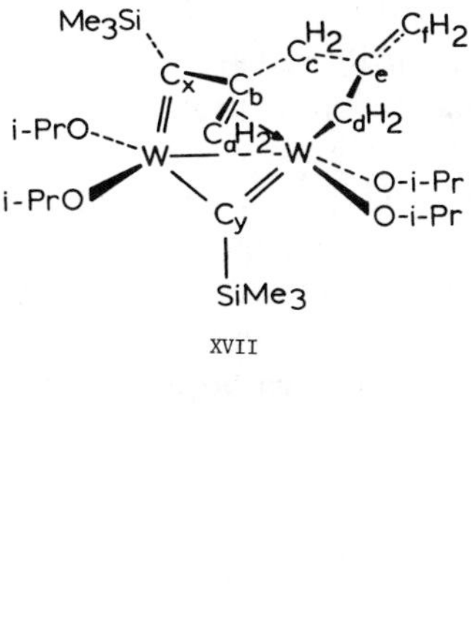

XVII

VII

REACTIONS INVOLVING ISOCYANIDE AND CARBON MONOXIDE LIGANDS

Isocyanide ligands and carbon monoxide ($C≡X$, where X = NBu^t, N-2,6-dimethylphenyl, and O) react in hydrocarbon solvents with $(Pr^iO)_4W_2$-$(\mu\text{-}CSiMe_3)_2$ to give compounds of formula $(Pr^iO)(X=)W(\mu\text{-}CSiMe_3)$-$(\mu\text{-}\eta^2,\eta^1\text{-}CCSiMe_3)W(OPr^i)_3$ (50). The generic VB pictorial description of these molecules is shown in **XVIII**.

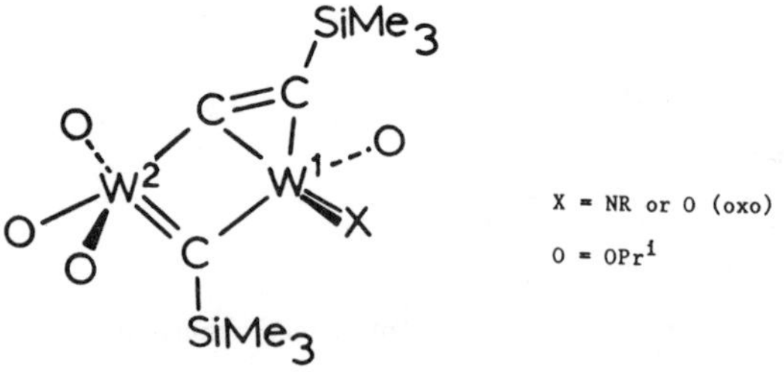

X = NR or O (oxo)

O = OPr^i

XVIII

The ^{1}H- and ^{13}C,^{1}H-NMR spectra of all three species are very similar. Specifically the ^{1}H-NMR spectra reveal four types of OPr^i ligands, each containing diastereotopic methyl groups and two types of $SiMe_3$ resonances. In the ^{13}C,^{1}H-NMR spectra there are three low-field signals of interest. One, at about 280 ppm, is assignable to a $\mu\text{-}CSiMe_3$ ligand bridging two tungsten atoms in an asymmetric manner $(J_{^{183}W-^{13}C} = 120$ and 100 Hz$)$. The other two signals occur at 250 and 170 ppm, with the resonance at 250 ppm showing a significant coupling to one tungsten atom $(J_{^{183}W-^{13}C} \sim 110$ Hz$)$. The other resonance at approximately 170 ppm shows a small coupling to one ^{183}W $(I = 1/2$, 14.5% natural abundance; $J_{^{183}W-^{13}C} \sim 45$ Hz$)$.

The data collectively are consistent with the proposed generic structure depicted by **XVIII** and the molecular structure of the derivative formed in the reaction involving 2,6-dimethylphenylisocyanide has been determined in the solid state (50) (see Fig. 7). If the $\mu\text{-}CCSiMe_3$ ligand can be counted as a -3 ligand with charge being partitioned as -1 to W(2) and -2 to W(1), which is η^2-bound, then each tungsten is in a formal oxidation state $+6$ and the W—W distance (2.91 Å) is nonbonding.

Reactions employing CO give a product which is an oil at room temperature but a waxy crystalline product at low temperatures. Reactions employing *CO, where *C represents 92 g atom % ^{13}C, show it is the resonance at approximately 250 ppm which is derived from CO. Moreover, there is ^{13}C—^{13}C coupling to the carbon signal at about 170 ppm of 35.4 Hz.

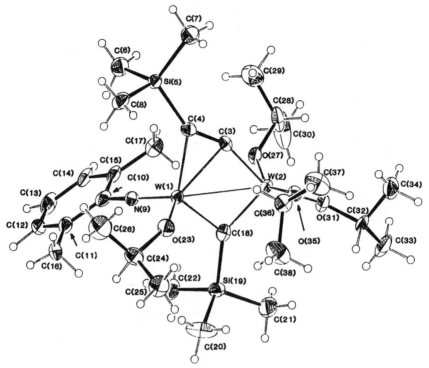

FIG. 7. ORTEP view of the $(Pr^iO)_4W_2(\mu\text{-}CSiMe_3)(\mu\text{-}\eta^1,\eta^2\text{-}CCSiMe_3)(NAr)$ molecule. Selected pertinent distances are W—W = 2.912(1) Å, W—N = 1.763(6) Å, W-1—C-4 = 2.109(7) Å, W-1—C-3 = 2.178(7) Å, W-2—C-3 = 2.022(7) Å.

Reactions employing C*O, where *O represents 98 g atom % ^{18}O, have allowed identification of $v(W{=}^{16}O)$ and $v(W{=}^{18}O)$ absorptions at 945 and 900 cm^{-1}, respectively. In the infrared spectra of the products derived from reactions involving CO (natural abundance) and *CO, bands at 1722 and 1690 cm^{-1}, respectively, can be assigned to $v(C{=}C)$ of the σ,π-alkynyl ligand (51).

Under analogous conditions reactions between $(Me_3SiCH_2)_4W_2$- $(\mu\text{-}CSiMe_3)_2$ and isocyanide and CO ligands proceed in a more complex manner. Perhaps this is due to the requirement of ligand transference between tungsten atoms in these reactions, which is easier for alkyoxy ligands than for alkyl ligands. This difference remains to be investigated, but what is evident is that the tantalum d^0—d^0 analogs are once again unreactive.

It is possible to propose a scheme, based on the alkyne insertion/addition reaction, for reactions involving these C≡X molecules. Whereas the alkynes give stable 1,3-dimetallated allyl derivatives, the analogs involving C—C bond formation with RNC and CO would be expected to be less stable and prone to C—X bond cleavage because of the thermodynamic favorability of forming the W=X bond which has triple bond character: W≡X. This proposed reaction pathway leading to cleavage of the C—X triple bond is shown in Scheme 5.

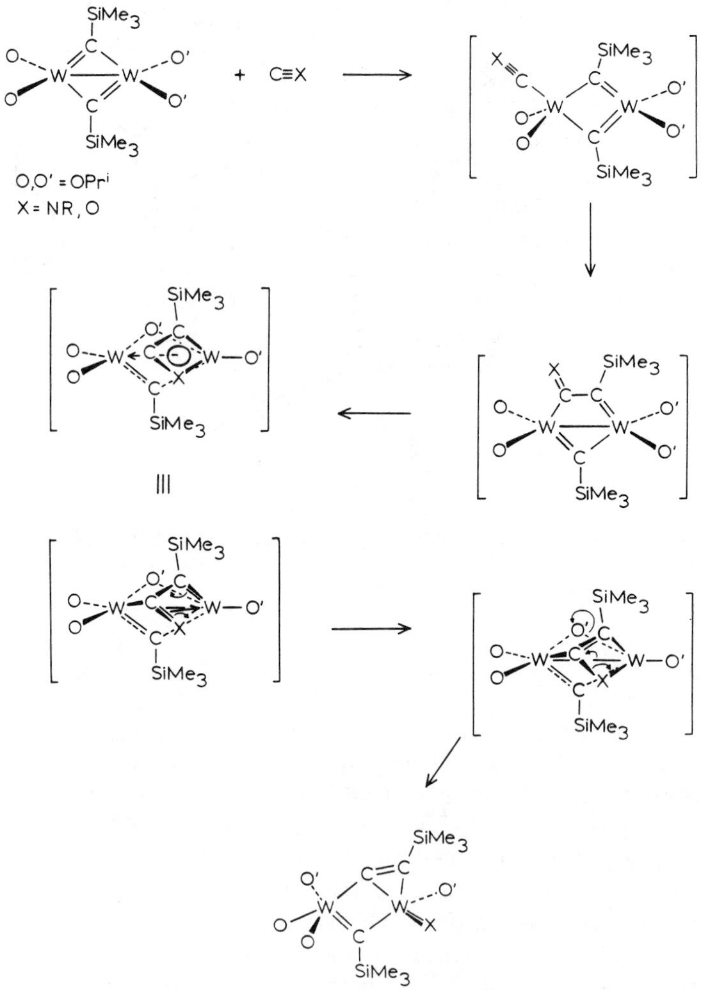

SCHEME 5. Proposed reaction pathway leading to C≡X bond cleavage (X = NAr and O) based on analogous reactions employing alkynes. The essential features of the reaction sequence are (i) 1:1 adduct (π complex) formation; (ii) C—C bond formation between coordinated C≡X and μ-CSiMe$_3$; (iii) OR group transference between tungsten atoms and C—X bond cleavage resulting in (iv) the W≡X and σ,π-CCSiMe$_3$ moieties.

VIII

REACTIONS INVOLVING DIPHENYLDIAZOMETHANE

Diazoalkanes have been used to form μ-CH$_2$- or μ-CRR^1-containing compounds in their reactions with M—M bonds (multiple or otherwise) (52). However, in our hands we have found that diphenyldiazoalkane reacts with the W$_2$(μ-CSiMe$_3$)$_2$ (d^1—d^1) compounds to form 1:1 adducts in which the coordinated Ph$_2$CN$_2$ ligand may be counted as a -2 ligand (53). The molecular structure of the compound (Me$_3$SiCH$_2$)$_2$(Ph$_2$CN$_2$)W(μ-CSiMe$_3$)$_2$W-(CH$_2$SiMe$_3$)$_2$ reveals fused trigonal–bipyramidal and tetrahedral tungsten centers in which the η^1-N$_2$CPh$_2$ ligand occupies the position akin to the η^2-alkyne adduct shown in Fig. 5. The geometry of the WN$_2$CPh$_2$ moiety is similar to that seen in Mo$_2$(OPr^i)$_6$(N$_2$CPh$_2$)$_2$(py) (54).

IX

CONCLUDING REMARKS

The reaction pathway interconverting a dimetallatetrahedane and two alkylidyne metal fragments still remains to be established. What we have shown is that two terminal W≡CSiMe$_3$ moieties are not readily formed in reactions employing X$_4$W$_2$(μ-CSiMe$_3$)$_2$ compounds; nor has it proved possible thus far to couple the two alkylidyne carbon atoms with formation of a W$_2${μ-C$_2$(SiMe$_3$)$_2$}-containing compound, or by elimination of TMSC≡CTMS to effect redox chemistry with accompanying formation of W$_2$X$_4$Y$_2$(M≡M) compounds. What has resulted is perhaps the first mechanistic glimpse of how a μ_2-CSiMe$_3$ ligand reacts with unsaturated molecules such as alkynes, allenes, isocyanides, and carbon monoxide and, in particular, how this is influenced by the presence or lack of a M—M bond [cf. the unreactive nature of the Ta$_2$(μ-CSiMe$_3$)$_2$ (d^0—d^0) moiety relative to the W$_2$(μ-CSiMe$_3$)$_2$ (d^1—d^1) group] and the substituents on the tungsten atoms, RO versus Me$_3$SiCH$_2$.

1. The reactions between the d^1—d^1 W$_2$(μ-CSiMe$_3$)$_2$-containing compounds and each of CO, RNC, Ph$_2$CN$_2$, allenes, and alkynes appear to proceed via a common route, namely initial 1:1 π complexation with concomitant cleavage of the W—W bond. It is evident that all of the reactive ligands listed above are two-electron σ-donors (i.e., Lewis bases) and π^*-acceptors. We should point out that the W—W bonding molecular orbital, 6a$_g$, depicted in Fig. 1 is amiably disposed to enter a $d\pi$-to-ligand π^* bonding interaction. A π-acid ligand attack at a tungsten center along a vector trans to

one of the alkylidyne bridges leads directly to the formation of an adduct having a structure common to **XII** or **XVI** and encounters the HOMO as an orbital of π symmetry. Without this π-backbonding interaction, stable 1:1 adducts cannot be formed and subsequent migratory insertion/addition to one of the μ-CSiMe$_3$ ligands does not occur.

2. Ligands such as alkynes, and diphenyldiazoalkanes can act as π-donor ligands (as well as π-acceptors). The 1:1 adducts of these derivatives are relatively stable compared with those of allenes, which, like olefins, are not generally viewed as π-acceptor and π-donor ligands.

3. The substitution of Me$_3$SiCH$_2$ by RO ligands destabilizes the 1:1 adducts. This can be understood in terms of the π-donating properties of the RO ligands and the available tungsten $d\pi$ acceptor orbitals on the metal. There is, however, a second effect which is not yet quantified but is apparent. The alkoxide ligands promote carbon–carbon bond forming reactions relative to Me$_3$SiCH$_2$ ligands. This can be inferred qualitatively from the fact that the rates of reactions involving X$_4$W$_2$(μ-CSiMe$_3$)$_2$ with Me$_3$SiC$\equiv$CH to give X$_4$W$_2$(μ-CSiMe$_3$)$\{\mu$-C(SiMe$_3$)C(H)C(SiMe$_3$)$\}$ compounds are the same for both X = CH$_2$SiMe$_3$ and OPr^i. For X = Me$_3$SiCH$_2$ the reaction proceeds through initial formation of the 1:1 alkyne adduct, while for X = OPr^i the reactions show approximately bimolecular kinetics for which the enthalpy of activation is lower but the entropy of activation is larger and more negative. Similarly, in the reaction between allene and X$_4$W$_2$(μ-CSiMe$_3$)$_2$ compounds, only when X = OPr^i do we observe a second insertion (a second C—C bond-forming reaction) even though the rate of reaction is slower: k_{obs} is influenced by both the rate of π complexation and the rate of migratory insertion/addition.

Related Work

In the organometallic literature there are structural analogs of the 1,3-dimetallaallyl insertion products, and these have been formed by reactions involving alkynes and μ_2-alkylidyne ligands, for example, Cp(CO)$_2$W-[μ-C(Me)C(Me)C(p-tolyl)]Fe(CO)$_3$ (55), which may be viewed as a metallacyclobutadiene analog of Pettit's (56) classic molecule (C$_4$H$_4$)Fe(CO)$_3$. The preferred ground-state geometry for these W(μ-C$_3$R$_3$)Fe- and W$_2$(μ-C$_3$R$_3$)-containing compounds may also be rationalized in terms of electron counting rules for clusters (57). The latter compounds may be viewed as isolobal analogs of Fe$_5$C(CO)$_{15}$ (58) having seven skeletal electron bonding pairs, and are thus predicted to exist as nido square-based pyramidal clusters (57).

Certain features of the W$_2$C$_3$ core of the alkyne insertion products bear striking resemblance to Schrock's (14,15) L$_3$W(η^2-C$_3$R$_3$) metallacyclobutadiene complexes which are prepared via reactions of alkynes with terminal alkylidyne complexes, $L_3W\equiv CR$. Also some of the C_3R_3 ligands in Schrock's compounds undergo facile rearrangements within the C_3R_3 core (*15*) apparently through reversible metallacyclobutadiene-to-metallatetrahedrane isomerizations, **I** $\rightleftharpoons$ **II**. Our proposal for the isomerizations within the $W_2(\mu\text{-}C_3R_3)$ ligand (Scheme 2) has again some parallel with the mononuclear systems.

To our knowledge the reactions described with allenes, isocyanides, and carbon monoxide are without precedent (*59*). The bridging π-allyl, σ-alkylidene ligand appears to be a new addition to bridging hydrocarbyl ligands (*60*), and the facile reactions leading to $C\equiv X$ triple bond cleavage (X = NR or O) are truly remarkable. The addition of the carbon atom to the μ-CSiMe$_3$ ligand to generate a σ,π-alkynyl$-M_2$-containing complex [the latter of which are well documented (*51*)] suggests that μ-alkylidyne ligands, which have been observed on metal surfaces (*61,62*), may be involved in C—C chain growth in heterogeneous catalyzed reactions employing either hydrocarbons or Fischer–Tropsch-type chemistry (*63–65*).

Several questions remain. In particular, why is it that only one alkylidyne ligand appears to be susceptible to attack (addition/insertion). Perhaps the answer to this question lies in a combination of steric and electronic factors associated with the initial insertion/addition products. In some cases these lack any formal W—W bond, the tungsten atoms being in a formal $+6$ oxidation state, whereas in others the nature of the W—W bond and its accessibility toward entering π-acid ligands is at present not well understood. Further studies are clearly warranted.

ACKNOWLEDGMENTS

We thank the National Science Foundation, the donors of the Petroleum Research Fund administered by the American Chemical Society, and the Wrubel Computing Center for financial support. We are also indebted to Drs. John Huffman, Kirsten Folting, and William Streib at the Molecular Structure Center for single-crystal X-ray studies, to Mr. David L. Clark for assistance with the Fenske–Hall calculations, and to Drs. Dennis Lichtenberger and Edward Kober for obtaining photoelectron spectra.

REFERENCES

1. D. S. Chappell and D. J. Cole-Hamilton, *Polyhedron* **1**, 739 (1982).
2. J. X. McDermott, J. F. White, and G. M. Whitesides, *J. Am. Chem. Soc.* **95**, 4451 (1973); *idem, ibid.* **98**, 6521 (1976).
3. J. X. McDermott and G. M. Whitesides, *J. Am. Chem. Soc.* **96**, 947 (1974).

4. J. X. McDermott, M. E. Wilson, and G. M. Whitesides, *J. Am. Chem. Soc.* **98,** 6529 (1976).
5. R. H. Grubbs, A. Miyashita, M-I.M. Liu, and P. L. Burk, *J. Am. Chem. Soc.* **99,** 3863 (1977).
6. R. H. Grubbs, and A. Miyashita, *J. Am. Chem. Soc.* **100,** 1300 (1978).
7. For a theoretical treatment of the relationship between metallacyclopentanes and bis(olefin) metal complexes see: A. Stockis and R. Hoffmann, *J. Am. Chem. Soc.* **102,** 2952 (1980).
8. T. J. Katz, *Adv. Organomet. Chem.* **16,** 283 (1977).
9. R. H. Grubbs, *Progr. Inorg. Chem.* **24,** 1 (1978).
10. R. J. Puddephatt, *Coord. Chem. Rev.* **33,** 149 (1980); *idem, ACS Symp. Ser.* **211,** 353 (1983).
11. See ref. 1 and work cited therein.
12. J. Sancho, and R. R. Schrock, *J. Mol. Catal.* **15,** 75 (1982).
13. R. R. Schrock, *ACS. Symp. Ser.* **211,** 369 (1983).
14. S. F. Pedersen, R. R. Schrock, M. R. Churchill, H. J. Wasserman, *J. Am. Chem. Soc.* **104,** 6808 (1982). For related MC_3R_3 compounds see ref. 15.
15. M. R. Churchill, J. W. Ziller, L. McCullough, S. F. Pedersen, and R. R. Schrock, *Organometallics* **2,** 1046 (1983); *idem, Organometallics* **3,** 1554, 1563 (1984).
16. M. R. Churchill, J. W. Ziller, S. F. Pedersen, and R. R. Schrock, *J. Chem. Soc., Chem. Commun.,* 485 (1984).
17. M. R. Churchill and H. J. Wasserman, *Organometallics* **2,** 755 (1983).
18. R. R. Schrock, M. L. Listemann, and L. G. Sturgeoff, *J. Am. Chem. Soc.* **104,** 4291 (1982).
19. M. L. Listemann and R. R. Schrock, *Organometallics* **4,** 74 (1985).
20. J. H. Freudenberger, S. F. Pedersen, and R. R. Schrock, *Bull. Soc. Chim. France,* 349 (1985).
21. D. M. Hoffman, R. Hoffmann, and C. R. Fisel, *J. Am. Chem. Soc.* **104,** 3858 (1982).
22. M. H. Chisholm, K. Folting, J. C. Huffman, and I. P. Rothwell, *J. Am. Chem. Soc.* **104,** 4389 (1982).
23. M. H. Chisholm, K. Folting, D. M. Hoffman, and J. C. Huffman, *J. Am. Chem. Soc.* **106,** 6794 (1984).
24. M. H. Chisholm, D. M. Hoffman, and J. C. Huffman, *Chem. Soc. Rev.* **14,** 69 (1985).
25. M. H. Chisholm, B. K. Conroy, D. M. Hoffman, and J. C. Huffman, results to be published.
26. T. Loerzer, R. Machineck, W. Lüttke, L. H. Franz, K.-D. Malsch, and G. Maier, *Angew. Chem. Int. Ed. Engl.* **22,** 878 (1983).
27. F. Huq, W. Mowat, A. C. Skapski, and G. Wilkinson, *J. Chem. Soc., Chem. Commun.,* 1477 (1971); W. Mowat and G. Wilkinson, *J. Chem. Soc., Dalton Trans.* 1120 (1973).
28. R. A. Andersen, A. L. Gayler, and G. Wilkinson, *Angew. Chim. Int. Ed. Engl.* **15,** 609 (1976).
29. M. H. Chisholm, F. A. Cotton, M. W. Extine, and C. A. Murillo, *Inorg. Chem.* **17,** 696 (1978); *idem, Inorg. Chem.* **15,** 2252 (1976).
30. M. Bochmann, G. Wilkinson, A. M. R. Galas, M. B. Hursthouse, and K. M. Abdul-Malik, *J. Chem. Soc., Dalton Trans.* 1797 (1980).
31. H. Taube, "Electron Transfer Reactions of Complex Ions in Solution," pp. 73–98. Academic Press, New York, 1970, and references therein.
32. This type of reaction may be compared to that in which Mo(6+) in MoS_4^{2-} is transformed to Mo(5+) in $Mo_2S_8^{2-}$ by reaction with RSSR (a formal oxidant) which in turn is reduced to 2 RS^-: W.-H. Pan, M. A. Harmer, T. R. Halbert, and E. I. Stiefel, *J. Am. Chem. Soc.* **106,** 459 (1984).
33. M. H. Chisholm, J. A. Heppert, E. M. Kober, and D. L. Lichtenberger, *Organometallics,* submitted for publication.
34. M. H. Chisholm, J. A. Heppert, and J. C. Huffman, *J. Am. Chem. Soc.* **107,** 5116.
35. F. A. Cotton, W. Schwotzer, and E. S. Shamshoum, *Organometallics* **2,** 1167.
36. For example, as seen in the structure of $[(t\text{-BuO})_3W\equiv CMe]_2$ which has a central planar $W_2(\mu\text{-OBu}^t)_2$ core: M. H. Chisholm, D. M. Hoffman, and J. C. Huffman, *Inorg. Chem.* **22,** 2903 (1983).

37. R. A. Andersen, M. H. Chisholm, J. F. Gibson, W. W. Reichert, I. P. Rothwell, and G. Wilkinson, *Inorg. Chem.* **20**, 3934 (1981).
38. M. H. Chisholm, K. Folting, J. C. Huffman, I. P. Rothwell, *Organometallics* **1**, 252 (1982).
39. M. H. Chisholm, J. A. Heppert, J. C. Huffman, and P. Thornton, *J. Chem. Soc., Chem. Commun.*, 1466 (1985).
40. The situation is not unlike the classic example of a spin singlet–triplet equilibrium in the dimeric carboxylates of copper, e.g., $[Cu(OAc)_2]_2$. See F. A. Cotton and G. Wilkinson, "Advanced Inorganic Chemistry: A Comprehensive Test," 4th Ed., pp. 817–818. Wiley (Interscience), New York, 1980.
41. M. H. Chisholm, D. M. Hoffman, and J. C. Huffman, *J. Am. Chem. Soc.* **106**, 6806 (1984).
42. M. H. Chisholm, B. W. Eichhorn, and J. C. Huffman, *J. Chem. Soc. Chem. Commun.*, 861 (1985).
43. M. H. Chisholm, J. A. Heppert, and J. C. Huffman, *J. Am. Chem. Soc.* **106**, 1151 (1984).
44. The term insertion is used in the context of the classical inorganic nomenclature "migratory insertion" which, in coordination chemistry, implies that one ligand bound to a metal center moves from its coordination site inserting into a neighboring metal–ligand bond to generate a new ligand. The formation of the μ-$C_3R_2(SiMe_3)$ ligand in this work can also be correctly defined as a cycloaddition or ring expansion reaction.
45. M. H. Chisholm, K. Folting, J. A. Heppert, and W. E. Streib, *J. Chem. Soc., Chem. Commun.*, 1755 (1985).
46. M. H. Chisholm, K. Folting, J. A. Heppert, and W. E. Streib, manuscript in preparation.
47. G. Wilkinson, F. G. A. Stone, and E. W. Abel, eds., *Comp. Organomet. Chem.* **6**, 409 (1982), and references cited therein.
48. S. A. R. Knox, R. F. D. Stansfield, F. G. A. Stone, M. J. Winter, and P. Woodward, *J. Chem. Soc., Chem. Commun.*, 221 (1978).
49. A. F. Dyke, S. R. Finnimore, S. A. R. Knox, P. J. Naish, A. G. Orpen, G. H. Riding, and G. E. Taylor, *ACS Symp. Ser.* **155**, 259 (1981).
50. M. H. Chisholm, J. A. Heppert, J. C. Huffman, and W. E. Streib, *J. Chem. Soc., Chem. Commun.*, 1771 (1985).
51. σ,π-Alkynyl complexes have been previously prepared by P—C bond cleavages in reactions between dinuclear polynuclear metal carbonyls and R_2P—C≡CR' compounds: A. J. Carty, *Pure Appl. Chem.* **54**, 113 (1982).
52. W. A. Herrmann, *Adv. Organomet. Chem.* **20**, 159 (1982).
53. M. H. Chisholm, K. Folting, and J. C. Huffman, results to be published.
54. M. H. Chisholm, K. Folting, J. C. Huffman, and A. L. Ratermann, *J. Chem. Soc., Chem. Commun.*, 1229 (1981).
55. J. C. Jeffrey, K. A. Mead, H. Razay, F. G. A. Stone, and M. J. Went, *J. Chem. Soc. Dalton Trans.*, 1383 (1983); idem, *J. Chem. Soc., Chem. Commun.*, 867 (1981).
56. G. F. Emmerson, L. Watts, and R. Pettit, *J. Am. Chem. Soc.* **87**, 131 (1965).
57. K. Wade, *Adv. Inorg. Chem. Radiochem.* **18**, 1 (1976).
58. E. H. Braye, L. F. Dahl, W. Hubel, and D. L. Wampler, *J. Am. Chem. Soc.* **84**, 4633 (1962).
59. It may be noted that compounds of formula $Fe_2M(\mu\text{-}CCR)(CO)_8Cp$ are formed in reactions between M≡CR(CO)$_2$Cp compounds (M = Mo, R = Me; M = W, R = Me and p-tolyl) and $[Et_3NH]^+[Fe_3(H)(CO)_{11}]^-$. Here the μ-CCR ligand is σ-bound to M and π-bonded (η^2) to each iron atom. Presumably in this reaction a CO ligand and an alkylidyne ligand react in a manner similar to that described herein, though the stoichiometry of the reaction involving the triiron carbonyl anion is not known [M. Green, K. Marsden, I. D. Salter, F. G. A. Stone, and P. Woodward, *J. Chem. Soc., Chem. Commun.*, 446 (1983)].
60. J. Holton, M. F. Lappert, R. Pearce, and P. I. W. Yarrow, *Chem. Rev.* **83**, 135 (1983).

61. L. L. Kesmodel, L. H. Dubois, and G. A. Somorjai, *Chem. Phys. Lett.* **56,** 267 (1978).
62. R. J. Koestner, J. C. Frost, P. C. Stair, M. A. Van Hove, and G. A. Somorjai, *Surface Science* **116,** 85 (1982).
63. P. C. Ford, (ed.) *ACS Symp. Ser.* **152** (1981).
64. R. L. Beanan, Z. A. Rahman, and J. B. Keister, *Organometallics* **2,** 1062 (1983).
65. For C—C bond-forming reactions involving the μ-CH moiety in $[Cp_2Fe_2(CO)_3(\mu\text{-CH})]^+$ see: C. P. Casey and P. J. Fagan, *J. Am. Chem. Soc.* **104,** 4950 (1982); C. P. Casey, P. J. Fagan, and V. W. Day, *J. Am. Chem. Soc.* **104,** 7360 (1982); C. P. Casey, M. W. Meszaros, S. R. Marder, and P. J. Fagan, *J. Am. Chem. Soc.* **106,** 3680 (1984).

Acyclic Pentadienyl Metal Complexes

P. POWELL

Department of Chemistry
Royal Holloway and Bedford New College
Egham Hill, Egham, Surrey TW20 0EX, England

I

INTRODUCTION

The cyclopentadienyl ligand has played a key role in the development of organometallic chemistry *(1–3)*, and herein we review the chemistry of complexes formed by the related acyclic pentadienyl group. Both ligands adopt a variety of bonding modes to main group and transition elements, the most important of which are shown in Fig.. 1. The cyclopentadienyl anion is present in the essentially ionic compounds of the alkali metals. η^1-Cyclopentadienyl groups are found both in main group derivatives such as $Me_3SiC_5H_5$ *(4)* and in transition metal complexes, $(\eta^5\text{-}C_5H_5)Fe(CO)_2(\eta^1\text{-}C_5H_5)$ *(5)*, for example. An η^3 linkage has been characterized crystallographically in $(\eta^5\text{-}C_5H_5)(\eta^3\text{-}C_5H_5)W(CO)_2$ *(6)*. The majority of cyclopentadienyls, however, contain η^5 bonding. This is found in metallocenes, in main group derivatives such as $Sn(\eta^5\text{-}C_5H_5)_2$ *(7)*, and in lanthanide and actinide *(8)* complexes.

In transition metal chemistry the cyclopentadienyl group is often kinetically rather inert and acts as a blocking ligand, enabling reactions to be carried out at other sites in the molecule. Methods of synthesis of cyclopentadienyls were developed during the 1950s and 1960s and are reviewed in an early volume of this series *(9)*. The same volume also contains a review of allyl complexes *(10)*. The allyl ligand also shows varied bonding characteristics, the most important being ionic, η^1 and η^3. Interconversion of η^1 and η^3 bonding modes is often facile and is the basis of some of the fluxional processes which are a feature of allyl chemistry *(11)*. Binary allyls are often thermally unstable, reactive compounds which are active catalysts, especially in olefin oligomerizations *(12–16)*.

The pentadienyl ligand would be expected to resemble the cyclopentadienyl group on the one hand and the allyl group on the other. Some possible bonding modes are illustrated in Fig. 1, many of which have been observed experimentally. Development of pentadienyl chemistry has occurred rather

125

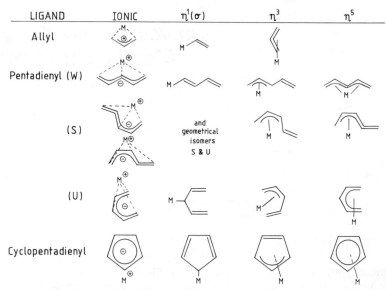

Fig. 1. Comparison of modes of bonding of allyl, pentadienyl, and cyclopentadienyl ligands. For pentadienyl the symbols refer to the shape of the group; the **W** and **U** conformations resemble the letter symbols, the **S** sickle.

belatedly compared with that of the other ligands, and it is only recently that the characteristics of the group are becoming understood and its potential realized. Acyclic pentadienyl derivatives only are discussed here, cyclic analogs such as cyclohexadienyls being omitted. Some aspects of the chemistry of these cyclic complexes, notably of η^5-cyclohexadienyltricarbonyliron salts, have been studied in great detail, and comprehensive accounts are available (17–20).

Ernst has recently reviewed structural and spectroscopic features of pentadienyl compounds. He discusses experimental and theoretical studies of pentadienyl anions, radicals, and cations and also the structures of transition metal pentadienyl complexes (21). He has also written an account of his own work in this field (22).

II

PENTADIENYL ANIONS

A. Preparation of Alkali Metal Derivatives (23,24)

The electron affinities of the pentadienyl and cyclopentadienyl radicals in the gas phase are 87 (25) and 173 kJ mol⁻¹ (26,27) and the first ionization energies 752 and 839 kJ mol⁻¹ (28,29), respectively. Cyclopentadienyl should therefore be more electronegative than pentadienyl, and cyclopentadiene is more acidic than 1,4-pentadiene in the vapor phase by about 29 kJ mol^{-1}. The pK_a values in solution for cyclopentadiene and for 1,4-pentadiene are about 18 and 30, respectively (30,31).

Cyclopentadiene is a fairly acidic hydrocarbon which can readily be deprotonated to give the cyclopentadienyl anion (30). Following a suggestion by Grignard (32), Paul and Tchelitcheff investigated the reaction of 1,4-pentadiene with strong bases and with alkali metals (33–35). While the hydrocarbon did not react at its boiling point with sodium, potassium, Grignard reagents, or sodium amide, phenylsodium in benzene gave a solution which reacted with carbon dioxide to give 2,4-pentadienoic acid and with water, 1,3-pentadiene.

Efficient methods for the preparation of pentadienyl compounds of the alkali metals have now been developed. Treatment of 1,4-dienes with butyllithium in the presence of tetrahydrofuran (thf) at −78° yields deep orange solutions which contain pentadienyllithiums (36,37). Any excess butyllithium may be destroyed by its reaction with thf by allowing the mixture to warm up briefly to room temperature (38). Similar results are obtained using potassium amide in liquid ammonia (39). 1,3-Dienes, however, do not yield pentadienyl anions under these conditions, unless the diene is conjugated with a phenyl (40), vinyl (41), trimethylsilyl (48), or similar stabilizing group. Unfortunately, 1,4-dienes are not very readily accessible. However, 1,3- as well as 1,4-dienes can be metallated using a 1:1 mixture of butyllithium and potassium *tert*-butoxide (49). Trimethylsilylmethylpotassium is also effective (44).

Yasuda found that either 1,3- or 1,4-dienes react with alkali metals dispersed in thf in the presence of triethylamine (45,46). An orange–red solution is produced without gas evolution. On addition of *n*-hexane a crystalline alkali metal derivative precipitates. With potassium the solvate KC_5H_7·thf is obtained, which loses thf above 50°C *in vacuo*. The reaction succeeds with all the alkali metals, although lithium gives a low yield even in the presence of 2 mol TMEDA. The polymerization of 1,3-dienes which occurs (47–50) in the absence of amine is thus suppressed. In addition to the pentadienyl derivative, dimers of pentadiene are obtained, probably through a radical anion (51). The method is apparently of general application; 2,4-hexadiene yields 1-methyl-pentadienylpotassium and 2,4-dimethyl-1,3-pentadiene the 2,4-dimethyl derivative. Cyclohepta- and cyclooctadienes also give the corresponding anions. Cyclohexadienes, however, yield benzene and cyclohexane only. Unstable cyclohexadienyl anions, however, have been prepared (52,53).

B. *Structures of Pentadienyl Anions (21,23,54)*

The structures of pentadienyl anions in solution have been studied either directly by NMR spectroscopy (55,56) or indirectly by analyzing product mixtures from trapping experiments (57,58). A delocalized pentadienyl group can assume a number of configurations (Fig. 1). The metal can be bonded η^3, which is usually favored for lithium, or η^5, which is adopted by the larger potassium ion. The results are listed in Table I. Pentadienyllithium in thf and pentadienylpotassium in liquid ammonia or in the solid state have the **W** configuration, but in thf the potassium compound is **U**. Introduction of 2,4-dimethyl substituents destabilizes the **W** form through nonbonded interactions, so that even the lithium derivative of the 2,4-dimethylpentadienyl anion assumes a **U** configuration in thf. Normally, however, the small lithium

TABLE I

PREDOMINANT CONFORMERS IN PENTADIENYLLITHIUM AND -POTASSIUM COMPOUNDS

Pentadienyl group	Metal	Medium	Configuration	Method[a]	Refs.
C_5H_7	Li	thf	**W**	A, B	37,55
C_5H_7	Li	Solid	**W**	C	60
C_5H_7	K	NH_3	**W**	A	39
C_5H_7	K	thf	**U**	C	57
C_5H_7	K	Solid	W	C	57
1-MeC_5H_6	Li	thf	cis **W**	B	55
1-MeC_5H_6	K	NH_3	cis **W**	A	59
1-MeC_5H_6	K	thf	trans **U** + trans **S** (4:1)	C	60
2-MeC_5H_6	Li	thf	**W** + **S** (1:1)	A	37
2-MeC_5H_6	K	thf	**U**	C	44
2-MeC_5H_6	K	Solid	exo **S**	C	60
3-MeC_5H_6	Li	thf	**W**	B	55
3-MeC_5H_6	K	NH_3	**W**	A	59
3-MeC_5H_6	K	thf	**U** (some **W**?)	C	60
3-MeC_5H_6	K	Solid	**W**	C	60
$3\text{-Et}_3SiOC_5H_6$	Li	thf	**W**	C	61
$1\text{-Me}_3SiC_5H_6$	Li	thf	**W**	C	42,62
$1\text{-Me}_3SiC_5H_6$	K	thf	trans **W** + trans **S** (7:3)	A, C	62
$1,1\text{-Me}_2C_5H_5$	Li	thf	**W** + **S** (4:1)	B	55
$2,4\text{-Me}_2C_5H_5$	Li	thf	**U**	C	44
$2,4\text{-Me}_2C_5H_5$	K	thf	**U**	C	44
$1,5\text{-Ph}_2C_5H_5$	Li	thf	**W**[b]	A	40
$1,5\text{-(Me}_3Si)_2C_5H_5$	Li	thf	**W**	A, C	62
$1,5\text{-(Me}_3Si)_2C_5H_5$	K	thf	**W**	A, C	62
$1,3,5\text{-Ph}_3C_5H_4$	Li	thf	**S** (two forms) + **W** 2.4:1	A, B	63
$1,3,5\text{-(Me}_3Si)_3C_5H_4$	Li	thf	**S**	A, B	63

[a] Key to methods: A, ^{1}H NMR; B, ^{13}C NMR; C, chemical.

[b] May contain rapidly equilibrating **S**.

atom is probably bonded in an η^3 fashion to three of the carbon atoms of a **W** pentadienyl unit. It is in rapid dynamic equilibrium with the other η^3 position along the chain. The introduction of bulky substituents such as Me_3Si or Ph at the ends of the pentadienyl group leads to the formation of some of the **S** isomers.

The complexes of the heavier alkali metals are probably essentially ionic in character. In thf the potassium and pentadienyl anions form contact ion pairs. The bonding is thought to be η^5 not only with the compact **U** form but also with the **W** and **S** forms. The preferred conformation is **U** except where there are bulky 1- and 5-substituents. In liquid ammonia, which is a better solvating medium than thf, solvent-separated ion pairs are present and the anion adopts a **W** form.

The electronic spectrum of 1,5-diphenylpentadienyllithium in thf shows the presence of contact (tight) and solvent-separated (loose) ion pairs. The smaller the cation and the more delocalized the anion, the greater is the tendency to form loose ion pairs. They are also favored as the temperature is lowered (*64,65*). ^{13}C-NMR studies (*66,67*) have been interpreted in terms of appreciable covalency, but the present view is that organolithiums are predominantly ionic (*68*). Schlosser and Stahle have analyzed coupling constants $^2J(C\!-\!H)$ and $^3J(H\!-\!H)$ in the ^{13}C- and ^{1}H-NMR spectra of allyl derivatives of Mg, Li, Na, and K and of pentadienyls of Li and K (*69*). They conclude that considerable pleating of the allyl and pentadienylmetal structures occurs. The ligand is by no means planar, and the metal binds to the electron-rich odd-numbered sites. The lithium is considered to be η^3 whereas the potassium is able to reach η^5 coordination with a **U**-shaped ligand.

Molecular orbital calculations of varying sophistication have been used to try to predict the relative stabilities of the **W**, **S**, and **U** conformers. The earliest using the EHMO method predicted that the **U** conformer is the most stable (*70*), but MNDO (*74*) suggests that the **W** form is of lowest energy for the parent C_5H_7 anion. Most of these calculations relate to the gaseous ion, but similar conclusions were reached by inclusion of lithium or BeH (*71–73*). The merits and demerits of the calculations are discussed by Dewar *et al.* (*74*).

Temperature variations of ^{1}H- and ^{13}C-NMR spectra of allyl and pentadienyl compounds of the alkali metals have given information about barriers to rotation about the C—C bonds. The endo and exo isomers of the allyl anions [Eq. (1)] are formed stereospecifically at low temperatures from Z- and E-alkenes, respectively (*75,76*).

$$\text{(1)}$$

The lithium compounds reach rotational equilibrium rapidly, but the sodium, potassium, and cesium compounds much more slowly (75). The free energies of activation for exchange of terminal allyl protons (kJ mol^{-1}), measured by ^{1}H NMR in thf solution are as follows: C_3H_5Li, 44.8; C_3H_5K, 69.9; C_3H_5Cs, 75.3 (77). The endo anion is thermodynamically the more stable. As the size of the substituent is increased the proportion of exo isomer increases owing to steric effects. Thus phenylallyllithium and diphenylallyllithium possess the exo structure (78). Similar results have been obtained for pentadienyl ions. The coalescence temperature of the NMR signals from the terminal protons of pentadienyllithium is 30°C, which indicates a rotation barrier of about 63 kJ mol^{-1}. The slow exchange low-temperature spectrum of pentadienylpotassium, however, is still sharp at 50°C. The indications are that generally the rotation barriers increase down group IA in the order Li < Na < K < Rb $\simeq$ Cs (77).

C. Rearrangements of Pentadienyl Anions (23)

The possibility of rearrangement in pentadienyl anions must be borne in mind when they are employed synthetically. When 1- or 5-alkyl groups are present, intramolecular 1,6-sigmatropic hydrogen shifts are possible and the stereochemistry follows Woodward–Hoffmann rules, being thermally antarafacial but photochemically suprafacial. Bates, for example, showed that the same equilibrium mixture of isomers results at 40°C from the deprotonation of either 5-methyl-1,4-hexadiene or 2-methyl-1,4-hexadiene (79). The tendency is to form isomers with fewer alkyl groups in the 1, 3, and 5 positions of the delocalized system (80).

Allyl, pentadienyl, and heptatrienyl anions can in principle undergo electrocyclic rearrangements (81). The thermal conversion of a pentadienyl into a cyclopentenyl anion is predicted to be a disrotatory process. The cyclooctadienyl anion cyclizes to the thermodynamically stable isomer of the bicyclo[3.3.0]octenyl ion having cis fused rings (52,82,83). The acyclic pentadienyl anions, however, do not normally cyclize. On the other hand, heptatrienyl anions cyclize readily at −30°C by a favorable conrotatory thermal process (41,84). This reaction sets a limit upon the synthetic utility of such anions.

D. Pentadienyl Compounds of Groups II and III

Bis(pentadienyl) derivatives of beryllium, magnesium, and zinc have been prepared from pentadienylpotassium (PlK) and the metal halides in thf [Eg. (2)].

$$PlK \xrightarrow[\text{thf}]{MX_2} Pl_2M \cdot 2thf \xrightarrow{TMEDA} Pl_2M \cdot TMEDA \qquad (M = Be, Mg, Zn) \qquad (2)$$

$$\qquad\qquad\qquad \mathbf{1} \qquad\qquad\qquad\qquad\qquad \mathbf{2}$$

The zinc compound **1** begins to decompose at $10°C$ to zinc and hydrocarbons, but the magnesium and beryllium complexes are more stable. Crystalline adducts **2** can be isolated on addition of TMEDA. The ^1H-NMR spectra of **2** $(M = Mg, Zn)$ show averaged AB_2X patterns on account of rapid fluxional processes. The spectra can be frozen out at low temperatures, indicating terminally bonded η^1-pentadienyl groups (85–87).

The crystal and molecular structures of $Pl'_2Mg \cdot TMEDA^1$ (**3**) (85) and of $PlZnCl \cdot TMEDA$ (**4**) (87) have been determined by X-ray diffraction. The dimethylpentadienyl groups in **3** retain a **U** shape, and the two double bonds are twisted relative to each other through $23°$. The **W**-shaped group in **4** is, however, essentially planar. Bond lengths alternate as expected for η^1-bonded pentadienyl groups.

Miginiac has obtained pentadienyl derivatives of lithium, magnesium, and zinc using methods developed for the allyl compounds (88). The reaction of 1-bromo-2,4-pentadiene with zinc in thf at or below $0°C$ gives good yields of $PlZnBr$. The analogous Grignard reagent can also be obtained in 75% yield, (89), but only with difficulty, by using very dilute solutions in diethyl ether, on account of coupling reactions (90). $(1\text{-}MeC_5H_6)_3Al_2Cl_3$ was similarly prepared from 1-chloro-2,4-hexadiene and aluminum in thf, although 1-bromo-2,4-hexadiene gave only 1,3,7,9-decatetraene (91).

III

ELECTROPHILIC ATTACK ON PENTADIENYL COMPOUNDS

A. *Derivatives of Groups I, II, and III*

^{13}C-NMR spectra show that the negative charge in the pentadienylmetal derivatives of groups I, II, and III resides preferentially at the terminal and central positions (21). Electrophiles can therefore attack either at the terminal or at the central position; Z and E isomers can result from terminal addition. For the reactions to be useful synthetically, high regiospecificity at one or other site is desirable.

$$C_5H_7M + E^+ \rightarrow H_2C=CHCH=CHCH_2E + H_2C=CHCH(E)CH=CH_2 + M^+ \qquad (3)$$

---

^{1} Pl = pentadienyl; Pl' = 2,4-dimethylpentadienyl.

Low-temperature protonation is essentially regiospecific at terminal sites for pentadienylpotassiums (46). PlK in thf (**U** configuration) yields 98% (Z)-1,3-pentadiene, while the solid (**W** configuration) affords 98% (E)-1,3-pentadiene. As the ionic character in the pentadienylmetal compound decreases, so does the regioselectivity. The percentage of 1,3 isomers in the 1,3, 1,4 mixture decreases as follows: K, 98; Li, 90; Mg, 80; Zn, 70; Be, 15; Al, 0 (46). With potassium and lithium reagents at any rate, reactions at low temperatures with Me_3SiCl (57) or with $BF(OMe)_2$ (44) (followed by alkaline hydrogen peroxide) occur exclusively at the terminal sites and have been used to determine the configurations of the species present. Alkyl halides, in general, however, couple with low regiospecificity (46).

There has been much interest in the addition of carbonyl compounds (CO_2, aldehydes, and ketones) to pentadienylmetals. The reactions of allyl (and by implication pentadienyl) Grignard and zinc reagents with carbonyl compounds is considered to follow a S_E2' mechanism via a six-membered transition state leading to allylic transposition (91). Lithium allyls, however, are thought to follow a S_E1 pathway. Aldehydes and ketones of low steric demand add preferentially to the central carbon of magnesium and zinc pentadienyls at or below room temperature (89,91). Lithium reagents are much less selective. Bulky ketones such as diisopropyl ketone give mainly terminal addition. Steric effects are thus important in determining the course of the reaction.

The nature of the counter ion and the solvent medium is also significant as these additions are reversible. The terminal adduct is thermodynamically the more stable, and the initial product mixture can in some cases be converted to it. Thus the zinc and lithium derivatives rearrange on prolonged heating in thf or thf–HMPA (91). The more ionic potassium compounds, however, which can be obtained by addition of potassium hydride, isomerize rapidly, especially in the presence of crown ethers and in polar solvents (92,93) [Eq. (4)].

Products from such reactions have been employed in the synthesis of terpenes via intramolecular Diels–Alder cyclizations (94–96). If the bulk of the carbonyl reagent is increased by inclusion of a large adjacent protecting group, much improved regioselectivity toward terminal attack is acheived (97,98). A complementary approach has been adopted by Oppolzer who prepared 3-triethylsilyloxypentadienyllithium (5) by the following route [Eq. (5)]:

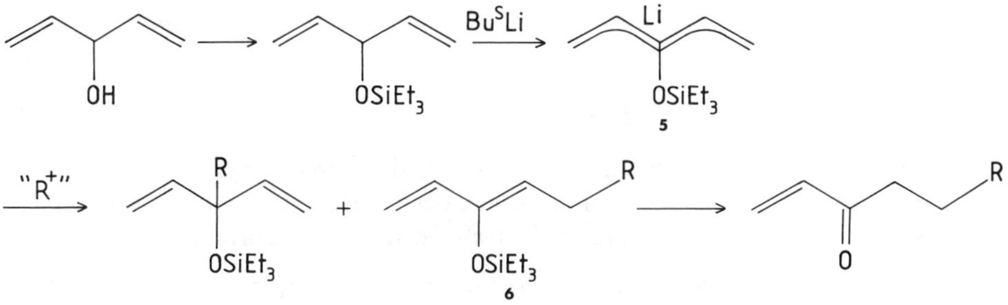

(5)

Terminal attack occurs with water, methyl iodide, and trimethylchlorosilane, whereas central attack was preferred with alkenyl halides, aldehydes, and ketones at low temperatures under kinetic control [Eq. (5)]. The Et_3SiO group is readily removed from 6 by potassium fluoride in isopropanol to give the vinyl ether $RCH_2CH_2COCH=CH_2$ (61). Some of these reactions have also been used in elegant syntheses of terpenes (99–102).

By introducing a thiomethyl substituent at the terminal position of 5 it was possible to direct attack to this position exclusively (61,103). Introduction of bulky 1,5-trimethylsilyl substituents into a pentadienyllithium provides a way of getting selective attack at C-3 (62). This approach worked with acetone, cyclohexanone, acetaldehyde, and 2-methylpropanal. With diisopropyl ketone, however, addition at C-3 still occurred, followed by Peterson elimination, giving a novel synthesis of 1,3,5-trienes (104) [Eq. (6)].

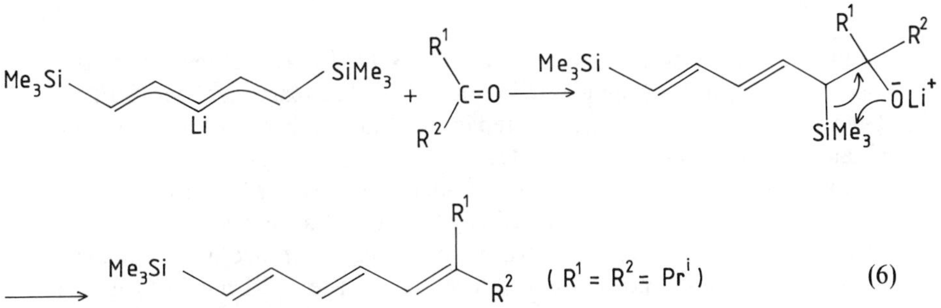

(6)

Yasuda has prepared magnesium, boron, and copper reagents by trans-metallation from the lithium derivatives. All these compounds are very regioselective toward aldehydes as well as toward ketones, giving essentially only C-3 addition (104). Diphenylpentadienylborane, prepared in situ from

P1Li and Ph_2BBr, behaves similarly (*105*). Pentadienylboranes, therefore, react as if the boron is terminally bonded, although, like allylboranes, they may consist of rapidly interconverting mixtures of 1- and 3-substituted isomers (*106*).

B. *Derivatives of Group IV*

Allyl silanes and stannanes are established as useful reagents in organic synthesis (*107–110*). Electrophilic attack occurs at the terminal carbon atom of the allyl system, leading to allylic transposition in the product [Eq. (7)].

$$N + Me_3Si—CH_2—CH\!\!=\!\!CHR + E \rightarrow N—SiMe_3 + H_2C\!\!=\!\!CH—CHRE \quad (7)$$

This idea has been extended to pentadienylsilanes and -stannanes and provides a useful complement to the reagents discussed in Section III,A, for the preparation of "homodienylic" alcohols. Reaction with aldehydes in the presence of a Lewis acid at low temperatures affords the alcohols 7 in good yields (*111,112*) [Eq. (8)].

$$Li^+C_5H_7^- + Me_3SiCl \rightarrow Me_3SiC_5H_7 \xrightarrow[\text{ii. hydrolysis}]{\substack{\text{i. RCHO/CH}_2\text{Cl}_2, \\ \text{TiCl}_4, \ -40°\text{C}}}$$

$$H_2C\!\!=\!\!CHCH\!\!=\!\!CHCH_2CH(OH)R \quad (8)$$
$$\mathbf{7}$$

Experiments with (*Z,E*)-2,4-hexadienyltrimethylsilane showed that an S_E2' process operates (*113*) [Eq. (9)].

$$MeCH\!\!=\!\!CHCH\!\!=\!\!CHCH_2SiMe_3 \xrightarrow[\text{ii. H}_2\text{O}]{\text{i. RCHO/TiCl}_4} RCH(OH)CH(Me)CH\!\!=\!\!CHCH\!\!=\!\!CH_2$$
$$(9)$$

Similarly, acetals were converted to ethers (*113,114*), ketones to tertiary alcohols (*111*), and acyl halides to ketones (*113*). The reaction of Me_3SiPl with 1-hexanal has been used as a key step in the total synthesis of the sperimidine alkaloid anhydrocannabisativine (*115*).

Allyl- and pentadienylstannanes undergo similar reactions. Attack of benzaldehyde on tributyl-2,4-hexadienyltin seems less regiospecific than that of the corresponding silicon derivative, as a mixture of alcohols arising from attack at both C-3 and C-5 is produced (*116*). Me_3SnPl has been used to introduce pentadienyl substituents into quinones in an approach to 11-deoxyanthracycline antibiotics (*117–119*). Protonation of R_3MPl (M = Si, Ge, Sn) gives only 1,3-pentadiene arising from terminal attack. This is also the predominant reaction with the hexadienyl derivatives. Insertion of sulfur dioxide into **8**, however, proceeds through an initial kinetic product **9** which rearranged within 30 minutes at 25°C to a mixture of **10** and **11** (*120*) [Eq. (10)].

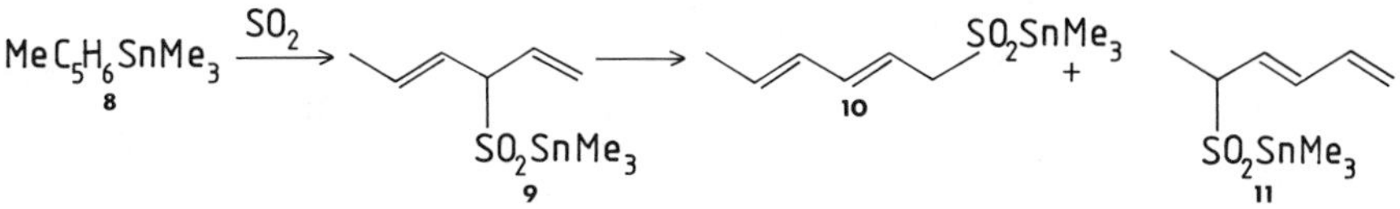

MeC5H6SnMe3
8
SO2
SO2SnMe3
9
10
SO2SnMe3
+
SO2SnMe3
11

(10)

IV

ACYCLIC PENTADIENYL COMPLEXES OF TRANSITION ELEMENTS

Methods of preparation of pentadienyl compounds are summarized in Table II.

A. *Binary Pentadienyls*

The preparation of pentadienylsodium was being studied at about the same time as ferrocene was discovered. It was not until 1968, however, that the first binary pentadienyl complex of a transition element, bis(pentadienyl)-chromium, was obtained from PlNa and $CrCl_2$ (*121*). This compound forms green, air-sensitive crystals, and like chromocene it has two unpaired electrons (μ_{eff} = 2.74 BM). This discovery was shortly followed by that of the curious complex, bis(pentadienyl)dinickel (**12**) which was prepared from $NiCl_2$ and triethylaluminium in 1,4-pentadiene (*122*). The pentadienyl ligands in **12**

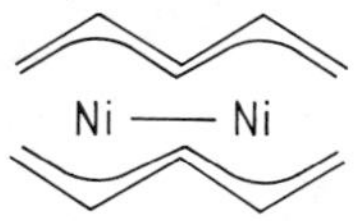

Ni — Ni

12

possess **W** configurations and are symmetrically bonded to the two nickel atoms, which are linked at a distance of 2.59 Å (*123*). In bridging two metal atoms the pentadienyl ligand resembles the allyl group in some binuclear palladium complexes (*124*). It would seem to be particularly suited to this purpose. Photoelectron spectra and MO calculations on **12** have been reported (*125*).

These compounds remained curiosities for over 10 years until R. D. Ernst obtained "open ferrocenes" from methyl-substituted pentadienylpotassiums and $FeCl_2$ in thf [Eq. (11)]. The orange–red complexes $FePl'_2$ can be sublimed

TABLE II

SUMMARY OF METHODS OF PREPARATION OF PENTADIENYL COMPOUNDS

Method, examples[a,b]	Section
Diene + base (M), e.g., 1,4-pentadiene + BuLi–thf, 1,4-pentadiene + KNH_2–NH_3, 1,3-pentadiene + BuLi–KOBu^t–thf	II,A
1,3- or 1,4-Diene + alkali metal + tertiary amine (M), e.g., PlM, Pl′M (M = Na, K, Cs)	II,A
Chloro- or bromodiene + metal (M), e.g., PlMgCl, PlZnBr, $Pl_3Al_2Cl_3$ (NOT alkali metals, coupling)	II,D
Addition of chloro- or bromodiene to metal complex (T)	IV,I
Pentadienyl group transfer (M, T): PlM′ + MX → PlM + M′X, e.g., 2 PlK + MCl_2 in thf → $Pl_2M\cdot 2thf$	II,D
PlK + Me_3MCl → Me_3MPl (M = Si, Ge, Sn)	III,B
2 Pl′K + MCl_2 → MPl_2 (M = Ti, V, Cr, Fe)	IV,A
4 Pl′K + MCl_4L_2 → Pl'_2ML (M = Zr, Nb, Mo; L = PEt_3)	IV,A
R_3SnPl + $BrM(CO)_5$ → $PlM(CO)_3$ (M = Mn, Re)	IV,F
R_3SnPl + $[Ru(CO)Cl_2L_2]_2$ → $PlRu(CO)ClL_2$ (L = PMe_2Ph)	
Electrophilic attack: Protonation of η^4-dienol complexes (T),	IV,D, IV,E
e.g., $PlFe(CO)_3{}^+$, $PlMCp^+$ (M = Rh, Ir)	IV,H
Protonation of η^4-triene complexes (T)	IV,D, IV,H
From alkynes (T)	IV,J

[a] Throughout this article, Pl = C_5H_7 (pentadienyl) and Pl′ = C_7H_{11} (2,4-dimethylpentadienyl).

[b] (M), Applicable to main group elements; (T), applicable to transition elements.

in vacuo and are essentially air stable. Spectroscopic and X-ray diffraction data are consistent with their formulation as sandwiches (*126,127*). The parent compound, bis(pentadienyl)iron, cannot be obtained from $FeCl_2$ and PlLi or PlK. It was prepared, however, using pentadienylmagnesium or -zinc reagents. It is rather less robust than the methylated analogs. Methyl substituents at the 2 and 4 positions of an η^5-pentadienyl group serve to stabilize the complex in a similar way to a 2-methyl substituent on an allyl ligand.

$$FeCl_2 + 2\ Pl'K \xrightarrow[-78°C]{thf} FePl'_2 + 2\ KCl \tag{11}$$

In similar reactions, $TiCl_2$, VCl_2 (prepared *in situ* by reduction of $TiCl_4$ or VCl_3 with magnesium or zinc, respectively), and $CrCl_2$ with 2 equiv of Pl′K

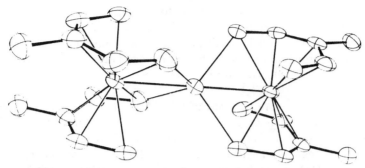

FIG. 2.   Molecular structure of tetrakis(3-methylpentadienyl)trimanganese. Reprinted with permission from the American Chemical Society, ref. *141*.

yield the acyclic sandwich compounds MPl′$_2$ (*128*). VPl′$_2$ has only one unpaired electron (*129*) whereas vanadocene has three (*130*). In view of the instability of titanocene and even of decamethyltitanocene, the isolation of the diamagnetic, 14-electron titanium complex is particularly interesting (*131*).

The reactions of MnCl$_2$ and of CoCl$_2$ do not yield simple monomeric pentadienyls. The former with 3-methylpentadienylpotassium affords an unusual trinuclear complex, Mn$_3$(C$_5$H$_6$Me)$_4$, (*129*) (Fig. 2). The complex is paramagnetic and has five unpaired electrons. The central manganese atom is bonded tetrahedrally to one of the terminal carbon atoms of each of the pentadienyl groups. The bonding has been discussed in the light of MO calculations (*132*).

CoCl$_2$ and Pl′K yield **13** in which the pentadienyl groups have dimerized. The monomer CoPl′$_2$ is a probable intermediate. In the decatetraene ligand the two butadiene units are conjugated (*133*) in contrast to the dimer formed on reduction of pentadienyltricarbonyliron cations (Section IV,D) in which the two units remain unconjugated (*134*).

13

Open ruthenocenes have been obtained by reaction of hydrated ruthenium trichloride with methylated pentadienes in ethanol in the presence of zinc dust

(135). This procedure was first used to prepare ruthenocene and other ruthenium complexes of cyclic olefin *(136)*. The acyclic ruthenocenes are pale yellow, air-stable materials, soluble in organic solvents and readily sublimed. One representative compound of the lanthanides, $NdPl'_3$ *(137)*, and one of the actinides, UPl'_3 *(138)*, have also been described.

The reactions of $[ZrCl_3L]_2$ or of MCl_4L_2 (M = Zr, Nb, Mo; L = PEt$_3$) with six or four equiv, respectively of Pl'K in thf at $-78°C$ afford complexes $[MPl'_2L]$ *(139)*. The Zr and Nb compounds are isomorphous. X-Ray diffraction studies confirm the syn-eclipsed conformation of the two η^5-pentadienyl ligands which is indicated by NMR for the diamagnetic Zr compound **14** (M = Zr). The molybdenum complex, which has an 18-electron

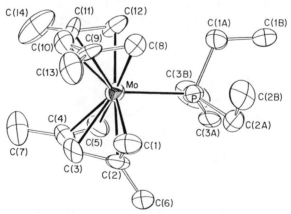

14

configuration, is virtually air stable. The pentadienyl ligands here, however, are not equivalent, the X-ray diffraction result showing that while both are bonded η^5, one adopts an **S** configuration (Fig. 3). This is the first structural

FIG. 3. Molecular structure of $(\eta^5\text{-Pl}')_2\text{Mo}(\text{PEt}_3)$, showing η^5-pentadienyl ligands in **U** and **S** conformations. Reprinted with permission from the American Chemical Society, ref. *139*.

characterization of an **S** pentadienyl, although others have been identified in solution by NMR (Section IV,D).

1. Molecular Structures in the Solid State

X-Ray diffraction studies have been carried out on MPl'_2 (M = V, Cr, Fe), $MPl'_2 \cdot PF_3$ (M = Ti, V), and $Ru(2,3,4-Me_3C_5H_4)_2$ (*127,135,140,141*). All these molecules possess sandwich structures, but the pentadienyl groups are rotated relative to each other about the axis joining the midpoint of the pentadienyl group and the metal atom (Fig. 4). In the iron compound this angle of twist (conformation angle) is 59.7°, very close to that required for the gauche-eclipsed conformation. Some bonding parameters for pentadienyls and cyclopentadienyls are listed in Table III. The metal–carbon distances for analogs are very similar except for the vanadium compounds.

The five-carbon skeletons of the pentadienyl groups are essentially planar. The C—C bond distances within the pentadienyl ligands (typically C-1—C-2, 1.415 Å; C-2—C-3, 1.39 Å) are shorter than in cyclopentadienyls (typically 1.435 Å) (*21*). The methyl substituents are bent out of the ligand planes toward the metal atom to a greater extent (between 6° and 11°) compared with methylferrocenes (3.7°). This tilt reflects the greater size of the open pentadienyl ligand compared with cyclopentadienyl. The C—C—C bond angles within the **U** pentadienyl groups are consistently greater than the "ideal" value of 120°. This has the effect of increasing the C-1—C-5 distance to about 2.8 Å. The pentadienyl ligand therefore had a larger spatial extent than cyclopentadienyl. Because the M—C bond distances are comparable in the two systems, the distances from the metal atom to the ligand center are markedly shorter in the pentadienyls. Consequently a pentadienyl ligand is more sterically demanding than C_5H_5 or even, perhaps, C_5Me_5. This may account, for example, for the formation of the 16-electron $TiPl'_2(CO)$ in contrast to $TiCp_2(CO)_2$ (*142,143*).

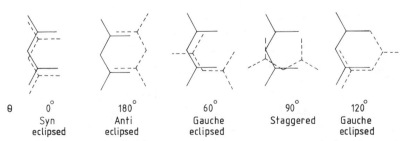

θ	0°	180°	60°	90°	120°
	Syn eclipsed	Anti eclipsed	Gauche eclipsed	Staggered	Gauche eclipsed

FIG. 4.   Relative orientations of pentadienyl ligands in bis(pentadienyl) metal complexes. θ is the conformation angle.

TABLE III

SOME BONDING PARAMETERS IN METALLOCENES AND 2,4-DIMETHYLPENTADIENYL COMPLEXES

	Ti	V	Cr	Mn	Fe
MPl$'_2$					
M—C$_{av}$ (Å)	2.236[a]	2.211(2)	2.165(4)	2.114(2)[b]	2.089(3)
M—center of mass (Å)	1.651	1.632(2)	1.594(4)	1.537(2)	1.508(2)
Conformation angle (°)	3.4	89.8	82.9	35.6	59.7
MCp$_2$					
M—C (Å)	—	2.280(5)	2.169(4)	2.112(2)[c]	2.064(3)
M—center of mass (Å)	—	1.9287(6)	1.798(4)	1.724(3)	1.660(10)

[a] TiPl$'_2$·PF$_3$.

[b] Mn$_3$Pl$'_4$.

[c] Mn(C$_5$Me$_5$)$_2$.

In a planar **U** pentadienyl group, even with internal bond angles greater than 120°, serious nonbonded interactions between endo substituents on C-1 and C-5 would occur, even when these are hydrogen. The endo hydrogen atoms are thus located away from the metal above the ligand plane, with the two exo hydrogens below this plane. In Ru(2,3,4-Me$_3$C$_5$H$_4$)$_2$ the average angles of tilt for endo and exo hydrogens are 42° and 17°, respectively (135). Similar features are to be expected in all **U** acyclic pentadienyl structures.

2. Structures in Solution

NMR spectroscopy affords evidence of dynamic processes which take place in solutions of acyclic ferrocenes and ruthenocenes. While the barriers to rotation of cyclopentadienyl rings about the metal–ring axis are very low (∼5 kJ mol^{-1}), significant barriers to oscillation of pentadienyl ligands have been observed and calculated (127). For example, the proton spectrum of Fe(3-MeC$_5$H$_6$)$_2$ at room temperature consists of four lines, but as the temperature is lowered the peaks broaden, collapsing at −69°C and reappearing as a seven-line spectrum. This indicates an unsymmetrical conformation which does not possess a vertical plane of symmetry, such as the gauche-eclipsed form which is present in the crystal. The free energy of activation for the oscillation is 33.5 kJ mol^{-1} (127). ^{13}C-NMR spectra also show parallel changes with temperature. Very complicated spectra are obtained for compounds with unsymmetrically substituted pentadienyl ligands on account of the presence of diastereoisomers which do not interconvert.

B. η^3-Pentadienyl Complexes

In principle pentadienyls can bond to transition elements in at least three basic ways, η^1, η^3, and η^5 (Fig. 1). These can be further subdivided when geometrical factors are considered. If η^5 coordination could be converted to η^3 or η^1, one or two coordination sites could become available at the metal center, and perhaps coordinate substrate molecules in catalytic processes. Little is known about the ability of pentadienyl complexes to act as catalysts. Bis(pentadienyl)iron derivatives apparently show "naked iron" activity in the oligomerization of olefins (144), resembling that exhibited by "naked nickel" (13). The pentadienyl groups are displaced from acyclic ferrocenes by PF_3 to give $Fe(PF_3)_5$ in a way reminiscent of the formation of $Ni(PF_3)_4$ from bis(allyl)nickel (144).

Bleeke and co-workers have studied the reactions of PlK with transition metal halides in the presence of trialkylphosphines. $FeCl_2(PMe_3)_2$ yields (η^3-Pl)Fe(PMe_3)_2$ in thf at $-78°C$. The molecular structure shows the presence of **W** pentadienyls which are η^3 bonded (145). The two pentadienyl groups remain as separate entities and do not link together as in the corresponding product obtained from $MnBr_2$, PMe_3, and PlK (146). The dimerization is avoided by using bis(dimethylphosphine)ethane as blocking ligand, as this permits the introduction of only one pentadienyl group (147), which is also **W** shaped (see Fig. 5). Reaction of $CoCl_2$ with PlK in the presence of zinc as reducing agent and PMe_3 as stabilizing ligand yields $Co(\eta^5\text{-Pl})(PMe_3)_2$. $CoCl(PMe_3)_3$, however, with PlK alone affords the complex $Co(\eta^3\text{-Pl})$-$(PMe_3)_3$ (**15**). Here the vinyl substituent on the allyl system adopts an anti configuration in the solid. The pentadienyl ligand, however, retains a **U** shape, but the vinyl group is bent away from the cobalt atom so that it does not coordinate. As with **16**, the anti isomer is in equilibrium with its syn isomer in solution. Spin saturation transfer NMR experiments suggest that the isomerization proceeds via a C-3-bonded η^1-pentadienyl intermediate (148a).

$(\eta^5\text{-Pl}')Co(PEt_3)_2$ reacts with $L = PMe_3$ or $P(OEt)_3$ to produce $(\eta^3$-Pl')CoL_3$. It is oxidized by $AgBF_4$ to the 17-electron complex $[(\eta^5$-Pl')Co(PEt_3)_2]^+BF_4^-$. The crystal structure and some reactions of the latter have been described (148b).

The molecular structures of three η^3-pentadienyl complexes having **W**, **S**, and **U** configurations, respectively, are shown in Fig. 5. Apart from showing the short uncoordinated C=C bond of the vinyl system a common feature is the unsymmetrical bonding of the η^3-allyl group with the longest M—C bond to the atom which bears the vinyl substituent (149).

Some η^3-pentadienyl complexes of nickel, $[CpNi(\eta^3\text{-}H_2CCMeCHCMe}=CH_2)]$ (syn and anti isomers) (**18** and **19**) (150) have arisen in the course of work on the insertion of alkenes into metal–alkyl bonds, relevant to Ziegler

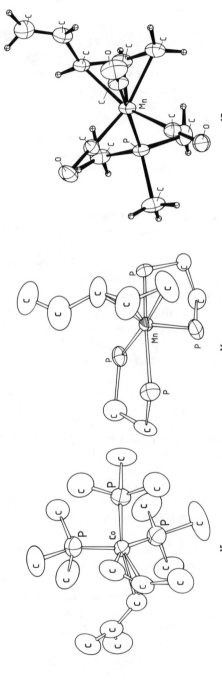

FIG. 5. Molecular structures of complexes **15**, **16**, and **17**, showing η^3-pentadienyl ligands in **U**, **W**, and **S** conformations, respectively. Reprinted with permission from the American Chemical Society, refs. *147, 148a, 149.*

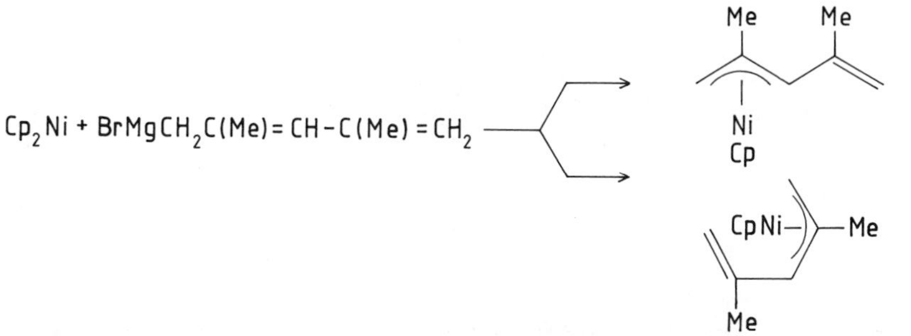
Cp2Ni + BrMgCH2C(Me)=CH-C(Me)=CH2
Me
Me
Ni
Cp
CpNi
Me
Me

syn
18

anti
19

(12)

catalysis (*151–153*) [Eq. (12)]. This research relates to a report of the reaction of bis(allyl)nickel with 1,5-hexadiene in the presence of difluorophosphoric acid. A very air-sensitive, red crystalline substance (**20**) was obtained. It was suggested that β-elimination is sterically disfavored in the chelated form of the C_9 chain (*154*).

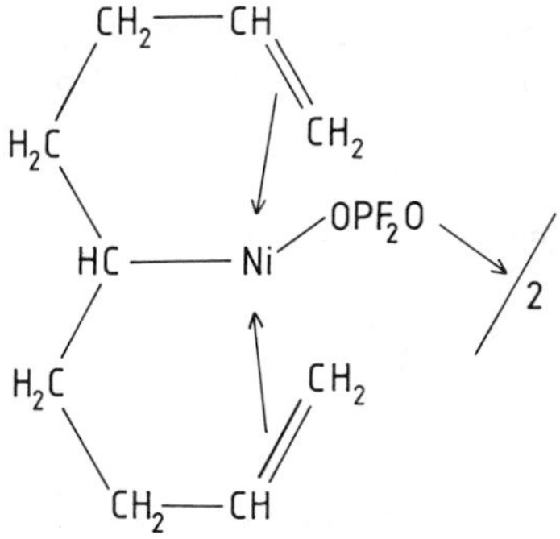
CH2
CH
H2C
CH2
OPF2O
HC
Ni
2
H2C
CH2
CH2
CH

20

C. η^1-*Pentadienyl Complexes*

There has been one report of an η^1-pentadienyl complex of a transition element, $Cp_2Zr(2\text{-}MeC_5H_6)_2$, prepared from Cp_2ZrCl_2 and 2-methylpentadienylpotassium (*155*). Cp_2ZrHCl, however, give η^4-complexes of 1,3-pentadiene and its higher homologs, when treated with pentadienylpotassiums. The reactions of these products with alkenes, alkynes, dienes, and carbonyl compounds are synthetically important (*156*).

D. η^5-*Pentadienyltricarbonyliron Cations (157)*

The first acyclic pentadienyl complexes to be prepared were pentadienyltricarbonyliron salts. They were discovered by Mahler and Pettit (*158*), who

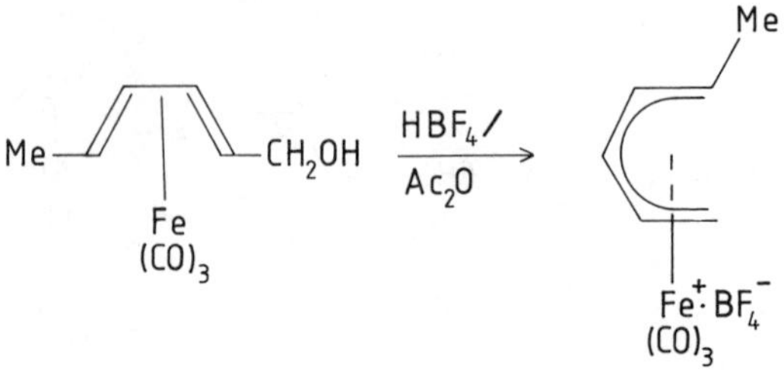

$$(13)$$

21

obtained them by protonation of η^4-dienol complexes [Eq. (13)]. The cis (**U**) structure of the pentadienyl group was established from NMR spectra and also from the observation that the parent ion $Fe(Pl)(CO)_3{}^+$ could be prepared by hydride abstraction from *cis*-pentadienetricarbonyliron, using $Ph_3C^+ \cdot BF_4{}^-$, but not from the *trans*-pentadiene complex (*159,160*). The cyclic analogs of **21**, the cyclohexadienyl- and cycloheptadienyltricarbonyliron cations, have been extensively investigated and applied in organic synthesis (*17–19,161*).

The diastereoisomeric alcohols **22** and **23** can be obtained preferentially using the methods indicated in Eq. (14) and (15) (*162–164*). Protonation of

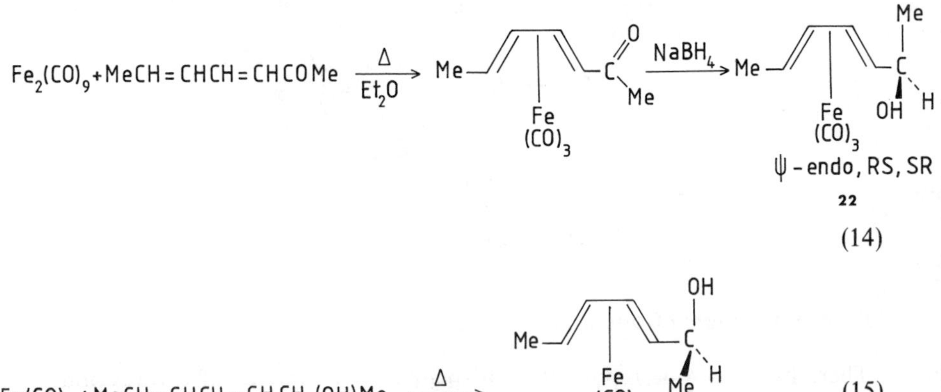

$$(14)$$

22

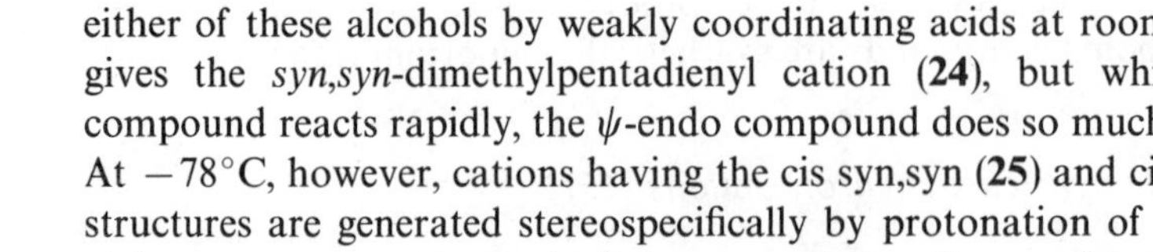

$$(15)$$

23

either of these alcohols by weakly coordinating acids at room temperature gives the *syn,syn*-dimethylpentadienyl cation (**24**), but while the ψ-exo compound reacts rapidly, the ψ-endo compound does so much more slowly. At $-78°C$, however, cations having the cis syn,syn (**25**) and cis syn,anti (**26**) structures are generated stereospecifically by protonation of ψ-exo and ψ-endo precursors, respectively. It is proposed that the reactions proceed via a "trans" ion which contains an **S**-shaped pentadienyl group. The syn,anti cation rearranges to the thermodynamically more stable syn,syn species on warming to room temperature [Eqs. (16) and (17)], (*165*). Compound **24** is

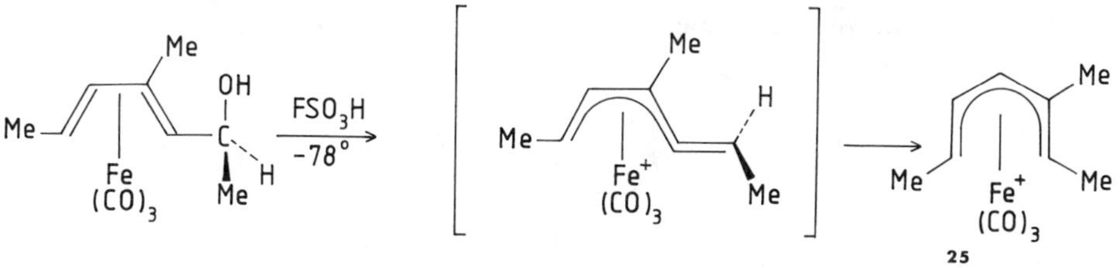

$$(16)$$

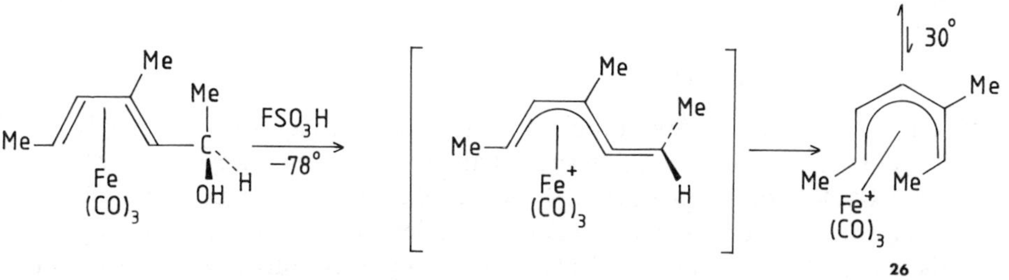

$$(17)$$

attacked by water stereospecifically to generate **23**, the ψ-exo alcohol. An X-ray diffraction study of **23** has shown it to be a mixture of enantiomers having the *R,R* and *S,S* configurations at the two chiral centers. The corresponding configurations in the racemate of **22** must therefore be *R,S* and *S,R* (*166*).

Evidence for the **S** pentadienyl ions has been obtained from kinetic and stereochemical studies on the solvolysis of dinitrobenzoate esters of diastereoisomeric alcohols like **22** and **23**. An S_N1 mechanism is suggested for the solvolyses, which proceed essentially with retention of configuration at C-1, (*162–164,167*). **S**-Shaped ions have also been observed directly by low-temperature NMR studies of the protonation of suitable alcohols. The normally rapid rearrangement of the **S** ion to the **U** ion is inhibited to some extent by destabilizing the **U** ion by introduction of an anti substituent (*168*) [Eq. (18)] or by constraining the **S** structure partly in a ring system (*169*). The

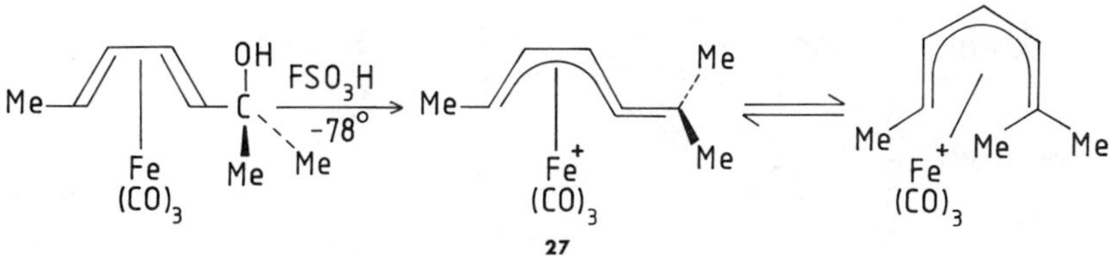

$$(18)$$

S (trans) ion (**27**) has a proton-NMR spectrum very similar to that of the protonated ketone (**28**) (*170,171*). The sorbaldehyde complex is protonated at low temperature to give a mixture of two hydroxypentadienyl ions with the **S** configuration which rearrange on warming to the **U** isomers (*172*).

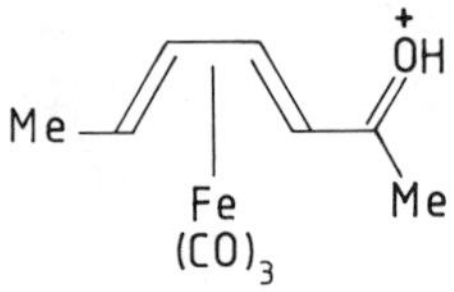

28

1. *Attack of Nucleophiles*

Pentadienyltricarbonyliron cations are readily attacked by nucleophiles. They are more reactive in this respect than the corresponding cyclic complexes (*160*). Cyclohexadienyltricarbonyliron tetrafluoroborate can be recrystallized from water, while the 1-methylpentadienyl derivative is rapidly converted to the alcohol. In reactions of 1-alkylpentadienyltricarbonyliron cations with water or methanol the nucleophile adds preferentially to the carbon atom bearing the substituent, even when this is bulky. It is suggested that the alkyl substituent stabilizes the **S**-ion intermediate. Some authors consider that this intermediate could have an η^3-pentadienyl structure, the 18-electron configuration at the metal center being maintained by coordination of a molecule of water or of solvent (*173*).

A common intermediate, possibly the **S** ion, formed in a reversible first step, is suggested by kinetic studies on the electrophilic substitution of di- and trimethoxybenzenes by pentadienyltricarbonyliron salts (*174,175*). Furthermore, weakly basic amines such as 4-nitroaniline yield products with a trans structure (**29**), whereas more strongly basic amines such as methylamine or dimethylamine afford **30** [Eq. (19)]. The former could be produced via the reactive **S** ion intermediate, whereas in the latter case direct attack on the **U** ion is likely (*176–178*).

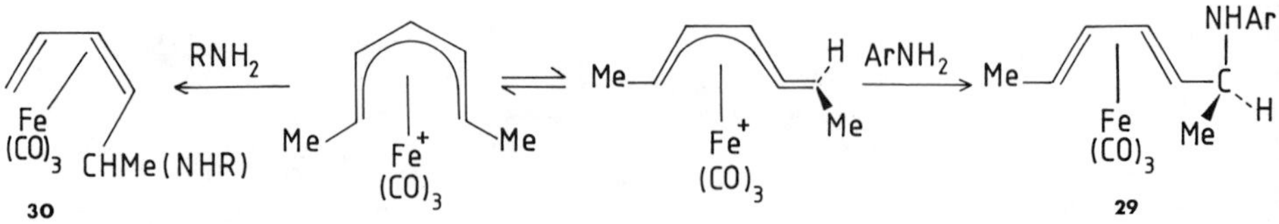

(19)

Reduction of 1-alkylpentadienyltricarbonyliron salts with various hydride donors such as $NaBH_4$, $NaBH_3CN$, and $NaBEt_3H$ has also been investigated (*179*).

The bulky ligands PPh_3 and $AsPh_3$ add to the unsubstituted end of **31** (*180,181*). The resulting phosphonium salt (**32**) is deprotonated by butyllithium at $-78°C$ to yield an ylid (**33**) which reacts with aldehydes in a Wittig reaction. Deprotonation with potassium *tert*-butoxide followed by addition of aldehyde **34** gives the *E* isomer (**35**) only [Eq. (20)] (*182*). Trimethylphosphite

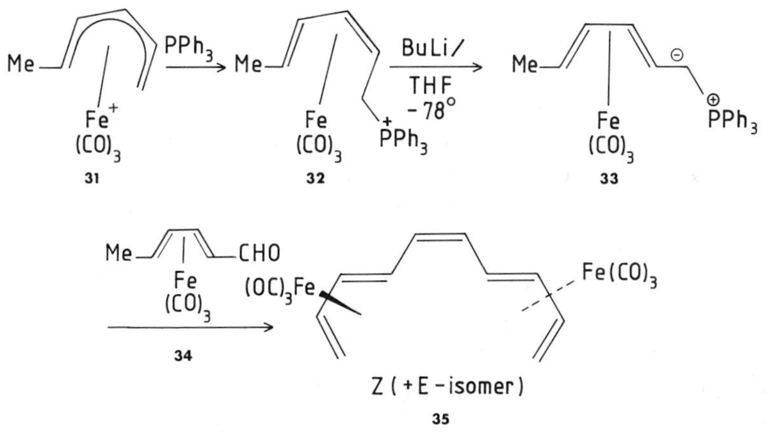

(20)

Z (+ E – isomer)
35

also adds, but the addition is followed by an Arbusov rearrangement. In this case the Wittig procedure leads to the *E* isomer exclusively. The dimethylphenylphosphonium salt derived from the cyclohexadienyltricarbonyliron cation also forms an ylid with base (*183*).

Pentadienyltricarbonyliron salts can also be prepared by protonation of η^4-triene complexes (*180,184,185*). The reverse transformation can be carried out by addition of trimethylamine followed by pyrolysis of the resulting quaternary ammonium salt [Eq. (21)] (*180*).

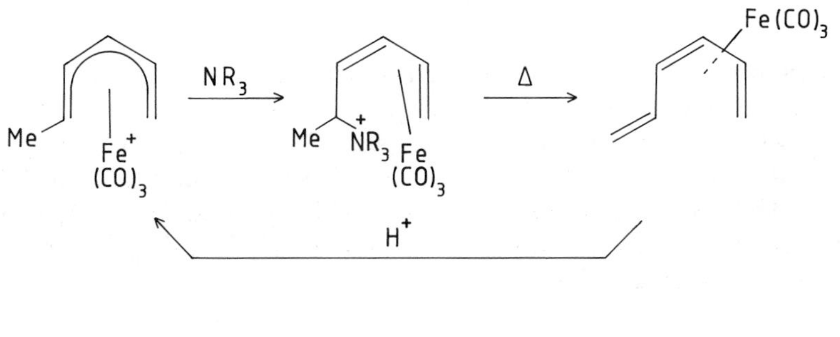

(21)

E. Cross-Conjugated Dienyltricarbonyliron Cations

Cross-conjugated dienyltricarbonyliron cations have been prepared as follows from trimethylenemethane complexes [Eq. (22)] (186,187):

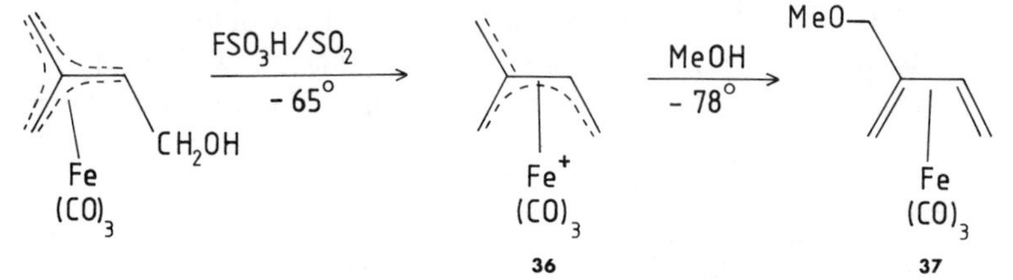

$$(22)$$

Quenching of the ion **36** with methanol at $-78°C$ gives the 2-methoxymethyl-butadiene complex (**37**). Alcohol **38**, when protonated at low temperature, first forms cation **39**, with an *anti*-methyl substituent, which rearranges to **40** on warming to room temperature [Eq. (23)]. Studies of trimethyl derivatives gave

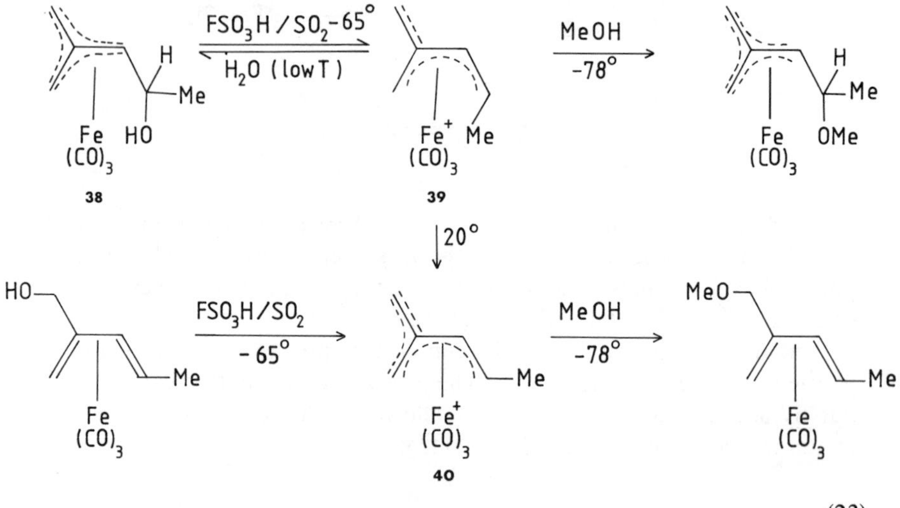

$$(23)$$

further information about isomerizations which occur in these systems (188). Protonation of **41** at $-65°C$ leads initially to a 2:1 mixture of two cationic species, which may possess structures **42** and **43**, respectively [Eq. (24)]. At $-50°C$ they isomerize to **44**, which probably has an agostic hydrogen (189), suggested by a peak at -14.1δ in the proton-NMR spectrum [Eq. (24)]. From studies of several ions the barrier to rotation about the C-2—C-3 bond in these systems is estimated to be about 55 kJ mol^{-1}.

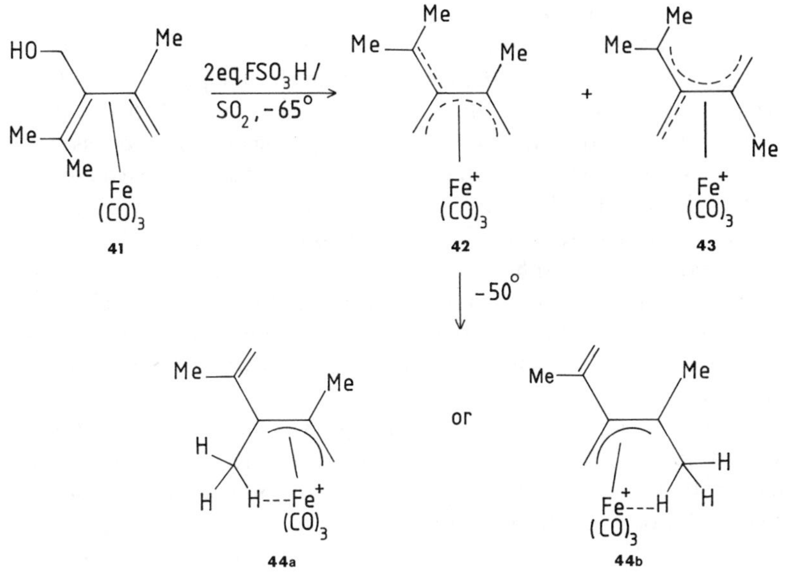

(24)

F. *Pentadienyltricarbonylmanganese and -rhenium*

Cyclopentadienyltricarbonylmanganese (cymantrene) has been extensively investigated (*190*). Typical reactions include photochemical substitution of carbonyl groups by other ligands and electrophilic substitution of the cyclopentadienyl ring. In 1973 a low yield of $Mn(\eta^5\text{-}1\text{-}MeC_5H_6)(CO)_3$ was obtained by heating $Mn(CO)_5Br$ with 1,5-hexadienyl-3-trimethyltin (*191*). Tropone and $Mn_2(CO)_{10}$ also yield an acyclic pentadienyl derivative (*192*). Subsequently the parent compound was obtained by refluxing trimethyl-pentadienyltin with $Mn(CO)_5Br$ in thf (*193*). It is a yellow crystalline solid, somewhat air sensitive, which sublimes readily under reduced pressure at room temperature. The rhenium analog was obtained by a similar route (*194*). ^1H- and ^{13}C-NMR spectra showed that both complexes contain the **U** pentadienyl ligand. They are isoelectronic with the pentadienyltricarbonyl-iron cation.

While $CpMn(CO)_3$ is quite inert to thermal substitution, on photo-excitation it loses one molecule of carbon monoxide dissociatively, yielding the 16-electron fragment $CpMn(CO)_2$, which has a high affinity for 2-electron ligands such as phosphines, phosphites, alkenes, alkynes, and dinitrogen (*190*). Thermal substitution of a carbonyl ligand in $CpCo(CO)_2$ follows an associative mechanism (*195*), probably via slippage from η^5 to η^3 coordination of the cyclopentadienyl group on approach of the nucleophile. Moreover η^3-CpRe(CO)$_3$PMe$_3$ is a common intermediate in the formation of fac-η^1-CpRe(CO)$_3$(PMe$_3$)$_2$ and of η^5-CpRe(CO)$_2$PMe$_3$ from CpRe(CO)$_3$ (196).

Carbon monoxide is more readily replaced in (η^5-indenyl)Mn(CO)$_3$ than in CpMn(CO)$_3$ because of much easier η^5 to η^3 slippage in the former case (197). Conversion of an η^5 to an η^3 pentadienyl should be easier still. Strongly basic phosphines (L = PMe$_3$, PMe$_2$Ph, PBu$_3$) react with PlMn(CO)$_3$ (45) at room temperature to give η^3-PlMn(CO)$_3$L (46). A single-crystal X-ray diffraction study of the trimethylphosphine adduct 17 reveals that the molecule contains an S-shaped η^3-pentadienyl group (see Fig. 5). In refluxing cyclohexane, however, 45 is converted by a variety of tertiary phosphines and phosphites to the substitution products η^5-PlMn(CO)$_2$L. As associative mechanism via 46 is likely.

In contrast to CpMn(CO)$_3$, no reaction was observed when 45 was photolyzed in thf. Compound 45 reacts with carbon monoxide under irradiation to form η^3-PlMn(CO)$_4$ (149). This compound has also been isolated by photolyzing Mn$_2$(CO)$_{10}$ with 1,3-pentadiene (47) [Eq. (25)]. Three

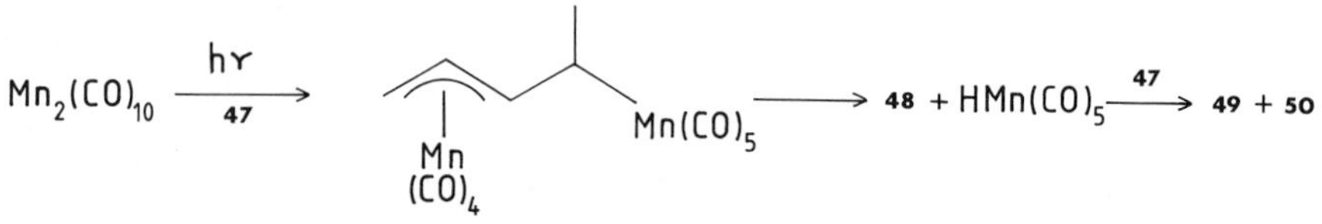

$$(25)$$

compounds [48, 49, and 50 (R = H)] were separated by chromatography (198). (Z,E)-Hexa-2,4-diene gives a similar mixture of products (R = Me). A 4% yield of 51, characterized by X-ray diffraction, was also formed in a side reaction (199).

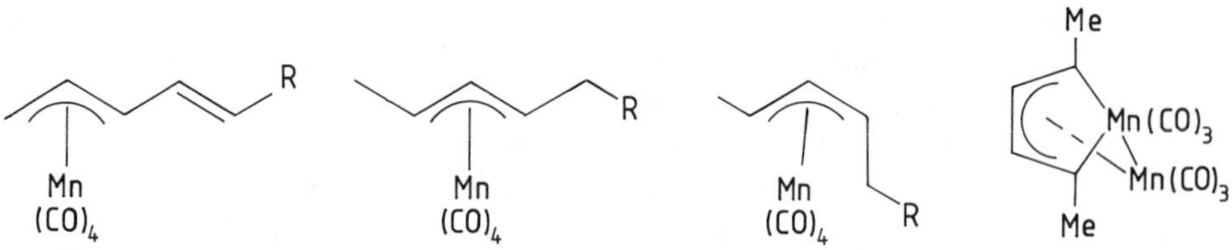

The tetracarbonyls lose carbon monoxide on heating in hexane to give (η^5-pentadienyl)Mn(CO)$_3$ derivatives (200). This provides a useful alternative route to these complexes. Two pathways for the decomposition of 48 can be envisaged. As (η^3-C$_3$H$_5$)Mn(CO)$_4$ undergoes substitution by a dissociative mechanism (201), route b may be the correct one [Eq. (26)].

Calculations predict a barrier to rotation of the M(CO)$_3$ group in 45 of about 50 kJ mol^{-1} (202). This has been confirmed from NMR studies on related compounds (203).

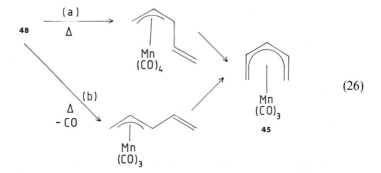

$$(26)$$

G. The η^5-Pentadienyltricarbonylmolybdenum Anion

A third member of the isoelectronic series to which PlMn(CO)$_3$ and [PlFe(CO)$_3$]$^+$ belong would be the anion [PlCr(CO)$_3$]$^-$. All three analogous cyclopentadienyls are well known. Apparently the 3-phenylpentadienyltricarbonylchromium anion has been prepared (204). Moreover treatment of (diglyme)Mo(CO)$_3$ in thf with 1 equiv Pl′K affords a solution which probably contains the anion [Pl′Mo(CO)$_3$]$^-$ (205). Addition of iodomethane at $-78°$C gives a dark solution from which a complex with the surprising structure **52** was isolated in 48% yield (Fig. 6).

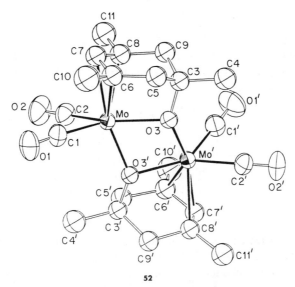

52

FIG. 6.　Molecular structure of complex **52**. Reprinted with permission from the American Chemical Society, ref. 205.

The formal electron configuration of each molybdenum atom is only 16. The Mo—Mo separation of 3.215 Å seems too long for a double bond [2.448 Å in $[CpMo(CO)_2I_2]$, cf. 3.235 Å in $[CpMo(CO)_3]_2]$]. The bond angles around oxygen suggest sp^2, hydridization, so that the molybdenum may attain an 18-electron configuration by π-bonding with the oxygen atom. The incorporation of carbon monoxide into the pentadienyl ligand as an alkoxide under such mild conditions is surprising and of considerable interest with regard to possible synthetic applications.

H. (Cyclopentadienyl)(pentadienyl)rhodium and -iridium Cations

Protonation of dienol complexes (53) with hexafluorophosphoric acid in diethyl ether affords air-stable salts (55) in high yield [Eq. (27)] (206). While the iron derivatives (21) are rapidly hydrolyzed, the cyclopentadienyl rhodium and iridium salts (55) are much less reactive (207).

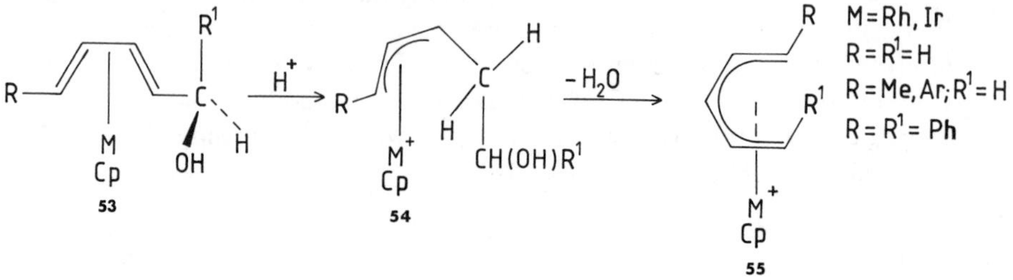

$$\tag{27}$$

NMR spectroscopic studies show that dienone complexes $[M = Fe(CO)_3]$ are protonated at oxygen (171,172). Protonation of 56 [M = CpRh, CpIr] takes place at carbon, yielding 57 (208) [Eq. (28)].

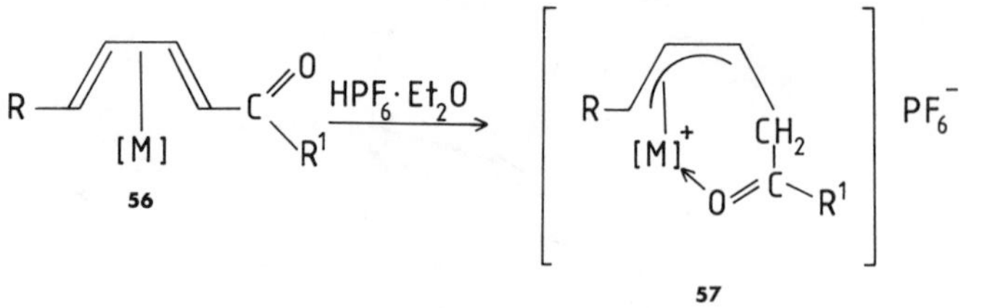

$$\tag{28}$$

Similar complexes can also be made from coordinated esters (R' = OMe) (209). On addition of CF_3COOH to a chloroform solution of 53 at $-60°C$, a deep red solution is produced which contains the protonated dienol complex 54. On warming to room temperature this eliminates water to form the η^5-

pentadienyl cations (**55**) [Eq. (27)]. Protonation of the ψ-endo dienol [**53** (R = R' = Ph)] proceeds stereospecifically endo, suggesting initial addition of H^+ to the metal center which rapidly transfers to the ligand (*210*). The substituent R' in **55** has the anti configuration, although it probably lies well out of the plane of the pentadienyl ligand.

Triene complexes of rhodium have been prepared by a crossed aldol condensation with acetophenone (*211*) or by Wittig reactions [Eq. (29)] (*212*).

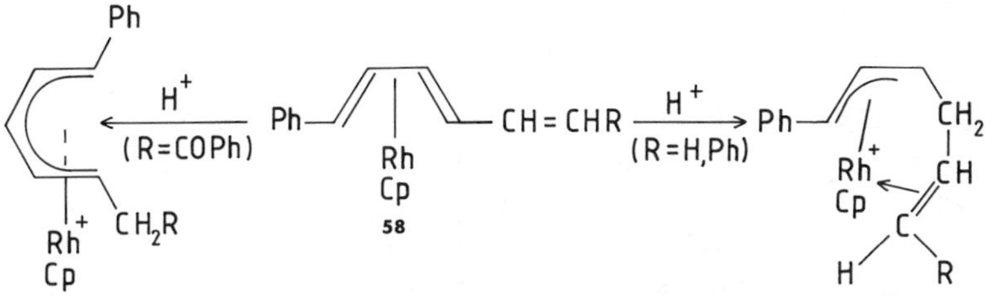

(29)

Protonation of the trienone [**58** (R = COPh)] yields an η^5-pentadienyl salt (**59**) but the trienes [**60** (R = H, Ph)] afford (1–3, 5–6-η-)pentadienyl complexes (**61**) in which addition of H^+ to C-4 of the coordinated diene system has occurred.

Some nucleophiles (OMe^-, H^-) add to the central carbon atom of the dienyl unit in **55**, yielding 1,4-pentadienyl complexes (*207*). Halide ions, however, attack the metal center to form η^3-pentadienyl complexes. A single-crystal X-ray diffraction study of $RhCp(\eta^5\text{-}1\text{-}PhC_5H_6)^+PF_6^-$ reveals an essentially planar pentadienyl ligand with the phenyl group twisted by 26° out of the plane (*213*).

I. Preparation of Pentadienyl Complexes by Oxidative Addition

A versatile method for the preparation of allyl complexes of transition elements involves addition of allyl halides or acetates to low oxidation state coordinatively unsaturated, or potentially unsaturated, complexes (*214*). A few attempts have been made to extend the method to halogenopentadienes. 1-Chloro-5-phenyl-2,4-pentadiene adds to $Pd_2(dba)_3$ (dba = dibenzoylac-etone) to give the η^3-pentadienyl complex **61** [Eq. (30)] (*215*). A similar palladium complex has also been obtained from $PdCl_2$, and 2,5-dimethyl-2,4-hexadiene in isopropanol (*216*).

$$Pd_2(dba)_3 + 2\, PhCH{=}CHCH{=}CHCH_2Cl \rightarrow$$

$$[(\eta^3\text{-}PhCH{=}CH{-}CH{=}CH{=}CH_2)PdCl]_2 \quad (30)$$

61

62

63

Hegedus and Varaprath studied the reactions of various bromodienes with $Ni(CO)_4$ and with bis(cyclooctadiene)nickel. 1-Bromo-2,5-hexadiene and 2-bromomethyl-1,3-butadiene give the stable products **62** and **63**, respectively, which resemble allyl nickel halides in their properties (*217*). Similar compounds had been prepared several years previously from geranyl halides (*218*). 1-Bromo-2,4-pentadiene and 1-bromo-2,4-hexadiene, however, formed intractable materials which could not be isolated and purified. In these cases the red color of the solution which was first produced faded and $NiBr_2$ was deposited. The desired compounds, however, could be generated *in situ* at $-30°C$ and used in coupling reactions with aryl, alkenyl, and allyl halides (*217*).

$Mo(CH_3CN)_3(CO)_3$ reacts with allyl chloride to form $MoCl(\eta^3\text{-}C_3H_5)$-$(CH_3CN)_2(CO)_2$ (*214*). 1-Bromo- and 1-chloro-2,4-hexadienes give analogous η^3-pentadienyl complexes (*219*).

J. *Complexes Derived from Alkynes*

Some unusual complexes that include pentadienyl or related ligands have been obtained from reactions of alkynes with various transition metal complexes. As the main thrust of this work is on oligomerization of alkynes and not toward the preparation and identification of pentadienyls, only a few key papers and results will be mentioned.

Reduction of $[Mo(CO)(Bu^tC{\equiv}CH)_2Cp]^+BF_4^-$ with $KBHBu_3(s)$ at $-78°C$ in an atmosphere of carbon monoxide yields a complex of a vinyl substituted γ-lactone linked η^3:η^2 (*220*). The allylidene ruthenium complex **64**, obtained by photochemical addition of one alkyne molecule to a μ-carbene derivative, is transformed into pentadienylidene complexes **65** and **66** on photolysis with more alkyne substrate. These reactions show clearly the stepwise growth of chains in alkyne oligomerizations at dimetal centers [Eq. (31)] (*221*). Similar reactions are also known for dinuclear iron (*222*), molybdenum (*223*), and tungsten (*224*) complexes.

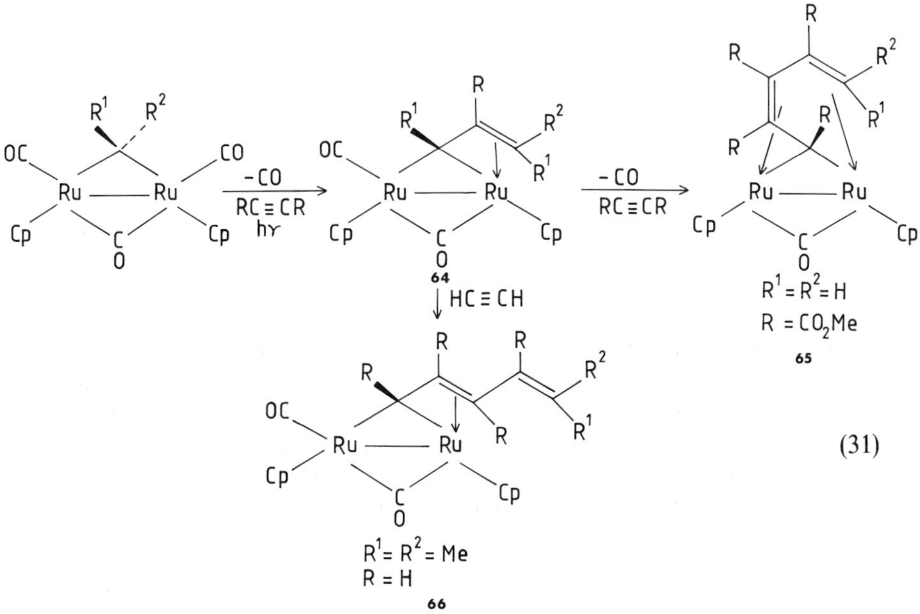

(31)

K. Oxapentadienyl Complexes

There have been a few reports of acyclic 1-oxapentadienyl complexes. 1-Phenylpropanone condenses with MeMn(CO)$_5$ to give **67** in 9% yield (*225*).

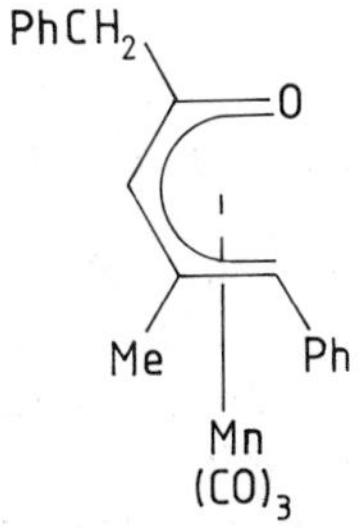

When PhMn(CO)$_5$ reacts with *trans*-1,3-pentadiene, and the resulting acyl substituted derivative **68** heated to 120°C (0.1 mm Hg), loss of carbon monoxide leads to a 73% yields of **69** [Eq. (32)] (*226*). Recently a derivative **70** of the parent ligand has been prepared from (Ph$_3$P)$_2$ReH$_7$ and furan under reflux in thf, in the presence of 3,3'-dimethyl-1-butene as hydrogen acceptor (*227*).

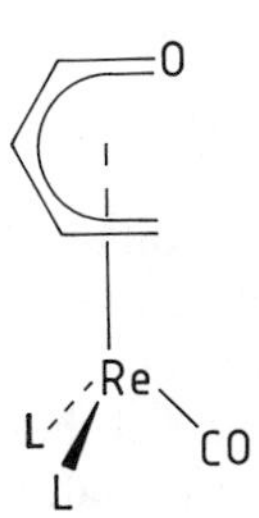

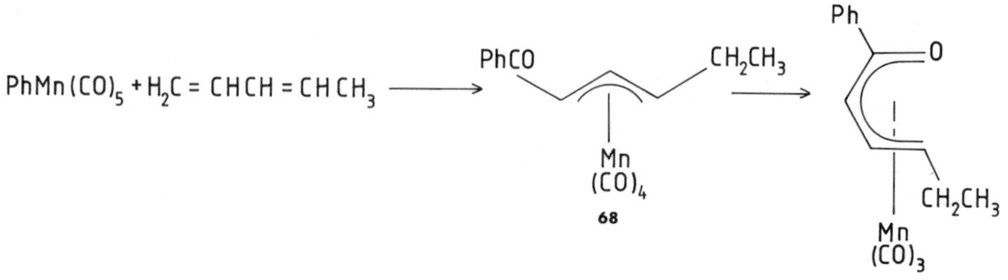

PhMn(CO)$_5$ + H$_2$C = CHCH = CHCH$_3$ $\longrightarrow$

(68) $\longrightarrow$ (69)

(32)

A cationic iridium complex (71), related to those discussed in Section IV,H, was obtained from [Ir(η^5-C$_5$Me$_5$)(Me$_2$CO)$_3$], (PF$_6$)$_2$, and mesityl oxide. If this or the original starting material is heated in solution in acetone, it is converted to the hydroxypentadienyl cation (72) (228). These synthetic procedures have not found general application. The results suggest, however, that oxapentadienyl complexes might repay further investigation.

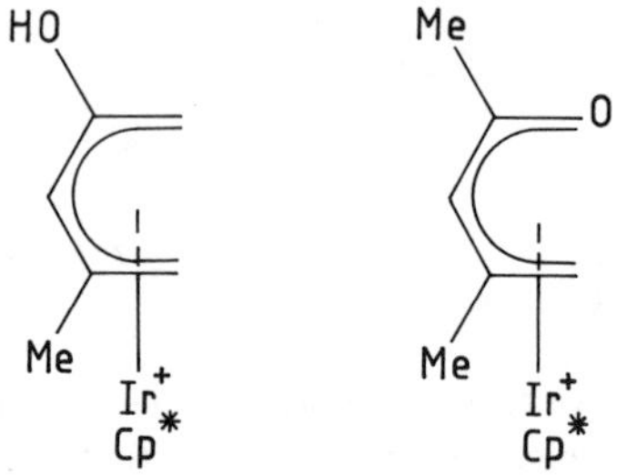

72   71

L. *Bonding in η^5-Pentadienyl Complexes of Transition Elements*

The π molecular orbitals of pentadienyl and cyclopentadienyl fragments, based on the simple Hückel molecular orbital treatment, are illustrated in Fig. 7. The basic similarity between the bonding characteristics of the two systems leads to the belief that many parallels in their coordination chemistry are to be expected. By analogy with cyclopentadienyl–metal bonds (229), it is likely that the most important interactions will be between metal d_{zx} and d_{yz} and ligand ψ_2 and ψ_3. The ψ_3 orbital lies considerably higher in energy than the corresponding pentadienyl one, so that C$_5$H$_7$– could be a better π donor than C$_5$H$_5$–. Also significant is the indication that ψ_4 is lower in energy than its counterpart. This leads to the prediction that δ bonding between this orbital and a filled metal $d_{x^2-y^2}$ orbital will be more important in pentadienyls than in cyclopentadienyls. This effect would be most noticeable

Metal Orbitals	Cyclopentadienyl Orbitals	Pentadienyl Orbitals

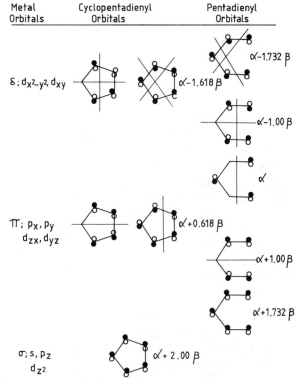

FIG. 7.   Comparison of Hückel molecular orbitals of cyclopentadienyl and pentadienyl ligands.

in complexes of metals late in the transition series which have a high number of d electrons. Under favorable circumstances, therefore, a pentadienyl ligand could be even more strongly bound than cyclopentadienyl (21,22).

The interactions between pentadienyl and $M(CO)_3$ fragments have been considered. Molecular orbital calculations using a Fenske–Hall semiempirical method were used to aid assignment of the photoelectron spectra of the cyclic complexes $(\eta^5\text{-}C_6H_7)Mn(CO)_3$, $(\eta^5\text{-}C_7H_9)Mn(CO)_3$, and $(\eta^5\text{-}C_7H_7)Mn(CO)_3$ (230), and a similar interaction diagram serves to explain the UV–PE spectra of $PlM(CO)_3$ (M = Mn,Re) (231). Here the effect on the metal ionizations of replacing C_5H_5 for C_5H_7 is only small. This is in marked contrast to the changes found on going from $FeCp_2$ to $FePl_2$ (232,233). Molecular orbital calculations show that there is considerable mixing between d orbitals in the open ferrocenes. One feature is an increase in the d_{xy} ionization energy which is attributable to mixing with the d_{xz} orbital and an interaction with the symmetric combination of the ψ_2 orbitals on both rings. This interaction is enhanced by means of a rotation of the terminal CH_2 groups so that their π orbitals point toward the iron atom, as noted above. Another conclusion is that δ bonding is much more important in the open complexes than in ferrocene itself. This means that the orbital populations of d_{xy} and $d_{x^2-y^2}$ are increased relative to those in ferrocene, and that those of d_{zx} and d_{yz} are decreased, a conclusion in accord with a Mossbauer study (234). The extent of δ bonding across the series $MP1_2$ (Ti to Fe) has been related to the conformation adopted in the ground states of the open metallocenes (233). These arguments have been reviewed and will not be repeated here (21).

A characteristic property of ferrocene is its reversible one-electron oxidation to the ferricinium cation (235). The mixed sandwich FeCpPl' and also $FePl'_2$ can likewise be oxidized reversibly at low temperatures, or irreversibly at room temperature on account of lability, to the corresponding cations. The potentials E^0 $(+/0)$ for the couples derived from $FeCp_2$, $Fe(MeC_5H_4)_2$, FeCpPl', and $FePl'_2/V$ vs SCE are 0.490, 0.304, 0.292, and 0.114 (extrapolated), respectively. The ESR spectrum of the open ferricinium cation indicates a nondegenerate ground state, in agreement with INDO calculations which predict that the unpaired electron is in the $3d_{z^2}$ orbital.

REFERENCES

1. G. Wilkinson, *J. Organomet. Chem.* **100**, 273 (1975).
2. J. P. Collman and L. S. Hegedus, "Principles and Applications of Organotransition Metal Chemistry." University Science Books, Mill Valley, California, 1980.
3. I. Haiduc and J. J. Zuckerman, "Basic Organometallic Chemistry." Walter de Gruyter, Berlin, 1985.
4. E. W. Abel, M. O. Dunster, and A. Waters, *J. Organomet. Chem.* **49**, 287 (1973).
5. C. H. Campbell and M. L. H. Green, *J. Chem. Soc. A*, 1318 (1970).
6. G. Huttner, H. H. Brintzinger, and L. G. Bell, *J. Organomet. Chem.* **145**, 329 (1978).
7. P. Jutzi, F. Kohl, P. Hofmann, C. Kruger, and Y.-H. Tsay, *Chem. Ber.* **113**, 757 (1980).
8. T. J. Marks and R. D. Ernst, *in* "Comprehensive Organometallic Chemistry", (G. Wilkinson, F. G. A. Stone, and E. W. Abel, eds.), Ch. 21. Pergamon, Oxford, 1982.
9. J. Birmingham, *Adv. Organomet. Chem.* **2**, 365 (1964).
10. M. L. H. Green and P. L. I. Nagy, *Adv. Organomet. Chem.* **2**, 325 (1964).
11. B. E. Mann, *in* "Comprehensive Organometallic Chemistry" (G. Wilkinson, F. G. A. Stone, and E. W. Abel, eds.), Ch. 20. Pergamon, Oxford, 1982.
12. H. Bonnemann, *Angew. Chem., Int. Ed. Engl.* **12**, 964 (1973).
13. P. W. Jolly and G. Wilke, "The Organic Chemistry of Nickel," Vols. I and II. Academic Press, New York, 1974 and 1975.
14. R. F. Heck, *Acc. Chem. Res.* **12**, 146 (1979).
15. B. M. Trost, *Chem. Soc. Rev.* **11**, 141 (1982).
16. P. W. Jolly, *Angew. Chem.* **97**, 279 (1985).
17. A. J. Pearson, *Transition Metal Chem.* (*Weinheim*) **6**, 67 (1981).
18. A. J. Pearson, "Metallo-organic Chemistry." Wiley, Chichester, 1985.
19. A. J. Pearson, *in* "Comprehensive Organometallic Chemistry" (G. Wilkinson, F. G. A. Stone, and E. W. Abel, eds.), Ch. 58. Pergamon, Oxford, 1982.

20. L. A. P. Kane-Maguire, E. D. Honig, and D. A. Sweigart, *Chem. Rev.* **84**, 525, (1984).
21. R. D. Ernst, *Structure and Bonding* **57**, 1 (1984).
22. R. D. Ernst, *Acc. Chem. Res.* **18**, 56 (1985).
23. R. B. Bates and C. A. Ogle, "Carbanion Chemistry." Springer-Verlag, New York, 1983.
24. D. H. O'Brien, *in* "Comprehensive Carbanion Chemistry" (E. Buncel and T. Durst, eds.), Part A, p. 271. Elsevier, Amsterdam, 1980.
25. A. H. Zimmerman, R. Gygax, and J. I. Brauman, *J. Am. Chem. Soc.* **100**, 5595 (1978).
26. J. H. Richardson, L. M. Stephenson, and J. I. Brauman, *J. Chem. Phys.* **59**, 5068 (1973).
27. P. C. Engelking and W. C. Lineberger, *J. Chem. Phys.* **67**, 1412 (1977).
28. S. Pignataro, A. Cassuto, and F. P. Lossing, *J. Am. Chem. Soc.* **89**, 3693 (1967).
29. A. G. Harrison, L. R. Honnen, H. J. Dauben, Jr., and F. P. Lossing, *J. Am. Chem. Soc.* **82**, 5593 (1960).
30. F. G. Bordwell, J. E. Bartmess, G. E. Drucker, Z. Margolia, and W. S. Matthews, *J. Am. Chem. Soc.* **97**, 3226 (1975).
31. M. Schlosser and P. Schneider, *Helv. Chim. Acta* **63**, 2404 (1980).
32. V. Grignard and L. Lapayre, *Compt. Rend.* **192**, 250 (1931).
33. R. Paul and S. Tchelitcheff, *Compt. Rend.* **224**, 1118 (1947).
34. R. Paul and S. Tchelitcheff, *Compt. Rend.* **232**, 1939 (1951).
35. J. E. Berger, E. L. Stogryn, and A. A. Zimmerman, *J. Org. Chem.* **29**, 950 (1964).
36. R. B. Bates, D. W. Gosselink, and J. A. Kaczynski, *Tetrahedron Lett.*, 199 (1967).
37. R. B. Bates, D. W. Gosselink, and J. A. Kaczynski, *Tetrahedron Lett.*, 205 (1967).
38. R. B. Bates, L. M. Kroposki, and D. E. Potter, *J. Org. Chem.* **37**, 560 (1972).
39. G. J. Heiszwolf and H. Kloosterziel, *Rec. Trav. Chim.* **86**, 807 (1967).
40. S. Brenner and J. Klein, *Isr. J. Chem.* **7**, 735 (1969).
41. R. B. Bates, W. H. Deines, D. A. McCombs, and D. E. Potter, *J. Am. Chem. Soc.* **91**, 4608 (1969).
42. W. Oppolzer, S. C. Burford, and F. Marazza, *Helv. Chim. Acta* **63**, 555 (1980).
43. J. J. Bahl, R. B. Bates, and B. Gordon, *J. Org. Chem.* **44**, 2290 (1979).
44. M. Schlosser and G. Rauchschwalbe, *J. Am. Chem. Soc.* **100**, 3258 (1978).
45. H. Yasuda, T. Narita, and H. Tani, *Tetrahedron Lett.*, 2443 (1973).
46. H. Yasuda, Y. Ohnuma, M. Yamauchi, H. Tani, and A. Nakamura, *Bull. Chem. Soc. Jpn.* **52**, 2036 (1979).
47. K. Ziegler, *Ann.* **542**, 90 (1940).
48. M. Newcomb and W. T. Ford, *J. Polym. Sci. B.* **10**, 17 (1972).
49. K. Suga and S. Watanabe, *Bull, Chem. Soc. Jpn.* **40**, 1257 (1967).
50. F. Schue, *Bull. Chem. Soc. Fr.* **4**, 980 (1965).
51. N. L. Bauld, *J. Am. Chem. Soc.* **84**, 4347 (1962).
52. H. Kloosterziel and J. A. A. Van Drunen, *Rec. Trav. Chim.* **89**, 368 (1970).
53. G. Fraenkel, *J. Organomet. Chem.* **197**, 249 (1980).
54. J. L. Wardell, *in* "Comprehensive Organometallic Chemistry" (G. Wilkinson, F. G. A. Stone, and E. W. Abel, eds.), Vol. 1, Ch. 2. Pergamon, Oxford, 1982.
55. W. T. Ford and M. Newcomb, *J. Am. Chem. Soc.* **96**, 309 (1976).
56. R. B. Bates, S. Brenner, C. M. Cole, E. W. Davidson, G. D. Forsythe, D. A. McCombs, and A. S. Roth, *J. Am. Chem. Soc.* **95**, 926 (1973).
57. H. Yasuda, M. Yamauchi, and A. Nakamura, *J. Organomet. Chem.* **202**, C1 (1980).
58. H. Bosshardt and M. Schlosser, *Helv. Chim. Acta* **63**, 2393 (1980).
59. G. J. Heiszwolf, J. A. A. van Drunen, and H. Kloosterziel, *Rec. Trav. Chim.* **88**, 1377 (1969).
60. H. Yasuda, M. Yamauchi, Y. Ohnuma, and A. Nakamura, *Bull. Chem. Soc. Jpn.* **54**, 1481 (1981).
61. W. Oppolzer, R. L. Snowden, and D. P. Simmons, *Helv. Chim. Acta* **64**, 2002 (1981).

62. H. Yasuda, T. Nishi, K. Lee, and A. Nakamura, *Organometallics* **2**, 21 (1983).
63. D. H. Hunter, R. H. Klinck, R. P. Steiner, and J. B. Stothers, *Can. J. Chem.* **54**, 1464 (1976).
64. H. M. Parkes and R. N. Young, *J. Chem. Soc., Perkin Trans.* 2, 1137 (1980).
65. R. J. Bushby and A. S. Patterson, *J. Chem. Res. Synop.* 306 (1980).
66. W. Neugebauer and P. von. R. Schleyer, *J. Organomet. Chem.* **198**, C1 (1980).
67. S. Bywater and D. J. Worsfold, *J. Organomet. Chem.* **159**, 229 (1978).
68. W. N. Setzer and P. von R. Schleyer, *Adv. Organomet. Chem.* **24**, 353 (1985).
69. M. Schlosser and M. Stahle, *Angew. Chem. Int. Ed. Engl.* **21**, 145 (1982).
70. R. Hoffmann and T. A. Olofson, *J. Am. Chem. Soc.* **88**, 943 (1966).
71. J. F. Sebastian, B. Hsu, and J. P. Grunwell, *J. Organomet. Chem.* **105**, 1 (1976).
72. R. J. Bushby and A. S. Patterson, *J. Organomet. Chem.* **132**, 163 (1977).
73. A. Bongini, G. Cainelli, G. Cardillo, P. Palmieri, and A. Umani-Ronchi, *J. Organomet. Chem.* **92**, C1 (1975).
74. M. J. S. Dewar, M. A. Fox, and D. J. Nelson, *J. Organomet. Chem.* **185**, 157 (1980).
75. M. Stahle, J. Hartmann, and M. Schlosser, *Helv. Chim. Acta* **60**, 1730 (1977).
76. M. Schlosser and J. Hartmann, *J. Am. Chem. Soc.* **98**, 4674 (1976).
77. T. B. Thompson and W. T. Ford, *J. Am. Chem. Soc.* **101**, 5459 (1979).
78. R. J. Bushby and G. J. Ferber, *Tetrahedron Lett.*, 3701 (1974).
79. R. B. Bates, *J. Am. Chem. Soc.* **92**, 6345 (1970).
80. J. Klein and S. Glily, *Tetrahedron* **27**, 3477 (1971).
81. R. Huisgen, *Angew. Chem. Int. Ed. Engl.* **19**, 947 (1980).
82. R. B. Bates and D. A. McCombs, *Tetrahedron Lett.*, 977 (1969).
83. R. B. Bates, S. Brenner, and C. M. Cole, *J. Am. Chem. Soc.* **94**, 2130 (1972).
84. H. Kloosterziel and J. A. A. van Drunen, *Rec. Trav. Chim.* **88**, 1084 (1969).
85. H. Yasuda, M. Yamauchi, A. Nakamura, T. Sei. Y. Kai, and N. Yasuoka, *Bull. Chem. Soc. Jpn.* **53**, 1089 (1980).
86. H. Yasuda and H. Tani, *Tetrahedron Lett.*, 11 (1975).
87. H. Yasuda, Y. Ohnuma, A. Nakamura, Y. Kai, N. Yasuoka, and N. Kasai, *Bull. Chem. Soc. Jpn.* **53**, 1101 (1980).
88. F. Gerard and P. Miginiac, *Bull. Soc. Chim. Fr.*, 1924 (1974).
89. P. Miginiac, *Ann. Chim.* **7**, 445 (1962).
90. G. F. Woods and H. F. Lederle, *J. Am. Chem. Soc.* **73**, 2245 (1954).
91. F. Gerard and P. Miginiac, *Bull. Soc. Chim. Fr.*, 2527 (1974).
92. S. R. Wilson, D. T. Mao, K. M. Jernberg, and S. T. Ezmirly, *Tetrahedron Lett.* **30**, 2559 (1977).
93. S. R. Wilson and D. T. Mao, *J. Org. Chem.* **44**, 3093 (1979).
94. S. R. Wilson and D. T. Mao, *J. Am. Chem. Soc.* **100**, 6289 (1978).
95. S. R. Wilson, A. Shedrinsky, and S. M. Haque, *Tetrahedron* **39**, 895 (1983).
96. S. R. Wilson, K. M. Jernberg, and D. T. Mao, *J. Org. Chem.* **41**, 3209 (1976).
97. S. R. Wilson, M. S. Hague, and R. N. Misra, *J. Org. Chem.* **47**, 747 (1982).
98. D. Seebach and M. Kolb, *Ann.*, 811 (1977).
99. W. Oppolzer and R. L. Snowden, *Tetrahedron Lett.*, 3505 (1978).
100. W. Oppolzer and R. L. Snowden, *Helv. Chim. Acta* **64**, 2592 (1981).
101. W. Oppolzer, *Angew. Chem. Int. Edn. Engl.* **16**, 10 (1977).
102. G. Brieger and J. W. Bennet, *Chem. Rev.* **80**, 63 (1980).
103. R. Grafing, A. V. E. George, and L. Brandsma, *Red. J. R. Neth. Chem. Soc.* **101**, 346 (1982).
104. H. Yasuda, T. Nishi, S. Miyanaga, and A. Nakamura, *Organometallics* **4**, 359 (1985).
105. M. G. Hutchings, W. E. Paget, and K. Smith, *J. Chem. Res., Synop.*, 31 (1983).
106. A. Pelter and K. Smith, *in* "Comprehensive Organic Chemistry" (D. H. R. Barton and W. D. Ollis, eds.), Vol. 3, p. 689. Pergamon, Oxford, 1979.

107. I. Fleming, *in* "Comprehensive Organic Chemistry" (D. H. R. Barton and W. D. Ollis, eds.), Vol. 3. Pergamon, Oxford, 1979.

108. E. W. Colvin, "Silicon in Organic Synthesis." Butterworths, London, 1981.

109. H. Sakurai, *Pure Appl. Chem.* **54**, 1 (1982).

110. H. Sakurai, A. Hosomi, M. Saito, K. Sasaki, H. Inguchi, J.-L. Sasaki, and Y. Ataki, *Tetrahedron* **39**, 883 (1983).

111. D. Seyferth, J. Pornet, and R. M. Weinstein, *Organometallics* **1**, 1651 (1982).

112. D. Seyferth and J. Pornet, *J. Org. Chem.* **45**, 1721 (1980).

113. A. Hosomi, M. Saito, and H. Sakurai, *Tetrahedron Lett.*, 3783 (1980).

114. J. Pornet, *Tetrahedron Lett.* **21**, 2049 (1980).

115. T. R. Bailey, R. S. Garigipati, J. A. Marton, and S. M. Weinreb, *J. Am. Chem. Soc.* **106**, 3240 (1984).

116. M. Koreeda and Y. Tanaka, *Chem. Lett.*, 1299 (1982).

117. Y. Naruta, Y. Arita, N. Nagai, H. Uno, and K. Maruyama, *Chem. Lett.*, 1859 (1982).

118. Y. Naruta, N. Nagai, Y. Arita, and K. Maruyama, *Chem. Lett.*, 1683 (1983).

119. Y. Naruta, M. Kashiwagi, Y. Nishigaichi, H. Uno, and K. Maruyama, *Chem. Lett.* 1687 (1983).

120. M. Jones and W. Kitching, *J. Organomet. Chem.* **247**, C5 (1983).

121. U. Giannini, E. Pellino, and M. P. Lachi, *J. Organomet. Chem.* **12**, 551 (1968).

122. R. Rienacker and Y. Yoshiura, *Angew. Chem. Int. Ed. Engl.* **8**, 677 (1969).

123. C. Krüger, *Angew. Chem. Int. Ed. Engl.* **8**, 678 (1969).

124. H. Werner, *Angew. Chem. Int. Ed. Engl.* **16**, 1 (1977).

125. M. C. Bohm and R. Gleiter, *Chem. Phys.* **64**, 183 (1982).

126. D. R. Wilson, A. A. DiLullo, and R. D. Ernst, *J. Am. Chem. Soc.* **102**, 5928 (1980).

127. D. R. Wilson, R. D. Ernst, and T. H. Cymbaluk, *Organometallics* **2**, 1220 (1983).

128. D. R. Wilson, L. Stahl, and R. D. Ernst, *Organomet. Syn.*, submitted for publication.

129. D. R. Wilson, J.-Z. Liu, and R. D. Ernst, *J. Am. Chem. Soc.* **104**, 1120 (1982).

130. K. R. Gordon and K. D. Warren, *Inorg. Chem.* **17**, 987 (1978).

131. G. P. Pez and J. N. Armor, *Adv. Organomet. Chem.* **19**, 1 (1981).

132. M. C. Bohm, R. D. Ernst, R. Gleiter, and D. R. Wilson, *Inorg. Chem.* **22**, 3815 (1983).

133. D. R. Wilson, R. D. Ernst, and M. S. Krailk, *Organometallics* **3**, 1442 (1984).

134. R. S. Sapienza, P. E. Riley, R. E. Davis, and R. Pettit, *J. Organomet. Chem.* **121**, C35 (1976).

135. L. Stahl and R. D. Ernst, *Organometallics* **2**, 1229 (1983).

136. P. Pertici, G. Vitalli, M. Paci, and L. Poni, *J. Chem. Soc., Dalton Trans.*, 1961 (1980).

137. R. D. Ernst and T. H. Cymbaluk, *Organometallics* **1**, 708 (1982).

138. T. H. Cymbaluk, J.-Z. Liu, and R. D. Ernst, *J. Organomet. Chem.* **255**, 311 (1983).

139. L. Stahl, J. P. Hutchinson, D. R. Wilson, and R. D. Ernst, *J. Am. Chem. Soc.* **107**, 5016 (1985).

140. R. D. Ernst, C. F. Campana, D. R. Wilson, and J.-Z. Liu, *Inorg. Chem.* **23**, 2732, (1984).

141. D. R. Wilson, J.-Z. Liu, and R. D. Ernst, *J. Am. Chem. Soc.* **104**, 1120 (1982).

142. R. D. Ernst, J.-Z. Lui, and D. R. Wilson, *J. Organomet. Chem.* **250**, 257 (1983).

143. D. J. Sikora, M. D. Rausch, R. D. Rogers, and J. L. Atwood, *J. Am. Chem. Soc.* **103**, 982 (1981).

144. S. J. Severson, T. H. Cymbaluk, R. D. Ernst, J. M. Higashi, and R. W. Parry, *Inorg. Chem.* **22**, 3833 (1983).

145. J. R. Bleeke and M. K. Heyes, *Organometallics* **3**, 506 (1984).

146. J. R. Bleeke and J. J. Kotyk, *Organometallics* **2**, 1263 (1983).

147. J. R. Bleeke and J. J. Kotyk, *Organometallics* **4**, 194 (1985).

148a. J. R. Bleeke and W.-J. Peng, *Organometallics* **3**, 1422 (1984).

148b. J. R. Bleeke and W.-J. Peng, *Organometallics* **5**, 635 (1986).

149. M. A. Paz-Sandoval, P. Powell, M. G. B. Drew, and R. N. Perutz, *Organometallics* **3**, 1026 (1984).

150. H. Lehmkuhl and C. Naydowski, *J. Organomet. Chem.* **240**, C30 (1982).
151. H. Lehmkuhl, C. Naydowski, R. Benn, A. Rufinska, and G. Schroth, *J. Organomet. Chem.* **246**, C9 (1983).
152. H. Lehmkuhl, F. Danowski, R. Benn, A. Rufinska, G. Schroth, and R. Mynott, *J. Organomet. Chem.* **254**, C11 (1983).
153. H. Lehmkuhl, A. Rufinska, R. Benn, G. Schroth, and R. Mynott, *Ann.*, 317 (1981).
154. R. Taube, U. Schmid, and H. Schwind, *Z. Anorg. Allg. Chem.* **458**, 273 (1979).
155. Y. Yasuda, K. Nagasuma, M. Akita, K. Lee, and A. Nakamura, *Organometallics* **3**, 1470 (1984).
156. H. Yasuda, K. Tatsumi, and A. Nakamura, *Acc. Chem. Res.* **18**, 120 (1985).
157. A. J. Deeming, in "Comprehensive Organometallic Chemistry" (G. Wilkinson, F. G. A. Stone, and E. W. Abel, eds.), Ch. 31.3. Pergamon, Oxford, 1982.
158. J. E. Mahler and R. Pettit, *J. Am. Chem. Soc.* **84**, 1551 (1962).
159. J. E. Mahler, and R. Pettit, *J. Am. Chem. Soc.* **85**, 3955 (1963).
160. J. E. Mahler, D. H. Gibson, and R. Pettit, *J. Am. Chem. Soc.* **85**, 3959 (1963).
161. A. J. Birch, *et al., Tetrahedron, Suppl.*, W289 (1981).
162. N. A. Clinton and C. P. Lillya, *J. Am. Chem. Soc.* **92**, 3058 (1970).
163. N. A. Clinton and C. P. Lillya, *J. Am. Chem. Soc.* **92**, 3065 (1970).
164. D. E. Kuhn and C. P. Lillya, *J. Am. Chem. Soc.* **94**, 1682 (1972).
165. D. G. Gresham, D. J. Kowalski, and C. P. Lillya, *J. Organomet. Chem.* **144**, 71 (1978).
166. P. E. Riley and R. E. Davis, *Acta Cryst.* **B32**, 381 (1976).
167. J. W. Burnill, B. R. Bonazza, D. W. Garnett, and C. P. Lillya, *J. Organomet. Chem.* **104**, C37 (1976).
168. T. S. Sorensen and C. R. Jablonski, *J. Organomet. Chem.* **25**, C62 (1970).
169. C. R. Jablonski and T. S. Sorensen, *Can. J. Chem.* **52**, 2085 (1974).
170. C. P. Lillya and R. A. Sahatjian, *J. Organomet. Chem.* **25**, C67 (1970).
171. C. P. Lillya and R. A. Sahatjian, *J. Organomet. Chem.* **32**, 371 (1971).
172. M. Brookhart and D. L. Harris, *J. Organomet. Chem.* **42**, 441 (1972).
173. R. S. Bayoud, E. R. Biehl, and P. C. Reeves, *J. Organomet. Chem.* **150**, 75 (1978).
174. T. G. Bonner, K. A. Holder, and P. Powell, *J. Organomet. Chem.* 77, C37 (1974).
175. T. G. Bonner, K. A. Holder, P. Powell, and E. Styles, *J. Organomet. Chem.* **131**, 105 (1977).
176. G. Maglio, A. Musco, R. Pulcino, and A. Striga, *J. Chem. Soc. Chem. Commun.*, 100 (1971).
177. G. Maglio, A. Musco, and R. Palumbo, *J. Organomet. Chem.* **32**, 127 (1971).
178. G. Maglio and R. Palumbo, *J. Organomet. Chem.* **76**, 367 (1974).
179. R. S. Bayoud, E. R. Biehl, and P. C. Reeves, *J. Organomet. Chem.* **174**, 297 (1979).
180. P. McArdle and H. Sherlock, *J. Chem. Soc., Dalton Trans.*, 1678 (1978).
181. A. Salzer and A. Hafner, *Helv. Chim. Acta* **66**, 1774 (1983).
182. A. Hafner, J. H. Bieri, R. Preivo, W. von Phillipsborn, and A. Salzer, *Angew. Chem. Int. Ed. Engl.* **22**, 713 (1983).
183. G. Jaouen, B. F. G. Johnson, and J. Lewis, *J. Organomet. Chem.* **231**, C21 (1982).
184. M. Anderson, A. D. H. Clague, L. P. Blaauw, and P. A. Couperus, *J. Organomet. Chem.* **56**, 307 (1973).
185. B. F. G. Johnson, J. Lewis, D. G. Parker, and S. R. Postle, *J. Chem. Soc., Dalton Trans.*, 794 (1977).
186. B. R. Bonazza, C. P. Lillya, E. S. Magyar, and G. Scholes, *J. Am. Chem. Soc.* **101**, 4100 (1979).
187. P. A. Dobosh, C. P. Lillya, E. S. Magyar, and G. Scholes, *Inorg. Chem.* **19**, 228 (1980).
188. B. R. Bonazza, C. P. Lillya, and G. Scholes, *Organometallics* **1**, 137 (1982).
189. M. Brookhart and M. L. H. Green, *J. Organomet. Chem.* **250**, 395 (1983).
190. K. G. Caulton, *Coord. Chem. Revs.* **38**, 1 (1981).
191. E. W. Abel and S. Moorhouse, *J. Chem. Soc., Dalton Trans.*, 1706 (1973).

192. K. J. Barrow, O. S. Mills, F. Haque, and P. L. Pauson, *J. Chem. Soc. D*, 1239 (1971).
193. D. Seyferth, E. W. Goldman, and J. Pornet, *J. Organomet. Chem.* **208**, 189 (1981).
194. M. A. Paz-Sandoval and P. Powell, *J. Organomet. Chem.* **219**, 81 (1981).
195. H. G. Schuster-Woldan and F. Basolo, *J. Am. Chem. Soc.* **88**, 1659 (1966).
196. C. P. Casey, J. M. O. Connor, W. D. Jones, and K. J. Haller, *Organometallics* **2**, 535 (1983).
197. K. N. Ji, M. E. Rerek, and F. Basolo, *Organometallics* **3**, 740 (1984).
198. M. Leyendecker and C. G. Kreiter, *J. Organomet. Chem.* **249**, C31 (1983).
199. M. Leyendecker, W. S. Sheldrick, and C. G. Kreiter, *J. Organomet. Chem.* **270**, C37 (1984).
200. C. G. Kreiter and M. Leyendecker, *J. Organomet. Chem.* **280**, 225 (1985).
201. G. T. Palmer and F. Basolo, *J. Am. Chem. Soc.* **107**, 3122 (1985).
202. T. A. Albright, *Acc. Chem. Res.* **15**, 149 (1982).
203. B. E. Mann, in "Comprehensive Organometallic Chemistry" (G. Wilkinson, F. G. A. Stone, and E. W. Abel, eds.), Vol. 3, p. 114. Pergamon, Oxford, 1982.
204. A. Ceccon, personal communication.
205. M. S. Kralik, J. H. Hutchinson, and R. D. Ernst, *J. Am. Chem. Soc.* **107**, 8296 (1985).
206. P. Powell and L. J. Russell, *J. Chem. Res., Synop.*, 283 (1978).
207. P. Powell, *J. Organomet. Chem.* **165**, C43 (1979).
208. P. Powell, *J. Organomet. Chem.* **206**, 229 (1981).
209. P. Powell, *J. Organomet. Chem.* **243**, 205 (1983).
210. P. Powell, *J. Organomet. Chem.* **206**, 239 (1981).
211. P. Powell, *J. Organomet. Chem.* **244**, 393 (1983).
212. P. Powell, *J. Organomet. Chem.* **266**, 307 (1984).
213. M. G. B. Drew and P. Powell, *J. Organomet. Chem.*, in press.
214. P. Powell, in "The Chemistry of the Metal Carbon Bond" (F. R. Hartley and S. Patai, eds.), Ch. 8. Wiley, New York, 1982.
215. T. Ukai, H. Kawazura, K. Ishii, J. J. Bonnet, and J. A. Ibers, *J. Organomet. Chem.* **65**, 253 (1974).
216. S. D. Robinson and B. L. Shaw, *J. Chem. Soc.*, 4806 (1963).
217. L. S. Hegedus and S. Varaprath, *Organometallics* **1**, 259 (1982).
218. E. J. Corey, M. F. Semmelhack, and L. S. Hegedus, *J. Am. Chem. Soc.* **90**, 2416 (1968).
219. M. A. Paz-Sandoval, Ph.D. thesis, University of London, 1983.
220. S. R. Allen, M. Green, N. C. Norman, K. E. Paddick, and A. G. Orpen, *J. Chem. Soc., Dalton Trans.*, 1625 (1983).
221. A. F. Dyke, S. A. R. Knox, P. J. Naish, and G. E. Taylor, *J. Chem. Soc., Chem. Commun.*, 803 (1980).
222. P. Q. Adams, D. L. Davies, A. F. Dyke, S. A. R. Knox, K. A. Mead, and P. Woodward, *J. Chem. Soc., Chem. Commun.*, 222 (1983).
223. M. Green, A. E. Orpen, C. J. Schaverien, and I. D. Williams, *J. Chem. Soc., Chem. Commun.*, 181 (1983).
224. J. Levisalles, F. Rose-Munch, H. Rudler, J. C. Daran, Y. Dromzee, and Y. Jeannin, *J. Chem. Soc., Chem. Commun.*, 1562 (1981).
225. R. L. Bennett and M. I. Bruce, *Aust. J. Chem.* **28**, 1141 (1975).
226. M. Green and R. I. Hancock, *J. Chem. Soc. A*, 109 (1968).
227. D. Baudry, J. C. Daran, Y. Dromzee, M. Ephritikhine, H. Felkin, Y. Jeannin, and J. Zakrzewski, *J. Chem. Soc., Chem. Commun.*, 813 (1983).
228. C. White, S. J. Thompson, and P. M. Maitlis, *J. Organomet. Chem.* **134**, 319 (1977).
229. D. M. P. Mingos, in "Comprehensive Organometallic Chemistry" (G. Wilkinson, F. G. A. Stone, and E. W. Abel, eds.), Vol. 3, Ch. 19. Pergamon, Oxford, 1982.
230. T. H. Whitesides, D. L. Lichtenberger, and R. A. Budnik, *Inorg. Chem.* **14**, 68 (1975).
231. J. C. Green, M. A. Paz-Sandoval, and P. Powell, *J. Chem. Soc., Dalton Trans.*, 2677 (1985).

232. M. C. Böhm, M. Eckert-Maksic, R. D. Ernst, D. R. Wilson, and R. Gleiter, *J. Am. Chem. Soc.* **104,** 2699 (1982).
233. R. Gleiter, M. C. Böhm, and R. D. Ernst, *J. Electron Spect. Relat. Phenom.* **33,** 269 (1984).
234. R. D. Ernst, D. R. Wilson, and R. H. Herber, *J. Am. Chem. Soc.* **106,** 1646 (1984).
235. C. Elschenbroich, E. Bilger, R. D. Ernst, D. R. Wilson, and M. S. Kralik, *Organometallics* **4,** 2068 (1985).

Computer Aids for Organometallic Chemistry and Catalysis

I. THEODOSIOU,
R. BARONE, and
M. CHANON

Laboratoire de Chimie Inorganique Moléculaire
Faculté des Sciences de St. Jérôme
13397 Marseille, Cédex 4, France

I

INTRODUCTION

When a chemist wishes to gain deeper insight in the understanding of a reaction mechanism, he usually proceeds according to a given sequence of steps. The first step is to objectively inventory all the feasible mechanisms that he should submit to experimental investigation. Then he proceeds to design the experiments which will, hopefully, allow him to select the mechanism which best fits the complete set of experimental data. If, at the preliminary stage, he overlooks one possibility, this will cast doubt on all his painstaking experiments as soon as someone realizes his oversight. Therefore, a tool to help the chemist in the preliminary stage would be useful.

Sometimes a pharmaceutical firm considers bypassing a long series of synthetic steps by using an elegant new reaction brought about through the use of homogeneous catalysis (see for example, ref. *1*). Before doing so, their first concern will be to determine whether any of the by-products expected in the new method may be toxic. A tool aiding them with an early diagnosis of possible by-products associated with a given process would probably be welcome.

A professor teaching the fundamentals of homogeneous catalysis based on transition metal complexes might be pleased to find a concise summary of the basic steps proposed up to the present in this area of chemistry. This would answer one worry many of us have, "What minimum basic set of reaction steps should I teach so that students can survive and compete in the fast moving field of homogeneous catalysis induced by transition metal complexes?" In these respects, and despite its imperfections (which we will discuss), TAMREAC has been designed to assist the chemist with the three foregoing questions. TAMREAC has to be distinguished from approaches which mathematically generate all mechanistic possibilities (*2c*) because it is supposed to embrace essentials of organometallic chemistry connected with catalysis.

165

TAMREAC (TrAnsition Metal REACtivity) (*2a,2b*), is a computer program whose aim has been to program into the computer the minimum basic set of reactions which are constantly encountered in homogeneous catalysis (from this point forward when this term is used in this article, it will specifically stand for homogeneous catalysis induced by transition metals), plus some rules concerning the relative articulation of these steps in a given catalytic cycle. It should therefore be able to provide the chemist with a set of sensible mechanisms to be expected for a given transformation and the set of possible by-products associated with a given mechanism. The first paper covering this program showed that its aid, even if imperfect, is not without value (*3*). Indeed, M. L. H. Green was able to select as sensible a mechanism not previously discussed from among the propositions that TAMREAC offered to rationalize the ethylene dimerization reaction. This new mechanism involved a metallacyclopentane intermediate, which was independently proposed as a possibility at about the same time by Schrock and co-workers (*4*).

In this article, we shall briefly describe the flowchart of TAMREAC, placing stress specifically on the matter of interest for the chemist, namely, basic reaction steps. Ionic as well as homolytic reaction steps will be considered. The possibility of using TAMREAC to propose not-yet-realized new reactions will be alluded to in the section dealing with the organization of reactions. In this section, the difficult problem of deciding which transition metal would best promote a given step will be addressed but of course, not definitively settled. The other set of rules taught to the computer deals with the successive steps; taking into account OS (oxidation state of the metal center), NVE (number of valence electrons), CN (coordination number of the complex), and Q (global charge of the complex). The last section will deal with results proposed by TAMREAC for selected transformations: ethylene dimerization, ethylene hydroformylation, and allyl–alkyl cross coupling.

TAMREAC is part of a wider study being developed by our group (*5a–5d*) and by others (*6*). This effort attempts to take advantage of the best characteristic of the computer: relieving the chemist of the burden of some repetitive tasks so that he may concentrate on tasks in which he excells, namely, creation of new steps and reactions, selection of the most sensible mechanistic schemes, and selection of the most lively materials to illustrate his lectures.

II

PROGRAM PRESENTATION

TAMREAC (TrAnsition Metal REACtivity) is written in FORTRAN and runs on a PDP 11/44 computer. The program is interactive and allows facile graphical communication of both input and output in a form most convenient and natural for chemists. First, the chemist transmits to the program the "starting materials," and then the program takes control to begin generating intermediates and products. Control is given back to the chemist for the final evaluation of the intermediates and for the choice of the next step. The main steps of the program are depicted in Fig. 1, and include the following:

INPUT of the compounds (complexes and organic reagents) is made by drawing them, using the keyboard of a VT125 display.

ANALYSIS of the catalytic system for its significant structural and electronic features is performed. The program perceives the nature of the ligands, the electronic and some structural features of the catalytic complex, and the functional groups of the organic reagents (Section II,B).

SEARCH for the substructures required to perform a given transformation, which corresponds to a possible reaction, is then performed by the program (Section II,C).

EVALUATION makes, when desirable, a rough estimation of the plausibility of the reaction, and either returns control to the search module if the reaction cannot take place, or proceeds to the construction of the intermediates or products.

CONSTRUCTION applies symbolic chemical modifications to create the structure of the products, which will be communicated to the chemist through

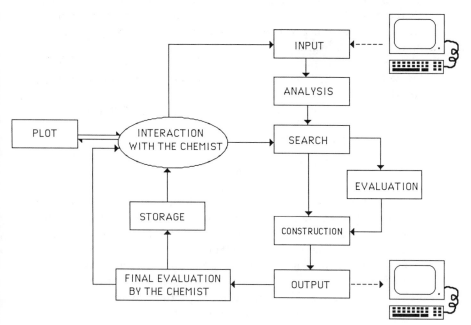

FIG. 1.   Main steps of the TAMREAC program.

the video display (*OUTPUT*) or the tracing table (*PLOT*), and waits for the chemist to decide whether to store or to discard the proposed intermediate (*FINAL EVALUATION*).

INTERACTION module allows the chemist to draw the intermediates on a tracing table (*PLOT*) and to decide about the following step. Generally he returns control to search until all stored reactions are treated. Then he chooses an intermediate which will be considered as the new starting material, and so on.

Limitations. At the moment TAMREAC is limited to mononuclear catalytic systems. It does not take into account stereochemistry, and the molecules or complexes are represented in only two dimensions. The number of nonhydrogen atoms is limited to 30 (see also Section V).

A. Catalyst Representation

TAMREAC is an interactive program, and the interface between chemist and program is provided by graphical communication. The chemist transmits the starting materials to the program by drawing them on the display, using the keyboard. The molecular drawing (external representation) is translated to the computer by means of a suitable numerical form in a connectivity table (internal representation) (*5d*), which informs the computer about the atoms in the molecules, the bonds between them, and the eventual charges (or oxidation states in the case of transition metals).

For output, the program uses a VT125 display which establishes the communication with the chemist and a Tektronix Plotter for tracing the graphics when desirable.

In the connectivity table, each atom is represented in the following way:

Atom no.	Nature	Connections	Bond nature	Charge
i	X	j, k, l, m, n, o	p, q, r, s, t, u	v

The atom i is bonded with the atom j (the nature of their bond is p), with the atom k (the nature of their bond is q), and so on. $p, q, r, s, t, u = 1, 2,$ or 3 (single, double, or triple bond, respectively). The nature of the atom i is X (X = C, H, O, Rh, Pd, etc.), and its charge is v (the oxidation state of the transition metal is represented as a charge, e.g., Rh^{I} is represented as Rh^{1+}). Hydrogen atoms are always represented by the atom 1, except when attached to a metal.

A problem which arises with catalyst representation by the connectivity tables, which were initially conceived for covalent bonds, is the representation

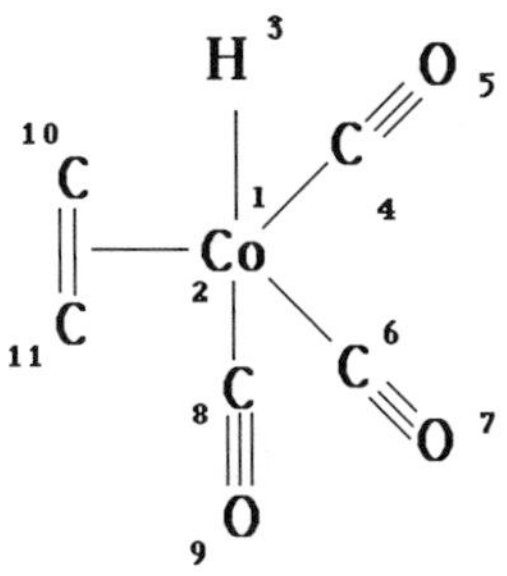

N°	Nature	Connections	Bond nature	Charges
1	H			0
2	Co	3 4 6 8	1 1 1 1	+1
3	H	2	1	0
4	C	2 5	1 3	0
5	O	4	3	0
6	C	2 7	1 3	0
7	O	6	3	0
8	C	2 9	1 3	0
9	O	8	3	0
10	C	11 1 1	2 1 1	0
11	C	10 1 1	2 1 1	0

N° of the atoms forming the unsaturated bond		N° of the metal
9	10	2
10	9	2

FIG. 2. Connectivity and π bond tables for the illustrated hydride complex. In the second line of the first table, atom 2 is a Co and is bonded with atom 3 with a single bond, with atom 4 with a single bond, etc. The charge of this atom is 1+ (in the case of the transition metals the charge corresponds to the oxidation state). In the first line of the second table atoms 9 and 10 form an unsaturated bond, which is associated to the metal (atom 2) by a π bond.

of donor bonds, such as in π association of unsaturated hydrocarbons with the transition metal. For example, in the case of η-bonded olefins, the question is still posed: Are these best represented as π complexes (Chatt–Dewar–Duncanson model), or as metallocyclopropane complexes (8)? In TAMREAC both of these representations are possible (2b). The π associations are specially dealt with in Fig. 2.

B. *Perception of Structural and Electronic Features*

The connectivity table provides only the topological features of the compounds and contains no chemical information about them. The ANALYSIS module recognizes relevant constituents, such as functional groups, electronic and structural features of the catalytic complex, and ligand properties. This module is of great importance, since the coding and evaluation of the reactions will be based on these perceptions. All the recognized constituents are given in Table 1. The functional groups perceived by this module are the nucleophilic and electrophilic centers most useful for the representation of organic reactions (7).

TABLE I

ATOMS AND BONDS ARE ANALYZED AS FOLLOWS BY TAMREAC IN THE ANALYSIS STEP.

C	Carbon atom
H	Hydrogen atom
O	Oxygen atom
N	Nitrogen atom
S	Sulfur atom
X	Halide atom (Cl, Br, F, I)
P	Phosphorus atom
L	Neutral ligand [H_2O, NH_3, $P(Ph)_3$ ...]
M	Transition metal (general case)
Nu	Nucleophilic atom
El	Electrophilic atom
T	All atoms
Z	All atoms except hydrogen atoms
Ht	Heteroatom(X, S, O, ...)
TL	All atoms except neutral ligands (L)
A	All atoms except carbon atoms
B	All atoms except carbon and hydrogen atoms
C2	sp and sp^2 carbon atoms
R	All carbon atoms except those which are bonded with the metal
Q	All atoms except those which are bonded with the metal
V	All atoms except carbons, hydrogens and atoms which are bonded with the metal
CM	Carbon atom in a metallacycle, bonded with the metal
CY	Carbon atom in a metallacycle, nonbonded to the metal
LS	Anionic σ-donor ligand
LA	σ-Donor and π-acceptor ligand
LP	σ-Donor and π-donor ligand
CS	"Schrock"-type carbene
CF	"Fischer"-type carbene
LA	Lewis acid
LB	Lewis base
—	Single bond
=	Double bond
≡	Triple Bond
⚌	Single or double bond
⚌	Double or triple bond
⚌	Single or double or triple bond

The ligand perceptions presented in Fig. 3 concern electronic characteristics. Donors, neutral or anionic ligands, π-donor neutral ligands, and so on are perceived. Carbene and carbyne ligands are considered as σ-donor neutral ligands, and a distinction between "Schrock"-type and "Fischer"-type carbenes is made (9).

A simplified electron-counting scheme (10) is also introduced in this module, providing a guide to likely products according to the 16- to 18-electron

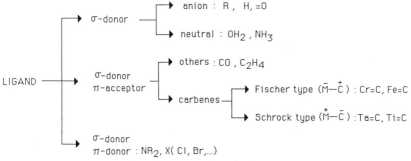

FIG. 3.   Ligand perceptions by the program.

rule (*11*). However, this only applies for "two-electron" reactions. We are aware of the limitations of this rule (*12a,12b*), but it may be valuable to the chemist who evaluates the intermediates. This electron-counting scheme is based on the following equation:

$$NVE = d - OS + 2(CN) \tag{1}$$

where NEV is the number of valence electrons, d the number of electrons in the d^n metal configuration, OS the formal oxidation state of the metal center, and CN the coordination number (σ bonds formed between a metal and its ligands). The oxygen ligand in the oxo form ($M = O$) is considered as a dianion (two σ bonds), and its coordination number will be two. In the case of metal carbenes ($M = C$), the coordination number retained is one.

The "global charge" of the catalysts derives from the oxidation state according to Eq. (2):

$$Q = OS - S \tag{2}$$

In this equation, Q is the global charge of the complex, S the number of the σ anionic ligands, and OS the formal oxidation state of the metal center. The perception of the global charge is needed for the establishment of some rules concerning the intermolecular interactions between the complex and organic reagents where acid–base interactions seem to be the driving forces (*13*). We will discuss these rules in Section III. Figure 4 shows the different type of complexes recognized by this module.

Another factor which is of importance for the choice of a mechanism is the coordination geometry of the complex. The electronic configuration of the complex depends on the geometry as well as on the spin state, and these are involved in the symmetry rules for chemical reactions (*14*). The ANALYSIS module makes a rough estimation in some cases (Fig. 5) according to the Angular Overlap and VSEPR models (*15a,15b*). Interaction with the chemist is sometimes necessary when the spin state or the geometry is not detected by the program.

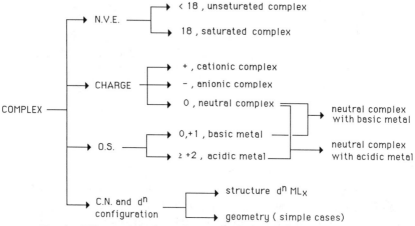

FIG. 4.   Different types of complexes recognized by the perception module.

d^n	spin	4-Coordination		5-Coordination	
		SPL	Td	SBY	TBY
0		0	1	0	1
2		0	1	0	1
4	ls	1	1	0	1
4	hs	1	1	1	1
6	ls	1	0	1	0
6	hs	0	1	0	1
8	ls	1	0	0	1
8	hs	1	1	1	1
10		1	1	1	1

FIG. 5.   Geometries detected by TAMREAC according to the Angular Overlap and VSEPR models for four- and five-coordinate complexes. Abbreviations: SPL = square planar, Td = tetrahedral, SPY = square pyramidal, TBP = trigonal bipyramidal, Is = low spin, hs = high spin, d^n = number of electrons in the d configuration of the metal.

C. Reaction Representation

The reaction research is based on the perception of the features in the ANALYSIS module, and a strategy is developed which includes the construction of possible intermediates or products. A reaction is described in three steps, each one corresponding to one module: (1) SEARCH for substructures within the molecules or complexes which signal the possibility of a reaction, (2) EVALUATION of this possibility, and (3) CONSTRUC-TION, which changes the substructure in order to build the product structure.

1. SEARCH

Let us take, as an example, the reductive elimination reaction in order to illustrate the approach. The general description of this reaction is given by Eq. (3).

$$A—M^n—B \rightarrow M^{n-2} + A—B \tag{3}$$

The first member in this equation represents the substructure to be recognized within the molecule. The symbols A, B, M, and their bonds derive from the perception module. M represents a transition metal, A and B all the anionic σ-donor ligands. The bonds between M—A and M—B can be single, double, or triple.

This method of coding the reactions, by identifying only the atoms which change, allows transformations to be applied to a wide range of molecules. So the reductive eliminations of Eqs. (3a), (3b), and (3c) will be found by this module, according to the mechanistic description of the reaction given by Eq. (3):

$$H—M—H \rightarrow M + H—H \tag{3a}$$

$$H—M{=}O \rightarrow M—OH \tag{3b}$$

$$CH_3—M—CH_3 \rightarrow M + CH_3CH_3 \tag{3c}$$

This kind of reaction representation cannot take into account, at this stage, the environment of the structure. These factors will be dealt with in the following module.

2. EVALUATION

The second stage of this approach is the evaluation of the reaction. This step is the most crucial one in our program, as well as in all other programs of computer-assisted synthesis (16a–16d). The evaluation by the computer is optional. If it is not desirable, the computer proceeds to the generation of all formal possibilities, otherwise the control passes to the EVALUATION module, which discards some solutions. The evaluation is based on the perception of various features of the catalyst and on some rules that we have introduced. These rules concern the two classes of two-electron reactions which will be presented in Section III. The first one contains key reactions of coordination chemistry, and the evaluation is based on the electronic configuration and the coordination number of the catalyst, rather than on the exact nature of the transition metal and the ligands. The evaluation matrices we addressed for some coordination reactions will be presented in Section III,A. The second class contains external attacks. These reactions are considered as generalized acid–base interactions, and rules for them are given in Section III,B.

The ultimate dream for an excellent evaluation would be to delineate the "dimensional space" (16e) of catalytic reactions. The three principal axes in such a space could be (1) the d^n configuration of the metal, (2) the coordination number, and (3) the element group (i.e., in the new ACS nomenclature 3, 4, ... 12). In terms of the dimensional space many of the transformations considered would occupy a limited volume. In a more elaborate n-dimensional space one could add ligand type and other considerations to better represent the whole set of transformations.

3. CONSTRUCTION

The CONSTRUCTION module performs modifications in the connectivity table, which consist of making and breaking bonds and the addition or loss of charges (or oxidation states). These modifications, specific for each reaction's substructure, lead to the product's structure.

In the case of reductive elimination [Eq. (3)], three modifications of the bonds and one of the oxidation state take place:

1. The order of the M—A bond decreases by 1 and becomes 0
2. The order of the M—B bond decreases by 1 and becomes 0
3. The order of the A—B bond increases by 1 and becomes a single bond
4. The oxidation state of the metal decreases by 2

These modifications lead to the product's substructure, which is given by the second member of Eq. (3).

III

STUDY AND CLASSIFICATION OF REACTIONS

Mechanistic studies are of great importance in homogeneous catalysis. The active catalyst, often prepared *in situ*, is not easily isolated and, even when it is, can prove misleading (17). The comprehension of the catalytic process is therefore not an easy matter. In addition to producing kinetic measurements, the study and identification of the intermediates of the reaction can provide some support for certain mechanisms. Hence, it would be convenient for the chemist to have an initial idea of possible intermediates to be expected before carrying out the mechanistic study. TAMREAC performs in this way, for it contains most of the key reactions of organotransition metal chemistry and generates most of the possible reactivity sequences. In this section we present and discuss these key reactions.

Reactions have been studied and classified according to their type of mechanism, rather than by organic functional group or by metal. We divided them into two classes, (1) two-electron equivalent concerted reactions and (2) one-electron equivalent reactions. The first class contains two groups of reactions that we have studied separately, namely, reactions involving the coordination sphere of the metal and intermolecular reactions (external attacks). The reactions of coordination chemistry studied are two-electron equivalent concerted reactions of mononuclear systems, such as oxidative additions reductive eliminations, β- or α-eliminations, etc. (Table II). The second group, intermolecular reactions, concerns external nucleophilic or electrophilic attacks on the complex. The remaining class, one-electron equivalent reactions, contains one-electron transfers and radical reactions. Each class and group will be discussed in the following section. We shall give an inventory of mechanisms and present the evaluation rules that we have adopted for the different types of reactions.

A. Reactions Involving the Coordination Sphere of the Metal

Table II shows the reactions in coordination chemistry coded in TAMREAC. For each we give a name, a general description, and an example. We also give the most common characteristics of each reaction: the variation of the number of valence electrons (NVE), of the coordination number (CN), of the oxidation state (OS), and of the global charge of the complex (Q).

Each reaction type (oxidative addition, reductive elimination, etc.) was studied according to the electronic configuration (at this time only the even d^n configurations have been considered), the coordination number, and the coordination geometry. The matrices that we have composed for the evaluation (Matrices 1, 2, 3, 4, 5, 6, and 7; see also Section I,C), show the structure $d^n\text{-}ML_X$ for which the reaction is allowed or forbidden. We must note that, in most of the cases, *the rules that we present derive from theoretical studies found in the literature and that exceptions certainly exist.* Another difficulty in this reaction evaluation is the importance of the coordination geometry (*15b*), related to the spin state (low or high), the choice of which is particularly difficult.

TABLE II
REACTIONS IN COORDINATION CHEMISTRY

Reaction name	General description[a]	Example	ΔOS, ΔNVE, ΔCN, ΔQ[b]	Ref.
OXIDATIVE ADDITION				
1. Oxidative addition of cyclopropane	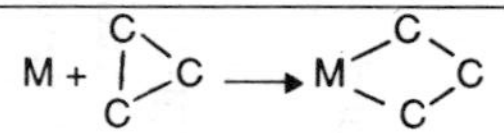 $M + \overset{C}{\underset{C}{\triangle}}C \rightarrow M\langle\overset{C}{\underset{C}{}}\rangle C$		+2,+2,+2,0	63,64
2. Oxidative addition of cyclobutane	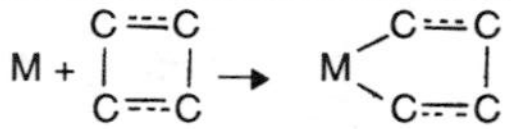 $M + \begin{matrix}C{=}C\\C{=}C\end{matrix} \rightarrow M\langle\begin{matrix}C{=}C\\C{=}C\end{matrix}$		+2,+2,+2,0	64,65
3. Oxidative addition of a polar addenda	 $M + \overset{A}{\underset{A}{\vDots}} \rightarrow M\langle\overset{A}{\underset{A}{}}$	 $M + \overset{O}{\underset{O}{\Vert}} \rightarrow M\langle\overset{O}{\underset{O}{}}$	+2,+2,+2,0	18a,66
4. Oxidative addition of a polar addenda	 $M + \overset{R}{\underset{V}{\vDots}} \rightarrow M\langle\overset{R}{\underset{V}{}}$	 $M + \overset{C}{\underset{Cl}{\vert}} \rightarrow M\langle\overset{C}{\underset{Cl}{}}$	+2,+2,+2,0	18a,66
5. Oxidative addition of allylic C-H	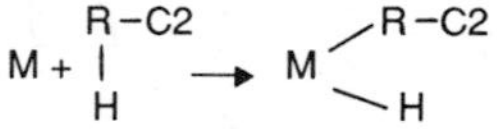 $M + \overset{R-C2}{\underset{H}{\vert}} \rightarrow M\langle\overset{R-C2}{\underset{H}{}}$	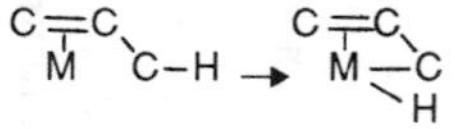 $\underset{M}{\overset{C{=}C}{\vert}}C{-}H \rightarrow \underset{M}{\overset{C{=}C}{\vert}}\langle\overset{C}{\underset{H}{}}$	+2,+2,+2,0	67
6. Oxidative addition of activated C-H	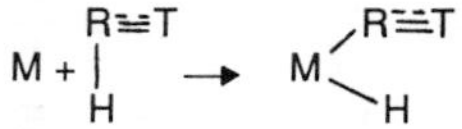 $M + \overset{R\equiv T}{\underset{H}{\vert}} \rightarrow M\langle\overset{R\equiv T}{\underset{H}{}}$	$M + H{-}C{\equiv}N \rightarrow H{-}M{-}C{\equiv}N$	+2,+2,+2,0	68
7. Oxidative addition of H_2	 $M + \overset{H}{\underset{H}{\vert}} \rightarrow M\langle\overset{H}{\underset{H}{}}$		+2,+2,+2,0	21

#	Reaction	Scheme		References
8.	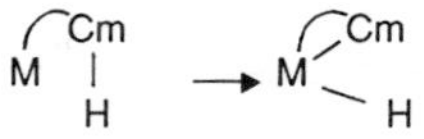 Cyclometallation	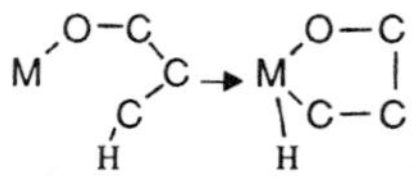 $M\overset{Cm}{\underset{H}{<}} \longrightarrow M\overset{Cm}{\underset{H}{<}}$	$+2,+2,+2,0$	69

REDUCTIVE ELIMINATION

#	Reaction	Scheme		References
9.	Reductive elimination	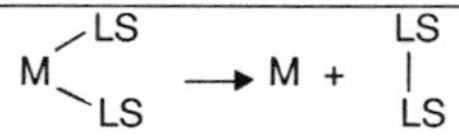 $M\overset{LS}{\underset{LS}{<}} \longrightarrow M + \overset{LS}{\underset{LS}{|}}$	$-2,-2,-2,0$	17,28

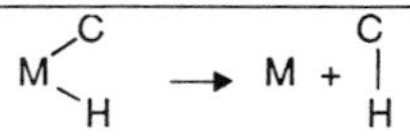

MIGRATORY- INSERTION

#	Reaction	Scheme		References
10.	Migratory insertion to unsaturated bond	$\overset{Z=Z}{\underset{M-LS}{}} \longrightarrow \overset{Z-Z}{\underset{M\ LS}{<}}$	$0,-2,-1,0$	44,45, 48,70
11.	Migratory insertion to carbonyl	$\underset{M-Z=Ht}{LS} \longrightarrow M-Z\overset{LS}{\underset{Ht}{<}}$	$0,-2,-1,0$	45e,46, 71
12.	Migratory insertion to carbene	$\underset{M=CF}{LS} \longrightarrow M-CF\overset{LS}{}$	$0,-2,-1,0$	56,72

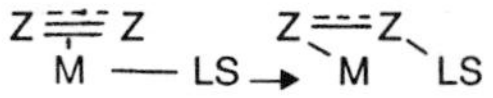

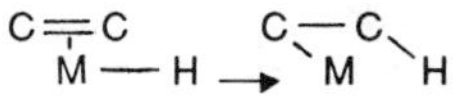

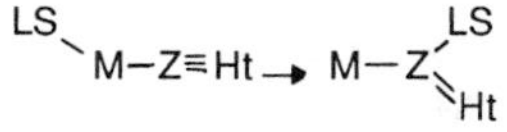

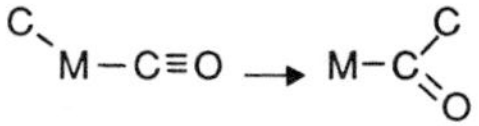

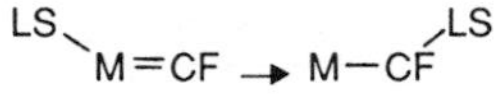

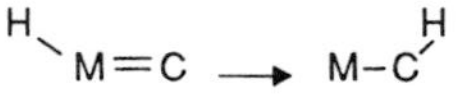

EXTRUSIONS

#	Reaction	Scheme		References
13.	α-Elimination	$\overset{A}{\underset{M-C}{}} \longrightarrow \overset{A}{\underset{M=C}{}}$	$0,+2,+1,0$	56,72
14.	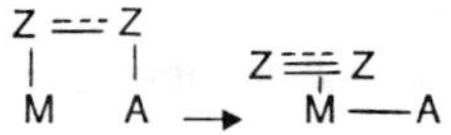 β-Elimination	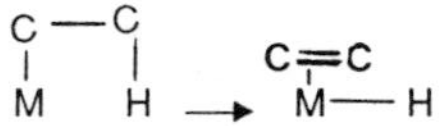 $\overset{Z=Z}{\underset{M\ A}{}} \longrightarrow \overset{Z=Z}{\underset{M-A}{}}$	$0,+2,+1,0$	24b,27, 48,52,53, 58,73

(continued)

177

TABLE II (continued)

Reaction name	General description[a]	Example	ΔOS, ΔNVE, ΔCN, ΔQ[b]	Ref.
15. α-Cleavage of C-C in metallacycles	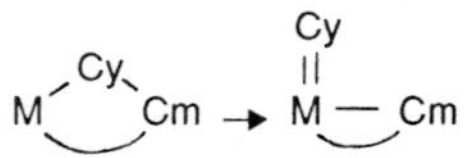 $M\diagdown\overset{Cy}{\underset{Cm}{C_y\text{-}Cm}} \rightarrow M-Cm$	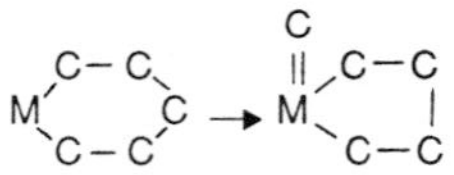 (structure)	0,+2,+1,0	29,39, 43,74
16. β-Cleavage of C-C in metallacycles	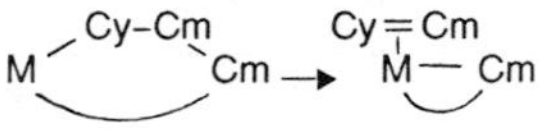 $M\diagdown\overset{Cy=Cm}{Cy\text{-}Cm} \rightarrow M-Cm$	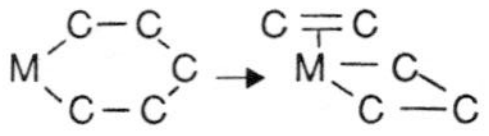 (structure)	0,+2,+1,0	29,35, 74
17. Decarbonylation	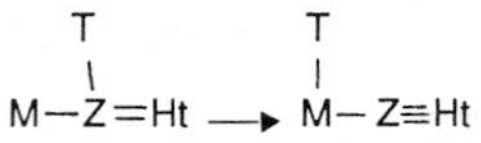 $M-Z=Ht \rightarrow M-Z\equiv Ht$	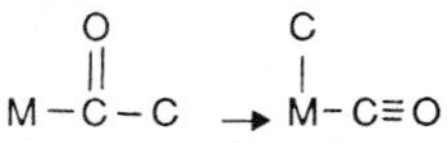 $M-C-C \rightarrow M-C\equiv O$	0,+2,+1,0	54,57, 75
COUPLING REACTIONS				
18. Oxidative coupling of unsaturated bonds	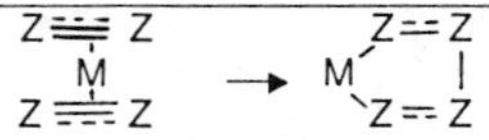 $\overset{Z=Z}{\underset{Z=Z}{M}} \rightarrow M\diagdown\overset{Z-Z}{\underset{Z-Z}{}}$	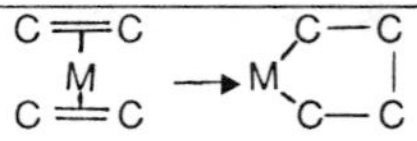 (structure)	+2,-2,0,0	29,35
19. Oxidative coupling affording a 4-membered metallacycle	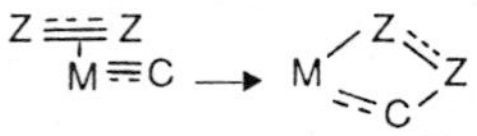 $\overset{Z=Z}{M=C} \rightarrow M\diagdown\overset{Z-Z}{\underset{C}{}}$	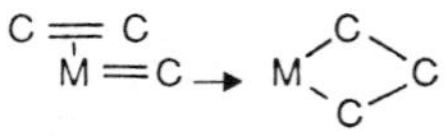 (structure)	+2,-2,0,0	39,43
20. Coupling of two carbene ligands	$M\overset{=C}{\underset{=C}{}} \rightarrow M-\overset{C}{\underset{C}{\|\|}}$	(structure)	0,-2,-1,0	59,76
21. Coupling of two carbyne ligands	 $M\overset{\equiv C}{\underset{\equiv C}{}} \rightarrow M-\overset{C}{\underset{C}{\|\|\|}}$	(structure)	0,-2,-1,0	59a

#	Reaction			
22. Coupling of two carbonyl ligands[c]	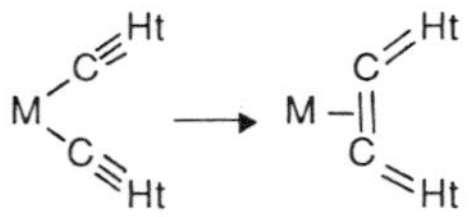 (structure)	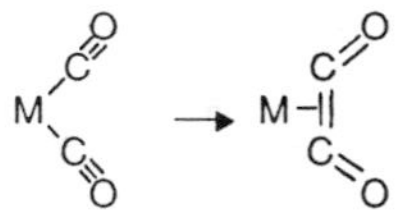 (structure)	0,-2,-1,0	59a,77
23. Oxidative coupling of two carbonyl ligands[c]	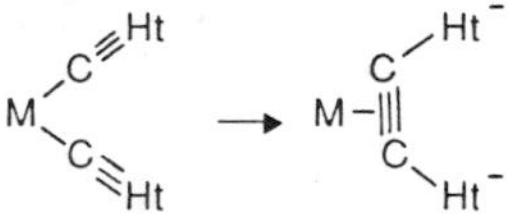 (structure)	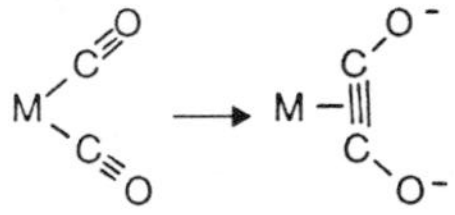 (structure)	+2,-2,-1,0	59a,77
24. Coupling between a carbonyl and a carbene (or carbyne)	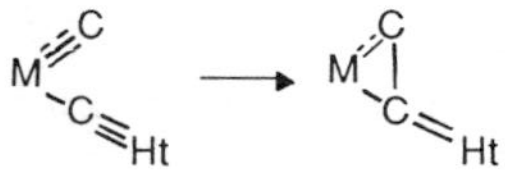 (structure)	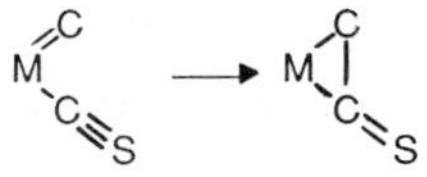 (structure)	+2,-2,-1,0	77a,77b

DECOUPLING REACTIONS

#	Reaction			
25. Reductive decoupling of a 5-membered metallacycle	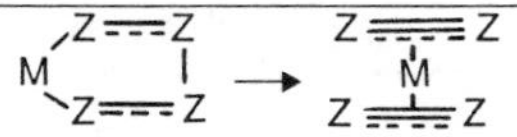 (structure)	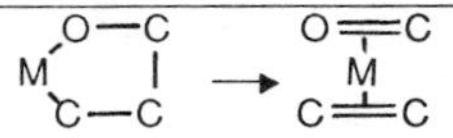 (structure)	-2,+2,0,0	29,38
26. Reductive decoupling of a 4-membered metallacycle	 (structure)	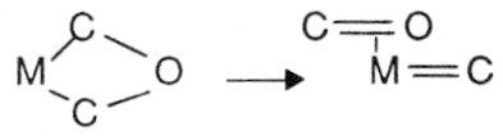 (structure)	-2,+2,0,0	39,43
27. Bis-alkylidene form of an olefin	 (structure)	(structure)	0,+2,+1,0	59,76
28. Bis-alkylidyne form of an alkyne	 (structure)	(structure)	0,+2,+1,0	59a

LIGAND ASSOCIATIONS

#	Reaction		
29. Neutral ligand association	$M + L \longrightarrow M{-}L \qquad M + H_2O \longrightarrow M{-}OH_2$	0,+2,+1,0	78

(continued)

179

TABLE II (continued)

Reaction name	General description[a]	Example	ΔOS, ΔNVE, ΔCN, ΔQ[b]	Ref.
30. Unsaturated bond association	$M + \overset{Z}{\underset{Z}{|||}} \longrightarrow M-\overset{Z}{|||}Z$	$M + \overset{O}{\underset{C}{||}} \longrightarrow M-\overset{O}{||}C$	0,+2,+1,0	79
31. Anion association	$M + T^- \longrightarrow M-T$	$M + Cl^- \longrightarrow M-Cl$	0,+2,+1,-1	61,80
32. Cation association	$M + T^+ \longrightarrow M-T$	$M + H^+ \longrightarrow M-H$	+2,0,+1,+1	81,60b,82
LIGAND DISSOCIATION				
33. Dissociation of unsaturated bond	$M-\overset{Z}{|||}Z \longrightarrow M + \overset{Z}{\underset{Z}{|||}}$	$M-\overset{C}{||}C \longrightarrow M + \overset{C}{\underset{C}{||}}$	0,-2,-1,0	83
34. Neutral ligand dissociation	$M-L \longrightarrow M + L$	$M-P(Ph)_3 \longrightarrow M + P(Ph)_3$	0,-2,-1,0	83
35. Hydride abstraction	$M-H + T^+ \longrightarrow M + T-H$	$M-H + R^+ \longrightarrow M + RH$	0,-2,-1,+1	60a,60b
36. Proton abstraction from a metal	$M-H + LB \longrightarrow M + LB-H$	$M-H + HO^- \longrightarrow M + H_2O$	-2,0,-1,-1	60
37. Anion dissociation	$M-Ht \longrightarrow M + Ht^-$	$M-NH_2 \longrightarrow M + NH_2^-$	0,-2,-1,+1	61,80
OTHER REACTIONS				
38. Oxidative form of an unsaturated bond[c]	$M-\overset{T}{\underset{T}{|||}} \longrightarrow M{\underset{T}{\overset{T}{\diagdown\diagup}}}$	$M-\overset{O}{\underset{O}{||}} \longrightarrow M{\underset{O}{\overset{O}{\diagdown\diagup}}}$	+2,0,+1,0	79
39. Reductive form of an unsaturated bond[c]	$M{\underset{T}{\overset{T}{\diagdown\diagup}}} \longrightarrow M-\overset{T}{|||}T$	$M{\underset{C}{\overset{C}{\diagdown\diagup}}} \longrightarrow M-\overset{C}{|||}C$	-2,0,-1,0	79

40. Reductive form of a carbonyl[c]	$M=Z=Ht \rightarrow M-Z\equiv Ht$	$M\equiv C=O \rightarrow M-C\equiv O$	-2,0,-1,0	84
41. Carbene to olefin (or alkyne) reaction	 $M=\overset{C\,F}{\underset{Z}{C}} \rightarrow M-\overset{CF}{\underset{Z}{\parallel C}}$	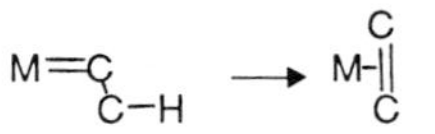 $M=\overset{C}{\underset{C-H}{C}} \rightarrow M-\overset{C}{\underset{C}{\parallel}}$	0,0,0,0	85
42. Olefin (or alkyn) to carbene reaction	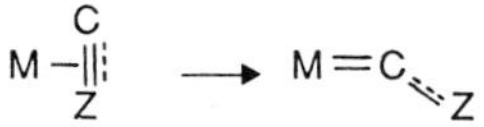 $M-\overset{C}{\underset{Z}{\parallel C}} \rightarrow M=C \overset{}{\underset{Z}{=}}$	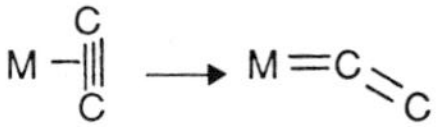 $M-\overset{C}{\underset{C}{\parallel}} \rightarrow M=C=C$	0,0,0,0	85
43. π-allyl rearrangement	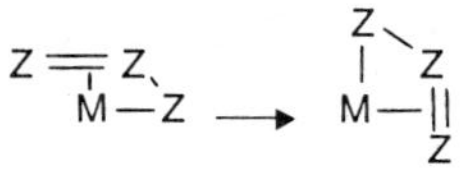 $Z=\overset{Z}{\underset{M-Z}{}} \rightarrow M-\overset{Z}{\underset{N}{\parallel}}$	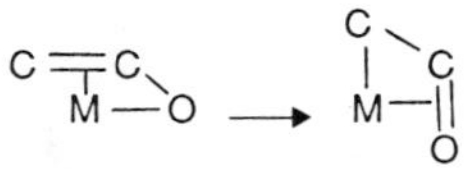 $C=\overset{C}{\underset{M-O}{}} \rightarrow M-\overset{C}{\underset{O}{\parallel}}$	0,0,0,0	86

[a] The meaning of the symbols A, V, C2,..., is given in Table I, p. 170.
[b] ΔOS = variation of the oxidation state of the metal, ΔNEV = variation of the number of valence electrons, ΔCN = variation of coordination number, ΔQ = variation of the global charge.
[c] Two different representations of the same reaction.
[d] This is not a reaction but a canonical representation.

1. *Oxidative Addition*

Reactions 1 through 8 of Table II represent different types of mononuclear two-electron equivalent oxidative additions. A first condition for the feasibility of these reactions is the presence of a vacant coordination site in the complex. Matrix 1 shows the structures for which the oxidative addition is possible.

d^n / ML_x	d^0	d^2	d^4	d^6	d^8	d^{10}
ML_1	▨	▨	▨	▨	▨	
ML_2	▨	▨	▨	▨	NO	YES
ML_3	▨	▨	▨	▨		YES
ML_4	▨	▨			if SPL	NO
ML_5	▨			if SPY	NO	▨
ML_6	NO	YES	YES	NO	▨	▨
ML_7	NO		NO	▨	▨	▨
ML_8	NO	NO	▨	▨	▨	▨
ML_9	NO	▨	▨	▨	▨	▨

MATRIX 1.   Evaluation matrix for oxidative addition. Each element of the matrix corresponds to a d^n-ML_x complex and answers the question, "Is such a reaction possible for this d^n-ML_x structure?" Empty boxes signify that no examples were found; shaded boxes correspond to complexes with less than 12 or more than 20 valence electrons. SPL, Td, SPY, and TBP (SPL = square planar, Td = tetrahedral, SPY = square pyramidal, TBP = trigonal bipyramidal) correspond to the geometry which is necessary in each case.

d^{10} **Configuration.** The oxidative addition is allowed for many complexes of this configuration (*18a–18d*). Apparently, the reaction is easier for ML_2 complexes than for ML_3 complexes (*19*), and leads to a d^8-ML_4 square planar complex.

d^8 **Configuration.** Most of the studies on oxidative additions were carried out on complexes of this configuration (*18a–18d*). In the case of four-coordinate complexes, only those having a low spin and a square planar geometry will react. We found no examples of such reactions on d^8-ML_4 tetrahedral complexes. Unsaturated ML_3 complexes react via an associative pathway. There is no example found of direct oxidative addition on ML_3 complexes.

d^6 **Configuration.** There are not many studies of oxidative addition on d^6 complexes (20). Apparently, for five-coordinated complexes, if the reaction occurs the geometry must be square pyramidal (low spin state) (21).

d^4 **Configuration.** The examples that we have found concern specifically the d^4-ML_2 complexes, where L is a cyclopentadienyl ligand. According to Pearson (22a), oxidative addition to ML_2 complexes is allowed only for the d^{10} configuration. This is true when the geometry is linear but bis-(cyclopentadienyl) unsaturated complexes have a bent geometry, which allows the formation, by oxidative addition, of stable tetrahedral complexes (22b–22d).

d^2 **Configuration.** As for the d^4 configuration, the bis(cyclopentadienyl) complexes allow oxidative additions in order to form d^0-ML_4 stable unsaturated complexes (NVE = 16) (22b–22d).

d^0 **Configuration.** Oxidative addition should be forbidden for these complexes, since the central metal does not have any electrons in its d orbitals.

2. Reductive Elimination

The general description of reductive elimination is given by Reaction 9 in Table II. The evaluation scheme is shown in Matrix 2. This matrix derives from

d^n / ML_x	d^0	d^2	d^4	d^6	d^8	d^{10}
ML_1						NO
ML_2					YES	NO
ML_3				YES	NO	NO
ML_4			NO	if SPL	if SPL	NO
ML_5		NO	NO	NO	if SPY	
ML_6	NO	YES	YES	YES		
ML_7	NO	NO	NO			
ML_8	NO	YES				
ML_9	NO					

MATRIX 2.   Evaluation matrix for the reductive elimination. For explanation, see Matrix 1.

a theoretical study of the concerted reductive elimination of two alkyl ligands
(23). We shall use this information for all concerted reactions of this type.

d^{10} **Configuration.** Reductive elimination is forbidden for the d^{10} systems.

d^8 **Configuration.** Many examples have been reported (24a–24c). Four-coordinate complexes must be square planar (24a–24c) and five-coordinate complexes must be square pyramidal (23,24a–24c).

d^6 **Configuration.** Complexes allowing this reaction are ML_3, ML_4 (square planar) (25), and ML_6 (octahedral) (26). As in most concerted reactions, the spin state and the spin conservation are of great importance. Symmetry restrictions are applied for d^n systems with $1 \leq n \geq 8$, and, when $n \geq 6$, the restrictions concern the products also. Norton (27a), has studied this problem in the case of d^6 osmium complexes: the product, $Os(CO)_4$, formed from the reductive elimination of ethane in the $Os(CO)_4(CH_3)_2$ complex, must be in a singlet state, otherwise the reaction is forbidden. Apparently, this complex is paramagnetic, as is $Fe(CO)_4$ (27b), and the triplet state is the most stable.

d^4 **Configuration.** The situation is quite confused for the d^4 systems. According to Akermark (23), reductive elimination is allowed only for octahedral complexes.

d^2 **Configuration.** Some d^2 examples, found in the literature, show that reductive elimination is allowed for tetrahedral (22b,28) and octahedral systems (23), but there is no evidence about the concertedness of the mechanism.

d^0 **Configuration.** For the d^0 systems, concerted reductive elimination is symmetry forbidden, since the ground-state product never possesses a d^2 low-spin configuration (23).

3. Oxidative Coupling Affording a Five-Membered Metallacycle

The general description of this reaction is given in Table II by Reaction 18. Many investigations, theoretical as well as experimental, have been carried out on this type of coupling (29a). The decrease in the number of valence electrons makes the reaction more plausible for saturated systems. The geometry of the complex is also very important, since the orientation of the olefins or alkynes is the factor which determines the feasibility of the reaction.

d^{10} **Configuration.** Coupling reactions have been found for the ML, ML_2, ML_3, and ML_4 systems (30). In the case of the ML_4 system the geometry must be tetrahedral (29b).

d^8 **Configuration.** Coupling takes place on five-coordinate, trigonal bipyramidal systems (29c,31). Four-coordinate saturated systems can give the coupling product if the geometry is tetrahedral (e.g., the case of d^8-ML_4, where one L is cyclopentadienyl) (32).

d^6 **Configuration.** Complexes of the d^6 configuration are often catalysts for metathesis reactions. We have not found any examples where formation of a five-membered intermediate was identified.

d^4 **Configuration.** An intermediate, derived from a d^4 coupling reaction, was proposed for the oligomerization of olefins (*33*).

d^2 **Configuration.** The d^2 reaction is symmetry allowed for saturated bis(cyclopentadienyl) complexes (*22b,29a,34*). The formation of a metallacycle was proposed for the Ziegler–Natta polymerization of ethylene with $Ti(Cl)_3(C_2H_4)_2Me$, which is a d^2-ML_6 complex (*35*).

d^0 **Configuration.** The reaction will be forbidden, for there are no electrons in the d orbitals of the metal.

Matrix 3 shows a simplified representation of the above discussion.

d^n / ML_x	d^0	d^2	d^4	d^6	d^8	d^{10}
ML_1						YES
ML_2						YES
ML_3						YES
ML_4					if Td	if Td
ML_5			YES	if TBY		
ML_6	NO	YES	YES			
ML_7	NO	YES				
ML_8	NO	YES				
ML_9	NO					

MATRIX 3. Evaluation matrix for the oxidative coupling reaction affording a five-membered metallacycle. For explanation, see Matrix 1.

4. *Fragmentation of a Five-Membered Metallacycle*

Fragmentation of a five-membered metallacycle is the reverse reaction of the previous coupling between olefins or alkynes. Reaction 25, in Table II, illustrates this transformation.

d^n / ML_x	d^0	d^2	d^4	d^6	d^8	d^{10}
ML_1						NO
ML_2						NO
ML_3						NO
ML_4					if Td	NO
ML_5				if TBY	NO	
ML_6	YES			NO		
ML_7	YES		NO			
ML_8	YES	NO				
ML_9	NO					

MATRIX 4. Evaluation matrix for the reductive decoupling of a five-membered metallacycle. For explanation, see Matrix 1.

d^{10} **Configuration.** No examples were found for the d^{10} system. If we consider this fragmentation as an extrusion [Eq. (5)], the reaction should be symmetry forbidden (*44*).

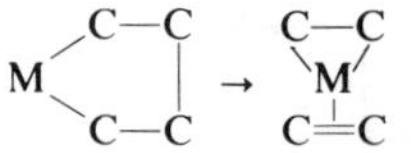

$$\begin{array}{cc} M\begin{array}{c} \diagup \overset{C-C}{\underset{C-C}{|}} \diagdown \end{array} & \rightarrow & \overset{C-C}{\underset{C=C}{M}} \end{array} \tag{5}$$

In order to respect the coherence of the evaluation scheme, this reaction is considered to be forbidden in the evaluation module.

d^8 **Configuration.** The d^8 systems have been studied (*29a–29h,36*), and the necessary geometry is the tetrahedral one. According to Hoffmann and co-workers (*36*), if the fragmentation is allowed, the reductive elimination (e.g., formation of cyclobutane) is forbidden. This rule is coherent with our evaluation scheme, since reductive elimination is forbidden for the tetrahedral complex.

d^6 **Configuration.** Apparently, the reaction is allowed for d^6-ML_5 complexes (*31,37*).

d^4 **Configuration.** No examples were found for these systems. Metallacyclopentane intermediates seem to be involved in metathesis processes where the fragmentation of the metallacycle occurs by other pathways (*38*) (see Reaction 15, Table II).

d^2 **Configuration.** No d^2 examples, were found.

d^0 **Configuration.** Most of the examples found concern the reverse reaction of the oxidative coupling in bis(cyclopentadienyl) complexes (*22b,34a*).

Matrix 4 represents the simplified scheme for the evaluation of this fragmentation reaction.

5. Oxidative Coupling Affording a Four-Membered Metallacycle

The general description of this oxidative coupling reaction is given by Reaction 19 in Table II. These reactions are involved in the metathesis of olefins or alkynes (*39*). We have not found any rules allowing or forbidding the reaction. Matrix 5 shows only the structures of complexes found in the literature which give such reactions: d^{10}-M (M = Ni) (*40*), d^8-ML$_5$ (*41*), d^6-ML$_6$ (*42*), and d^2-ML$_5$ (*43*). For the d^0 and d^4 systems we were unable to find any examples.

d^n / ML$_x$	d^0	d^2	d^4	d^6	d^8	d^{10}
ML$_1$						
ML$_2$						
ML$_3$						
ML$_4$						YES
ML$_5$		YES			YES	
ML$_6$	NO	YES	YES	YES		
ML$_7$	NO	YES				
ML$_8$	NO	YES				
ML$_9$	NO					

MATRIX 5. Evaluation matrix for the oxidative coupling affording a four-membered metallacycle. For explanation, see Matrix 1.

6. Fragmentation of Four-Membered Metallacycles

The general description of the fragmentation reaction is given by Reaction 26 in Table II. We have come across few data in the literature concerning studies on this mechanism and thus we do not give any evaluation matrix. The examples found are given in the references on the reverse coupling reaction (*39–43*). The reaction increases the number of valence electrons and would take place only for unsaturated systems.

7. *Migratory Insertion*

Different types of migratory insertions are exemplified by reactions 10, 11, and 12 in the Table II. An evaluation scheme is given in Matrix 6. In the present study we consider this reaction according to its general mechanism, shown in Eq. (6), where L is a neutral ligand (olefin, carbon monoxide, carbene, etc.) which becomes X' (anionic σ-donor ligand). X is also an anionic σ-donor ligand.

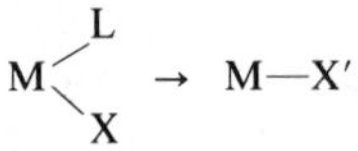

$$M\overset{\displaystyle L}{\underset{\displaystyle X}{\big<}} \;\rightarrow\; M\text{---}X' \tag{6}$$

d^n / ML_x	d^0	d^2	d^4	d^6	d^8	d^{10}
ML_1	░	░	░	░	░	NO
ML_2	░	░	░	░		NO
ML_3	░	░	░			NO
ML_4	░	░	░	if Td	YES	NO
ML_5	░				YES	░
ML_6				YES	░	░
ML_7			YES	░	░	░
ML_8	YES	YES	░	░	░	░
ML_9	YES	░	░	░	░	░

MATRIX 6. Evaluation matrix for the migratory insertion reaction. For explanation, see Matrix 1.

d^{10} **Configuration.** Migratory insertion is a symmetry-forbidden reaction in all d^{10} systems (*44*).

d^8 **Configuration.** The reaction is allowed for the d^8 systems. The ML_4 complexes often give such reactions via an associative pathway (*45*).

d^6 **Configuration.** The migration occurs in d^6-ML_6 complexes (*46*). The presence of an incoming ligand is often a necessary condition in order to stabilize the square pyramidal product. Moreover, the formation of a five-coordinate saturated trigonal bipyramidal complex is forbidden by the spin conservation rules (*47*). The reaction also occurs in tetrahedral saturated d^6-ML_4 complexes where one L is cyclopentadienyl (*48*).

d^4 **Configuration.** Examples of migrations to carbonyl involving the d^4 configuration have been found, for example, in $CpM(CO)_3R$ complexes where M is Cr, Mo, or W (*49*).

d^2 **Configuration.** The saturated systems of the d^2 configuration are often the bis(cyclopentadienyl) complexes. With these systems, the migratory insertion is slow, and could provide an explanation concerning the relative stability of complexes such as $Cp_2Mo(C_2H_4)H^+$, $Cp_2Nb(C_2H_4)H$, and $Cp_2Ta(CO)R$ (*29d*). As for insertions in d^6 complexes, the starting materials and the products may have different ground spin states, and this could be the reason why the reaction is slow (*47*). Sometimes the reaction occurs in the presence of an incoming ligand, involving a second-order mechanism.

d^0 **Configuration.** Complexes of d^0 configuration undergo migratory insertions more easily than d^2 systems. Indeed, such unsaturated d^0 complexes (16 valence electrons) have a smaller activation energy than the d^2 saturated complexes (*22b,29d*). Some of the soluble catalysts of the Ziegler–Natta type, $(Cp)_2M$, seem to involve d^0 intermediates, such as Cp_2TiR^+, which give fast migrations to olefins, providing polymerization products (*29a,50*).

8. Extrusion Reactions

Extrusions are the reverse of migratory insertion reactions. These are α- and β-eliminations, decarbonylations, and fragmentations of metallacycles with more than five members. Their general descriptions are given in Table II by Reactions 13 through 17. The evaluation table is given by Matrix 7.

d^{10} **Configuration.** As for the migrations, the extrusions are symmetry forbidden in d^{10} systems.

d^8 **Configuration.** Complexes of this configuration are often olefin isomerization catalysts, for example, $HNiL_3^+$ (*51*), where insertions and extrusions are involved in the mechanism. Many other examples of β-eliminations are known with Ni(II) (*52*) and Pt(II) (*53*).

d^6 **Configuration.** Studies on the d^6 systems especially concern decarbonylation reactions (*54*) and β-eliminations have also been studied (*48*).

d^4 **Configuration.** Apparently, complexes of d^4 configuration favor migration reactions. An explanation could be provided by the concept of hard and soft acids and bases. Most of the metals of these systems can be considered hard acids, and, if all the other parameters are of comparable size (*55a*), could

d^n / ML_x	d^0	d^2	d^4	d^6	d^8	d^{10}
ML_1						NO
ML_2						NO
ML_3					YES	NO
ML_4					YES	NO
ML_5				YES	NO	
ML_6		YES		NO		
ML_7	YES	YES	NO			
ML_8	YES	NO				
ML_9	NO					

MATRIX 7. Evaluation matrix for the extrusion reaction (reverse of the migratory insertion). For explanation, see Matrix 1.

favor migration reactions rather than extrusion when there is formation of a soft ligand from a hard one (55b).

d^2 **Configuration.** As for the d^4 systems, d^2 extrusion reactions are unfavorable. Therefore, they must be involved in polymerization reactions. Some α-eliminations are also known for these systems (56).

d^0 **Configuration.** Certain reversible carbonyl insertions are known for d^0 complexes (57). Also, the fragmentation of tantalacyclopentanes involves a β-elimination before the reductive elimination which leads to olefins (58).

9. Coupling Reactions Between Two π-Ligands

These reactions concern coupling between two methylene ligands (Reaction 20, Table II), two methyne ligands (Reaction 21, Table II), or two other π systems (Reactions 22–24 in Table II). According to a theoretical study (59a–59d), the methylene coupling would be allowed for the d^2 and d^{10} configuration systems and the methyne coupling for d^6, d^8, and d^{10} systems. However, the nature of the metal, the ligands, and the substituent groups on the methylene or methyne carbon atoms could affect the energy barriers of these reactions and make them possible for other electron configuration systems as well. The program proposes these couplings without any special evaluation.

10. *Ligand Associations and Dissociations*

The different types of ligand interactions with the metal are given in Table II (Reactions 29–37). The only evaluation carried out by the program concerns the number of valence electrons. If the complex is saturated, association reactions are forbidden. Some of these reactions, such as cation (e.g., proton) or anion (e.g., hydride) associations and dissociations (Reactions 31, 32, 35, 36, and 37), modify the global charge of the complex. However, they are not considered as acid–base interactions but as coordination reactions (*60a–60e,61*).

11. *Other Reactions*

In order to cover all the possible reactions and structures of the complexes, we have also introduced canonical forms (Reactions 38–40 in Table II) and rearrangements (Reactions 41–43) of some ligands. These reactions do not modify the number of valence electrons, and their choice is left up to the chemist's decision.

B. *Intermolecular Reactions*

The second class of TAMREAC's inventory includes the reactions between the coordinated ligands and external organic reagents. We divide these reactions into nucleophilic and electrophilic attacks and consider them as acid–base interactions. Table III presents their general description. The nucleophilic attacks are either addition reactions to unsaturated coordinated ligands (Reactions 44–46) or abstraction reactions (usually a proton abstraction, Reactions 47–50). The electrophilic attacks are similarly addition reactions (Reactions 51 and 52) and abstraction reactions (usually a hydride abstraction, Reactions 53–59). Reactions 60 to 63 represent some other intermolecular reactions.

The evaluation is based on two rules which correspond to extreme situations (*87*) for which many exceptions are known (*88a,88b*):

Rule 1: Cationic complexes or complexes with a metal in a high oxidation state (≥ 2) are susceptible only to nucleophilic attack.

Rule 2: Anionic complexes or complexes with a metal in a low oxidation state (0, +1) are susceptible only to electrophilic attack.

TABLE III
INTERMOLECULAR REACTIONS

Reaction name	General description[a]	Example	ΔOS, ΔNVE, ΔCN, ΔQ[b]	arr[c] Ref.
NUCLEOPHILIC ADDITIONS				
44. External nucleophilic attack to unsaturated bond	$M-\overset{Z}{\underset{Z}{\|}} + Nu \rightarrow M-Z\cdots Z-Nu$	$M-\overset{C}{\underset{C}{\|}} + MeO^{-} \rightarrow M-C\cdots C-OMe$	0,0,0,-1	88,89
45. External nucleophilic attack to carbonyl	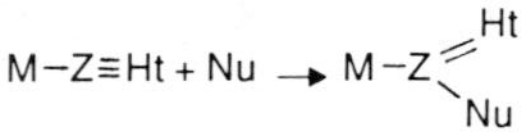 $M-Z\equiv Ht + Nu \rightarrow M-Z{=}Ht(Nu)$	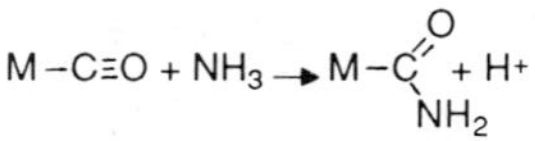 $M-C\equiv O + NH_3 \rightarrow M-C(\!\!\overset{O}{\underset{NH_2}{}}\!\!) + H^{+}$	0,0,0,-1	90
46. External nucleophilic attack to carbene or carbyne	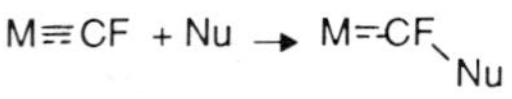 $M\equiv CF + Nu \rightarrow M{=}CF(Nu)$	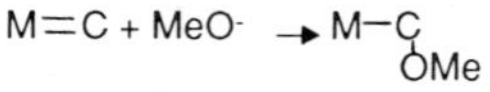 $M{\equiv}C + MeO^{-} \rightarrow M-C(OMe)$	0,0,0,-1	91
NUCLEOPHILIC ABSTRACTIONS				
47. Proton abstraction from a β position	$M-\overset{Z}{\underset{Z-H}{}} + LB \rightarrow M-\overset{Z}{\underset{Z}{\|}} + LB\text{-}H$	$M-\overset{C}{\underset{C-H}{}} + HO^{-} \rightarrow M-\overset{C}{\underset{C}{\|}} + H_2O$	-2,-2,0,-1	92
48. Proton abstraction from an α position	$M-\overset{C}{\underset{H}{}} + LB \rightarrow M{=}C + LB\text{-}H$	$M-\overset{C}{\underset{H}{}} + HO^{-} \rightarrow M{=}C + H_2O$	-2,-2,0,-1	93
49. Allylic proton abstraction	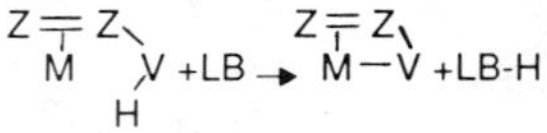 $M(\overset{Z=Z}{\underset{H}{\diagdown\diagup}}) + LB \rightarrow M\text{-}V + LB\text{-}H$	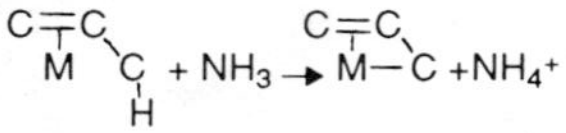 $M(\overset{C\equiv C}{\underset{H}{\diagdown\diagup}}) + NH_3 \rightarrow M-C(C\equiv C) + NH_4^{+}$	0,+2,+1,-1	94
50. Proton abstraction from a neutral ligand	$M-L-H + LB \rightarrow M-L + LB\text{-}H$	$M-OH_2 + HO^{-} \rightarrow M-OH + H_2O$	0,0,0,-1	95

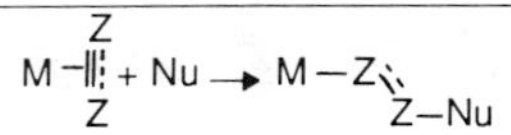

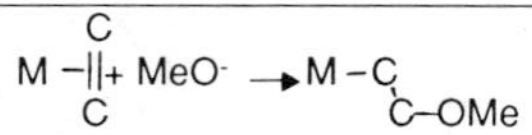

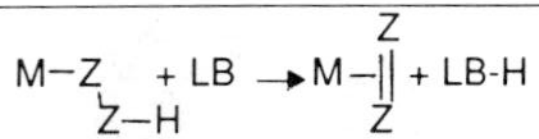

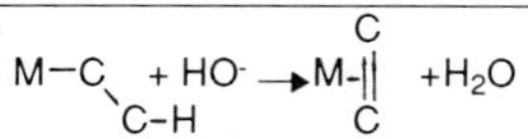

ELECTROPHILIC ADDITIONS

No.	Reaction	Example	Oxidation state	Ref.
51.	Electrophilic addition to an unsaturated bond	$M-\overset{Z}{\underset{Z}{\|\|\|}} + T^+ \longrightarrow M-Z\overset{}{\underset{Z-T}{\|\|}}$	+2,-2,0,+1	96
52.	Electrophilic addition to carbene or carbyne	$M\equiv CS + T^+ \longrightarrow M\approx CS-T$ $M=\overset{C}{\|\|}_{C} + H^+ \longrightarrow M-\overset{}{\underset{C-H}{C}}$	+2,-2,0,+1	97,102

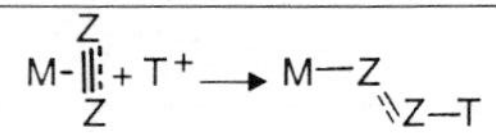

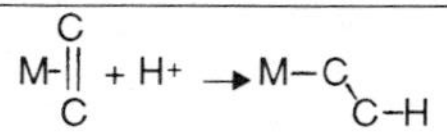

ELECTROPHILIC ABSTRACTIONS

No.	Reaction	Example	Oxidation state	Ref.
53.	Protonolysis	$M-LS + H^+ \longrightarrow M + LS\text{-}H$ $M-CH_3 + H^+ \longrightarrow M + CH_4$	0,-2,-1,+1	98
54.	Hydride or heteroatom abstraction from a β-position	$M-\overset{Z}{\underset{C-A}{}} + T^+ \longrightarrow M-\overset{Z}{\underset{C}{\|\|}} + T\text{-}A$ $M-C\overset{Ph_3C^+}{\underset{C-H}{\longrightarrow}} M-\overset{C}{\underset{C}{\|\|}} + Ph_3CH$	0,0,0,+1	99
55.	Hydride or heteroatom abstraction from an α-position	$M-\overset{}{\underset{A}{C}} + T^+ \longrightarrow M=C + T\text{-}A$ $M-\overset{}{\underset{OEt}{C}} + H^+ \longrightarrow M=C + EtOH$	0,0,0,+1	100
56.	Abstraction of an allylic hydride or allylic heteroatom	$M\overset{Z=Z}{\underset{C}{\longrightarrow}}\overset{Z=Z}{\underset{A}{}} + T^+ \longrightarrow M-C + T\text{-}A$ $M\overset{C=C}{\underset{OEt}{\longrightarrow}}C + H^+ \longrightarrow M-C + EtOH$	+2,0,+1,+1	101
57.	Proton abstraction from a "Fischer"-type carbene	$M\equiv CF + LB \longrightarrow M\equiv CF + LB\text{-}H$ $M=C\overset{}{\underset{C-H}{}} + R^- \longrightarrow M=C\overset{}{\underset{C^-}{}} + RH$	0,0,0,0	97,102

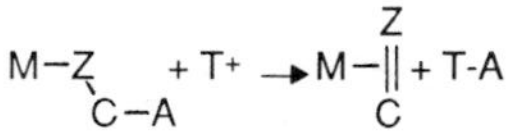

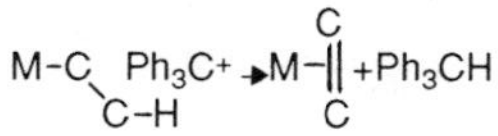

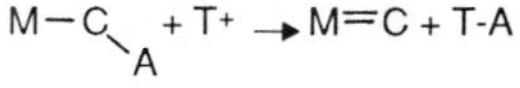

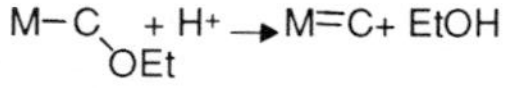

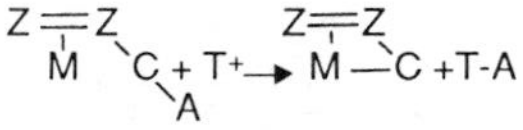

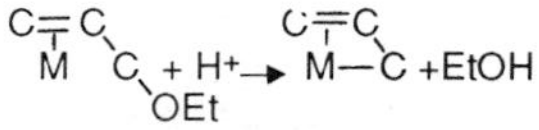

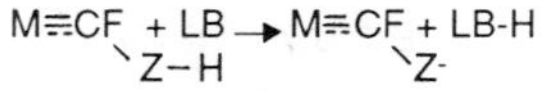

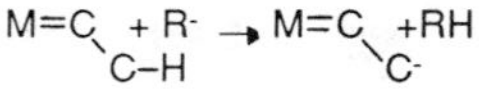

(continued)

TABLE III (continued)

Reaction name	General description[a]	Example	ΔOS, ΔNVE, ΔCN, ΔQ[b]	Ref.
58. Proton abstraction from a "Schrock" type carbene	$M=CS(H) + LB \rightarrow M\equiv CS + LB\text{-}H$	$M=C(H) + R^- \rightarrow M\equiv C + RH$	0,+2,0,0	97,102
59. Heteroatom abstraction from a "Fischer"-type carbene	$M=CF(Ht) + T^+ \rightarrow M\equiv C + T\text{-}Ht$	$M=C(OEt) + H^+ \rightarrow M\equiv C + EtOH$	0,0,0,0	97,102
OTHER REACTIONS				
60. Nucleophilic substitution by a metal	$M + El\text{-}Ht \rightarrow M\text{-}El + Ht$	$M\text{-}C(=O) \rightarrow M=C(O^-)$	+2,0,+1,+1	103
61. Solvolysis of acyl complexes	$M\text{-}C(=O) + Nu\text{-}H \rightarrow M\text{-}H + Nu\text{-}C(=O)$	$M\text{-}C(=O) + MeOH \rightarrow M\text{-}H + MeO\text{-}C(=O)$	0,0,0,0	104
62. Nucleophilic substitution on a "Fischer" type carbene	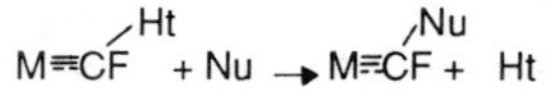 $M\equiv CF(Ht) + Nu \rightarrow M\equiv CF(Nu) + Ht$	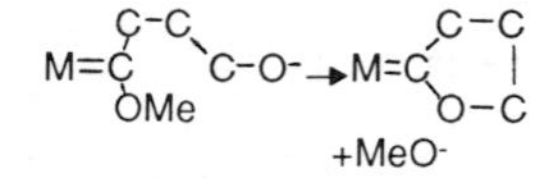 $M=C(OMe)\text{-}C\text{-}C\text{-}O^- \rightarrow M=C(O\text{-}C)\text{-}C\text{-}C + MeO^-$	0,0,0,0	97,102
63. Diels-Alder with a vinyl-carbene	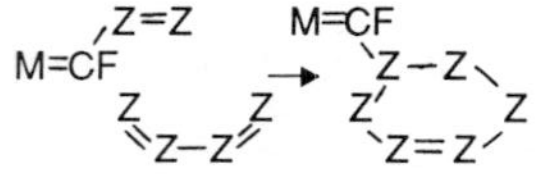 $M=CF(Z=Z)(Z\text{-}Z) + (Z=Z)(Z=Z) \rightarrow M=CF(\cdots)$	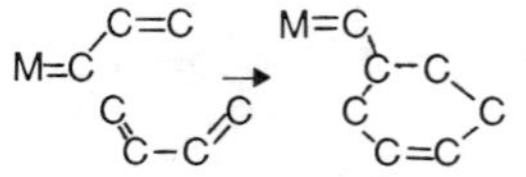 $M=C\text{-}C=C + C=C\text{-}C=C \rightarrow M=C\cdots$	0,0,0,0	105

[a] The meaning of the symbols A, V, C2..., is given in Table I.
[b] ΔOS = variation of the oxidation state of the metal, ΔNEV = variation of the number of valence electrons, ΔCN = variation of the coordination number, ΔQ = variation of the global charge.

The effect of the charge is stronger than the effect of oxidation state. The chemist must be aware of the simplicity of these rules which, however, provide an indication of the reactivity of the coordinated ligands. In some external attacks it appears that other factors may be at the origin of the driving forces (*88a*).

C. One-Electron Transfers and Radical Reactions

Free radical mechanisms are involved in homogeneous catalysis and play a considerable role in some transformations (*106a–106d,107a–107d*). We have developed an inventory of the key reactions in this field (Table IV). Reactions 64–67 represent different possibilities of radical formation from a metal complex: electron transfer reactions (Reactions 64 and 66), atom transfers, which within inorganic nomenclature could be called inner sphere electron transfers (*106d*) (Reaction 65), and homolytic bond breaking (Reaction 67). For the present, we have introduced only electron transfer to polar bonds, such as C—X (X = halide), and to heteroatomic bonds such as O—O (Reaction 64, Table IV). Electron transfer activation of nonpolar bonds such as C—H or H—H has not yet been introduced into the program.

Reactions 68–73 (Table IV) represent the radical interactions with metal complexes: radical addition (Reaction 68), electron transfer from a radical to a metal (Reaction 70) or from a metal to a radical (Reaction 69), homolytic displacements at the metal or at a ligand (Reactions 71 and 72), and β-radical fragmentations (Reaction 73). The final reactions (74–79) concern the reactivity of radicals with organic reagents.

IV

EXAMPLES

In this last section we wish to illustrate the use of TAMREAC with three examples. The first, ethylene dimerization, is of importance since it is involved in the oligomerization and polymerization reactions of olefins. The second example concerns the hydroformylation of ethylene, and the last example is the allyl–alkyl coupling reaction induced by a Ni(II) complex.

A. Ethylene Dimerization

In a preliminary article (*3*), we discussed the ethylene dimerization reaction starting from a hydride complex of general formula MLH. Formal generation

TABLE IV
One-Electron Transfers and Radical Reactions

Reaction name	General description[a]	Example	ΔOS, ΔNVE, ΔCN, ΔQ[b]	Ref.
FORMATION OF RADICALS WITH METAL COMPLEXES				
64. Electron transfer	$M + Z\text{–}Ht \rightarrow M + Z\text{–}Ht^{\overline{\cdot}}$	$M + C\text{–}Cl \rightarrow M + C\text{–}Cl^{\overline{\cdot}}$	+1,-1,0,+1	108
65. Atom transfer	$M + T\text{–}Ht \rightarrow M\text{–}Ht + T^{\cdot}$	$M + C\text{–}Cl \rightarrow M\text{–}Cl + C^{\cdot}$	+1,+1,+1,0	108,109
66. Electron transfer from an unsaturated bond	$M + T{\equiv}T \rightarrow M + T^{+}{\equiv\!\!\overline{\cdot}\,}T^{\cdot}$	$M + C{=}C \rightarrow M + C^{+}\text{—}C^{\cdot}$	-1,+1,0,-1	110
67. Homolytic bond breaking	$M\text{–}LS \rightarrow M + LS^{\cdot}$	$M\text{–}C \rightarrow M + C^{\cdot}$	-1,-1,-1,0	111
REACTIONS OF RADICALS WITH METAL COMPLEXES				
68. Radical addition to the metal	$M + T^{\cdot} \rightarrow M\text{—}T$	$M + H^{\cdot} \rightarrow M\text{—}H$	+1,+1,+1,0	112
69. Electron transfer to a radical	$M + T^{\cdot} \rightarrow M + T^{-}$	$M + Cl^{\cdot} \rightarrow M + Cl^{-}$	+1,-1,0,+1	113,108b
70. Electron transfer from a radical to the metal	$M + T^{\cdot} \rightarrow M + T^{+}$	$M + H^{\cdot} \rightarrow M + H^{+}$	-1,+1,0,-1	114
71. Homolytic displacement at the metal	$M\text{–}TL + T^{\cdot} \rightarrow M\text{–}T + TL^{\cdot}$	$M\text{–}C + Cl^{\cdot} \rightarrow M\text{–}Cl + C^{\cdot}$	0,0,0,0	115

72. Homolytic displacement at a ligand	M–LS + T· → M + LS–T	M–Cl + C· → M + Cl–C	-1,-1,-1,0	115
73. β-Metalloalkyl radical fragmentation	$M-Z' \overset{Z}{\underset{Z}{\diagup}} \rightarrow M-\overset{Z}{\underset{Z}{\parallel}}$	$M-C \overset{C}{\underset{C·}{\diagup}} \rightarrow M-\overset{C}{\underset{C}{\parallel}}$	-1,+1,0,0	116

REACTIONS OF RADICALS WITH ORGANIC SUBSTRATES

74. SH2 with organic substrates	El–Ht + Z· → El–Z + Ht·	C–Cl + C· → C–C + Cl·	0,0,0,0	117
75. Radical addition to an unsaturated bond	$T· + Z{\equiv}Z \rightarrow Z'{=}Z'^{\diagdown T}$	$Cl + C{=}C \rightarrow C·-C{\diagdown}^{Cl}$	0,0,0,0	118
76. Radical recombination	T· + T· → T–T	C· + C· → C–C	0,0,0,0	119
77. Anion-radical dissociation	Z–Ht·⁻ → Z· + Ht⁻	C–Cl·⁻ → C· + Cl⁻	0,0,0,0	120,108a
78. Radical allyl rearrangement	$Z{\diagdown}^{Z}{-}Z· \rightarrow Z·{-}Z'^{\diagdown Z}$	$C{\diagdown}^{C}{=}C· \rightarrow C·{-}C'^{\diagdown C}$	0,0,0,0	121
79. Cyclopropylmethyl radical rearrangement	$Z·{-}Z{\diagup}^{\underset{Z}{\diagdown}}Z{=}Z'{-}Z$	$C·{-}C{\diagup}^{\underset{\overset{··}{C}}{\diagdown}}C{=}C{-}C$	0,0,0,0	122

[a] The meaning of the symbols A, V, C2,..., is given in Table I.

[b] ΔOS = variation of the oxidation state of the metal, ΔNEV = variation of the number of valence electrons, ΔCN = variation of the coordination number, ΔQ = variation of the global charge.

197

of all possible structures was carried out, but no geometrical or electronic considerations were included. However, all the intermediates which had been postulated up to that point in time were found by the program, which also proposed a new metallacyclopentane intermediate. This possibility was reported at about the same time by Schrock for a similar catalytic system (4).

We have again studied this reaction [Eq. (7)] with the current version of TAMREAC in order to compare the information and the processing of the program. In the new study, TAMREAC did not find any other novel intermediates, but it allowed us to follow the reactivity sequences in a more rational way.

$$2\,C_2H_4 + [Rh] + HCl \rightarrow C_4H_8 + [Rh] + HCl \tag{7}$$

The starting materials given to the computer were, according to Cramer's scheme (123), a bis(ethylene)rhodium(I) complex, an HX molecule (X = chloride atom) and a solvent molecule (L). TAMREAC finds, in the initial step, two possible product-intermediates (2 and 5, Fig. 6). We store them in the disk and we continue the generation of the synthetic sequences starting from the intermediate 2 or 5 in order to find the intermediates of the second step, which will be also stored. The same operation is repeated with all the intermediates until the generation of the initial catalyst (1, Fig. 6).

We shall discuss only the first steps of this reaction, shown in Fig. 6, in order to indicate the new information provided by TAMREAC (for more information concerning all the possible intermediates, see ref. 3). The information concerning the number of valence electrons, the charge, and the geometry of the intermediates allows a first evaluation of the different reaction pathways. For example, the metallacyclopentane intermediate (6, Fig. 6) derives from intermediate 5 (or from the sequence involving a hydride intermediate, 4) only if the geometry is trigonal bipyramidal. Direct coupling from 1 is not found by the program. In contrast with what was proposed in the preliminary approach (3), according to the evaluation matrix (Section III, Matrix 3) the coupling reaction in the case of a four-coordinate d^8 complex occurs when the geometry is a tetrahedral one, while in our case the monovalent rhodium complex (1, Fig. 6) has a square planar geometry.

Another possibility found by TAMREAC is the electrophilic addition of the proton to the coordinated ethylene. The product, intermediate 7, is a neutral one (here also the initial coordination by a solvent ligand is necessary). According to Cramer's study, all the intermediates are anionic ones, so the chemist could discard this possibility.

The above considerations (geometry and the ligand's role) are helpful in the understanding of mechanistic sequences. They also aid the chemist in the selection of the most probable reactivity pathways.

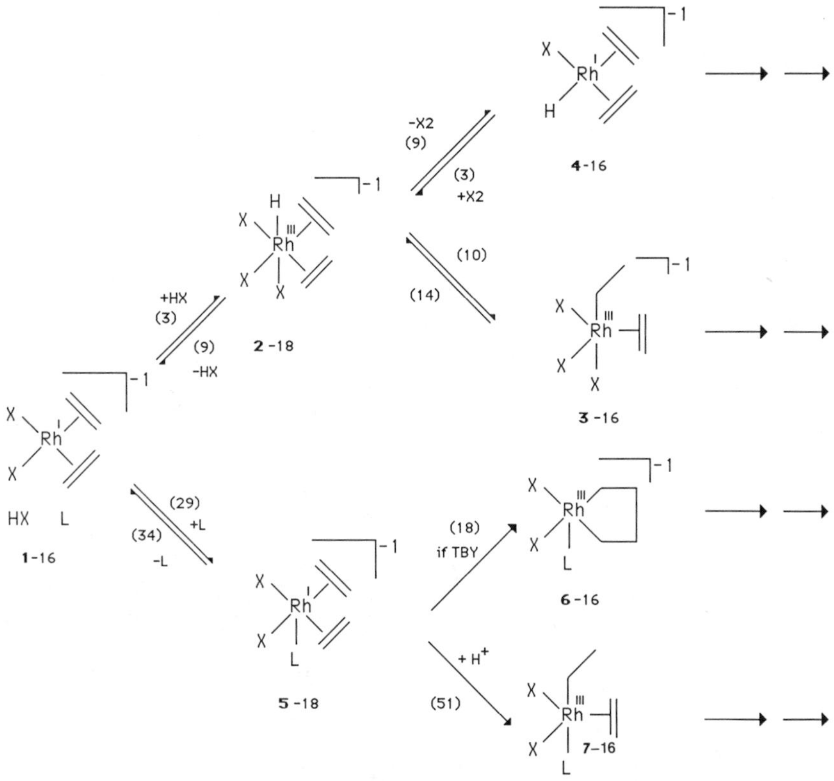

FIG. 6. Intermediates proposed by TAMREAC for the two initial steps of the ethylene dimerization. The two numbers below each complex correspond to the complex number and to the valence electron number, respectively. Numbers in parentheses on, or below, the arrows correspond to the reaction numbers (Tables II, III, and IV). TBY = trigonal bipyramidal geometry.

B. *Ethylene Hydroformylation*

As a second example, we generated possible intermediates and synthetic sequences for the hydroformylation of olefins [Eq. (8)] with the hydride complex $HRh(CO_2)L_2$ (L = PPh_3).

$$C_2H_4 + [Rh] + CO + H_2 \rightarrow CH_3CH_2CHO + [Rh] \qquad (8)$$

Beyond the usually discussed intermediates, anionic and radical ones (124,125), TAMREAC proposes new sequences involving alkylidene and ketene intermediates (2a). Such sequences had never been reported before the appearance of our article, and we were pleased to note a recent article (126) discussing this possibility with a $W(CO)_4Cl_2$ catalyst.

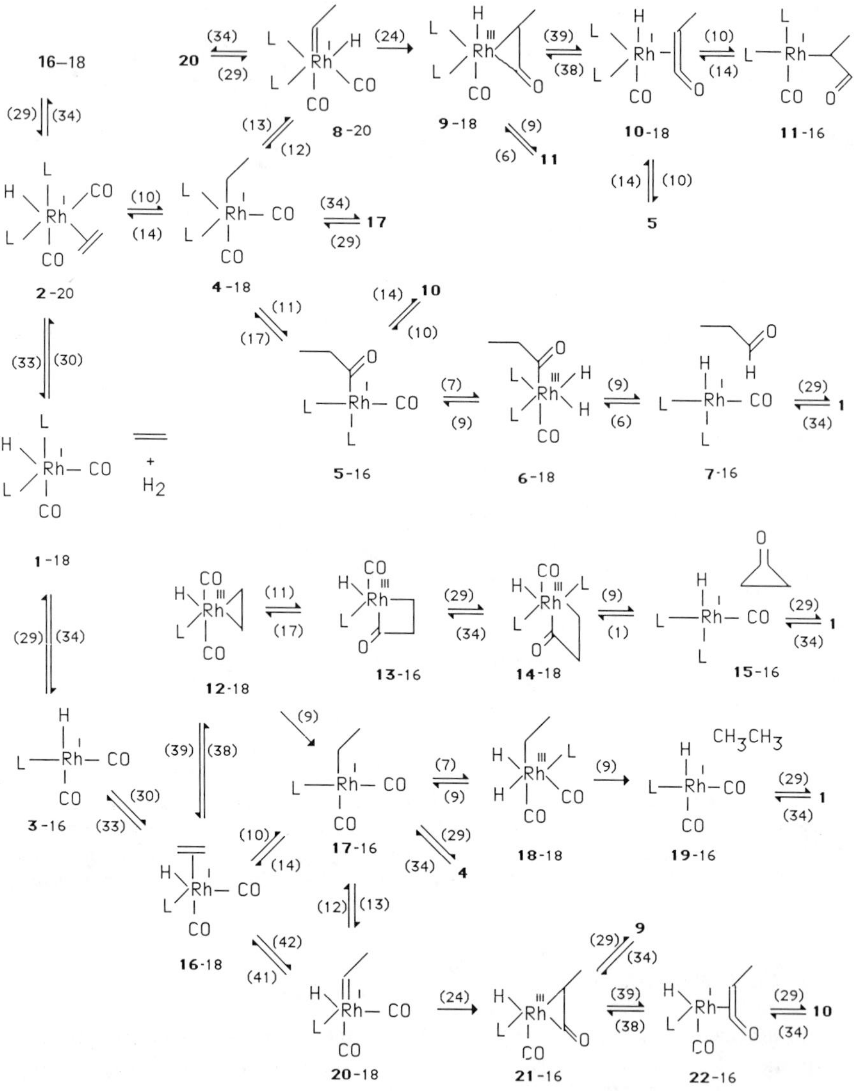

FIG. 7.   Synthetic sequences for the hydroformylation reaction. The two numbers below each complex correspond to the complex number and to the valence electron number, respectively. Numbers in parentheses on, or below, the arrows correspond to the reaction numbers (Tables II, III, and IV). L = PPh$_3$.

Figures 7 and 8 show the intermediates generated by TAMREAC. It should be noted that we have not discarded 20-electron intermediates (intermediates **2** and **8**, Fig. 7) in order to find the associative pathway proposed by Wilkinson and co-workers (*124a,124b*). Figure 9 shows the most probable catalytic cycle, which proceeds via a dissociative pathway (*124a*), and Figs. 10 and 11 the novel cycles found by the program involving an intramolecular carbonylation.

Even if these intermediates appear improbable in this present example of the rhodium complex, the interest in our approach lies in the possibility of the generation of most of the formally conceivable reactions (with no evaluation

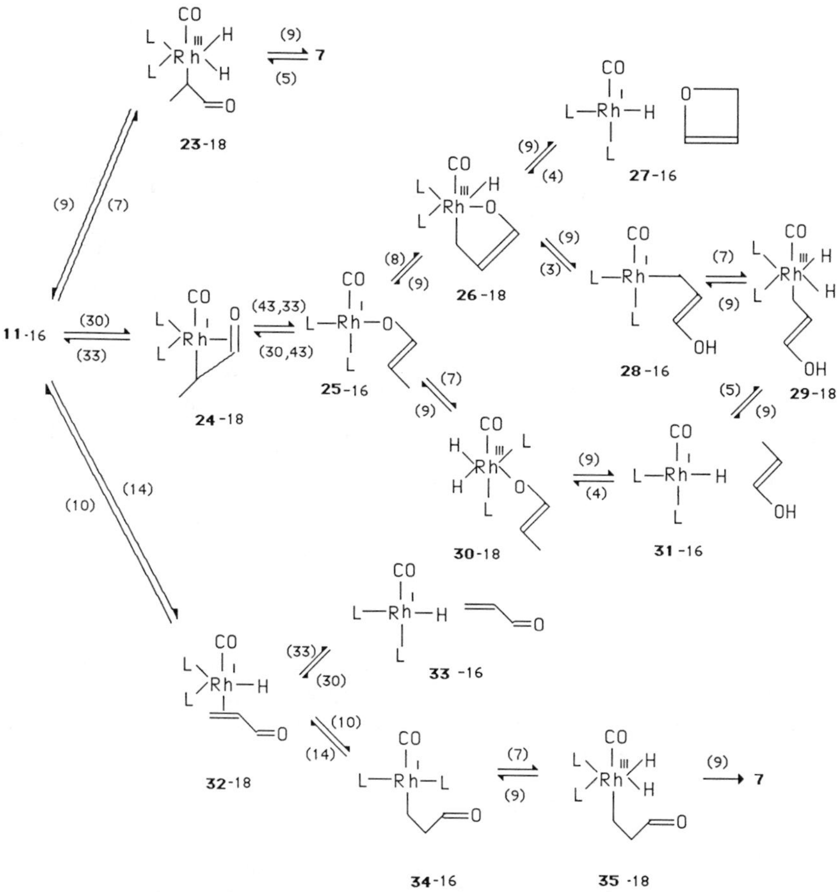

FIG. 8.   Synthetic sequences for the hydroformylation reaction starting from intermediate **11** (Fig. 7). Numbers in parentheses on, or below, the arrows correspond to the reaction numbers (Tables II, III, and IV). L = PPh₃.

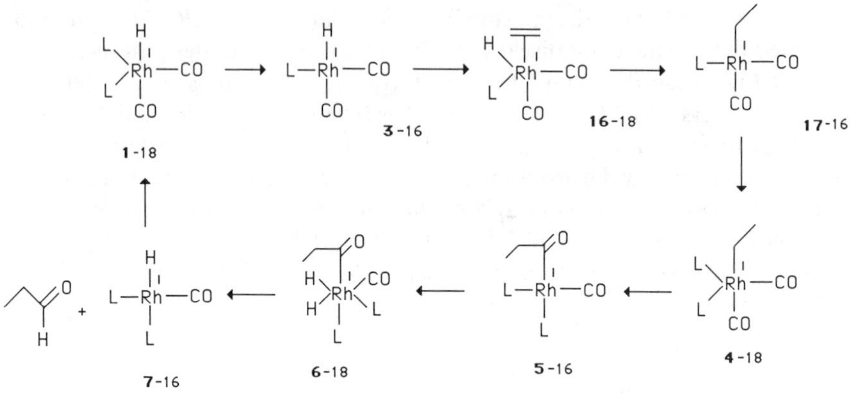

FIG. 9. Dissociative catalytic cycle.

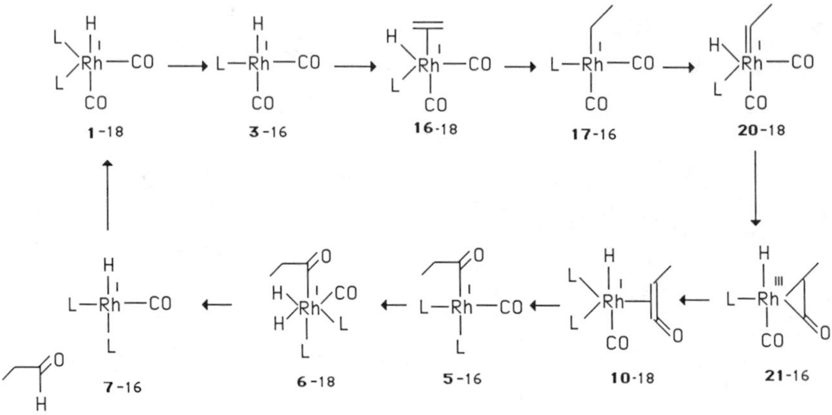

FIG. 10. Dissociative catalytic cycle involving alkylidene–ketene intermediates.

by the computer) without important interference by the chemist–user, who then carries out the selection of the intermediates. The recent evidence concerning such a possible alkylidene–ketene mechanism with a tungsten complex underscores the interest of developing these kinds of programs in organometallic chemistry.

C. Allyl–Alkyl Cross Coupling

The final example concerns the cross-coupling between an alkyl halide and a π-allylnickel(II) halide complex [Eq. (9)]. The mechanisms of this reaction have been extensively studied, and several schemes, essentially radical ones,

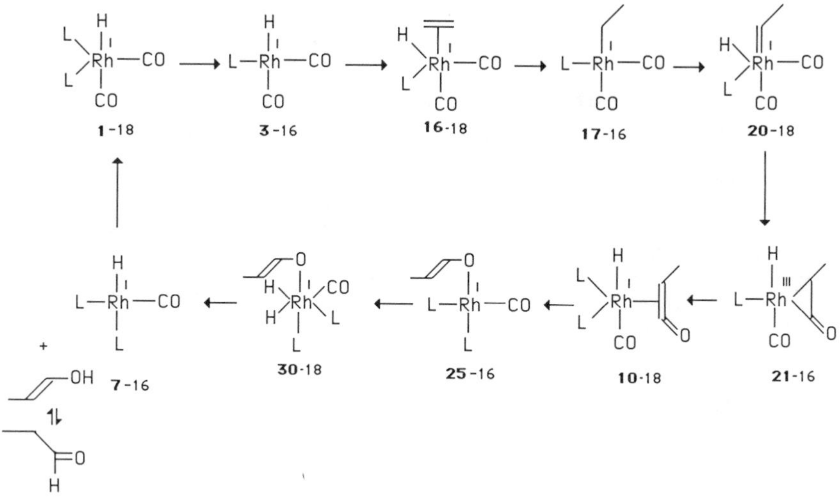

FIG. 11.   Dissociative catalytic cycle involving alkylidene–ketene intermediates.

are reported (*127a,127b*). The starting materials given to the computer were a mononuclear π-allylnickel(II) complex, a ligand molecule, L (L = solvent), and an alkyl halide (RX).

$$\text{(allylNiX)}_2 + RX \rightarrow R\text{—allyl} + NiX_2 \qquad (9)$$

Figure 12 shows two-electron equivalent mechanisms proposed by TAMREAC. The product of the coupling (7, Fig. 12) is formed from the sequences $1 \rightarrow 4 \rightarrow 8 \rightarrow 7$, which involve an oxidative addition followed by a reductive elimination (or migration of the alkyl to the allyl, the step $4 \rightarrow 10$). Such two-electron sequences seem to be unlikely since a wide range of organic halides (aryl, alkyl, and vinyl), undergo this reaction (*127a*). Several pieces of evidence suggest an electron-transfer-catalyzed mechanism (*127a,127b*), and we carried out a study of this possibility with TAMREAC. In a first stage, TAMREAC generated one-electron nonchain sequences, shown in Fig. 13. Electron transfer (step $1 \rightarrow 25$, Fig. 13), atom transfer (step $1 \rightarrow 27$, Fig. 13), and homolytic breaking of Ni—X bonds (step $1 \rightarrow 18$), Fig. 13) have been found. This last reaction, homolytic breaking, corresponds to the intramolecular electron transfer within the dimeric π-allylnickel halide, proposed in the literature (*127b*) [Eq. (10)].

$$\text{allylNi}^{II}XX\text{Ni}^{II}\text{allyl} \rightleftarrows \text{allylNi}^{III}X + XNi^{I}\text{allyl} \rightleftarrows \text{allylNi}^{I} + \text{allylNi}^{III}X_2 \qquad (10)$$

In order to have TAMREAC proposing radical-catalyzed mechanisms, we proceeded in the following way: we introduced successively in the program a

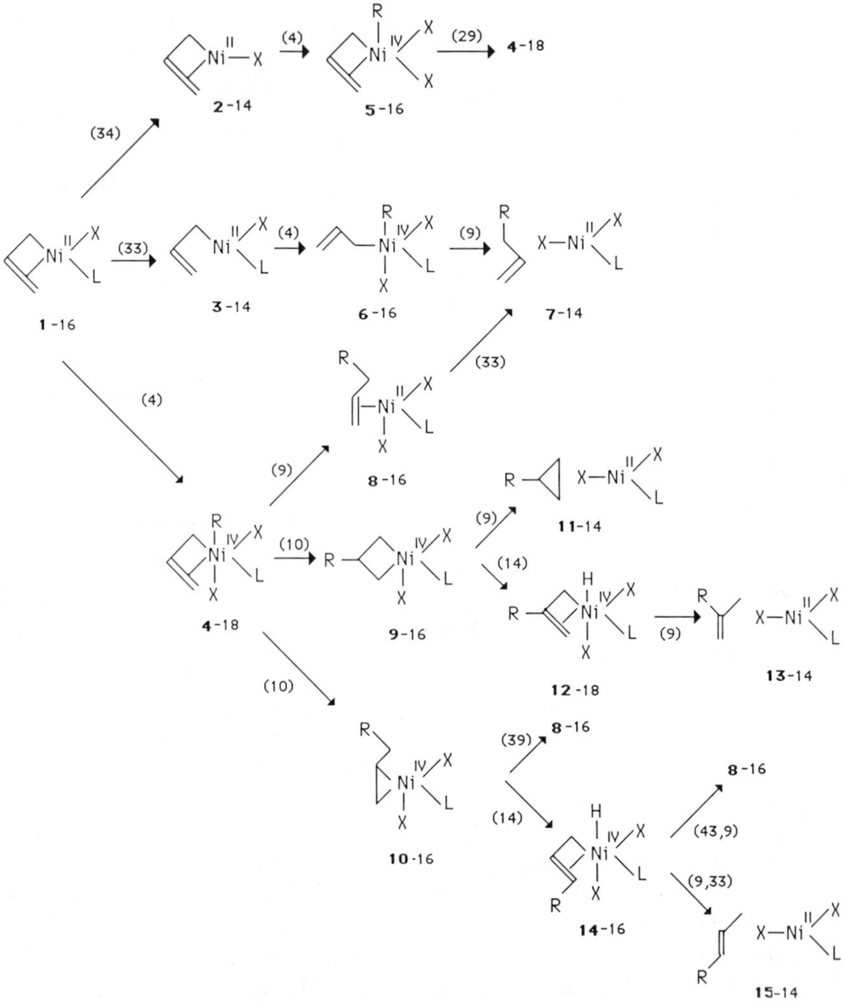

FIG. 12. Two-electron mechanisms proposed by TAMREAC for the allyl–alkyl cross-coupling. The two numbers below each complex correspond to the complex number and to the valence electron number, respectively. Numbers in parentheses by the arrows correspond to the reaction numbers (Tables II, III, and IV). R = alkyl, X = halide, L = solvent molecule.

paramagnetic species [alkyl radical, Ni(I), or Ni(III)] with a diamagnetic one and followed the sequences proposed by TAMREAC. Figure 14 shows the result of the first operation: interaction of the alkyl radical with the allylNi(II) complex afforded two paramagnetic species, a Ni(I)XL complex (B, Fig. 14) or a Ni(III) intermediate (C, Fig. 14). In a second step, these two complexes were

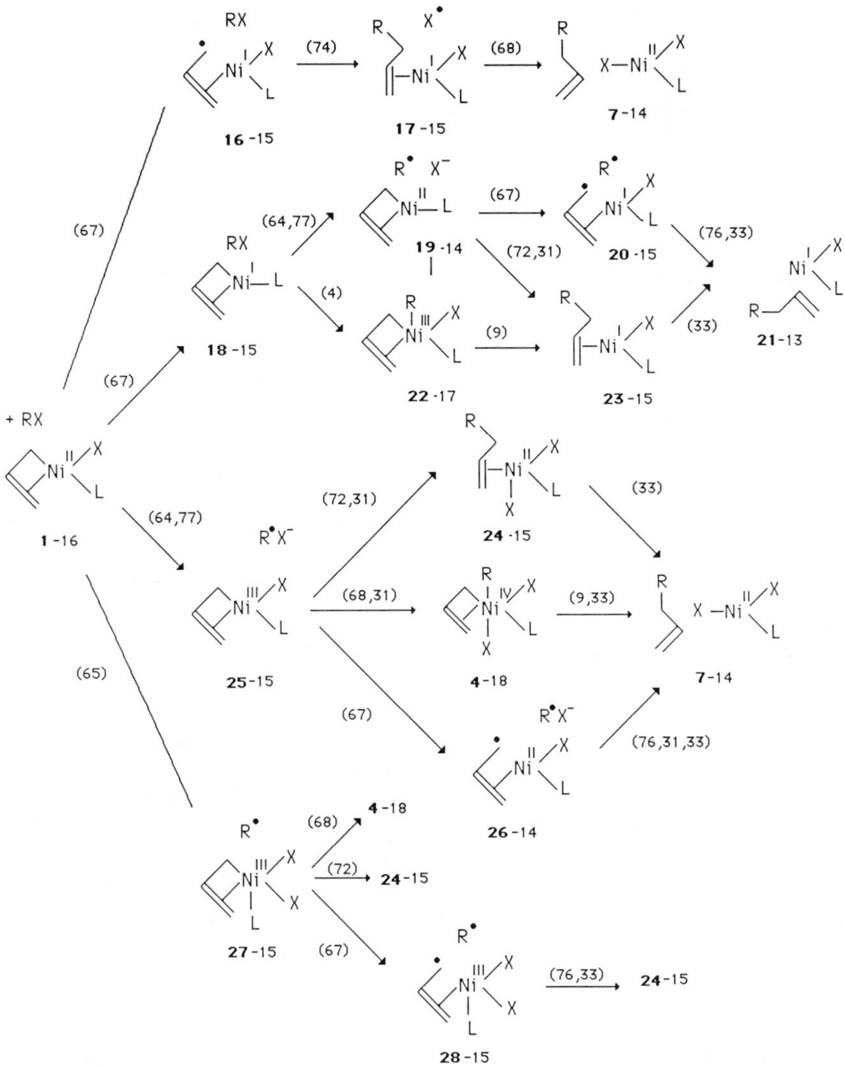

FIG. 13.   One-electron nonchain mechanisms proposed by TAMREAC for the allyl–alkyl cross-coupling.

given to the program with an alkyl halide substrate (Fig. 15). TAMREAC generated the alkyl radical (A, Fig. 15), which will be chain carrying in this case. Complex E in Fig. 15 derives from an electron-transfer reaction between Ni(I)XL and allylNi(II)XL complexes (Fig. 16). TAMREAC is not able to find such reactions, since it deals only with mononuclear systems. All the reactions

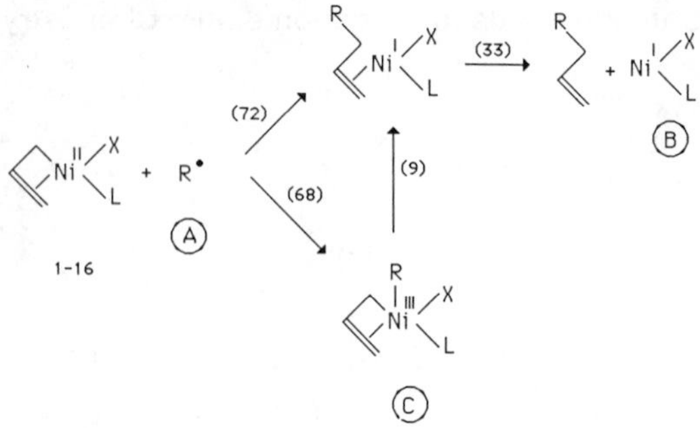

FIG. 14. Possible reactions between an alkyl radical and a π-allylnickel(II) complex.

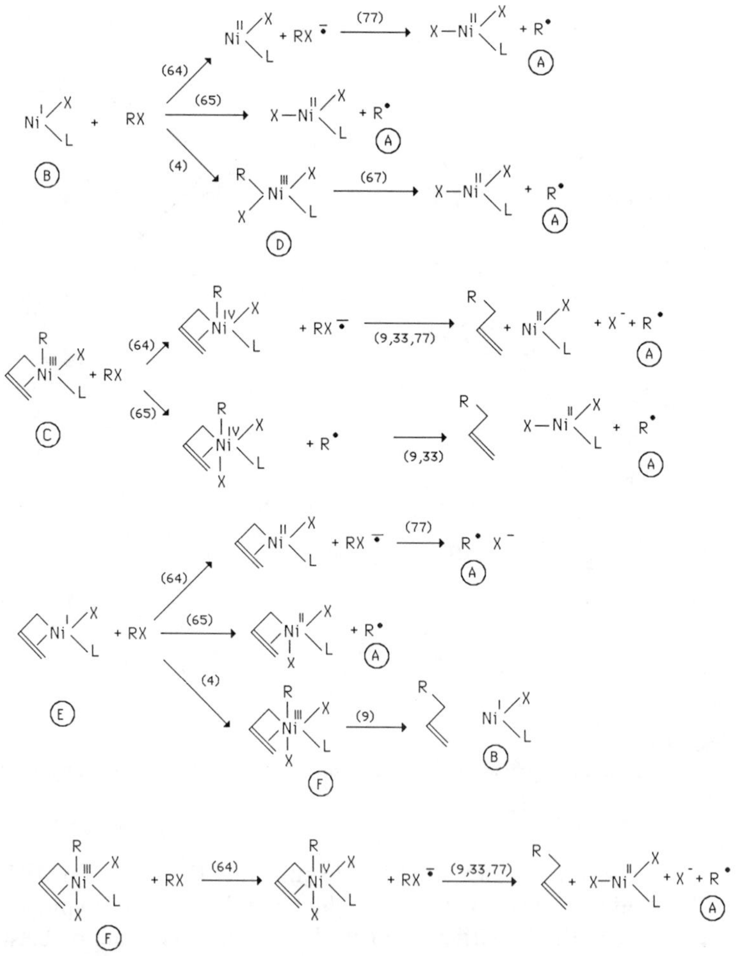

FIG. 15. Possible reactions between paramagnetic nickel species and the halide substrate. Formation of the complex E is shown in Fig. 16.

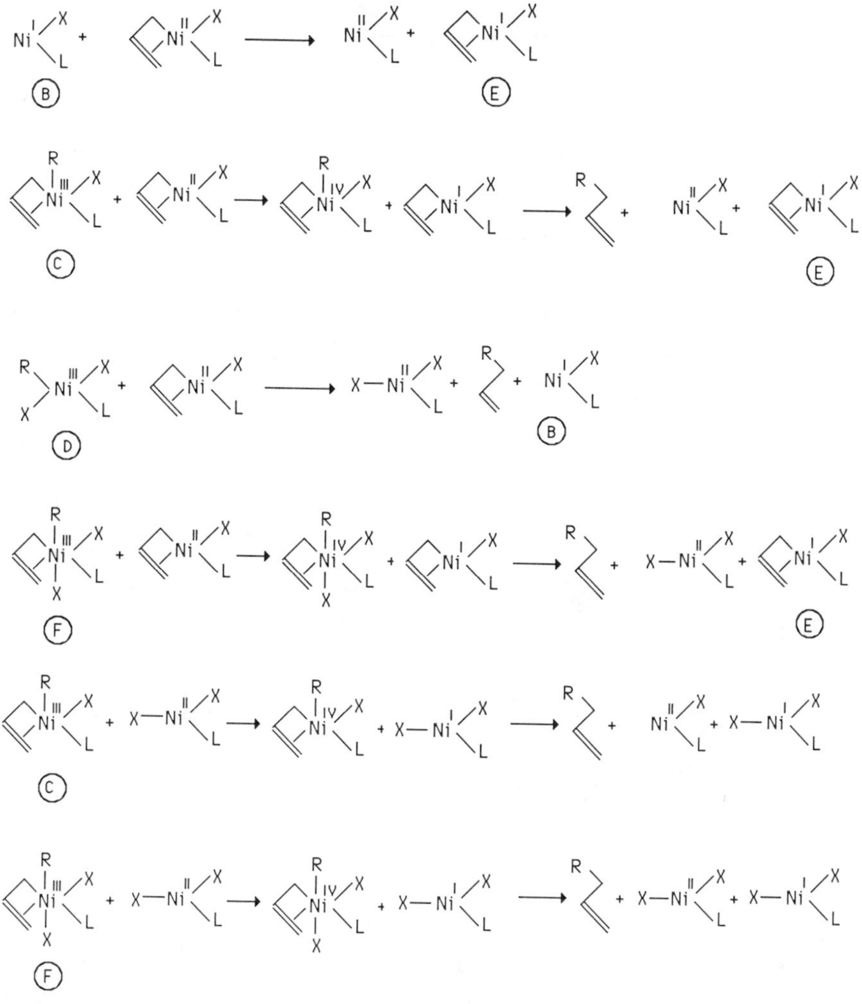

FIG. 16. Electron-transfer sequences between paramagnetic and diamagnetic nickel species. These reactions cannot be found by the program as presently designed, but were proposed by Hegedus (*127b*). At this stage, the chemist has to "help" TAMREAC.

in Fig. 16, have been formally proposed by us, according to the principle of the electron-transfer reaction between a paramagnetic and a diamagnetic species. This example shows that for some cases, the interaction TAMREAC–chemist leads to a situation where the chemist gives up TAMREAC's aid and follows his own intuition because he is aware of the present limitation of the program (see Section V).

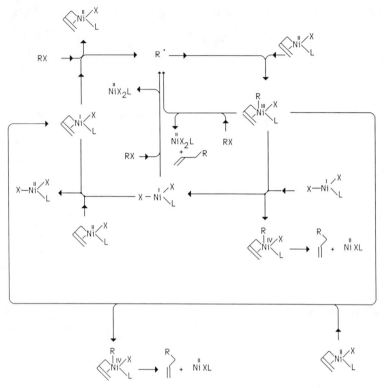

FIG. 17. Electron-transfer catalytic cycle, where the chain-carrying species is an alkyl radical.

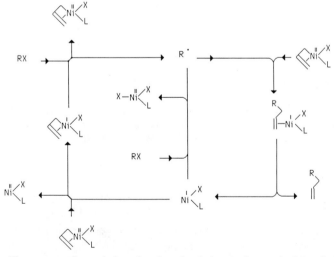

FIG. 18. Electron-transfer catalytic cycle, where the chain-carrying species is an alkyl radical.

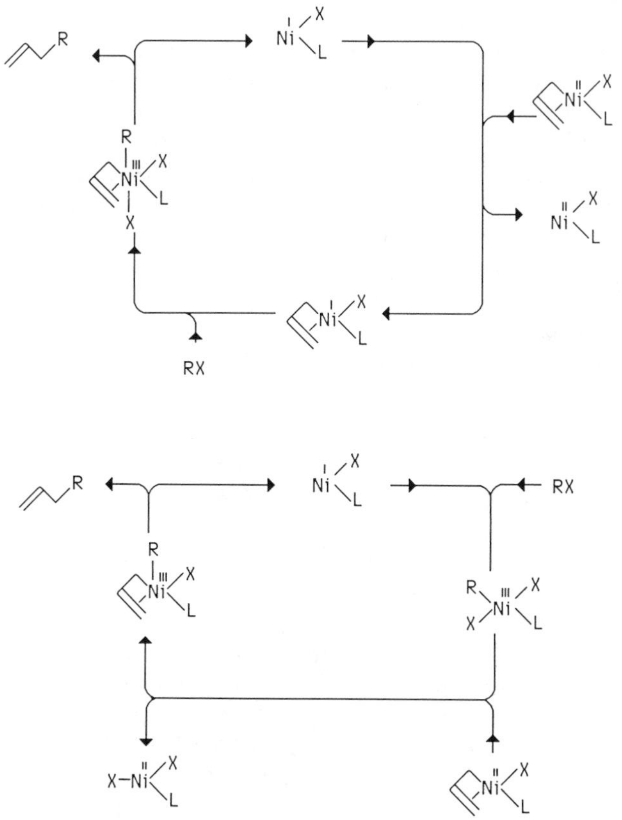

FIG. 19.   Electron-transfer catalytic cycles. The possible chain-carrying species are Ni^IXL, allylNi^IXL, or $RNi^{III}X_2L$.

Figures 17, 18, and 19 show the different catalytic cycles derived from the foregoing approach. The chain-carrying species of the cycles in Figs. 17 and 18 is an alkyl radical, whereas in the cycles of Fig. 19 it is a Ni(I) or an allylNi(I) complex. We may note that, for this example where both paramagnetic and diamagnetic intermediates were considered, TAMREAC did not find more solutions than Hegedus and Thompson did (*127b*).

V

CONCLUSION

Although the computer proposed, in two cases at least (dimerization, hydroformylation), innovative mechanistic schemes before the experimental chemist, and although it provides interesting hints for by-products to be expected in a catalytic transformation, we believe that this approach is still in its infancy. We shall therefore conclude by outlining what is presently beyond the grasp of TAMREAC, delineating by the same token directions for future improvements. At this point, if TAMREAC uses the fact that intermediates may be square planar, tetrahedral, or any other shape to select allowed and forbidden pathways (see Section III) it totally overlooks the stereochemical features which allow the design of enantioselective catalysis (128). In this direction, the improvement of ligand abilities in enantioselective discrimination could possibly profit from molecular mechanistic calculations extended to organometallic complexes (129). One other field presently out of the scope of TAMREAC approach is polynuclear complexes and clusters. This would deserve a treatment on its own, particularly since the isolobal approach seems to provide a new tool for rationalizing and designing new experiments (130). We selected 79 basic mechanistic schemes to cover organometallic reactivity connected with homogeneous catalysis. More could have been included, but our experience of heterocyclic chemistry (5d) and statistical studies of other groups of reactions (131) show that, in a given field of reactivity, there exists a hard core of a small number of important reactions surrounded by concentric circles of reactions of decreasing importance and use. In this direction, however, it would be important to reach a tighter evaluation of which metal–ligand couples allow a given transformation of Tables II, III, and IV to occur. This can be done only at the price of an extensive bibliographical compilation and should lead to a Mendeleevian chart of transition metals for each of the 79 reactions that we retained as data base for TAMREAC.

REFERENCES

1. B. M. Trost, *Acc. Chem. Res.* **13**, 385 (1980); J. Tsuji, *Acc. Chem. Res.* **6**, 8 (1973).
2a. I. Theodosiou, R. Barone, and M. Chanon, *J. Mol. Catal.* **32**, 27–50 (1985).
2b. I. Theodosiou, Ph.D. dissertation, Marseille, 1985.
2c. J. Happel and P. H. Sellers, *Adv. Catal.* **32**, 273 (1983).
3. R. Barone, M. Chanon, and M. L. H. Green, *J. Organomet. Chem.* **85**, 185 (1980).
4. J. D. Fellman, G. A. Rupprecht, and R. R. Schrock, *J. Am. Chem. Soc.* **101**, 5099 (1979).
5a. R. Barone, M. Chanon, P. Cadiot, and J. M. Cense, *Bull. Soc. Chim. Belg.* **91**, 333 (1982).
5b. R. Barone, M. Chanon, and J. Metzger, *Tetahedron Lett.* **32**, 2751 (1974).
5c. R. Barone and M. Chanon, "The Chemistry of Heterocyclic Flavoring and Aroma Compounds," (G. Vernin ed.), p. 249. Ellis Horwood Ltd., Chichester, 1982.
5d. R. Barone, Ph.D. dissertation, Marseille, 1976.
6. I. Ugi and P. Gillespie, *Angew. Chem., Int. Ed. Engl.* **10**, 914 (1977); A. Weisse, *Z. Chem.* **13**, 155 (1973); K. K. Agarwal, D. L. Larsen, and H. L. Gelernter, *Comput. Chem.* **2**, 75 (1978); T. D. Salatin and W. L. Jorgensen, *J. Org. Chem.* **45**, 2043 (1980); B. L. Roos Kozel and W. L. Jorgensen, *J. Chem. Inf. Comput. Sci.* **21**, 101 (1981); T. D. Salatin, D. McLaughlin, and W. L. Jorgensen, *J. Org. Chem.* **46**, 5284 (1981); J. Schmidt Burnier and W. L. Jorgensen, *J. Org. Chem.* **48**, 3923 (1983); C. E. Peishoff and W. L. Jorgensen, *J. Org. Chem.* **50**, 3174, 1056

(1985); E. J. Corey, A. K. Long, and S. D. Rubenstein, *Science* **228,** 408 (1985); R. Barone and M. Chanon, in "Computer Aids to Chemistry" (G. Vernin and M. Chanon, eds.). Ellis Horwood, Ltd., Chichester, 1982.

7.   E. J. Corey and W. T. Wipke, *Science* **166,** 178 (1969).
8.   M. J. S. Dewar, *Bull. Soc. Chim. Fr.* **C79,** 18 (1951); M. J. S. Dewar and G. P. Ford, *J. Am. Chem. Soc.* **101,** 783 (1979); J. Chatt and L. A. Duncanson, *J. Chem. Soc.,* 2339 (1953).
9.   K. H. Dötz, *Angew. Chem., Int. Ed. Engl.* **23,** 579 (1984).
10. J. P. Collman and L. S. Hegedus, "Principles and Applications of Organotransition Metal Chemistry," pp. 17–19, 43–45. University Science Books, Mill Valley, California, 1980.
11. C. A. Tolman, *Chem. Soc. Rev.* **1,** 137 (1972).
12a. P. R. Mitchell and R. V. Parish, *J. Chem. Educ.* **46,** 811 (1969).
12b. J. K. Burdett, "Molecular Shapes. Theoretical Models of Inorganic Stereochemistry," p. 222. Wiley, New York, 1980.
13. D. A. White, *Organometal. Chem. Rev., A* **3,** 497 (1968); R. Ugo, *La chimica e l'industria* **51,** 1319 (1969); R. G. Pearson, "Hard and Soft Acids and Bases," p. 318. Dowden, Hutchinson & Ross, Inc., Stroudsberg, Pennsylvania, 1973.
14. R. G. Pearson, "Symmetry Rules for Chemical Reactions." Wiley, New York, 1976.
15a. Ref. 12b, p. 171.
15b. K. F. Purcell and J. C. Kotz, "Inorganic Chemistry," pp. 543–545. Saunders, Philadelphia, 1977.
16a. E. J. Corey, *Pure Appl. Chem.* **14,** 19 (1967).
16b. E. J. Corey and W. T. Wipke, *Science* **166,** 178 (1969).
16c. J. Gasteiger and C. Jochum, *Topics Curr. Chem.* **74,** 94 (1978).
16d. J. Gasteiger, *Comp. Chem.* **2,** 85 (1978).
16e. M. Chastrette, M. Rajzmann, M. Chanon and K. Purcell, *J. Am. Chem. Soc.* **107,** 1 (1985).
17. J. Halpern, *Pure Appl. Chem.* **55,** 99 (1983).
18a. J. P. Collman, *Acc. Chem. Res.* **1,** 136 (1968).
18b. J. Halpern, *Acc. Chem. Res.* **3,** 386 (1970).
18c. L. Vaska, *Acc. Chem. Res.* **1,** 335 (1968).
18d. R. Ugo, *Coord. Chem. Rev.* **3,** 319 (1968).
19. J. P. Birk, J. Halpern, and A. L. Pickard, *J. Am. Chem. Soc.* **90,** 4441 (1968).
20. M. L. H. Green, *Ann. N.Y. Acad. Sci.,* 229 (1980).
21. J-Y. Saillard and R. Hoffmann, *J. Am. Chem. Soc.* **106,** 2006 (1984).
22a. Ref. 14, pp. 286–288.
22b. T. A. Albright, *Tetrahedron* **38,** 1339 (1982).
22c. H. H. Brintzinger, L. L. Lohr, and K. J. T. Wong, *J. Am. Chem. Soc.* **97,** 5146 (1975).
22d. G. P. Pez, *J. Am. Chem. Soc.* **98,** 8072 (1976).
23. B. Åkermak and A. Ljungqvist, *J. Organomet. Chem.* **182,** 59 (1979).
24a. S. Komiya, T. A. Albright, R. Hoffmann, and J. K. Kochi, *J. Am. Chem. Soc.* **98,** 7255 (1976).
24b. M. C. Baird, *J. Organomet. Chem.* **64,** 289 (1974).
24c. J. A. Osborn, F. H. Jardine, J. F. Young, and G. Wilkinson, *J. Chem. Soc., A,* 1711 (1966).
25. E. R. Evitt and R. G. Bergman, *J. Am. Chem. Soc.* **100,** 3288 (1978).
26. B. L. Shaw and J. D. Ruddick, *J. Chem. Soc. A,* 2969 (1969).
27a. J. R. Norton, *Acc. Chem. Res.* **12,** 139 (1979).
27b. T. J. Barton, R. Grinter, A. J. Thompson, B. Davies, and M. Poliakoff, *J. Chem. Soc., Chem. Commun.,* 841 (1977).
28. J. E. Barcaw, *J. Am. Chem. Soc.* **96,** 5087 (1974).
29a. Ref. 14, p. 424.
29b. A. Stockis and R. Hoffmann, *J. Am. Chem. Soc.* **102,** 2952 (1980).
29c. J. W. Lauther and R. Hoffmann, *J. Am. Chem. Soc.* **98,** 1729 (1976).

29d. R. H. Grubbs and A. Miyashita, *J. Am. Chem. Soc.* **100**, 7416, 7418 (1978).

29e. R. H. Grubbs, A. Miyashita, M. M. Liu, and P. L. Burk, *J. Am. Chem. Soc.* **99**, 3863 (1977).

29f. M. Ephritikhine, J. Levisalles, H. Rudler, and D. Villemin, *J. Organomet. Chem.* **124**, C1–C4 (1977).

29g. R. Taube and K. Seyferth, *J. Organomet. Chem.* **249**, 365 (1983).

29h. C. P. Casey and H. E. Tuinsta, *J. Am. Chem. Soc.* **100**, 2270 (1978).

30. P. W. Jolly and G. Wilke, "The Organic Chemistry of Nickel," Vol. 2. Academic Press, New York, 1975.

31. P. S. Braterman, *J. Chem. Soc., Chem. Commun.*, 70 (1979).

32. K. P. C. Vollhardt and R. G. Bergman, *J. Am. Chem. Soc.* **96**, 4996 (1974); K. P. C. Vollhardt, *Acc. Chem. Res.* **10**, 1 (1977); K. P. C. Vollhardt, in "Strategies and Tactics in Organic Synthesis" (T. Lindberg, ed.), p. 303. Academic Press, Orlando, 1984.

33. J. D. Fellmann, G. A. Rupprecht, and R. R. Schrock, *J. Am. Chem. Soc.* **101**, 5099 (1979).

34. G. Erker, K. Engel, U. Dorf, J. L. Atwood, and W. E. Hunter, *Angew. Chem., Int. Ed. Engl.* **21**, 914 (1982); M. Pascali, C. Floriani, A. Chiesi-Villa, and C. Cuastini, *J. Am. Chem. Soc.* **101**, 4740 (1979).

35. R. J. McKinney, *J. Chem. Soc., Chem. Comm.*, 491 (1980).

36. R. J. McKinney, D. L. Thorn, R. Hoffmann, and A. Stockis, *J. Am. Chem. Soc.* **103**, 2595 (1981).

37. L. Cassar, P. E. Eaton, and J. Halpern, *J. Am. Chem. Soc.* **92**, 3515 (1970).

38. J. Levisalles, H. Rudler, and D. Villemin, *J. Organomet. Chem.* **193**, 69 (1980).

39. T. J. Katz, *Adv. Organomet. Chem.* **17**, 449 (1979); N. Calderon, E. A. Ofstead, and W. A. Judy, *Angew. Chem., Int. Ed. Engl.* **15**, 401 (1976); J-L. Herisson and Y. Chauvin, *Makromol. Chem.* **141**, 161 (1970); O. Eisenstein, R. Hoffmann, and A. R. Rossi, *J. Am. Chem. Soc.* **103**, 5582 (1981); R. H. Grubbs, D. D. Carr, C. Hoppin, and P. L. Burk, *J. Am. Chem. Soc.* **98**, 3478 (1976); D. A. Straus and R. H. Grubbs, *Organometallics* **1**, 1658 (1982); J. B. Lee, K. C. Ott, and R. H. Grubbs, *J. Am. Chem. Soc.* **104**, 7491 (1982); T. J. Katz and R. Rothchild, *J. Am. Chem. Soc.* **98**, 2519 (1976).

40. Ref. 10, p. 517.

41. E. R. Evitt and R. G. Bergman, *J. Am. Chem. Soc.* **101**, 3973 (1979).

42. T. J. Katz, *Adv. Organomet. Chem.* **16**, 283 (1977); K. H. Dötz, *Pure Appl. Chem.* **55**, 1689 (1983); M. F. Semmelhack, *J. Am. Chem. Soc.* **106**, 5363 (1984); M. F. Semmelhack and R. Tamura, *J. Am. Chem. Soc.* **105**, 6750 (1983).

43. R. R. Schrock, *Acc. Chem. Res.* **12**, 98 (1979).

44. R. G. Pearson, *Chem. Br.* **12**, 160 (1976).

45. D. L. Thorn and R. Hoffmann, *J. Am. Chem. Soc.* **78**, 2079 (1978); G. Booth and J. Chatt, *J. Chem. Soc., A*, 2403 (1969); L. Succoni, P. Dapporto, and P. Stoppioni, *J. Organomet. Chem.*, C33, 116 (1976); G. Yagupsky, C. K. Brown, and G. Wilkinson, *J. Chem. Soc., A*, 1392 (1970); G. K. Anderson and R. J. Cross, *Acc. Chem. Res.* **17**, 67 (1984).

46. F. Calderazzo, *Angew. Chem., Int. Ed. Engl.* **16**, 299 (1977); A. L. Lapidus and N. M. Saveler, *Russian Chem. Rev.* **53**, 535 (1984).

47. Ref. 10, pp. 275, 281.

48. E. R. Evitt and R. G. Bergman, *J. Am. Chem. Soc.* **101**, 3973 (1979); H. Werner and R. Feser, *Angew. Chem., Int. Ed. Engl.* **18**, 157 (1979).

49. K. W. Barnett, D. L. Beach, S. P. Gaydos, and T. F. Pollmann, *J. Organomet. Chem.* **69**, 121 (1974).

50. F. G. Dyachkowskii, A. K. Shilova, and A. E. Shilov, *J. Polym. Sci., Part. C* **16**, 2336 (1967).

51. C. A. Tolman, *J. Am. Chem. Soc.* **94**, 2994 (1972).

52. T. Yamamoto, A. Yamamoto, and S. Ikeda, *J. Am. Chem. Soc.* **93**, 3350 (1971).

53. G. M. Whitesides, J. F. Gaasch, and E. R. Stedronsky, *J. Am. Chem. Soc.* **94**, 5258 (1972).

54. C.-H. Cheng, B. D. Spivack, and R. Eisenberg, *J. Am. Chem. Soc.* **99**, 3003 (1977); M. Kubota, D. M. Blake, and S. A. Smith, *Inorg. Chem.* **10**, 1430 (1971).

55a. M. Arbelot and M. Chanon, *Nouv. J. Chimie* **7**, 499 (1983).

55b. R. G. Pearson, "Hard and Soft Acids and Bases," p. 318. Dowden, Hutchinson & Ross, Inc., Stroudsberg, Pennsylvania, 1973.

56. N. J. Cooper and M. L. H. Green, *J. Chem. Soc., Dalton*, 1121 (1979).

57. Ref. 10, p. 283.

58. S. J. McLain, C. D. Wood, and R. R. Schrock, *J. Am. Chem. Soc.* **101**, 4558 (1979).

59a. C. N. Wilker, R. Hoffmann, and O. Eisenstein, *Nouv. J. Chimie* **7**, 535 (1983).

59b. M. F. Lappert, P. L. Pye, and G. M. McLaughlin, *J. Chem. Soc., Dalton Trans.*, 1272 (1977).

59c. D. J. Cardin, B. Cetinkaya, and M. F. Lappert, *J. Chem. Soc., Dalton Trans.*, 541 (1973).

59d. J. C. Jeffery, R. Navarro, H. Razay, and F. G. A. Stone, *J. Chem. Soc., Dalton Trans.*, 2471 (1981).

60a. B. E. Burster and M. G. Gatter, *Organometallics* **3**, 941 (1984).

60b. B. E. Burster and M. G. Gatter, *Organometallics* **3**, 895 (1984).

60c. M. L. H. Green, *Endeavour*, **XXVI**, 129 (1967).

60d. W. Hieber and W. Hybel, *Z. Elektrochem.* **57**, 235 (1953).

60e. M. Orchin, *Acc. Chem. Res.* **14**, 259 (1981).

61. R. S. Drago, *J. Chem. Ed.* **51**, 300 (1974).

63. T. J. Katz, *J. Am. Chem. Soc.* **91**, 2405 (1969).

64. K. C. Bishop, *Chem. Rev.* **76**, 467 (1976).

65. L. Cassar, P. E. Eaton, and J. Halpern, *J. Am. Chem. Soc.* **92**, 3515 (1970).

66a. W. S. Knowels, *Acc. Chem. Res.* **16**, 106 (1983).

66b. I. Ojima, *Pure Appl. Chem.* **56**, 99 (1984).

66c. J. P. Collman and W. R. Roper, *Adv. Organomet. Chem.* **7**, 53 (1968).

66d. H. B. Kagan, *Ann. N.Y. Acad. Sci.* **1**, 333 (1980).

66e. P. N. Rylander, *Cat. Org. Synth.*, **7**, 155 (1980).

67. B. M. Trost, P. E. Strege, L. Weber, T. J. Fullerton, and T. J. Dietsche, *J. Am. Chem. Soc.* **105**, 3929 (1983).

68a. G. W. Parshall, *Acc. Chem. Res.* **8**, 113 (1975).

68b. M. E. Thompson and J. E. Bercaw, *Pure Appl. Chem.* **56**, 1 (1984).

68c. A. E. Shilov, *Chemistry Rev.* **4**, 71 (1982).

69a. A. H. Janowicz and R. G. Bergman, *J. Am. Chem. Soc.* **105**, 3929 (1983).

69b. M. I. Bruce, *Angew. Chem., Int. Ed. Engl.* **16**, 73 (1977).

70. M. F. Lappert and B. Prokai, *Adv. Organomet. Chem.* **5**, 225 (1967).

71a. G. P. Chiusoli, *Acc. Chem. Res.* **6**, 422 (1973).

71b. A. Wojciski, *Adv. Organomet. Chem.* **11**, 88 (1983).

72a. M. L. H. Green, *Pure Appl. Chem.* **56**, 47 (1984).

72b. S. J. Holmes, D. N. Clark, H. W. Turner, and R. R. Schrock, *J. Am. Chem. Soc.* **104**, 6322 (1982).

72c. R. R. Schrock, *Acc. Chem. Res.* **12**, 98 (1979).

72d. Ref. 10, pp. 285–292.

72e. E. L. Muetterties, *Inorg. Chem.* **14**, 951 (1975).

73a. Ref. 10, pp. 246–261.

73b. P. J. Davidson, M. F. Lappert, and R. Pearce, *Chem. Rev.* **76**, 214 (1976).

73c. R. R. Schrock and G. W. Parshall, *Chem. Rev.* **76**, 243 (1976).

74a. Ref. 10, p. 506.

74b. G. Wilke, *Pure Appl. Chem.* **50**, 677 (1978).

75. Ref. 10, pp. 277, 497–502.

76a. F. A. Cotton and W. T. Hall, *J. Am. Chem. Soc.* **101**, 5094 (1979).

76b. H. G. Ala and H. I. Hayen, *Angew. Chem., Int. Ed. Engl.* **22**, 1008 (1983).

76c. J. L. Templeton and B. C. Ward, *J. Am. Chem. Soc.* **102**, 3288 (1980).

77a. J. C. Jeffery, J. C. V. Laurie, and F. G. A. Stone, *J. Organomet. Chem.* **2**, 258 (1983).

77b. M. R. Churchill, H. J. Wasserman, S. J. Holmes, and R. R. Schrock, *Organometallics* **1**, 766 (1982).

77c. R. Hoffman, C. N. Wilker, S. J. Lippard, J. L. Templeton, and D. C. Brower, *J. Am. Chem. Soc.* **105**, 146 (1983).

78a. Ref. 10, p. 43.

78b. R. F. Heck, "Organotransition Metal Chemistry," pp. 11–12. Academic Press, New York, 1974.

78c. G. W. Parshall, "Homogeneous Catalysis," pp. 8–12. Wiley, New York, 1980.

79a. Ref. 10, pp. 44–45.

79b. M. J. S. Dewar and G. P. Ford, *J. Am. Chem. Soc.* **101**, 783 (1979).

79c. M. J. S. Dewar, *Bull. Soc. Chim. Fr.*, **C79**, 18 (1951).

80a. Ref. 10, p. 43.

80b. Ref. 78b, pp. 11–12.

80c. Ref. 78c, pp. 8–12.

81. D. F. Shriver, *Acc. Chem. Res.* **3**, 231 (1970).

82a. H. Werner, *Angew. Chem., Int. Ed. Engl.* **22**, 927 (1983).

82b. J. K. Kotz and D. G. Pedrotty, *Organomet. Chem. Rev.* **A4**, 479 (1969).

82c. H. Werner, *Pure Appl. Chem.* **54**, 177 (1982).

82d. P. J. Staples, *Coord. Chem. Rev.* **11**, 277 (1973).

82e. H. J. Reich and C. S. Cooperman, *J. Am. Chem. Soc.* **95**, 5078 (1973).

82f. P. Meakin, R. A. Schun, and J. P. Jesson, *J. Am. Chem. Soc.* **96**, 277 (1974).

82g. C. A. Tolman, *J. Am. Chem. Soc.* **92**, 4217 (1970).

82h. H. Werner, H. Neukomm, and W. Klavi, *Helv. Chim. Acta* **60**, 326 (1977).

83. Ref. 10, pp. 16–25.

84a. I. Tkatchenko "Genie Chimique" (J. E. Leger, ed.), "Techniques de l'ingénieur," J1 187-1. Paris, 1979 .

84b. J. C. Hayes, P. Jernakoff, G. A. Miller, and N. J. Cooper, *Pure Appl. Chem.* **56**, 25 (1984).

85a. B. A. Dolgoplosk, K. L. Malkovetsky, T. G. Golenko, Y. U. Korshak, and E. I. Timyakova, *Eur. Polym. J.* **10**, 901 (1974).

85b. F. Garnier, P. Krausz, and H. Rudler, *J. Organomet. Chem.* **186**, 77 (1980).

85c. N. Calderon, J. P. Lawrence, and E. A. Ofstead, *Adv. Organomet. Chem.* **17**, 449 (1979).

85d. N. Calderon, E. A. Ofstead, and W. A. Judy, *Angew. Chem. Int. Ed. Engl.* **15**, 401 (1976).

85e. J. Silvestre and R. Hoffmann, *Helv. Chim. Acta* **68**, 1461 (1985).

86a. M. Tsutsui and A. Courtney, *Adv. Organomet. Chem.* **16**, 241 (1977).

86b. B. Gorewit and M. Tsutsui, *Adv. Catal.* **27**, 227 (1978).

86c. B. A. Dolgoplosk and E. I. Tinyakova, *Russian Chem. Rev.* **53**, 22 (1984).

87. D. A. White, *Organomet. Chem. Rev., A,* **3**, 497 (1968); R. Ugo, *Chim. Ind.* **51**, 1319 (1969).

88a. O. Eisenstein and R. Hoffmann, *J. Am. Chem. Soc.* **103**, 4308 (1981).

88b. O. Eisenstein and R. Hoffmann, *J. Am. Chem. Soc.* **102**, 6148 (1980).

89. J. D. Munro and P. L. Panson, *J. Chem. Soc.,* 3475, 3484 (1961); P. Lennon, A. M. Rosan, and M. Rosenblum, *J. Am. Chem. Soc.* **99**, 8426 (1977); M. F. Semmelhack, *Pure Appl. Chem.* **53**, 2379 (1981); J. E. Bäckwall, *Acc. Chem. Res.* **16**, 335 (1983); S. G. Davies, M. L. H. Green, and D. M. P. Mingos, *Tetrahedron* **34**, 3047 (1978); M. H. Chisholm and H. C. Clark, *Acc. Chem. Res.* **6**, 202 (1973).

90. E. O. Fischer and A. Maasbol, *Angew. Chem. Int. Ed. Engl.* **3**, 580 (1964); J. P. Collman, *Acc. Chem. Res.* **8**, 342 (1975); R. J. Angelici, *Acc. Chem. Res.* **5**, 335 (1972); T. L. Brown and P. A. Bellus, *Inorg. Chem.* **17**, 3726 (1978); D. Drew, M. Y. Darensbourg, and D. J. Darens-

bourg, *J. Organomet. Chem.* **85**, 73 (1975); W. Tam, W. K. Wong, and J. A. Gladysz, *J. Am. Chem. Soc.* **101**, 1589 (1979); J. A. Gladysz and W. Tam., *J. Am. Chem. Soc.* **100**, 2545 (1978); C. P. Casey, M. A. Andrews, and J. E. Rinz, *J. Am. Chem. Soc.* **101**, 741 (1979).

91. Ref. 10, p. 307; M. I. Bruce and A. G. Swincer, *Adv. Organomet. Chem.* **22**, 60 (1983); K. H. Dötz, H. Fischer, P. Hofmann, F. R. Kreissl, U. Schubert, and K. Weib, "Transition Metal Carbene Complexes," p. 151. Verlag Chemie, Weinhein, 1983; N. J. Copper and M. L. H. Green, *Chem. Commun.,* 761 (1974); M. L. H. Green, *Pure Appl. Chem.* **50**, 27 (1978).

92. R. G. Schultz, *J. Organomet. Chem.* **6**, 435 (1966); J. Tsuji and H. Takahashi, *J. Am. Chem. Soc.* **87**, 3275 (1965).

93. Ref 78c, p. 173; R. R. Schrock and P. R. Sharp, *J. Am. Chem. Soc.* **100**, 2389 (1978).

94. W. P. Giering, S. Raghn, M. Rosenblum, A. Cutler, D. Ehntholt, and R. W. Fish, *J. Am. Chem. Soc.* **94**, 8251 (1972).

95. Ref 78b, p. 106.

96. M. L. H. Green and P. L. I. Nagy, *J. Chem. Soc.,* 189 (1963); M. L. H. Green and P. L. I. Nagy, *Adv. Organomet. Chem.* **2**, 325 (1964); S. Raghy and M. Rosenblum, *J. Am. Chem. Soc.* **95**, 3060 (1973); M. A. Bernett, K. Hoskins, W. R. Kneen, R. S. Nyholm, P. B. Hitchock, R. Mason, G. B. Robertson, and A. D. C. Towl, *J. Am. Chem. Soc.* **93**, 4591 (1971).

97. J. Ushio, H. Nakatsuji, and T. Yonezawa, *J. Am. Chem. Soc.* **106**, 5892 (1984).

98. M. Cousins and M. L. H. Green, *J. Chem. Soc.,* 889 (1963); M. L. H. Green and A. N. Stear, *J. Organomet. Chem.* **1**, 230 (1964); J. Kochi and D. M. Singleton, *J. Org. Chem.* **33**, 1027 (1968); S. Komiya and J. K. Kochi, *J. Am. Chem. Soc.* **98**, 7599 (1976).

99. J. E. Mahler and R. Pettit, *J. Am. Chem. Soc.* **84**, 1511 (1962); M. Brookhart and G. O. Nelson, *J. Am. Chem. Soc.* **99**, 6099 (1977); J. M. Jerkunica and T. G. Traylor, *J. Am. Chem. Soc.* **93**, 6278 (1971).

100. J. K. Kochi, "Organometallic Mechanisms and Catalysis," p. 285. Academic Press, New York, 1978; A. Sanders, L. Cohen, W. P. Giering, D. Kennedy, and C. V. Magatti, *J. Am. Chem. Soc.* **95**, 5430 (1973).

101. Ref. 87a; B. M. Trost, *Acc. Chem. Res.* **13**, 385 (1980).

102. K. H. Dötz, *Angew. Chem., Int. Ed. Engl.* **23**, 587 (1984).

103. H. Werner, *Angew. Chem., Int. Ed. Engl.* **22**, 927 (1983); K. Öfele, *Angew. Chem., Int. Ed. Engl.,* 916 (1969); J. P. Collman, *Acc. Chem. Res.* **8**, 342 (1975); J. P. Collman, R. G. Finke, J. N. Cawse, and J. I. Brauman, *J. Am. Chem. Soc.* **99**, 2515 (1977); R. G. Pearson, H. Sobe, and J. Songstad, *J. Am. Chem. Soc.* **90**, 319 (1968).

104. R. Ugo, "Catalysis in C1 Chemistry" (W. Keim, ed.), pp. 135–167 Reidel, Dordrecht, 1983.

105. W. D. Wulff and D. C. Yang, *J. Am. Chem. Soc.* **105**, 6726 (1983).

106a. J. K. Kochi, "Organometallic Mechanisms and Catalysis." Academic Press, New York, 1978.

106b. J. Halpern, in "Fundamental Research in Homogeneous Catalysis" (M. Tsutsui and R. Ugo, eds.), p. 25. Plenum, New York, 1977.

106c. M. F. Lappert and P. W. Lendor, *Adv. Organomet. Chem.* **14**, 374 (1976).

106d. M. Chanon, *Bull. Soc. Chim. Fr.,* 197 (1982).

107a. A. V. Kramer, *J. Am. Chem. Soc.* **96**, 9145 (1974).

107b. M. P. Brown, *J. Chem. Soc. Dalton Trans.* 2457 (1974).

107c. J. Evans, *J. Organomet. Chem.,* C37, 81 (1974).

108a. M. Chanon, *Bull. Soc. Chim. Fr.,* 197 (1982).

108b. M. Chanon and M. L. Tobe, *Angew. Chem., Int. Ed. Engl.* **21**, 1 (1980).

108c. Ref. 106a, pp. 15–20.

108d. J. J. Eisch, *Pure Appl. Chem.* **56**, 35 (1984).

108e. E. C. Chukovskaya, R. K. Freidlina, and N. A. Kuzmina, *Synthesis*, 773 (1983).

108f. H. L. M. Van Gaal and J. G. M. Van der Linden, *Coord. Chem. Rev.* **47**, 41 (1982).

109. J. K. Kochi and J. W. Powers, *J. Am. Chem. Soc.* **92**, 13 (1970); Ref. 106b, p. 25.
110. R. M. Dessan, *J. Am. Chem. Soc.* **92**, 6356 (1970).
111. Ref. 106a, pp. 33, 350; I. Y. Levitin, A. L. Sigan, R. M. Bodnar, R. G. Gasanov, and M. E. Volpin, *Inorg. Chim. Acta* **76**, 169 (1983).
112. Ref. 106a, p. 17; J. A. Labinger, J. A. Osborn, and N. J. Coville, *Inorg. Chem.* **19**, 3236 (1980); J. A. Osborn, *J. Am. Chem. Soc.* **94**, 4043 (1972).
113. Ref. 106a, p. 446.
114. K. Jonas, *Pure Appl. Chem.* **56**, 63 (1984); P. R. Jones, *Adv. Organomet. Chem.* **15**, 273 (1977); Ref. 106a, p. 18.
115. M. F. Lappert and P. W. Lendor, *Adv. Organomet. Chem.* **14**, 374 (1976); M. D. Johnson, *Acc. Chem. Res.* **16**, 343 (1983).
116. Ref. 106a, pp. 33–42.
117. Ref. 106a, p. 532.
118. Ref. 106a, pp. 197–203.
119. D. J. Cram and G. Hammond, "Organic Chemistry," p. 609. McGraw-Hill, New York, 1959; H. M. Feder and J. Halpern, *J. Am. Chem. Soc.* **97**, 7186 (1975).
120. Ref. 106a, pp. 172–173. M. O. Albers and N. J. Coville, *Coord. Chem. Rev.* **53**, 227 (1984); J. A. S. Howel and P. M. Burkinshaw, *Chem. Rev.* **83**, 557 (1983); T. L. Brown, *Ann. N.Y. Acad. Sci.* **331**, 80 (1980).
121. Ref. 106a, p. 62.
122. Ref. 106a, p. 148; B. Maillard and K. U. Ingold, *J. Am. Chem. Soc.* **98**, 7024 (1976).
123. R. Cramer, *J. Am. Chem. Soc.* **87**, 4717 (1965).
124a. C. K. Brown and G. Wilkinson, *J. Chem. Soc., A,* 2733 (1970).
124b. G. Yagupsky, C. K. Brown and G. Wilkinson, *J. Chem. Soc., A,* 1392 (1970).
125. R. L. Pruett and J. A. Smith, *J. Org. Chem.* **34**, 327 (1969); H. Seigel and W. Himmele, *Angew. Chem., Int. Ed. Engl.* **19**, 178 (1980); P. Pino, *J. Organomet. Chem.* **200**, 223 (1980); R. L. Pruett, *Adv. Organomet. Chem.* **17**, 1 (1979).
126. L. Bencze and L. Prokai, *J. Organomet. Chem.* **294**, C5 (1985).
127a. Ref. 106a, p. 401.
127b. L. S. Hegedus and D. H. P. Thompson, *J. Am. Chem. Soc.* **107**, 5663 (1985).
128. T. P. Dang and H. B. Kagan, *Chem. Commun.,* 481 (1971); W. Dumont, J. C. Poulin, T. P. Dang, and H. B. Kagan, *J. Am. Chem. Soc.* **95**, 8295 (1973); H. B. Kagan, in "Comprehensive Organometallic Chemistry" (G. Wilkinson, F. G. A. Stone, and E. W. Abel, eds.), Vol. 8, p. 463. Pergamon, Oxford, 1982; B. Bosnich, *Chem. Scripta* **25**, 45 (1985).
129. J. F. Labarre, *Structure Bonding* **35**, 2 (1978); G. R. Brubaker and D. W. Johnson, Coord, Chem. Rev., **53**, 1 (1984); J. I. Seeman and S. G. Davies, *J. Chem. Soc., Chem. Commun.,* 1019 (1984).
130. R. Hoffmann, *Angew. Chem., Int. Ed. Engl.* **21**, 711 (1982); F. G. A. Stone, *Angew. Chem., Int. Ed. Engl.* **23**, 89 (1984).
131. E. Garagnani and J. C. J. Bart, *Z. Naturforsch* **32b**, 455, 465 (1977); J. T. J. Bart and E. Garagnani, *Z. Naturforsch* **32b**, 678 (1977).

π *Bonding to Main-Group Elements*

PETER JUTZI

Fakultät für Chemie
Universität Bielefeld
D-4800 Bielefeld 1, Federal Republic of Germany

I

INTRODUCTION

To most chemists the term "π complex" is connected with compounds in which π ligands are polyhapto bonded to a transition metal. Indeed, this class of compounds is highly important and has experienced fantastic development during the last three decades. On the other hand, polyhapto bonding of π liagnds is not restricted to transition elements. Main-group elements can also serve as central atoms, as documented by many relatively recent examples.

π Complexes of main-group elements have a long history that goes back to the nineteenth century. The first such compounds to be obtained were probably the benzene complexes to a gallium center, described in 1881 by Lecoq de Boisbaudran as crystals which precipitate from solutions of gallium dihalides in benzene (see Section IX,D), and arene complexes to an antimony center, already described in 1879 by Smith as crystals arising from antimony trichloride and different aromatic hydrocarbons (see Section XVI,C).

This article is a comprehensive treatment of π complexes of all main-group elements including their synthesis, structure, and bonding, and covers the literature up to 1985. Compounds described here have been characterized by X-ray or electron diffraction, or at least by conclusive IR or NMR data. No matrix chemistry will be mentioned.

II

π COMPLEXES OF LITHIUM

Many examples of the interaction between lithium and diverse π systems are documented in the literature. In the first review in this field Stucky (*1*) described the structure and bonding in π complexes of N-chelated lithium units. More recently, the synthesis and structure of organolithium compounds, including those containing unsaturated organic systems, have been reviewed by Wardell (*2*). A collection of X-ray crystal structure data of

217

organolithium species exhibiting lithium π interactions has been presented by Schleyer and Setzer (3) in a comprehensive review covering the literature up to 1983. Structures of the most relevant π complexes, including very new results in this field, are schematically represented in Fig. 1 and will be briefly described together with theoretical considerations for some selected π systems.

A. *Alkene Complexes*

To the best of our knowledge, X-ray structural data of complexes with simple dihapto interactions between a lithium atom and the π system of an alkene or alkyne ligand are unknown, but there is some spectroscopic evidence for weak π interactions in solutions of 3-alkenyllithium compounds from ^{7}Li- and ^{1}H-NMR data (4). Interactions of this sort are presumably important in addition (polymerization) reactions between organolithium compounds and alkenes or alkynes.

On the other hand, crystal structures of rather complicated mixed-metal organometallic π complexes are known; the common feature in these species is the interaction of a lithium atom with a π ligand which is further complexed to a transition metal. Strong interactions between the lithium atoms and the transition metals are also indicated. Compounds of this type have previously been reviewed by Jonas (5,6) and by Schleyer and Setzer (3) and will not be discussed here.

B. *Allyl Complexes*

Recently, Boche *et al.* (8) have reported the synthesis of a highly interesting crystalline adduct of 1,3-diphenylallyllithium with diethyl ether. The crystal structure of this complex (**I** in Fig. 1) shows symmetrical π bonding between lithium atoms and allylic fragments. Each lithium atom interacts with two allylic π systems and further with the oxygen atom of a diethyl ether molecule. An exo, exo orientation of the phenyl ligands has been observed in this coordination polymer.

An alternative bonding situation without π interaction has been found by Weiss and Köster (9) for the allyllithium–tetramethylethylenediamine (TMEDA) complex. A low-resolution X-ray structure shows an endless polymer with TMEDA-solvated lithium atoms attached to both terminal allyl carbons.

π-Allyl interactions are also present in the crystal structures of some other hydrocarbon π complexes with LiN_2 units (1–3). A trihapto coordination has been observed in the benzyllithium-1,4-diaza[2.2.2]bicyclooctane complex (10), as well as in the TMEDA adduct of indenyllithium (11) and in the

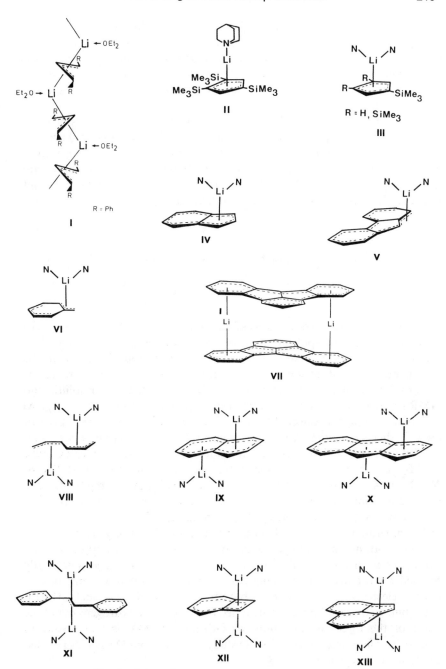

FIG. 1. Schematic structures of π complexes of lithium.

bisquinuclidine complex of fluorenyllithium (12) (structures **VI**, **IV**, and **V** in Fig. 1). Finally, the crystal structure of the complex dilithiohexadiene (TMEDA)$_2$ (13) (structure **VIII** in Fig. 1) can be roughly approximated as two C—C connected allylic systems each forming a π complex with a lithium–TMEDA unit. Some further delocalization energy is very likely contributed by the coplanarity of the two π systems.

The question concerning the nature of allyllithium in solution is of great interest (2). The interpretation of experimental studies is complicated because the bonding situation is considerably influenced by solvent and temperature effects (14,15). Possible structures for allyllithium species in solution are aggregates (**A**), contact ion pairs (**B** and **C**), solvent separated ion pairs (**D**) and free ions (**E**) (bases coordinated to the lithium atoms have been omitted for clarity).

Recent ^{13}C-NMR investigations of allyllithium in tetrahydrofuran (THF) solution based on the application of Saunders' isotopic perturbation method rule out a dynamic equilibrium between species of type **B**. More or less symmetrically π-bridged species (type **C**) are generally concluded to represent the structure of allyllithium in this solvent (16–19). In dilute THF solution no NMR spectroscopic evidence is found for the presence of aggregates (type **A**) (19), though at higher concentrations association has been observed, also with diethyl ether as solvent (20). Allyllithium in THF solution gives rise to an **AA′BB′C**-type ^{1}H-NMR spectrum at low temperatures; at room temperature an **AB**$_4$ spectrum is observed due to topomerization processes (20). Recent NMR investigations of some substituted allyllithium compounds give detailed information concerning the conformations, the barriers to rotation and the spatial position of the hydrogen atoms (2,21,22).

A comparative *ab initio* study of the geometry and bonding in allyllithium, -sodium, and -magnesium hydride has been presented by Schleyer et al. (23). Monomeric, unsolvated allyllithium is predicted to have a symmetrically bridged π structure. The bonding can be best interpreted as consisting of mainly electrostatic interactions, but the Li 2p and 2sp functions also contribute to the bonding. The calculated bending of the hydrogens out of the plane of the three carbon atoms as indicated in **XIV** is mainly due to electrostatic interactions. An MNDO study by Boche and Decher (24) has led to the same conclusions.

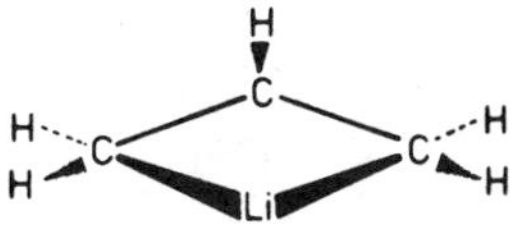

XIV

C. Cyclopentadienyl, Indenyl, and Fluorenyl Complexes

Since the discovery of ferrocene and the development of metallocene chemistry, the compound cyclopentadienyllithium, C_5H_5Li, has been of considerable interest to both experimentalists and theoreticians. In its physical properties cyclopentadienyllithium qualitatively behaves like a typical salt. It is nearly insoluble in noncoordinating solvents and possesses a high melting (decomposition) point and a very low volatility. To our knowledge the crystal structure of C_5H_5Li is unknown.

A drastic change in physical behavior can be observed when three of the hydrogens in the cyclopentadienyl unit are substituted by trimethylsilyl groups. The resulting tris(trimethylsilyl)cyclopentadienyllithium, $(Me_3Si)_3$-H_2C_5Li, behaves like a covalent species, soluble even in nonpolar solvents like hexane and existing as a monomer in the gas phase (25). Crystalline adducts with some nitrogen donors have been obtained (25); the most interesting is the 1:1 adduct with quinuclidine (complex II), which exists as a monomeric species in the solid state, in solution, and in the gas phase. An X-ray structure determination show that the metal is unusually coordinated. The cyclopentadienyl unit is pentahapto bonded to the lithium atom, which—in contrast to all other comparable π-complexes (see Fig. 1)—is further stabilized by only one monodentate ligand. This complex fulfills the criteria for a nido cluster. The distance of the lithium atom to the centroid of the planar cyclopentadienyl ring is found to be 1.79 Å and is therefore considerably shorter (~0.2 Å) than in the corresponding Li–TMEDA complex; the same is

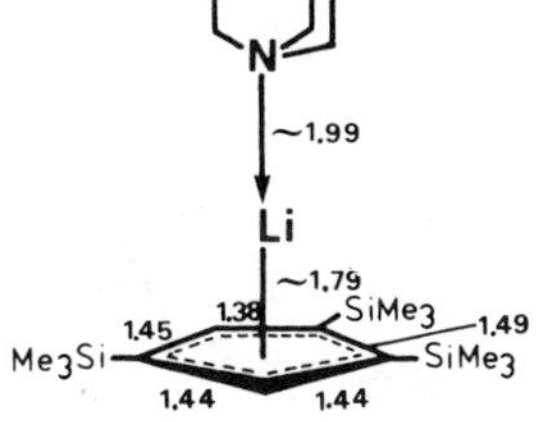

II 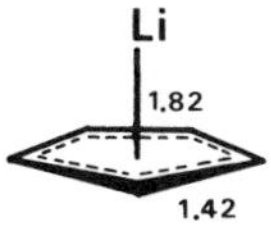   XV

true of the lithium–nitrogen distance which is found to be 1.99 Å in comparison with ∼ 2.2 Å in the TMEDA complex.

Many calculations on cyclopentadienyllithium at different levels have been published during the last decades (26–32). On the basis of *ab initio* MO calculations Streitwieser *et al.* (29) have predicted that in the most stable configuration of C_5H_5Li the lithium is situated symmetrically above the cyclopentadienyl ring; the authors thus felt justified in describing cyclopentadienyllithium as the simplest metallocene (see **XV**). The bond length between lithium and the centroid of the cyclopentadienyl ring is in a very shallow well and may be considered optimal at 1.82 Å, a value very similar to that found experimentally in compound **II**. According to these calculations the hydrogens in cyclopentadienyllithium should be bent out of the C_5 plane away from the lithium for simple electrostatic reasons: such bending puts more negative charge on the side of the ring facing the lithium cation (32). There is still some controversy over the extent of multicenter covalent bonding in these lithium π complexes, where ionic interactions clearly dominate. According to calculations by Schleyer and Jemmis (30) some degree of covalency must be implicated to explain the bending of the hydrogens, by which better overlap between ring π orbitals and the lithium p orbitals is obtained. According to the above calculations and those by Cowley and Lattman (31), the HOMO in cyclopentadienyllithium is doubly degenerate and localized predominantly at the C_5H_5 ring, but some contribution of the lithium $2p_{x,y}$ AO's is evident from the charge density data. Schleyer and Jemmis (30) have treated cyclopentadienyllithium as an aromatic six-membered nido polyhedron belonging to the same family as other cyclopentadienyl systems with isolobal main-group or transition metal fragments. "Aromaticity in three dimensions" describes the bonding in these compounds.

In his classical review, Stucky (1) has already mentioned that in many π complexes the position of the lithium–base fragment is determined by the interaction of the frontier orbitals in the π fragment with the relevant orbitals at the lithium atom. This is nicely demonstrated by a series of cyclopentadienyl-, indenyl-, and fluorenyllithium complexes containing two further nitrogen atoms coordinated to the lithium as portrayed in Figs. 1 and 2.

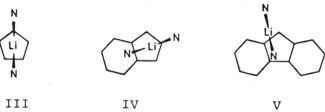

III   IV   V

FIG. 2.   Position of the LiN_2 unit in the complexes **III**, **IV**, and **V**.

In tris(trimethylsilyl)cyclopentadienyllithium · TMEDA (25) as well as in trimethylsilylcyclopentadienyllithium · TMEDA (33) (compounds **III**), a symmetrical pentahapto coordination is observed; in indenyllithium · TMEDA (11) (**IV**) there is a tendency toward orbital-controlled trihapto coordination; finally, in fluorenyllithium bis(quinuclidine) (12) (V) an allylic trihapto coordination is favored.

D. Other π Complexes

One of the most interesting structures in the series of π organolithium compounds has been published by Rewicki and co-workers (34). Metallation of indenofluorene with butyllithium yields ruby-red crystals of the first organolithium sandwich, having two lithium atoms in a linear coordination with two hexahapto-bonded six-membered ring ligands (see **VII**, Fig. 1). The molecular geometry in this metallocenophane is consistent with both a simple electrostatic and a molecular orbital model. There are several examples in the literature where two protons in a dihydroolefinic or dihydroaromatic hydrocarbon are abstracted by N-chelated organolithium reagents to give the corresponding dilithium compounds (1–3). As portrayed in Fig. 1, for some of these compounds [**XI** (35), **XII** (36), and **XIII** (37)] inverse sandwich structures have been found, with two lithium atoms coordinating to opposite sides of a π system. In other cases, as, for example, in **VIII** (13), **IX** (38), and **X** (39), the two lithium atoms are not situated directly opposite to each other. The location of the lithium atoms in some situations is consistent with simple electrostatic interactions; in other situations some covalent bonding is necessary to explain the observed structure.

Very recently an X-ray structure of a π-conjugated trimethylenemethane dianion derivative complexed to two TMEDA-solvated lithium cations has been described (40). In this π complex the most symmetrical form, with the lithium atoms lying along the threefold axis, is not lowest in energy; instead, a less symmetrical structure was found.

III

π COMPLEXES OF THE HEAVIER ALKALI METALS

There are many examples known of π bonding between the heavier alkali metals and carbanionic fragments. In most cases evidence for π complexation stems from spectroscopic studies in solution (2); here we restrict ourselves to those situations where π interactions have been proved by X-ray structural

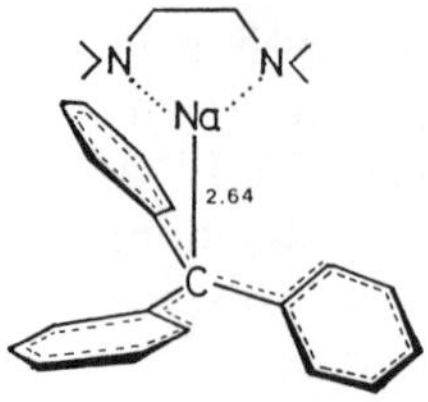

FIG. 3.   Simplified structures of π complexes of the heavier alkali metals.

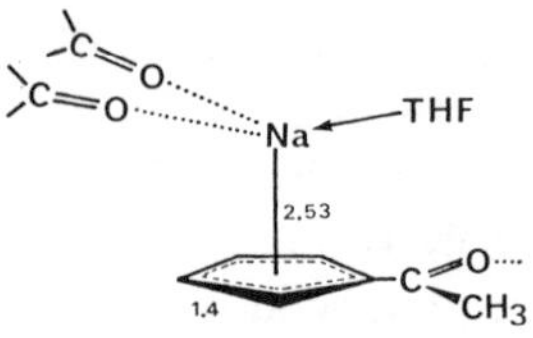

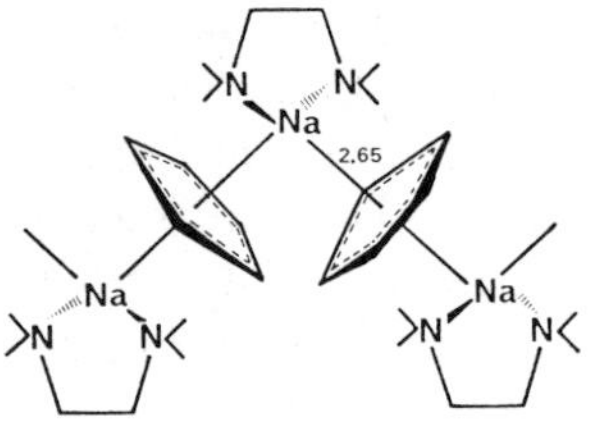

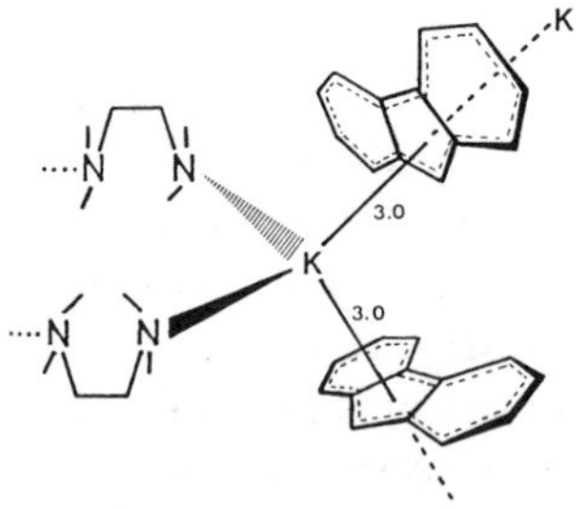

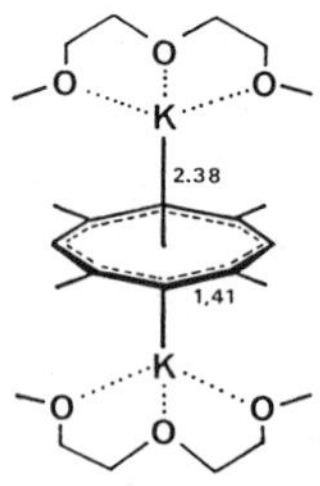

data. In Fig. 3 the simplified structures of those π complexes of the heavier alkali metals so far known in the literature are illustrated.

The crystal structure of the red triphenylmethylsodium·TMEDA complex (compound **XVI** in Fig. 3), published by Weiss and Köster (41), resembles that of the red triphenylmethyllithium·TMEDA complex (42) and can be described as a π complex between a triphenylmethyl carbanion with an sp^2-hybridized central carbon atom and a sodium cation coordinated to the bidentade ligand TMEDA. The sodium atom has close contacts to several carbon atoms of the triphenylmethyl ligand, which possesses twisted phenyl groups. An additional short distance exists between sodium and a p-C (phenyl) atom of a neighboring π system.

The crystal structure of the acetylcyclopentadienylsodium–tetrahydro-furan adduct (compound **XVII** in Fig. 3) has been obtained only recently (43). In this π complex the coordination sphere around each sodium consists of the oxygen atoms of two neighboring acetyl groups, the oxygen of a THF mole-cule, and a rather symmetrical contact between the sodium and the π system of the cyclopentadienyl unit. Several adducts of cyclopentadienylsodium and Lewis bases have been synthesized by Wade et al. (44); the complex with TMEDA (compound **XVIII** in Fig. 3) is structurally characterized. The struc-ture consists of puckered chains of sodium atoms, each with a chelating TMEDA molecule attached, linked by bridging cyclopentadienyl groups each bisecting the Na—Na axes. The degree of puckering of the chain reflects the distorted tetrahedral coordination of the sodium atoms.

The crystal structure of the fluorenylpotassium·TMEDA complex (compound **XIXa** in Fig. 3) has been solved by Stucky et al. (45). The coordination sphere of the potassium atom is made up of two tertiary amines and two unsaturated groups instead of two tertiary amines and one unsaturated organic group as found for the lithium atom in the fluorenyllithium complex **VI** (12).

The structure of a potassium salt of a cyclooctatetraene dianion has been determined by Raymond and co-workers (269). In this bright yellow complex (**XIXb** in Fig. 3) a diglyme molecule is coordinated to each of the potassium atoms through all three oxygen atoms. Two potassium–diglyme units lie on either side of the planar carbocyclic ring equidistant from the ring center.

Finally, two other types of π coordination to sodium documented by crystal structure data will be mentioned. π-Type bonding interactions between bis(THF)sodium units and the benzene rings of complex aluminate anions derived from naphthalene or anthracene have been found in the compounds $[Na(THF)_2]_2[Me_2AlC_{10}H_8]_2$ and $[Na(THF)_2]_2[Me_2AlC_{14}H_{10}]_2$ (46). Even more complex coordination patterns between sodium, transition metals, and π systems have been reported by Jonas and Krüger (5).

The different type of π bonding in the lithium and in the heavier alkali metal complexes can be deduced from the metal–ligand centroid distances, which are collected for some comparable alkali metal π complexes in Table I. In the complexes with the heavier metals these distances are greater by the value $\delta\Delta$ than expected from the differences in the ionic radii ΔM^+. As already pointed out by Stucky (1), the bonding in the sodium and potassium π complexes can

TABLE I

METAL–LIGAND CENTROID DISTANCES FOR SOME ALKALI METAL π Complexes

π Complexes	Metal–ligand centroid distances (Å)	Δ (Å)	ΔM^+ (Å) (Pauling)	$\delta\Delta$ (Å)
Ph$_3$CNa·TMEDA (**XVI**)	2.64	0.42	0.20	0.22
Ph$_3$CLi·TMEDA	2.22			
(CH$_3$CO)C$_5$H$_4$Na·THF (**XVII**)	2.53	0.55	0.20	0.35
(Me$_3$Si)$_3$C$_5$H$_2$Li·TMEDA (**III**)	1.98			
C$_5$H$_5$Na·TMEDA (**XVIII**)	2.65	0.67	0.20	0.47
(Me$_3$Si)$_3$C$_5$H$_2$Li·TMEDA (**III**)	1.98			
(Fluorenyl)K·TMEDA (**XIX**)	3.00	1.00	0.55	0.45
(Indenyl)Li·TMEDA (**IV**)	2.00			

be best explained by the assumption of nondirectional, predominantly electrostatic interactions between the anionic π system and the solvated alkali metal cation.

IV

π COMPLEXES OF BERYLLIUM

A. Alkyne Complexes

To our knowledge there is no experimental evidence for π bonding between a BeR_2 unit and a simple alkene or alkyne ligand. However a π interaction has been found in the X-ray crystal structure of the dimeric dipropynylberyllium trimethylamine adduct (48). The unit cell of this molecule contains two independent centrosymmetric dimers in which the alkynyl groups exhibit different types of interaction with the beryllium atoms, one of them forming an effectively electron-precise dimer unit by π interaction as portrayed in **XX**.

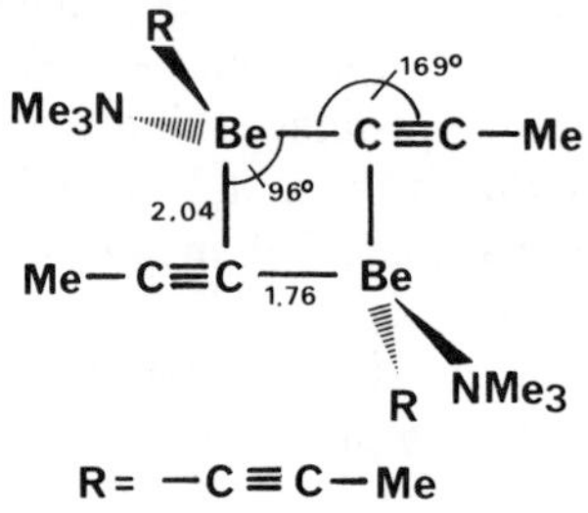

$$R = -C \equiv C - Me$$

B. Cyclopentadienyl Complexes

The synthesis of the first cyclopentadienylberyllium compound was reported by Fischer and Hofmann in 1959 (49). Reaction of cyclopentadienyl alkali-metal compounds with $BeCl_2$ in diethyl ether or benzene leads to dicyclopentadienylberyllium (beryllocene), a rather volatile, colorless, diamagnetic complex, easily soluble in benzene, diethyl ether, and hydrocarbons, and very sensitive to air and moisture.

The structure of this interesting π complex has stimulated controversial discussion and is still a subject of debate. From the large dipole moment ($\mu_{25°C} = 2.46 \pm 0.06$ D in benzene, 2.24 ± 0.09 D in cyclohexane), it is evident that the structure must be unsymmetrical. An X-ray structural investigation at low temperature ($-120°C$) shows that the molecule assumes a sandwich structure

F (*50*), in which one of the two nearly parallel rings has slipped sideways in its plane by about 1.2 Å. The distance between the planes of the rings is 3.34 Å. The beryllium atom is 1.53(3) Å from the plane of one ring on the fivefold symmetry axis with all beryllium–carbon distances equal [1.94(4) Å] and is therefore considered to be π bonded. The second ring is essentially parallel to the first and is lying at a perpendicular distance of 1.81(5) Å from beryllium, but has slipped sideways and may therefore be considered to be weakly π bonded. An alternative bonding description regards the structure as a pentahapto–trihapto constellation (*51*), in agreement with Raman spectral data (*52*). At room temperature the large average thermal motion prevents a precise structure determination (*53*).

The electron diffraction pattern (*54*) of beryllocene is consistent with a structure **G**, which contains two planar and parallel cyclopentadienyl rings and a beryllium atom, which moves freely between two positions on the fivefold rotation axis. This C_{5v} structure, although highly unusual, would provide a nice explanation for the dipole moment of the molecule, but a recent reinvestigation of the electron diffraction data (*55*) indicates that a slip-sandwich model similar to the crystal structure also agrees adequately with the observed data. A slip-sandwich model is also in agreement with the PE spectrum of beryllocene. A model of type **G** is compatible with the PE data only if an exceptionally large Jahn–Teller splitting is allowed (*56*). The ^{1}H-NMR spectrum of beryllocene consists of a single sharp singlet even at $-135°C$ and is temperature independent (*57*), typical for a highly fluxional molecule. Very recently, a structure similar to that of beryllocene has been found for the isovalent bis(pentamethylcyclopentadienyl)zinc in the gas phase (*58*).

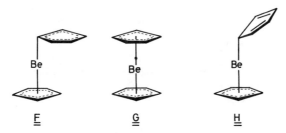

In several theoretical papers at different degrees of sophistication attempts have been made to determine the most favorable structure for beryllocene (*59–66*). *Ab initio* molecular orbital calculations indicate that in the lowest energy form one pentahapto- and one monohapto-bonded cyclopentadienyl ring are present, as portrayed in structure **H**. Neither the slip sandwich **F** nor the off-center double well potential structure **G** suggested experimentally are indicated to be favorable. Even the symmetrical D_{5d} sandwich structure [found experimentally for $Mg(C_5H_5)_2$] is calculated to be more stable than structure **G**. Comparing the experimental and theoretical results, there is still some disagreement concerning the most stable structure of beryllocene. The rather flat energy surface indicated both experimentally and by various calculations undoubtedly contributes to this situation.

During the last decade a great number of beryllium compounds containing only one cyclopentadienyl ligand have been described, mainly by the groups of Morgan, Coates, and Gaines. The structures for most of these complexes are displayed graphically in Fig. 4. Most of these compounds have been synthesized by the reaction between beryllocene and the corresponding disubstituted beryllium compound BeR_2 leading to the mixed products C_5H_5BeR with R = H (*67,81*), Cl (*68*), Br (*69*), I (*71*), CH_3 (*70,81*), t-C_4H_9 (*72*), C_6F_5 (*72*), C_6H_5 (*73*), C≡CH (*74*), and C≡C—CH_3 (*74*) (see structures **XXIa–e** in Fig. 4). The cyclopentadienyl complexes **XXIf–i** with different borate ligands have been synthesized by other routes. Reaction of cyclopentadienyl beryllium chloride with $LiBH_4$ and KB_5H_8 leads to complexes **XXIf** (*75*) and **XXIi** (*76*), respectively. The synthesis of compounds **XXIg** (*77*) and **XXIh** (*78*) has been achieved by the reaction of NaC_5H_5 with $Be(B_3H_8)_2$ and $B_5H_{10}BeBr$, respectively.

All the complexes **XXIa–i** are air-sensitive colorless solids or liquids. Structural parameters are available mainly from X-ray structure data (*76,266*) and from electron diffraction and microwave studies (*67–69,71,79,80*). In nearly all cases the molecules are monomeric in the solid state, in solution, and

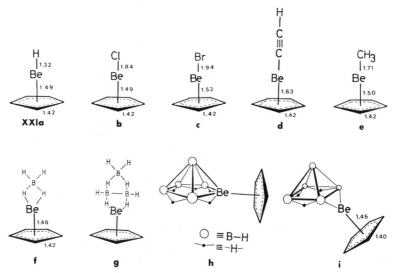

FIG. 4.   Structures of monocyclopentadienyl complexes of beryllium.

in the gas phase, with pentahapto coordination of the cyclopentadienyl ligand. The beryllium to ring centroid distances are all similar to that to the nearest ring in beryllocene. The Be—CH_3 bond distance in $C_5H_5BeCH_3$ (**XXIe**) is not significantly different from the Be—C bond distance in dimethylberyllium [1.698(5) Å], but the Be—Cl bond distance in C_5H_5BeCl (**XXIb**) is significantly longer in the gas phase [1.837(6) Å] and in the solid state [1.869(3) Å] than in monomeric beryllium dichloride (1.75 Å). It seems reasonable to assume that there is less dative Cl—Be π bonding in C_5H_5BeCl than in $BeCl_2$. The hydrogen atoms of the cyclopentadienyl ring in **XXIb** show a slight tendency to bend toward the beryllium atom (266), in agreement with theoretical predictions (63). In the π complex **XXIi** the cyclopentadienylberyllium moiety resides in a nonvertex bridging position between two adjacent basal boron atoms in a square-pyramidal framework. In complex **XXIh** the beryllium appears to be incorporated as a vertex in a six-atom nido framework that is structurally and electronically similar to that of the pentagonal pyramidal B_6H_{10} molecule (82).

Some NMR data of the above complexes are worth mentioning. It is interesting for the general discussion of bonding in these π complexes that in the ^{13}C-NMR spectrum of cyclopentadienylberyllium bromide (**XXIc**) a ^{13}C—^{9}Be coupling has been observed (83). Rapid intramolecular hydrogen exchange takes place in the borate units of the compounds **XXIf** and **g** (75,77).

Theoretical calculations at different levels have been performed for monocyclopentadienylberyllium compounds, especially for cyclopentadienyl-beryllium hydride (63). It has been shown that C_5H_5BeH strongly prefers C_{5v} symmetry and pentahapto bonding with a high degree of covalent character. The direction of the dipole moment [experimentally 2.08 D (67)], revealed by the calculations to have the negative end toward the BeH group, argues against ionic bonding of the type $C_5H_5^-$ BeH^+ and implies considerable electron donation from the ring to the beryllium orbitals.

The bonding in the C_5H_5BeR compounds can easily be rationalized by considering the linear combination of C_5H_5 and BeR fragment orbitals, as indicated in Fig. 5a. The degenerate C_5H_5 π orbitals (e_1) interact with the p_x and p_y orbitals of an sp-hybridized BeR fragment, while the lowest C_5H_5 π orbital (a_1) interacts with the BeR sp hybrid. The resulting three bonding molecular orbitals are occupied by six electrons, giving rise to a "three-dimensional aromaticity" (30). This type of bonding is characteristic for nido clusters with a pentagonal pyramidal structure (84). In C_5H_5BeR compounds where the ligand R possesses a degenerate set of orbitals with the proper symmetry (lone pairs or π orbitals) further interactions can take place leading to a stabilization of the e_1 set in the C_5H_5Be unit and a destabilization of the degenerate orbital set centered mainly at the ligand, as indicated schematically

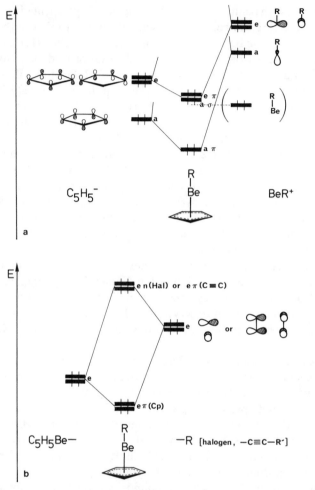

FIG. 5. Qualitative diagram between the fragment orbitals of (a) a C_5H_5 and a BeR unit and (b) a C_5H_5Be and a halogen or acetylenic unit.

in Fig. 5b. Molecular orbital energies have been calculated for the compounds **XXIb–e** and for C_5H_5Be—$C\equiv C$—CH_3 using a modified CNDO procedure. The data obtained are in close agreement with the measured vertical ionization potentials derived from PE spectra of these complexes (see Table II) (74). In contrast to beryllocene (56), where the split of the first band in the PE spectrum is found to be 1 eV, the C_5H_5BeR compounds show a small Jahn–Teller splitting of the order of 0.1–0.2 eV, supporting the conclusion that ionization occurs from a weakly bonding molecular orbital. The splitting of

TABLE II

MEASURED VERTICAL IONIZATION POTENTIALS (I_v) AND CALCULATED ORBITAL ENERGIES ($-\varepsilon_j$)
FOR SOME C_5H_5BeR COMPOUNDS[a]

		e_π (C≡C)	e_π (Cp)	e_n (Hal)	σ (Be—Hal)	σ (Be—C)
C_5H_5BeCl	$-\varepsilon_j$	—	11.57	9.88	11.83	—
	I_v	—	11.15; 11.5	9.60; ~9.9	12.45	—
C_5H_5BeBr	$-\varepsilon_j$	—	11.30	9.23	11.77	—
	I_v	—	10.79; ~11.0	9.52; 9.78	12.00	—
$C_5H_5BeC≡CH$	$-\varepsilon_j$	9.69	11.21	—	—	12.12
	I_v	9.40; ~9.6	10.30; ~10.5	—	—	12.36
$C_5H_5BeC≡C—CH_3$	$-\varepsilon_j$	9.33	10.98	—	—	12.11
	I_v	8.82; ~9.1	9.85; ~10.15	—	—	11.95
$C_5H_5BeCH_3$	$-\varepsilon_j$	—	10.14	—	—	10.03
	I_v	—	9.43; ~9.8	—	—	~10.3

[a] All values in eV (73).

0.26 eV found in the first band of the PE spectrum of **XXIc** is typical for the spin–orbit coupling in bromides.

C. Dicarbollyl Complexes

There are two examples in the literature where a beryllium atom is π complexed to the open face of the (3)-1,2-dicarbollide ion. Dimethyl- or diethylberyllium dietherate reacts with $7,8-C_2H_2B_9H_{11}$ in benzene/ether solution at room temperature to produce one equivalent of alkane; a second equivalent is evolved during removal of the solvent at 50°C, and the crystalline complex $Et_2O \cdot BeC_2H_2B_9H_9$ (**XXIIa**) is formed (85,86). This compound is very sensitive to moist air and hydrolyzes with the formation of $7,8-C_2H_2B_9H_{10}^-$. Treatment of the ether complex with trimethylamine yields the much more stable amine complex $Me_3N \cdot BeC_2H_2B_9H_9$ (**XXIIb**). The spectroscopic data are in accord with a closo icosahedral structure for these beryllium compounds with the ether or amine ligand coordinated to the beryllium atom, and therefore they are formulated as π complexes.

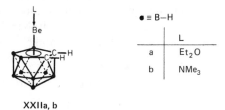

XXIIa, b

● ≡ B—H

	L
a	Et_2O
b	NMe_3

V

π COMPLEXES OF MAGNESIUM

A. Alkene Complexes

To our knowledge there is no example in the literature of a stable alkene (or alkyne) complex with magnesium as central atom. π Complexes have, however, been postulated as intermediates in the rearrangement reactions of some alkenyl Grignard reagents (86).

B. Cyclopentadienyl Complexes

During the genesis of metallocene chemistry the first π complex of magnesium was synthesized in the groups of Fischer and Wilkinson (87–89). Dicyclopentadienylmagnesium (**XXIII**, magnesocene) can be prepared by the thermal decomposition of cyclopentadienylmagnesium bromide as a colorless and very air-sensitive, crystalline compound, isomorphous with other metallocenes. Direct synthesis starting from cyclopentadiene and magnesium metal is also possible at higher temperatures (560–600°C) (90). A refined X-ray diffraction study (91) shows a typical sandwich structure (**XXIIIa**) for magnesocene with Mg—C and C—C distances of 2.304(8) and 1.39(2) Å, respectively. In the crystalline state the two parallel rings have a staggered conformation, whereas the electron-scattering pattern is consistent with an eclipsed conformation (**XXIIIb**) in the gas phase (see Fig. 6) (92). An asymmetric structure similar to that of beryllocene can be ruled out. 1,1'-Dimethyl-, decamethyl- and 1,2,4,1',2',4'-hexakis(trimethylsilyl)magnesocene have been obtained by the reactions shown in Eqs. (1), (2) and (3) (93,94,111). The π structures of these complexes have been assigned from spectroscopic data.

$$MeH_4C_5MgBr \xrightarrow{\Delta} (MeH_4C_5)_2Mg + MgBr_2 \tag{1}$$

$$Me_5C_5MgBr \xrightarrow{dioxane} \tfrac{1}{2} Mg(C_5Me_5)_2 + \tfrac{1}{2} MgBr_2 \cdot dioxane \tag{2}$$

$$MgMe_2 + 2 C_5H_3(SiMe_3)_3 \xrightarrow[\text{TMEDA}]{Et_2O} Mg[C_5H_2(SiMe_3)_3]_2 + 2 MeH \tag{3}$$

Mono-π-cyclopentadienylmagnesium compounds have so far only been isolated in the form of adducts with oxygen or nitrogen bases. The bis-(tetrahydrofuran) complexes of cyclopentadienyl(phenyl)- and -(methyl)-magnesium have been synthesized by Whitesides and co-workers (96) according to Eq. (4).

$$\tfrac{1}{2}(H_5C_5)_2Mg + \tfrac{1}{2} R_2Mg \rightarrow C_5H_5MgR \xrightarrow{2L} C_5H_5MgR \cdot L_2 \tag{4}$$

$$R = Me, Ph; L = THF$$

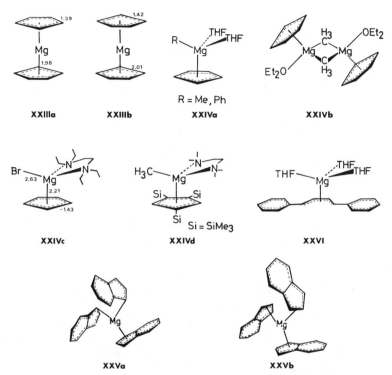

FIG. 6. Structures of π complexes of magnesium.

These compounds are monomeric in solution and are thought to possess the π structures as indicated in **XXIVa** (Fig. 6). Reaction of cyclopentadiene with dimethylmagnesium in diethyl ether leads to an ether adduct of cyclopentadienyl(methyl)magnesium (*97*) [Eq. (5)].

$$C_5H_6 + Me_2Mg \xrightarrow{\text{Et}_2\text{O}} \tfrac{1}{2}[C_5H_5MgMe \cdot OEt_2]_2 + MeH \qquad (5)$$

From IR data it appears that in the solid state the magnesium atoms are associated via methyl bridge bonds and that the cyclopentadienyl rings are approximately pentahapto bonded as indicated in structure **XXIVb**. When dissolved in benzene, the methyl bridge bonds are retained.

The TMEDA complex of tris(trimethylsilyl)cyclopentadienyl(methyl)-magnesium (**XXIVd** in Fig. 8), which is monomeric in benzene solution, has been obtained similarly (*94*). Two Grignard reagents with a π-bonded cyclopentadienyl ligand have been isolated so far. The cyclopentadienyl-magnesium bromide tetraethylethylenediamine adduct **XXIVc** was crystallized from a solution of C_5H_5MgBr in diethyl ether by slow addition of base (*98*). A crystal structure analysis has confirmed the π structure. The closest approach of the magnesium atom to the plane of the cyclopentadienyl group is 2.21 Å with an average Mg—C distance of 2.55 Å. The terminal Mg—Br distance (2.63 Å) is the same as for the six-coordinate magnesium in $MgBr_2(THF)_4$, but significantly longer than that for the four-coordinate magnesium in $EtMgBr(OEt_2)$. An adduct of tris(trimethylsilyl)cyclopenta-dienylmagnesium bromide and TMEDA has been isolated from the reaction of methylmagnesium bromide and tris(trimethylsilyl)cyclopentadiene in the presence of TMEDA [Eq. (6)] (94). The π structure of this complex has been confirmed by spectroscopic data. Considerable disagreement still exists as to whether the bonding between the cyclopentadienyl unit and the magnesium atom in cyclopentadienylmagnesium compounds should be classified as ionic or covalent.

$$MeMgBr + C_5H_3(SiMe_3)_3 + TMEDA \xrightarrow{Bu_2O} [C_5H_2(SiMe_3)_3]MgBr \cdot TMEDA + MeH$$

$$(6)$$

Magnesocene itself and the substituted derivatives occupy an interesting position at the lighter end of the series of dicyclopentadienyl sandwich compounds and are distinguished by the fact that the central atom has no d electrons available for bonding (100–102). A qualitative molecular orbital scheme for $Mg(C_5H_5)_2$ with the assumption of D_{5d} symmetry is depicted in Fig. 7; only the interactions between the metal valence AOs and the ligand π orbitals are considered. The highest occupied MOs (e_{1g}) are supposed to be essentially nonbonding orbitals belonging to the ligand π framework. The subsequent MOs (e_{1u}) should have the most significant bonding character. Assuming that Koopman's theorem is valid, vertical ionization energies obtained from the PE spectra of some magnesocenes have been assigned to the molecular orbitals from Fig. 7 (101,102). The relevant data and assignments are collected in Table III. In the permethylated magnesocene ionization energy shifts of about 1 eV have been observed compared to the unsubstituted compound (102).

A comprehensive summary of theoretical arguments in favor of covalent or ionic bonding in magnesocene has been presented recently in a paper by Lüthi and co-workers (100). Ab initio MO–LCAO investigations predict the charge separation in magnesocene to be slightly larger than in ferrocene, though not sufficiently so to justify the classification of one as ionic and the other as covalent. The metal–ring bond in magnesocene is much weaker than in ferrocene, as indicated by the lower force constant of the metal–ring stretch and by the longer metal–ring distance (91). From the chemical shifts and line widths observed in ^{25}Mg-NMR spectra, it has been concluded that mainly covalent interactions determine the bonding in magnesocene and its derivatives (103).

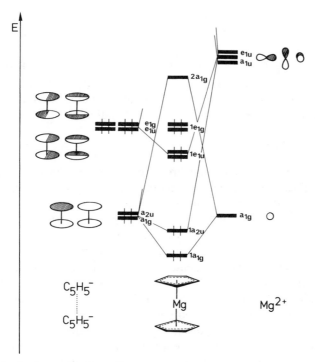

Fig. 7. Qualitative molecular orbital scheme for $Mg(C_5H_5)_2$ in the staggered D_{5d} configuration.

TABLE III

VERTICAL IONIZATION ENERGIES[a] AND THEIR ASSIGNMENTS FOR SOME
MAGNESOCENES

	$e_{1g}(\pi)$	$e_{1u}(\pi)$	a_{1g}, a_{2u}, σ C—H
$Mg(C_5H_5)_2$ (101)	8.11	9.03	12.2
	8.23	9.26	12.5
	8.44	—	13.5
$Mg(C_5H_4Me)_2$ (101)	7.78	8.62	11.7
	7.90	8.86	12.4
	8.10		13.0
$Mg(C_5Me_5)_2$ (102)	7.06	7.75	11.09
			12.38
			13.98

[a] All values in eV.

Some changes in bonding may occur going from the di- to the mono-cyclopentadienyl magnesium compounds. For instance, in the crystal structure of the Grignard complex **XXIVc** longer Mg—C and cyclo-pentadienyl ring C—C distances than in magnesocene have been observed. Judging from these data and from the similarity in the coordination pattern of the monocyclopenatdienyl π complexes of magnesium and sodium, essentially ionic interactions appear to determine the bonding in these π complexes.

C. Other π Complexes

Reaction of ethylmagnesium bromide with indene yields indenylmagne-sium bromide, which on heating decomposes to diindenylmagnesium (**XXV**) (*104,105*), an air-sensitive, colorless, crystalline solid, soluble in ethers and slightly soluble in aromatic hydrocarbons. The crystal structure analysis (*105*) reveals the magnesium atom in two different environments with both terminal and bridging indenyl groups. One magnesium is bonded to the five-membered ring of an indenyl ligand in a pentahapto fashion and to two others through essentially only one carbon. The second magnesium is coordinated to one ring in a pentahapto fashion and to two others through two carbon atoms. The closest carbon to magnesium approach is 2.26(1) Å, the average pentahapto ring carbon to magnesium distance is 2.43 Å. Each magnesium atom is coordinated to three indenyl moieties, as indicated in **XXVa,b** in Fig. 6; the substance therefore exists in an infinite polymeric arrangement. The complex-ity of the structure is probably determined by the most effective packing of indenyl groups in the lattice.

Several magnesium–diene complexes prepared from 1,3-dienes and acti-vated magnesium have already been described in the literature (*106*), but only recently has the first X-ray structure analysis been presented (*107*). The compound tris(tetrahydrofuran)magnesium–s-cis-1,4-diphenylbutadiene can be regarded as containing a magnesium dication coordinated by three oxygens of the THF ligands and the π system of a dianionic butadiene fragment, as indicated schematically in **XXVI** (Fig. 6). However, a bonding situation similar to that discussed for the related tris(tetrahydrofuran)magnesium–anthracene (*110*) cannot be excluded.

VI

π COMPLEXES OF THE HEAVIER ALKALINE-EARTH METALS

Many compounds of the heavier alkaline-earth metals with organic li-gands are described in the literature, for which from spectroscopic data π-interactions can be anticipated. Recent reviews by Gowenlock (*108*) and Lindsell (*86,108*) provide comprehensive information in this field of chemistry.

Organic fragments which can serve as π ligands include cyclopentadienyl, indenyl, and fluorenyl systems as well as polynuclear aromatic anionic and dianionic species. From IR and NMR data it has been concluded that the bonding in these complexes may be regarded as fully ionic.

To our knowledge dicyclopentadienylcalcium, prepared by the reaction of calcium metal with cyclopentadiene in THF, is the sole example whose structure in the crystalline state has been determined. This structure shows interesting features (see **XXVII** *108–110*). In the crystal lattice of $(C_5H_5)_2Ca$ each Ca atom is associated with four planar cyclopentadienyl rings which are bridging to varying degrees. Three of the rings are disposed about the Ca atom in a roughly trigonal manner, two being pentahapto and one trihapto bonded; the fourth ring is monohapto bonded, all the other carbon atoms in this ring being too far away for significant interaction. The bonding is presumably mainly ionic as evidenced by the long mean Ca—C distances. The mass spectrum of $(C_5H_5)_2Ca$ shows a high abundance of the molecular ion which is probably derived from a simple sandwich molecule in the gas phase.

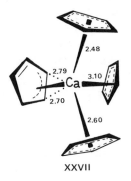

XXVII

VII

π COMPLEXES OF BORON

A. *Alkene Complexes*

π Complexation is believed to be involved in the first step of the hydroboration of alkenes and alkynes; rearrangement reactions of organoboranes most likely involve intermediates of a π-complex type. However, the stability of such complexes is generally too low to allow their isolation (*112*). However, evidence for π-complex formation has been obtained by the device of anchoring the alkene function to the metal atom in question by means of a suitable alkyl chain. In the IR spectrum of 1-(3-methylbuten-4-yl)-3-methylborolane (**XXVIII**) a shift to lower frequencies for the C=C vibrations has been observed, which disappears completely upon addition of donor molecules (*113*). This shift suggests an interaction between the C=C double bond and the boron atom.

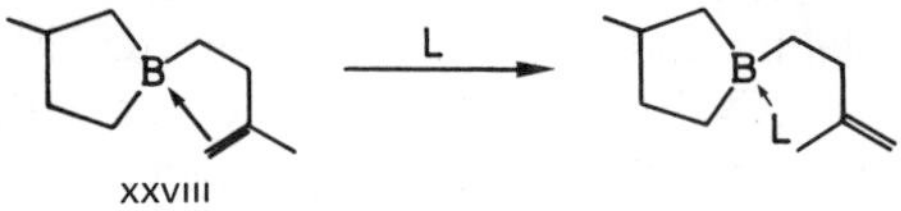

XXVIII

B. *Cyclopentadienyl Complexes*

Cationic complexes of boron with a cyclopentadienyl unit can be obtained from neutral pentamethylcyclopentadienylboranes containing a good leaving group in a process involving a $\eta^1-\eta^5$ rearrangement of the cyclopentadienyl system. Halide abstraction according to Reaction (7) leads to the compounds **XXIXa–h** (*114,115*); treatment of pentamethylcyclopentadienylbis-(dimethylamino)borane with tetrafluoroboric acid yields the boron compound **XXIXi** [Reaction (8)] (*116*). The exchange of the tin atom in the complex $Me_5C_5Sn^+ \ CF_3SO_3^-$ for a BI moiety yielding the compound **XXIXj** has also been described (*123*). The π complexes **XXIXa–j** are colorless, very air-sensitive, crystalline compounds with varying thermal stability.

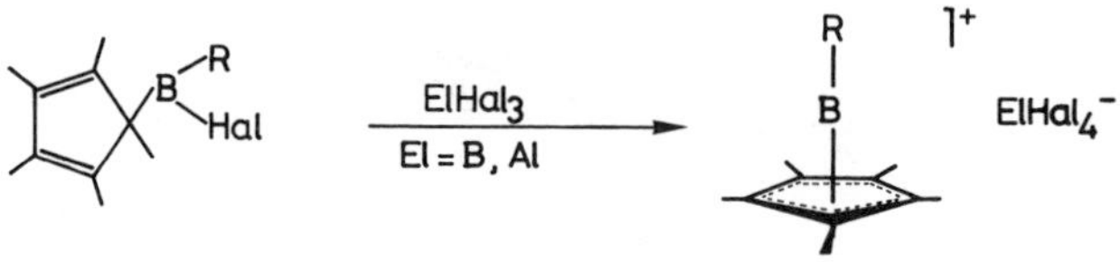

XXIXa–j

XXIXa	b	c	d	e	f	g	h	j
R	Cl	Cl	Br	I	I	C_5Me_5	CMe_3	$C_6H_2Me_3$	I
$ElHal_4$	BCl_4	$AlCl_4$	BBr_4	BI_4	BF_4	BCl_4	BCl_4	BCl_4	CF_3SO_3

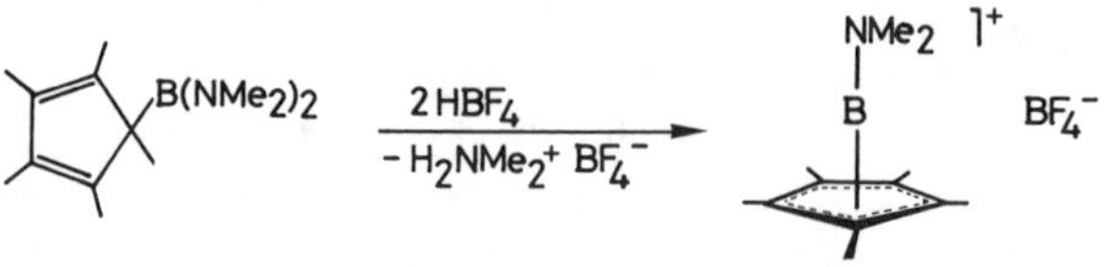

XXIXi

The structure of these complexes has been unambiguously proved by ^{11}B-, ^{1}H-, and ^{13}C-NMR spectroscopic and other analytical data. Spin coupling is observed between the boron and the five equivalent carbon atoms of the cyclopentadienyl ring as well as between the boron and the hydrogens of the ring-bonded methyl groups, indicating a highly symmetrical environment around the boron atom in these complexes. In accord with that, a comparably long spin-lattice relaxation time of 650 mseconds has been measured for the ^{11}B nucleus in the cation of compound **XXIXd**. The ^{11}B-NMR data (114–116) of the complexes **XXIXa–i** are collected in Table IV. The resonances for the π-complexed boron are found at rather high field, indicating that the positive charge in the cationic species is not concentrated at the boron atom; charge density is transferred mainly from the cyclopentadienyl system. The bonding in the $Me_5C_5BR^+$ cations can be rationalized in similar fashion to that in the neutral, isostructural cyclopentadienylberyllium compounds, H_5C_5BeR, namely, by the linear combination of $Me_5C_5^-$ and BR^{2+} fragment orbitals (114).

An interesting fluxional behavior has been observed for the cation in compound **XXIXf** by D-NMR techniques. At higher temperatures rapid sigmatropic rearrangements (115) in the cyclopentadiene unit take place, finally leading to a situation where the σ-bonded cyclopentadienyl ligand becomes π bonded and vice versa. The transition state for such a σ/π interchange may be represented by a symmetrical sandwich structure, which has been shown to be the ground state for the isoelectronic decamethylmagnesocene.

C. Dicarbollyl Complexes

The electron counting rules of Wade (83), Williams (117), and Rudolph (118) can serve as a useful concept to explain structure and bonding in a variety of systems which at first glance are very different: Zintl phases, boranes and carboranes, transition metal π complexes and carbonyl clusters, nonclassical carbocations, and also π complexes of main-group elements. According to

TABLE IV

^{11}B-Chemical Shifts^a of Complexes **XXIXa–i**

	Compound								
	XXIXa	**b**	**c**	**d**	**e**	**f**	**g**	**h**	**i**
Cation	−42.1	−42.1	−39.0	−50.8	−50.8	−44.1	−43.6	−41.4	−1.3
Anion	+6.7	—	−24.1	−127.6	−0.3	+6.7	+17.2	+7.4	−0.5

^{a} Chemical shifts, δ, in ppm, $BF_3 \cdot OEt_2$ reference compound.

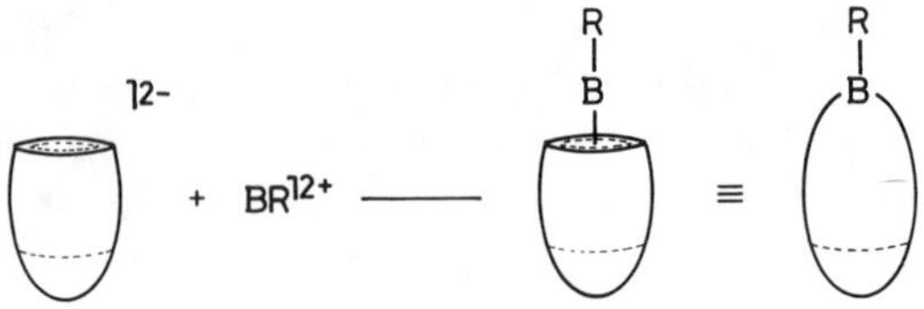

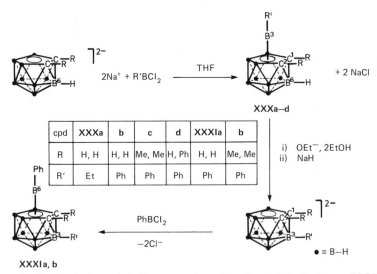

SCHEME 1.   Formation of a π complex or a closo or nido cluster.

these bonding considerations closo or nido carboranes can be regarded as π complexes, composed of a nido (or arachno) dianion as π ligand and a BR^{2+} fragment as central atom in a vertex position, as indicated schematically in Scheme 1.

In this article all known boranes, carboranes, and metallacarboboranes cannot be treated, and the reader is referred to recent reviews in this field (119–121). Only those cases will be described where the reaction principle of Scheme 1 has been used experimentally. Some 7,8-dicarbollide dianions were employed in reactions with phenyl- or ethylboron dichloride to reconstitute the boron-substituted dicarbaclosodecaborane icosahedrons **XXXa–d** (122). These carboranes were further degraded to 7,8-dicarbollide dianions, whereby the substituted boron remained within the product. Further reactions with phenyl- or ethylboron dichloride led once more to reconstruction of dicarbaclosoborane species, as indicated in Scheme 2. Further degradation reactions with the carbaboranes **XXXIa,b** were unsuccessful.

cpd	XXXa	b	c	d	XXXIa	b
R	H, H	H, H	Me, Me	H, Ph	H, H	Me, Me
R'	Et	Ph	Ph	Ph	Ph	Ph

i) OEt⁻, 2EtOH
ii) NaH

● ≡ B—H

SCHEME 2.   Synthesis of π-carbollylboron complexes from dicarbollide dianions and RBHal₂.

VIII

π COMPLEXES OF ALUMINUM

A. Alkene and Alkyne Complexes

It has been suggested that alkene or alkyne complexes are mechanistically important intermediates in hydroalumination and carboalumination reactions (124–126). Clear spectroscopic evidence for π interactions stems from investigations of alkenylaluminum compounds having a suitable intramolecular separation between the aluminum center and the double bond (127). IR and NMR data of these compounds show comparably lower alkene stretching frequencies and deshielded vinylic protons. Furthermore, these molecules are monomeric in solution; this indicates that the tendency toward π-complex formation is stronger than that toward the dimeric bonding usual in aluminum alkyls.

Structural data of alkynylaluminum compounds have supported the likelihood of π complexation. The X-ray crystal structure of dimeric diphenyl(phenylethynyl)aluminum (**XXXII**), prepared from phenylacetylene and triphenylaluminum (128), shows an interesting mode of phenylethynyl bridging (129), which suggests that the alkynyl group is largely σ bonded to one aluminum and π bonded to the other, as portrayed in Fig. 8. The bridging aluminum and carbon atoms form a rectangular array where the aluminum atoms are π bonded only to the α-acetylenic carbon atoms. The unusual π bonding in the dimer can be explained by overlap between one carbon $2p\pi$ orbital and the aluminum $3p_z$ orbital. A similar bonding situation has been revealed for the dimeric dimethyl(1-propynyl)aluminum (**XXXIII**) (131) by gas phase electron diffraction data (see Fig. 8) (130). Alkynylaluminum compounds with a symmetrical bridged structure are also known (132). It appears that there is a competition between two types of bridge bonds with similar energies: a symmetrical (type **J**) and an unsymmetrical (π type, **K**) one. Which one will be preferred depends on the circumstances.

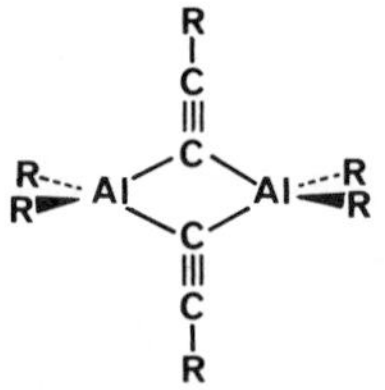

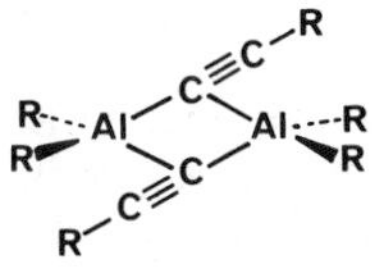

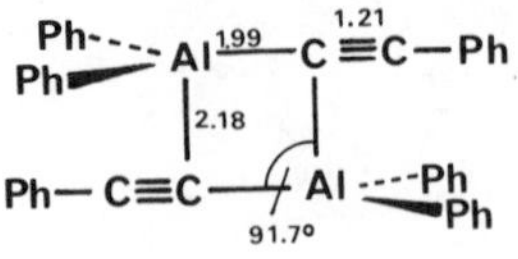

XXXII

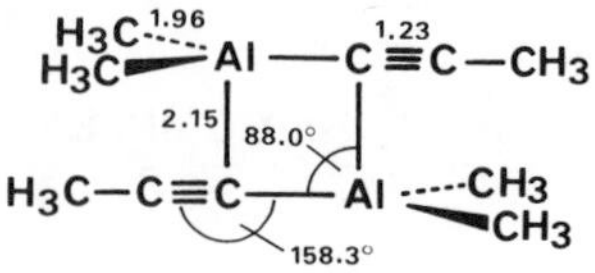

XXXIII

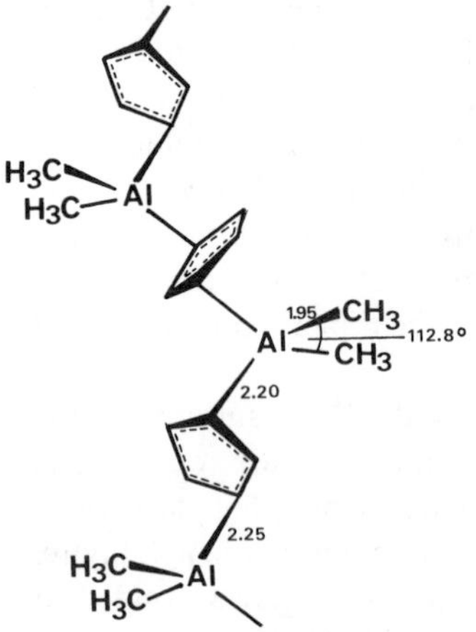

XXXIV, gas phase

XXXIV, solid state

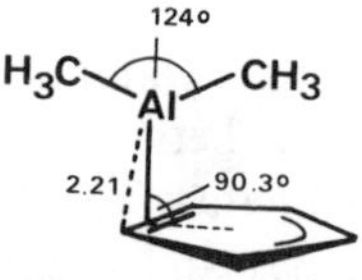

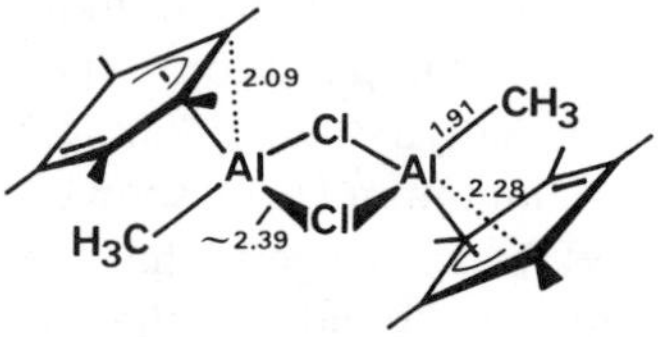

XXXV

● ≡ B—H

XXXVIII

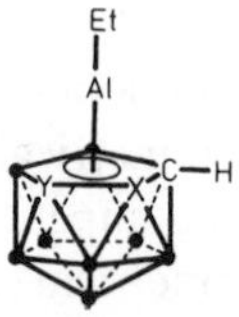

XXXVI
XXXVII

	X	Y
XXXV	CH	BH
XXXVI	BH	CH

FIG. 8. Schematic structures of π complexes to aluminum.

B. Cyclopentadienyl Complexes

The known dialkyl(cyclopentadienyl)aluminum compounds are all highly fluxional (124,133); IR and NMR data are therefore not conclusive evidence for their ground-state structures. We restrict ourselves to those compounds where X-ray crystal structure or gas phase electron diffraction data are available.

Dimethyl(cyclopentadienyl)aluminum **XXXIV** has been prepared by the direct reaction of Al_2Me_6 and excess cyclopentadiene (*134,135*). It is a colorless solid with a rather high melting point (above 140°C) and a low vapor pressure (about 0.001 mm at 60°C). Its solubility in hydrocarbon solvents is very poor; the degree of association in freezing benzene is 1.4 (*136*). In the solid state **XXXIV** consists of infinite chains composed of dimethylaluminum groups bridged by cyclopentadienyl rings (*137*), as portrayed in Fig. 8. The bridging units are unsymmetrically bound to the metal and show significant distortion within the ring. One bridge–bond distance has the length anticipated for an electron-deficient bond while the other is significantly longer. These results may be interpreted as indicating the presence of essentially molecular units, which are held together by secondary interactions. In the gas phase only the monomeric dimethyl(cyclopentadienyl)aluminum has been observed. From ED data (*138*) this species is best described as containing a cyclopentadienyl ring π bonded in a dihapto fashion (see Fig. 8). The dihapto bonding is also favored by MO calculations (*139*).

Chloro(pentamethylcyclopentadienyl)methylaluminum (**XXXV**) has been prepared by reaction of a toluene solution of pentamethylcyclopentadienyl-magnesium chloride with $[Me_2AlCl]_2$, that is, by the exchange of a pentamethylcyclopentadienyl group for one methyl group in dimethylaluminum chloride. This complex forms colorless crystals and is extremely air and moisture sensitive (*140*). The X-ray crystal structural study reveals a dimeric unit in which each aluminum atom is trihapto bonded to a pentamethylcyclopentadienyl ring (see Fig. 8). The rather long Al—Cl bridge bond distances and the Al_2Cl_2 ring bond angles are consistent with a weak interaction between two monomeric units. The bonding in the above π complexes can be rationalized by the linear combination of C_5R_5 and AlR_2 or AlR_2·donor fragment orbitals (*133,139*). As a result di-, tri-, or pentahapto configurations are all very similar in energy.

C. *Dicarbollyl Complexes*

Mixing the nido-carborane $7,8\text{-}C_2H_2B_9H_{11}$ with triethylaluminum in benzene at room temperature generates the fluxional nido cluster $Et_2AlC_2H_2B_9H_{10}$, in which the Et_2Al group is in a bridging position on the open face of a C_2B_9 fragment (*142*). Further reflux of the benzene solution causes the formation of the icosahedral 1-ethyl-1-alumino-2,3-dicarba-*closo*-dodecaborane (**XXXVI**), here formulated as a π complex (see Scheme 3) (*141*).

Compound **XXXVI** undergoes thermal rearrangement in the vapor phase at 410°C to the isomer **XXXVII** in high yield (*141*). When **XXXVI** is dissolved in THF the crystalline complex $EtAlC_2H_2B_9H_9 \cdot 2THF$ (**XXXVIII**) is obtained.

$$7,8\text{-}C_2H_2B_9H_{11} + Et_3Al \xrightarrow[-\text{EtH}]{} 7,8\text{-}C_2H_2B_9H_{10}(AlEt_2)$$

$$\downarrow -\text{EtH} \; \Delta$$

$$1,2,3\text{-}AlEtC_2H_2B_9H_9\cdot2THF \xleftarrow{2\ THF} 1,2,3\text{-}AlEtC_2H_2B_9H_9 \; (\textbf{XXXVI})$$

$$(\textbf{XXXVIII})$$

$$\uparrow THF \qquad\qquad\qquad\qquad \downarrow 410°C$$

$$Na_2[7,8\text{-}C_2H_2B_9H_9] + EtAlCl_2 \qquad 1,2,4\text{-}AlEtC_2H_2B_9H_9 \; (\textbf{XXXVII})$$

SCHEME 3. Formation of dicarbollyl complexes to aluminum.

This adduct is also formed by the salt-elimination method shown in Scheme 3 (*144*). The structure of **XXXVI** has been confirmed by a crystallographic study (*143*). The structure of **XXXVIII** is unknown so far, but following the isolobal relationship between the fragments **L** and **M** it is not unreasonable to suggest that a structure similar to that of the stannacarborane bipyridyl adduct (*145*) will be present, as indicated in Fig. 8.

IX

π COMPLEXES OF GALLIUM

A. *Alkyne Complexes*

Two gallium compounds are known in which π complexation to alkyne units has been confirmed by crystallographic or electron diffraction studies. Tetramethylbis(μ-phenylethynyl)digallium (**XXXIX**) was prepared as colorless, air-sensitive crystals by the reaction of trimethylgallium and phenylacetylene (*146*). The crystal and molecular structure of **XXXIX** (*147*) reveals that the molecular unit is held together by bridging ethynyl groups (see Fig. 9) similar to the situation in the aluminum compound **XXXII**. Dimerization of the gallium complex does not occur in THF, due no doubt to the formation of solvated monomers (*146*).

The colorless, crystalline tetramethylbis(μ-propynyl)digallium (**XL**) was prepared by the reaction of dimethylgallium chloride and sodium propynide

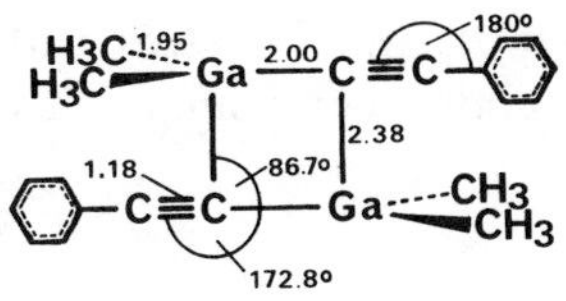

XXXIX

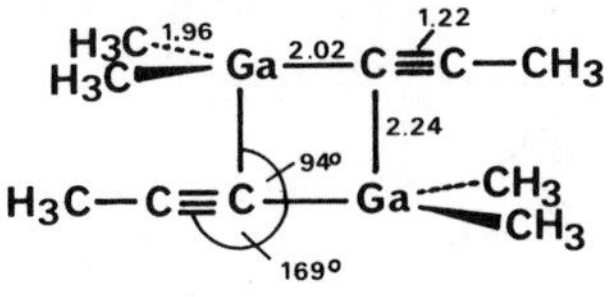

XL

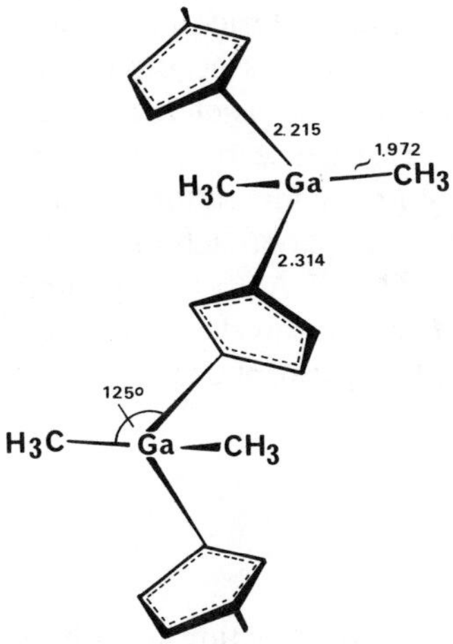

XLIa

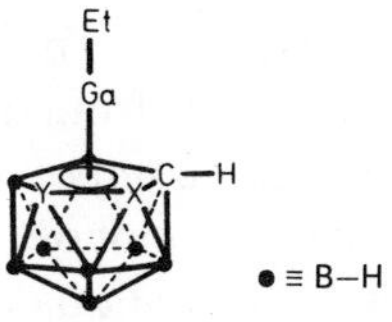

	X	Y
XLIII	CH	BH
XLIV	BH	CH

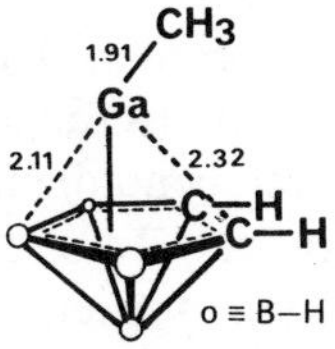

XLII

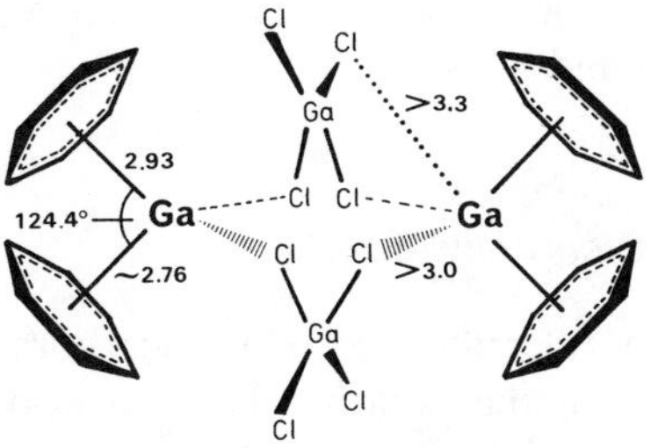

XLV

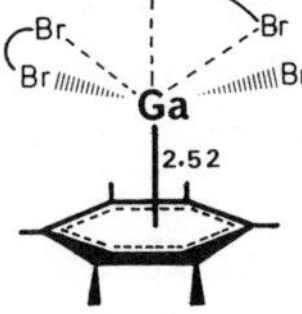

XLVIIb

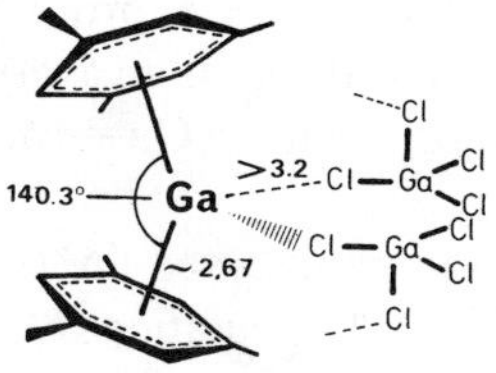

XLVI

FIG. 9. Schematic structures of π complexes of gallium.

(*148*). The electron diffraction data of this complex are consistent with a structure similar to that of the analogous aluminum compound **XXXIII** (see Fig. 9).

B. *Cyclopentadienyl Complexes*

A small number of cyclopentadienylgallium(III) compounds are known in the literature, and only in the $R_2GaC_5H_5$ species is there some indication of π interaction. The very air-sensitive dimethyl- and diethylcyclopentadienyl-gallium(**XLIa,b**) can be prepared from the dialkylgallium chloride and sodium cyclopentadienide (*150*). Both compounds show degrees of association in the range 1.2–1.5 in benzene or cyclohexane. A detailed analysis of the IR and Raman spectra in both solid and liquid phases gives evidence for the presence of di- or trihapto-bonded cyclopentadienyl rings. Low-temperature ^{1}H- and ^{13}C-NMR spectra failed to give any indication of the approach of the exchange limit in these highly fluxional molecules (*82,133*). A crystallographic study (*151*) of the dimethylgallium compound **XLIa** confirms the existence of a polymeric structure (see Fig. 9) similar to that of the isomorphous aluminum compound **XXXIV**.

C. *Dicarbollyl Complexes*

Two different types of gallacarboranes with a closo structure have been reported in the literature. They can also be regarded as π complexes of a nido-carborane dianion and a RGa^{2+} center.

Trimethylgallium reacts with $2,3\text{-}C_2H_2B_4H_6$ in the gas phase at 180–215°C to form the closo metallacarborane $1,2,3\text{-}MeGaC_2H_2B_4H_4$ (**XLII**) (*152*). The single-crystal X-ray diffraction study (*152*), while confirming the essential features of the structure, produced two unexpected findings: (1) substantial displacement of the gallium away from the framework carbon atoms and (2) a large angle between the Ga—Me bond and the Ga—B-7 axis, as indicated in Fig. 9.

The bending observed in the above complex can be explained using extended Hückel calculations (*153*). The orbital **N** is strongly concentrated on the gallium and on the central boron. Bending of the Ga—Me bond increases overlap between the carborane and the Ga—Me fragments, and consequently the orbital **N** is stabilized. The orbital **O** can be described as $\pi_{C=C} + \sigma_{GaMe}$. As a consequence of the bending the overlap is decreased, but the destabilization is less than the stabilization of orbital **N**. In summary, the distortion increases the skeletal bonding.

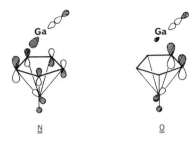

$\underset{=}{N}$ $\underset{=}{O}$

In analogy to the synthesis of the aluminum compounds **XXXVI** and **XXXVII** the closo gallium carboranes **XLIII** and **XLIV** have been prepared (*141*), but crystal structures are not yet available. From NMR spectroscopic data it appears that they possess icosahedral structures as shown in Fig. 9.

D. *Arene Complexes*

Very intriguing and well known from the beginning of gallium chemistry is the appreciable solubility of $GaCl_2$ and $GaBr_2$ in benzene and other aromatic solvents. The question of whether gallium forms π complexes with arenes has therefore been of long-standing interest (*154–156*), but only very recently have several arene complexes of gallium(I) been isolated and unambiguously characterized, by Schmidbaur's group (*157–159*). Under carefully controlled conditions discrete phases can be crystallized from aromatic solvents. From solutions of $GaCl_2$ in benzene a phase of the composition $(C_6H_6)_{3.5}Ga_2Cl_4$ has been isolated (*157*). The structure of the transparent, air-sensitive crystals can be described as consisting of cyclic centrosymmetrical $[(C_6H_6)_2Ga(I)Ga-(III)Cl_4]_2$ units (**XLV**) and isolated benzene molecules. In the bis(benzene)-gallium(I) moieties the two benzene rings are hexahapto bonded to the gallium atom forming an interplane angle of 124.4°. Two weakly distorted tetrachlorogallate tetrahedra provide singly and doubly bridging chlorine atoms to the gallium(III) centers. The gallium(III)–chlorine distances range from 2.152 to 2.183 Å, whereas the Ga(I)—Cl distances are much larger, thus indicating only weak interactions between the cationic π complex and the $GaCl_4^-$ units, as indicated in the schematic representation of this complex in Fig. 9.

From solutions of $GaCl_2$ in hot 1,3,5-trimethylbenzene the colorless crystalline bis(arene) complex $[C_6H_3(CH_3)_3]_2Ga^+$ $GaCl_4^-$ (**XLVI**) has been isolated (*158*). The structural study reveals a bent sandwich structure for the cationic moiety, as indicated schematically in Fig. 9. The tetrachlorogallate tetrahedra are once more only weakly coordinated, this time via one chlorine atom each to the Ga(I) centers of two bis(arene)gallium units, forming a chain-like coordination polymer. The steric requirement of the mesityl ligands is presumably the reason for the greater interplane angle of 140.3° and also for the coordination of only two chlorine atoms to the cation. The solubility of this complex suggests that a degradation to smaller units takes place in solution (158).

Suspensions of $GaCl_2$ or $GaBr_2$ in toluene yield clear solutions after addition of hexamethylbenzene and warming to 80–90°C, from which colorless compounds of the type $[C_6(CH_3)_6]Ga^+ GaHal_4^-$ (**XLVII**) crystallize on cooling (159). X-Ray diffraction data show a polymeric structure for the bromo compound **XLVII**, containing bridging tetrabromogallate units, but indicating in principle the presence of hexamethylbenzene complexed gallium(I) cations and $GaBr_4^-$ anions, as portrayed in Fig. 9. The bromine atoms in the coordination sphere of the Ga(I) cation come from three different nearly undistorted $GaBr_4^-$ units, and their distances to Ga(I) vary between 3.204 and 3.585 Å. They are therefore only weakly coordinated. The complexed hexamethylbenzene shows systematic deviations from planarity.

The most important differences between the structures of the above complexes are manifested in the different gallium(I) arene ring centroid distances, which shorten with increasing methylation of the arene ring system. The bonding in a benzene-complexed gallium(I) cation has already been qualitatively described by Rundle and Corbett (155), who discussed the symmetry-allowed interactions of a_1- and e-type π orbitals in the aromatic system with sp- and p-type atomic orbitals at the Ga(I) cation. Applying the isolobal principle to the species C_6H_6 and $C_5H_5^-$, and to the fragments Ga(I)$^+$ and Sn(II)$^{2+}$, the bonding in cations of the type $C_6H_6Ga^+$ and $C_5H_5Sn^+$ and also in species of the type $(C_6H_6)_2Ga^+$ and $(C_5H_5)_2Sn$ must be very similar.

X

π COMPLEXES OF INDIUM

A. Alkyne Complexes

Dimethylpropynylindium (**XLVIII**) can be prepared by the reaction of dimethylindium chloride or bromide with sodium propynide as a colorless, crystalline substance (148). Whereas indium trimethyl is monomeric in benzene, the alkynyl compound is dimeric, but only sparingly soluble in this and other aprotic noncoordinating solvents. Improved solubility has been observed in diethyl ether due to the formation of a weak adduct (148). According to crystallographic data (148), compound **XLVIII** can be regarded as a coordination polymer in the solid state, as indicated in Fig. 10. The axial positions on the trigonal bipyramidal coordination polyhedron of each indium atom are occupied by π-coordinating C≡C fragments of two neighboring units. The indium atom is located almost at the midpoint of the carbon–carbon triple bond with the two metal–carbon distances nearly equivalent at 2.93 and 2.99 Å. The gas phase structure (149) of **XLVIII** is similar to that of the analogous aluminum and gallium compounds. Once more there is a favored interaction with the α-carbon atom of the alkyne π system, as indicated in Fig. 10.

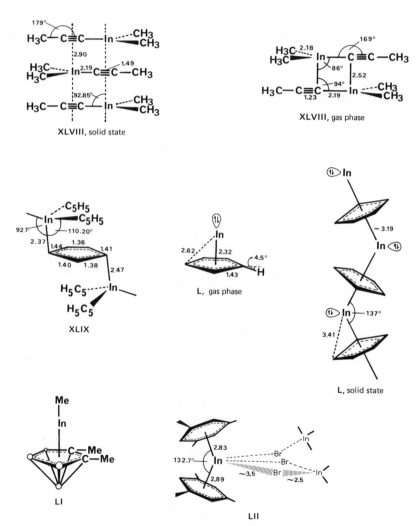

FIG. 10. Structures of π complexes of indium.

B. Cyclopentadienyl Complexes

Cyclopentadienyl compounds are known for In(III) as well as for In(I), and in both cases π interaction has been observed. Tris(cyclopentadienyl)-indium(III) (**XLIX**) was first reported by Fischer and Hofmann (*160*) as a side product of the preparation of cyclopentadienylindium(I) (**L**), in the reaction of indium trichloride with sodium cyclopentadienide. Later it was shown that the use of lithium cyclopentadienide produces **XLIX** in good yields as light yellow crystals (*161*). In the solid state this complex consists of infinite polymeric chains, with each chain unit consisting of an indium atom linked to two σ-bonded and, with different distances, to two π-bonded bridging cyclopentadienyl groups, as portrayed in Fig. 10. This constellation gives rise to a slightly distorted tetrahedral environment around the indium atom (*162*). It has been inferred from spectroscopic data that cyclopenta-dienylindium(III) compounds are highly fluxional in solution (*150*).

As mentioned earlier, cyclopentadienylindium(I) (**L**) was prepared nearly 30 years ago (*160*) as a pale yellow compound, sublimable in a high vacuum. In a later publication, Tuck and Poland (*161*) suggested the reaction sequence in Eqs. (9) and (10) for the synthesis of **L**. The thermal decomposition of

$$InCl_3 + 3\,NaC_5H_5 \rightarrow In(C_5H_5)_3 + 3\,NaCl \qquad (9)$$

$$In(C_5H_5)_3 \rightarrow InC_5H_5(L) + C_{10}H_{10} \qquad (10)$$

In $(C_5H_5)_3$ to **L** *in vacuo* above 100°C was confirmed by these authors, and the compound $InC_5H_4(CH_3)$ was obtained in a similar way (*161*). The preparation of **L** by condensation of indium vapor onto a matrix of C_5H_6 at 77 K was subsequently reported (*169*).

Cyclopentadienylindium(I) (**L**) has been the subject of many investigations, including electron diffraction (*163*) and crystal structure (*164*) studies. The compound is rather stable in the solid state, sensitive to air, but not affected by water. It is insoluble in aprotic and nonpolar solvents and sublimable in a vacuum. In the gas phase **L** possesses a half-sandwich structure as originally suggested by Fischer (see Fig. 10) (*160*). The cyclopentadienyl hydrogens are bent away from the cyclopentadienyl plane by approximately 4°. In the solid state **L** has a polymeric structure forming zigzag chains of alternating metal atoms and planar cyclopentadienyl rings, as sketched in Fig. 10. The indium–cyclopentadienyl centroid distances in the polymer (3.19 Å) are equal and considerably larger than those in the monomer (2.32 Å); the interplane angle is 137°.

The extensive controversy in the literature concerning the bonding in cyclopentadienylindium is summarized in a recent paper by Eisenstein *et al.* (*153*), which also contains a theoretical analysis of the bonding in other monomeric and polymeric C_5H_5M compounds. It is suggested that the bonding between the indium atom and the cyclopentadienyl unit has a large covalent component. The modest difference in electronegativity between carbon and indium does not imply high ionic character, nor does the relatively short 2.62 Å In—C distance in the gaseous molecule. The 1.3 Å ionic radius of In^+ and the 1.7-Å van der Waals radius of carbon imply an ionic bond length of roughly 3.0 Å. The long distances in the polymeric solid-state structure of L thus indicate an enhancement of ionic character, but the zigzag orientation of the chain still suggests some covalent metal–ligand interaction.

A highly covalent monomeric cyclopentadienylindium(I) species may be regarded as a nido cluster with a pentagonal pyramidal structure. It is isoelectronic with cyclopentadienylberyllium compounds of the type C_5H_5BeR, with cyclopentadienyl boron cations of the type $C_5H_5BR^+$, and with cyclopentadienylgermanium and -tin species of the type $C_5H_5El^+$. The bonding situation is characterized by three bonding molecular orbitals, arising from interactions of a_1- and e_1-type orbitals on the π-cyclopentadienyl fragment and on the indium atom, and by one further occupied orbital which represents the "lone pair" at indium. According to SCF calculations (153) the highest occupied orbitals in L are a pair of degenerate e_1 orbitals, the next highest being the "lone pair." The results of this theoretical analysis are in complete agreement with the interpretation of the PE spectra by Cradock and Duncan (165) and by Fragala et al. (166). The ionization energies and their assignments from the PE spectrum are as follows: 8.28 eV ($e_1\pi$); 9.23 eV (a_1); 12.89 eV ($a_1\pi + \sigma$ CH $+ \sigma$ C—C). The MO sequence described here disagrees with that calculated by the CNDO method (167).

The polymeric cyclopentadienylindium(I) has also been analyzed theoretically (153). The zigzag structure of the polymer is shown to be due to a minimization of repulsion. Bending of the linear structure changes the character of the highest energy band to make it essentially nonbonding.

Very recently, the synthesis and structure of pentamethylcyclopentadienylindium(I) have been reported (277). The golden yellow, very air-sensitive compound (CII) crystallizes in the form of monomeric units (indium–cyclopentadienyl centroid distance, 2.3 Å) which are held together by an octahedral arrangement of indium atoms indicating weak indium–indium interactions (3.96 Å).

C. Dicarbollyl Complexes

In the gas phase at 95–110°C trimethylindium reacts with the carborane 2,3-$C_2H_2B_4H_6$ to form the closo compound $MeInC_2H_2B_4H_4$ (LI) in 50% yield as a colorless, slightly volatile crystalline solid, which is slowly degraded by air (152). This carborane, which can also be regarded as a π complex (see Fig. 10), has been characterized by ^{11}B- and ^1H-NMR data. The gross structure of **LI** is that of a pentagonal bipyramid with the MeIn group occupying an apex position, comparable to that of the gallium compound (**XLII**).

D. Arene Complexes

The first arene π complexes of indium have been described only recently. Mesitylene takes up $In(InBr_4)$ on warming; from these solutions on cooling colorless, air- and moisture-sensitive crystals of the compound $(1,3,5-Me_3C_6H_3)_2In_2Br_4$ (**LII**) are formed (*168*). An X-ray structure analysis shows that this compound is a coordination polymer containing bis(mesitylene)-indium(I) cations and tetrabromoindate(III) anions linked by halogen bridges, as illustrated in Fig. 10. In the plane which bisects the angle between the arene rings, three only weakly coordinating bromine atoms are located, belonging to one chelating and one monodentate $InBr_4$ ligand. The compound **LII** easily loses mesitylene, especially under reduced pressure. An indium complex with unsubstituted benzene as π ligand has not yet been isolated (*168*).

XI

π COMPLEXES OF THALLIUM

A. Alkyne Complexes

In contrast to the behavior of comparable aluminum, gallium, and indium compounds, the hitherto known alkynyl thallium(III) compounds (*170*) of the type $R_2TlC{\equiv}CR$ show no tendency to dimerize via π complexation. Molecular weight determinations performed in aniline show the monomeric character of these species, which behave as weak electrolytes due to partial dissociation into R_2Tl^+ and $C{\equiv}CR^-$ (*170*).

B. Cyclopentadienyl Complexes

Many cyclopentadienylthallium(I) compounds have been described in the literature; the first one, cyclopentadienyl thallium (**LIIIa**), was synthesized by Fischer and Hofmann (*160*) nearly 30 years ago. The colorless solid can be best prepared from cyclopentadiene and thallium hydroxide or thallium(I) salts in the presence of alkali in aqueous solution (*171*) [Eq. (11)]. The ring-substituted species **LIIIb–i** (*172,173,268,270*) (Table V) as well as hydro-

TABLE V

CYCLOPENTADIENYLTHALLIUM(I) COMPOUNDS **LIIIb–l**

Compound	**LIIIb**	c	d	e	f	g
R¹	Me	Me$_3$C	Me$_3$Si	3-FC$_6$H$_4$	4-FC$_6$H$_4$	CMe(H)(Ph)
R²	H	H	H	H	H	H
R³	H	H	H	H	H	H

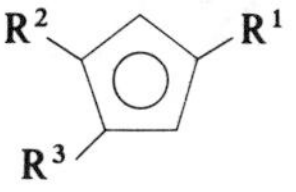

	h	i	j	k	l
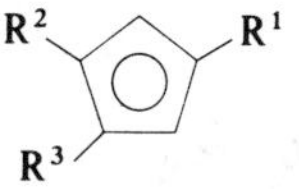 R¹	Hal	P(C$_6$H$_5$)$_2$	Me$_3$Si	Me$_3$Si	C(CN)C(CN)$_2$
R²	H	H	Me$_3$Si	Me$_3$Si	H
R³	H	H	H	Me$_3$Si	H

pentalenethallium (**LIIIn**) and isodicyclopentadienylthallium (**LIIIo**) can be synthesized in a similar way. The pentamethylcyclopentadienylthallium (**LIIIp**) and the silylated compounds **LIIIj,k** are best prepared by the reaction of the corresponding cyclopentadienyllithium species with thallium(I) chloride (*175*) [Eq. (12)]. The dark red thallium compound **LIIIl** has been synthesized from **LIIIa** and tetracyanoethylene (*176*) [Eq. (13)]. Reaction of hexachlorocyclopentadiene with thallium amalgam results in the formation of **LIIIm**, the pentachloro analog of **LIIIa** [Eq. (14)]. This compound may also be prepared by the reaction of pentachlorocylopentadiene with thallium ethoxide (*177*). A new method for the preparation of **LIIIa** uses the reaction of thallium vapor with cyclopentadiene (*169*).

$$C_5H_6 + TlX \xrightarrow[-KX]{KOH/H_2O} C_5H_5Tl \ (\textbf{LIIIa}) \tag{11}$$

$$[C_5R_{5-x}R'_x]Li + TlX \xrightarrow[-LiX]{} [C_5R_{5-x}R'_x]Tl \ (\textbf{LIIIj,k,p}) \tag{12}$$

$$C_5H_5Tl + (CN)_2C{\equiv}C(CN)_2 \xrightarrow[-HCN]{} C_5H_4[(CN)C{=}C(CN)_2]Tl \ (\textbf{LIIIl}) \tag{13}$$

$$C_5Cl_6 + TlHg \xrightarrow[-Hg, \, -TlCl]{} C_5Cl_5Tl \ (\textbf{LIIIm}) \tag{14}$$

Many investigations have been performed in order to elucidate the structure and bonding in cyclopentadienylthallium(I) compounds. Various bonding situations are found in this class. The parent compound, cyclopentadienylthallium (**LIIIa**) is nearly unsoluble in polar solvents. It is rather stable to air and water and sublimable in a vacuum. The microwave spectrum (*178*) suggests a half-sandwich structure in the gas phase with a thallium to ring centroid distance of 2.41 Å (see Fig. 11). The thallium–carbon distances of

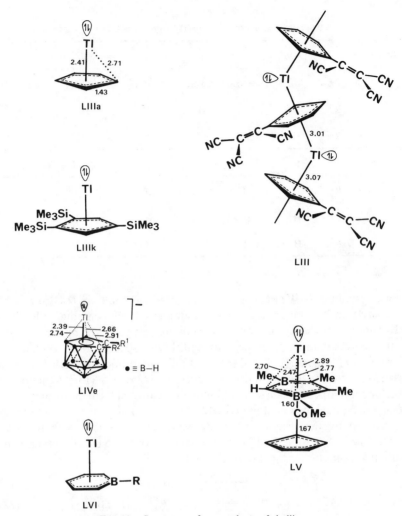

FIG. 11.   Structures of π complexes of thallium.

2.71 Å are smaller than the sum of the van der Waals radius of carbon and the ionic radius of Tl⁺ (3.1 Å). The structure (164) of **LIIIa** in the solid state contains linear polymeric chains, which are composed of alternating cyclopentadienyl rings and thallium atoms, a situation similar to that observed for cyclopentadienylindium(I). The difference between the thallium–ring distances in the gas phase and in the solid state is very great (171).

In the ¹H- and ¹³C-NMR spectra of **LIIIa** no spin–spin coupling between ²⁰⁵Tl and ¹H or ¹³C nuclei has been observed (172). Apparently there is considerable ionic character in the bonding in solution and in the solid state. Similar conclusions can be drawn regarding the bonding in the substituted cyclopentadienylthallium(I) compounds **LIIIb–i, l,** and **m** (*172,173*).

The behavior of the trimethylsilyl-substituted cyclopentadienylthallium(I) compound **LIIIk** in solution differs from that of the so far described thallium complexes. The air-stable species **LIIIk** is fairly soluble even in aromatic solvents and behaves as a covalent monomer. Furthermore, spin–spin coupling between the ^{205}Tl nucleus and the ring proton nuclei and the ^{13}C and ^{1}H nuclei of the trimethylsilyl groups has been observed (*175*). A half-sandwich structure in solution and in the gas phase has been suggested for this π complex.

The tricyanovinylcyclopentadienyl-substituted thallium(I) compound **LIIIl** shows properties characteristic of ionic species (*176*). The solution conductivity in dimethylformamide is much higher than that of the unsubstituted cyclopentadienylthallium(I) (**LIIIl,** 6.5×10^{-2}; **LIIIa,** $5.1 \times 10^{-4}\ \Omega^{-1}\ cm^2\ equiv^{-1}$). A low-temperature single-crystal X-ray investigation confirmed a polymeric chain structure similar to that of **LIIIa** (see Fig. 11). The rather long ring centroid to thallium distances of 3.014 and 3.065 Å reveal a high extent of ionic character, but the zigzag structure of the chain still suggests some covalent metal–ligand interaction.

Concerning the structure of the pentachlorocyclopentadienyl-substituted thallium(I) compound **LIIIm** it has been suggested that in the solid state the thallium cation nestles in the $C_5Cl_5^-$ anion and that some charge is transferred from the anion to the empty thallium orbital (*177*), a situation typical for an ion pair with a low degree of covalent interaction.

According to very recent theoretical calculations based on the SCF pseudopotential method (*153*), the bonding between thallium and the cyclopentadienyl fragment in **LIIIa** has a large covalent component. The influence of *d* orbitals has been found to be negligible, contrary to the result of an earlier SCF calculation using single basis sets (*179*). The calculated molecular orbital sequence is comparable to that for C_5H_5In and can be correlated with the PE spectroscopic data for **LIIIa** [7.96 eV ($e_1\pi$); 10.12 eV (a_1)] (*165,166*).

C. Complexes with Other Anionic 6π Ligands

The first thallium(I) π complexes with carbollide ligands were synthesized by Stone *et al.* (*180*); addition of an aqueous solution of thallium(I) acetate to an aqueous alkaline solution of the dicarbaundecaborane anion, 7,8-$C_2B_9H_{12}^-$, or its *C*-alkyl derivatives causes the formation of the new compounds (**LIVa–c**) [Eq. (15)]. One of the thallium atoms in the air-stable,

$$Tl(I)OAc + 7,8\text{-}C_2R^1R^2B_9H_{10}]^-Na^+ \longrightarrow Tl_2C_2R^1R^2B_9H_9$$

$$\text{(LIVa–c)}$$

	LIVa	b	c	d	e
Cation	Tl	Tl	Tl	Ph$_4$As	Ph$_3$PMe
R^1	H	H	Me	H	H
R^2	H	Me	Me	H	H

$$\downarrow M^+Hal^- \qquad (15)$$

$$M[1,2,3\text{-}TlC_2R^1R^2B_9H_9]$$

$$\text{(LIVd,e)}$$

pale to bright yellow compounds **LIVa–c** can be readily replaced by the tetraphenylarsonium or the triphenyl(methyl)phosphonium (*181*) cation, suggesting the presence of icosahedral anions of the type 1,2,3-TlC$_2$R^{1}R^{2}B$_9$H$_9^-$. This has been proved by an X-ray structure analysis of **LIVe** (see Fig. 11) (*181*). The thallium atom in the anion of this salt is not symmetrically located with respect to the cage, but shifted away from the two carbons by about 0.3 Å.

The formation of a π complex between a thallium atom and a transition metal sandwich π ligand has been reported by Siebert *et al.* (*182*). Reaction of (η^5-cyclopentadienyl)cobalt(η^5-1,3,4,5-tetramethyl-2,3-dihydro-1,3-diborole) with cyclopentadienylthallium in tetrahydrofuran leads to the orange–brown complex **LV**, which is rather air sensitive, easily soluble in nonpolar solvents, and sublimable in a high vaccum. The compound is a monomer in the solid state, in the gas phase, and presumably also in solution. According to the X-ray crystal structure, the position of the thallium atom as well as the distances from thallium to the ring carbon and boron atoms are similar to those found in the anion of compound **LIV**. The acceptor qualities of the B$_2$C$_3$ ring are held to be responsible for the fact that **LV** does not dissociate into ions in the solid state with concomitant formation of a polymeric structure. The bonding in the π complexes **LIV** and **LV** can qualitatively be related to the bonding in **LIIIa** because the open face of the carbaborane fragment in **LIV**, the heterocyclic site of the sandwich in **LV**, and the cyclopentadienyl π system are nearly isolobal.

Borabenzene anions can serve as π ligands to a thallium(I) center, as demonstrated by the work of Herberich *et al.* (*183*). Reaction of alkali metal borinates with thallium(I) chloride yields the pale yellow, sublimable complexes **LVIa,b** [Eq. (16)] (see Fig. 11), which are only sparingly soluble in nonpolar aprotic solvents but easily soluble in pyridine and dimethylsulfoxide (*183*).

$$C_5H_5BR^-Na^+ + TlCl \xrightarrow[-NaCl]{} C_5H_5BRTl \qquad (16)$$

$$\text{(LVIa,b)}$$

	LVIa	b
R	Me	Ph

In the NMR spectra of **LVIa,b** no spin–spin coupling has been observed between ^{205}Tl and ^1H, ^{13}C, and ^{11}B nuclei. According to mass spectroscopic data the compounds are monomeric in the gas phase. From the similarity between the physical–chemical data of **LVIa,b** and **LIIIa** similar bonding in both classes of compounds is anticipated (*183*).

D. *Arene Complexes*

Compounds of the composition $C_6H_6 \cdot 2TlAlCl_4$ and $(C_6H_6)_2 \cdot TlAlCl_4$ have been mentioned in a publication by Amma (*184*), but no structural assignments were presented. A crystalline adduct of composition $[Tl \cdot 2C_6H_5(CH_3)]^+ C_5Cl_5^-$ has been described by West and Wulfsberg (*177*), but the structure of this compound has also not been identified. In a very recent paper, Schmidbaur *et al.* have presented the first unambiguously characterized arene complex containing $(mesitylene)Tl^+$ and $(mesitylene)_2Tl^+$ moieties weakly coordinated to $GaBr_4^-$ anions (*185*).

XII

π COMPLEXES OF CARBON

Using the isolobal relationship between CH or CH_2 units and corresponding transition metal or main-group element fragments, the majority of the nonclassical carbocations can be theoretically treated as π complexes (*30,186–188*). This is demonstrated for some representative examples in Scheme 4. The description of these species, which have not been synthesized from positively charged CH or CH_2 fragments and organic π systems, is beyond the scope of this review.

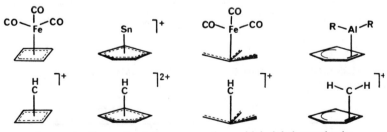

SCHEME 4.   Connection between π complexes with isolobal central units.

XIII

π COMPLEXES OF GERMANIUM

A. Cyclopentadienyl Complexes

The first π complex of germanium was synthesized by Curtis and Scibelli (*189*). Reaction of sodium cyclopentadienide with germanium dibromide leads to dicyclopentadienylgermanium(II) (**LVIIa**, germanocene), a colorless compound, quite stable in its pure state. It sublimes under reduced pressure and can be easily handled under an inert gas atmosphere (*190*). Following the general reaction scheme in Eq. (17), the substituted germanocenes **LVIIb–f** and the germanocenophane **LVIIj** (Table VI) have also been prepared. The mixed substituted germanocenes **LVIIg,h,i** have been synthesized by a metathesis reaction between pentamethylcyclopentadienylgermanium(II) chloride and the appropriate potassium cyclopentadienide [Eq. (18)] (Table VI).

$$\text{GeHal}_2 + 2\ \text{MC}_5\text{R}_5 \xrightarrow[-2\ \text{MHal}]{} (\text{R}_5\text{C}_5)_2\text{Ge}\ (\textbf{LVIIa–fj}) \qquad (17)$$

$$\text{Me}_5\text{C}_5\text{GeCl} + \text{KC}_5\text{H}_2\text{R}_3 \xrightarrow[-\text{MCl}]{} (\text{Me}_5\text{C}_5)\text{GeC}_5\text{H}_2\text{R}_3\ (\textbf{LVIIg,h,i}) \qquad (18)$$

$$\text{R} = \text{H, SiMe}_3$$

TABLE VI

GERMANOCENES **LVIIa–j**, INTERPLANE ANGLES (α), AND AVERAGE Ge—C DISTANCES (a)

	LVIIa (*189,190*)	**b** (*191*)	**c** (*192*)	**d** (*193*)	**e** (*193*)
R_5C_5	H_5C_5	MeH_4C_5	Me_5C_5	$(\text{Me}_3\text{Si})\text{H}_4\text{C}_5$	$(\text{Me}_3\text{Si})_2\text{H}_3\text{C}_5$
R_5C_5	H_5C_5	MeH_4C_5	Me_5C_5	$(\text{Me}_3\text{Si})\text{H}_4\text{C}_5$	$(\text{Me}_3\text{Si})_2\text{H}_3\text{C}_5$
α (°)	50.4[a] (*190*)	34 ± 7[b] (*195*)	22 ± 2[b] (*196*)	—	—
a (Å)	2.52	2.53	2.52	—	—

	f (*193*)	**g** (*194*)	**h** (*194*)	**i** (*194*)	**j** (*267*)
R_5C_5	$(\text{Me}_3\text{Si})_3\text{H}_2\text{C}_5$	Me_5C_5	Me_5C_5	Me_5C_5	$\text{Me}_2\text{Si—C}_5\text{Me}_4$ $\mid$ $(\text{CH}_2)_2$ $\mid$
R_5C_5	$(\text{Me}_3\text{Si})_3\text{H}_2\text{C}_5$	$(\text{Me}_3\text{Si})\text{H}_4\text{C}_5$	$(\text{Me}_3\text{Si})_2\text{H}_3\text{C}_5$	$(\text{Me}_3\text{Si})_3\text{H}_2\text{C}_5$	$\text{Me}_2\text{Si—C}_5\text{Me}_4$
α (°)	20.8[a] (*193*) 14.9	—	—	—	—
a (Å)	2.56	—	—	—	—

[a] X-Ray crystal structure.
[b] GED.

The germanocenes are all air-sensitive compounds, stable under ordinary conditions, with the most stable being the permethylated species **LVIIc**. They are easily soluble in aprotic nonpolar and polar solvents and according to present data exist as monomers in the solid and in the gaseous state as well as in solution. The common structure of these complexes can be described as a more or less distorted sandwich allowing the presence of a stereochemically active lone pair on the metal, as indicated in **LVII** (see Fig. 12). The distortion can be regarded as a small bending of the cyclopentadienyl centroid—Ge—cyclopentadienyl centroid angle from linearity and as small tilts of the cyclopentadienyl ring planes relative to the germanium–centroid vectors, as portrayed in **P**. As a consequence of these tilts the Ge—C distances are spread between 2.34 and 2.73 Å; the germanium atom is no longer required to reside over the center of the cyclopentadienyl rings. The angles between the cyclopentadienyl planes (α) and the average Ge—C distances (a) known so far are collected in Table VI.

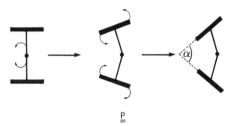

P

The angle α in **LVIIb,c** could not be determined accurately by electron diffraction, partly due to large amplitude ring–metal–ring bending vibrations. Even a qualitative comparison of the values for α in Table VI shows that bulky substituents at the cyclopentadienyl rings force the two rings into a more parallel conformation. According to NMR spectroscopic data, the germanocenes **LVIIa–i** all are rather flexible molecules in solution. The barrier to internal rotation of the cyclopentadienyl rings is very small. Even in the case of the sterically crowded germanocene **LVIIf** no evidence for hindered rotation (gear effect) could be found; a steric energy program was applied to explain the experimental observations (*193*).

The bonding in the germanocenes **LVII** (and more generally in all group IV metallocenes) can be rationalized qualitatively by inspection of molecular orbital schemes constructed for species with parallel (D_{5d} symmetry, magnesocene-type MO sequence) and with bent (C_{2v} symmetry) cyclopentadienyl rings as portrayed in Fig. 13. Some important conclusions can be drawn from these MO diagrams. First, in the germanocene with D_{5d} symmetry, interaction between the e_{1g} set in the Cp$\cdots$Cp unit and a germanium centered orbital is symmetry forbidden; thus 4 of the overall 14

PETER JUTZI

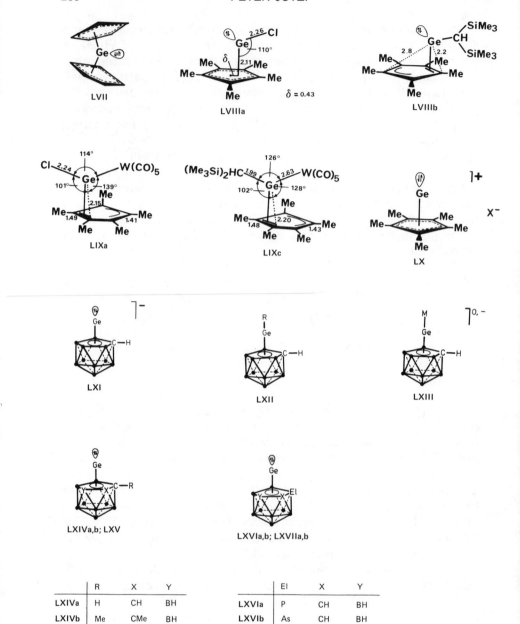

	R	X	Y
LXIVa	H	CH	BH
LXIVb	Me	CMe	BH
LXV	H	BH	CH

	El	X	Y
LXVIa	P	CH	BH
LXVIb	As	CH	BH
LXVIIa	P	BH	CH
LXVIIb	As	BH	CH

FIG. 12.   Structures of π complexes of germanium.

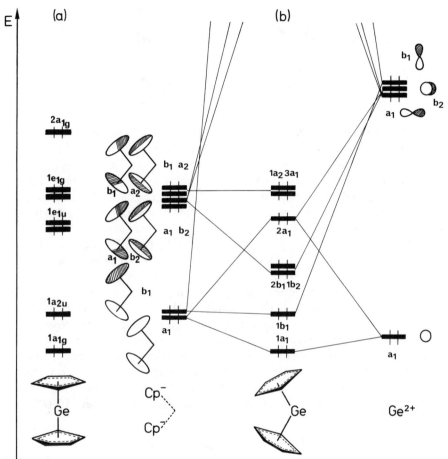

FIG. 13.   Molecular orbital schemes for a germanocene with D_{5d} (a) and C_{2v} (b) symmetry.

electrons in the germanocene are located in nonbonding molecular orbitals $(1e_{1g})$; this situation remains qualitatively unchanged going to the molecule with C_{2v} symmetry. Second, in the coparallel species the highest occupied molecular orbital $(2a_{1g})$ is highly destabilized and antibonding with respect to germanium cyclopentadienyl ring interactions (195). Bending of the cyclopentadienyl rings has no great effect on the other molecular orbitals but drastically reduces the energy of the former $2a_{1g}$ orbital. Admixing of p character from germanium to the wave function now leads to a situation in which the germanium–cyclopentadienyl antibonding is reduced (see **Q**). The resulting orbital $2a_1$ represents to a great extent the nonbonding electron pair (lone pair) at the germanium atom.

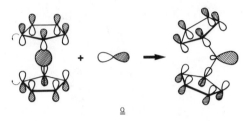

$\underline{\underline{\text{o}}}$

In the molecular orbital diagram for a bent germanocene, the three highest orbitals are very close in energy. Which one will be the HOMO thus varies with the particular situation. The energy of the "lone pair" orbital $2a_1$ depends drastically on the interplane angle α; the energies of the nonbonding orbitals $1a_2$ and $3a_1$ and of the bonding orbitals $2b_1$ and $1b_2$ depend greatly on the energies of the parent cyclopentadienyl orbitals. For example, the π-fragment orbitals of a permethylated cyclopentadienyl system are about 1 eV higher in energy than those of an unsubstituted one. In the PE spectrum of decamethylgermanocene (**LVII**), the two highest molecular orbitals have been assigned π character (197) (see Table X and the discussion in Section XIV,A).

In a second class of germanium(II) complexes, there is only one cyclopentadienyl ligand present. Compounds of the type Me_5C_5GeR (**LVIII**) have been synthesized during the last few years. The pentamethylcyclopentadienyl (chloro) germylene (**LVIIIa**) is prepared by the reaction of pentamethylcyclopentadienyllithium with germanium dichloride·dioxane or by coproportionation of decamethylgermanocene with germanium dichloride (198); further substitution reactions of **LVIIIa** with some alkyllithiums or lithium amides lead to the germylenes **LVIIIb–d** [Eqs. (19)–(21)] (194).

$$Me_5C_5Li + GeCl_2\cdot dioxane \xrightarrow[\substack{-LiCl \\ -dioxane}]{} Me_5C_5GeCl \textbf{ (LVIIIa)} \qquad (19)$$

$$(Me_5 \ (Me_5C_5)_2Ge + GeCl_2\cdot dioxane \xrightarrow[-dioxane]{} \textbf{LVIIIa} \qquad (20)$$

$$\textbf{LVIIIa} + RLi \xrightarrow[-LiCl]{} Me_5C_5GeR \textbf{ (LVIIIb–d)} \qquad (21)$$

	LVIIIa	**b**	**c**	**d**
R	Cl	CH(SiMe₃)₂	N(SiMe₃)₂	

These germylene complexes are light yellow to orange, rather air sensitive but thermostable, crystalline compounds, easily soluble in aprotic polar and nonpolar solvents. According to X-ray investigations, gas phase electron diffraction studies, and molecular weight determinations, the complexes are monomeric species in the solid and gaseous state and also in solution. An

X-ray structure analysis of **LVIIIb** (*194*) has shown that the planar cyclo-pentadienyl ring is dihapto bonded to the germanium atom (see Fig. 12). The observed carbon–germanium distances are similar to those found in the germanocenes **LVIIa** and **c**. The orientation of the bulky bis(trimethylsilyl)methyl group allows for the presence of a sterically active lone pair at the germanium(II) center. A similar structure has been found for the germylene **LVIIIa** in the gas phase (see Fig. 12) (*196*). The electron diffraction pattern is consistent with an angular model of C_s symmetry in which the cyclo-pentadienyl ring is bonded to the germanium atom in an asymmetric, polyhapto fashion. The germanium atom is displaced from the fivefold symmetry axis of the cyclopentadienyl ring towards the midpoint of a C—C bond. The perpendicular germanium to ring distance is significantly shorter than in decamethylgermanocene (**LVIIc**); the Ge—Cl distance is longer than in gaseous $GeCl_2$. According to NMR measurements the compounds **LVIIIa**–**e** are highly fluxional in solution (*194*). Even at low temperatures it is not possible to observe the static structure. The dynamic behavior of these compounds is believed to be similar to that of the closely related bicyclo[3.1.0]hexenyl cation, investigated in detail by Winstein and Childs (*199*).

Ab initio MO calculations on the molecule H_5C_5GeCl have been carried out by Haaland and Schilling (*200*). The minimum of the total energy is represented by a structure similar to that found for the compound **LVIIIa**. The orbital energies obtained for H_5C_5GeCl and the measured ionization energies for the species **LVIIIa** (*197*) together with their assignments are compared in Table VII; the differences in the MO sequences may be caused by the different energies of the π orbitals in C_5H_5 and C_5Me_5 fragments.

In a more qualitative and general description, the π bonding in compounds of type **LVIII** is represented by the linear combination of GeR and Me_5C_5 fragment orbitals. The germanium atom in the GeR species can be regarded as approximately sp^2 hybridized, thus offering one sp^2-hybrid and a further p orbital for interaction with the relevant cyclopentadienyl π orbitals, as indicated in the orbital pictures **R** and **S**. It is obvious from these con-siderations that the bonding situations of differing hapticity possess very

TABLE VII

CALCULATED ORBITAL ENERGIES ($-\epsilon$), VERTICAL IONIZATION ENERGIES
(IE_v), AND THEIR ASSIGNMENTS FOR H_5C_5GeCl AND Me_5C_5GeCl

H_5C_5GeCl	$-\epsilon$ (eV)	Me_5C_5GeCl (**LVIIIa**)	IE_v (eV)
Ge lone pair	9.49	Cp eπ + Ge 4p	8.13
Cp eπ + Ge 4p	9.56	Cp eπ + Ge 4p	8.32
Cp eπ + Ge 4p	10.09	Ge lone pair	9.07
Cl lone pairs	11.69	Cl lone pairs	10.47

$\underline{R}$ $\underline{S}$

similar total energies. A dihapto situation placing the germanium lone pair in the electron void above the ring center seems to be most favorable. The observed high degree of fluxionality is consistent with a rather soft calculated energy surface. Haptotropic shifts in cyclopentadienyl systems by fragments isolobal to a GeR moiety have been discussed in the same fashion (28).

In a third class of π-cyclopentadienyl complexes, compounds of type **LVIII** are bound to a transition metal via the lone pair at the germanium atom. The first compounds in this series, the pentacarbonyl-[chloro(pentamethylcyclopentadienyl)germylene]chromium(0) (**LIXa**) and -tungsten(0) (**LIXb**) were synthesized by alkylation of the ylide complexes $(CO)_5M—GeCl_2 \cdot THF$ according to Eq. (22) (201). Nucleophilic substitution reactions of **LIXb** lead to the alkyl- and amino-substituted complexes **LIXc–e** [Eq. (23)] (202).

$$(CO)_5M—GeCl_2 + Me_5C_5SnMe_3 \xrightarrow[-THF]{Me_3SnCl} (CO)_5W \leftarrow Ge\begin{array}{c} Cl \\ C_5Me_5 \end{array} \quad (22)$$
$$\uparrow$$
$$THF$$
$$(\textbf{LIXa,b})$$

$$\textbf{LIXb} \qquad + RLi \xrightarrow{-LiCl} (CO)_5W \leftarrow Ge\begin{array}{c} R \\ C_5Me_5 \end{array} \quad (23)$$
$$(\textbf{LIXc,d,e})$$

	LIXa	b	c	d	e
M	Cr	W	W	W	W
R	Cl	Cl	CH(SiMe_3)_2	CH_3	N(SiMe_3)_2

The bright yellow to orange compounds **LIXa–e** are extremely air sensitive and easily soluble in benzene, diethyl ether, di- and trichloromethane. They are monomeric in the solid and gaseous state and also in solution. NMR investigations indicate that all these complexes are highly fluxional. X-Ray crystal structure analyses of **LIXb** and **c** (see Fig. 12) show very similar features. The germanium atom is dihapto bonded to the cyclopentadienyl ring. In these highly symmetric structures the $(CO)_5W$ unit is in a position above the plane of the cyclopentadienyl ring. The molecular parameters show some small but significant differences between the two structures, which probably reflect the changes resulting from replacement of the electron-withdrawing chlorine by the electron-releasing and bulky alkyl ligand. The π bonding in these complexes can be explained in similar fashion to that in the uncomplexed compounds **LVIII**.

In a further class of cyclopentadienyl complexes, a cyclopentadienyl π system is pentahapto bonded to an isolated germanium cation. Cations of the composition $R_5C_5Ge^+$ were first observed in the mass spectra of various cyclopentadienylgermanium compounds. The isolation of compounds containing such an unusual cation has so far been successful only in pentamethylcyclopentadienyl chemistry. Reaction of decamethylgermanocene with electrophiles (192,203) proceeds by attack at the cyclopentadienyl π system, followed by abstraction of cyclopentadiene and formation of ionic species **LX**, which contains the pentamethylcyclopentadienylgermanium(II) cation [Eq. (24)]. In an alternative procedure, a chloride anion is abstracted from pentamethylcyclopentadienylgermanium chloride by halide acceptors like $AlCl_3$ or $GeCl_2$ [Eq. (25) and (26)] (204). Finally, a synthesis starting from pentamethylcyclopentadienyllithium and two equivalents of $GeCl_2 \cdot$dioxane was also successful (198). The presence of anions with low nucleophilicity promotes the stability of the cationic π complexes. The colorless, air-sensitive salts **LXa–e** are easily soluble in aprotic polar solvents. Their structure (see Fig. 12) was established by NMR spectroscopic data, molecular weight determinations, and conductivity measurements.

$$(Me_5C_5)_2Ge + EX \xrightarrow[-Me_5C_5E]{} Me_5C_5Ge]^+ X^- \text{ (LXa,b,c)} \qquad (24)$$

	LXa	**b**	**c**
X^-	BF_4	$C_5(COOMe)_5$	CF_3SO_3

$$Me_5C_5GeCl + AlCl_3 \longrightarrow Me_5C_5Ge]^+ AlCl_4^- \text{ (LXd)} \qquad (25)$$

$$Me_5C_5GeCl + GeCl_2 \cdot dioxane \xrightarrow[-dioxane]{} Me_5C_5Ge]^+ GeCl_3^- \text{ (LXe)} \qquad (26)$$

The π bonding in these cations can be easily understood using the isolobal relationship between the fragments portrayed in **T** and the linear combination of those fragment orbitals with the relevant orbitals of a cyclopentadienyl π system (192). Thus, the bonding is represented by three bonding MO's, arising from interactions of a_1- and e_1-type orbitals on the π-cyclopentadienyl fragment with those of the same symmetry at the germanium center. A further occupied MO mainly represents the lone pair at germanium, as illustrated in the orbital picture **U**. *Ab initio* MO calculations (200) performed on the cation $H_5C_5Ge^+$ yield an optimum germanium to ring distance of 1.99 Å, about 0.10 Å shorter than the experimental distance in Me_5C_5GeCl and about 0.20 Å shorter than the experimental distance in $(Me_5C_5)_2Ge$.

The cationic pentamethylcyclopentadienylgermanium unit reacts with pyridine, pyrazine, and 2,2'-bipyridine to form the corresponding azine complexes (205). In analogy to the comparable tin complexes, a distortion of the η^5 bonding toward η^2 and η^3 structures may be expected.

B. Carbollyl Complexes

Germanium atom insertion into a vacant site of an icosahedral carborane fragment was first reported by Todd et al. (206,207). Reaction of $Na_3[7\text{-}CHB_{10}H_{10}](THF)_2$ with methyl trichlorogermane in tetrahydrofuran produces the icosahedral 1-germa-2-carba-closo-dodecaborane (**LXIIa**, R = CH_3) in very low yield as well as salt-like species containing the anion $1,2\text{-}GeCHB_{10}H_{10}^-$ (**LXI**). In an improved procedure, the closo compound **LXIIa** can be prepared by the reaction of the carbollide ion with germanium tetrachloride and by subsequent methylation [Eq. (27)]. The ethyl-substituted compound **LXIb** was also synthesized in an analogous way.

$$Na_3[7\text{-}CHB_{10}H_{10}](THF)_2 \xrightarrow{\text{GeCl}_4} [1,2\text{-}GeCHB_{10}H_{10}]Na \xrightarrow{\text{RI}} 1,2\text{-}GeRCHB_{10}H_{10} \quad (27)$$

$$\textbf{(LXI)} \qquad\qquad\qquad \textbf{(LXIIa,b)}$$

Compounds **LXIIa,b** are sublimable solids which can be stored in air for months with negligible decomposition. Interestingly, basic reagents can easily demethylate the cluster: treatment of **LXIIa** with piperidine generates the anion **LXI** and, in addition, the anion $7\text{-}CHB_{10}H_{12}^-$. The closo-cluster structures of **LXI** and **LXII**, which here are regarded as π complexes, are consistent with the spectroscopic data, especially those from ^{11}B-NMR investigations (see Fig. 12).

Transition metal complexes of the type **LXIIIa–e** have been synthesized starting from **LXI** and neutral or cationic transition metal reagents according to the equations in Scheme 5 (207,208). The structures of the complexes **LXIIIa–e** (see Fig. 12) have been mainly inferred from IR and NMR spectroscopic data.

$$\xrightarrow[-\text{CO}]{\text{M(CO)}_6,\ \text{THF},\ h\nu\ (\text{M}=\text{Cr, Mo, W})} Me_4N[1,2\text{-}GeCHB_{10}H_{10}M(CO)_5]$$

$$\textbf{(LXIIIa–c)}$$

$$Me_4N[1,2\text{-}GeCHB_{10}H_{10}] \xrightarrow[-\text{cyclohexene}]{[(\pi\text{-}C_5H_5Fe(CO)_2(\text{cyclohexene})]PF_6} (\pi\text{-}C_5H_5)Fe(CO)_2GeCHB_{10}H_{10}$$

$$\textbf{(LXIIId)}$$

$$\xrightarrow[-\text{CO}]{[(\pi\text{-}C_7H_7)MO(CO)_3]BF_4} (\pi\text{-}C_7H_7)MO(CO)_2GeCHB_{10}H_{10}$$

$$\textbf{(LXIIIe)}$$

SCHEME 5.   Synthesis of transition metal complexes with the ligand $1,2\text{-}GeCHB_{10}H_{10}^-$.

Germanium atom insertion into the vacant site of an isolobal icosahedral fragment has been reported by Rudolph *et al.* (*209*). Treatment of the 7,8-$C_2H_2B_9H_9^{2-}$ dianion with germanium diiodide afforded the 1-germa-2,3-dicarba-*closo*-dodecaborane (**LXIVa**). The methyl derivative (**LXIVb**) has been similarly obtained (*210*). The compounds, here formulated as π complexes to a germanium center, are colorless, sublimable, air-sensitive solids, monomeric in benzene. Their structure (see Fig. 12) has been surmised by analogy with other metallacarborane structures and is consistent with the ^{11}B-NMR data. On heating to 600°C, the compound **LXIVa** rearranges to the 1,2,4-isomer (**LXV**), which can be synthesized independently starting from the 7,9-$C_2H_2B_9H_9^{2-}$ dianion and germanium diiodide (*211*).

A further extension of the known isoelectronic series in the class of icosahedral species has been reported by Todd and Beer (*212*). Substitution of a C—H fragment in the complexes **LXIV** and **LXV** by a "bare" phosphorus or arsenic atom leads to the closo-cluster compounds **LXVI** and **LXVII**, which are here also described as π complexes with regard to their synthesis. Treatment of the disodium salt of the 7,8-$ElCHB_9H_9^{2-}$ dianion (El = P, As) with a germanium(II) species results in the formation of the 1,2,3-$GeElCHB_9H_9$ complexes (**LXVIa,b**) [Eq. (28)]. The isomeric species 1,2,4-$GeElCHB_9H_9$ (**LXVIIa,b**) can be prepared starting from the 7,9-$ElCHB_9H_9^{2-}$ dianion [Eq. (29)].

$$Na_2[7,8\text{-}ElCHB_9H_9] + GeI_2 \rightarrow 1,2,3\text{-}GeElCHB_9H_9 \text{ (\textbf{LXVIa,b})} \quad (28)$$

$$Na_2[7,9\text{-}ElCHB_9H_9] + GeI_2 \rightarrow 1,2,4\text{-}GeElCHB_9H_9 \text{ (\textbf{LXVIIa,b})} \quad (29)$$

The complexes are colorless, sublimable, crystalline solids and can be stored in air for several weeks with negligible decomposition. In both series of compounds, the 1,2,4-isomers are thermally more stable than the 1,2,3-isomers owing to the greater charge separation in the former. The structural assignment (see Fig. 12) is once more based mainly on NMR spectroscopic data. It is suggested that the germanium atom in the compounds **LXVIa,b** is displaced toward the boron atoms on the open face of the π system.

XIV

π COMPLEXES OF TIN

A. *Cyclopentadienyl Complexes*

The situation in the cyclopentadienyl chemistry of tin is similar to that described for germanium; several classes of cyclopentadienyl compounds are now known. The first compound in this series was reported by Fischer and

Grubert (*213*) as early as 1956. Reaction of sodium cyclopentadienide with tin(II) chloride results in the formation of dicyclopentadienyltin(II) (**LXVIIIa**), trivially named stannocene. This compound, together with bis(methylcyclopentadienyl)tin(II) (**LXVIIIb**) (*214*), synthesized some years later, enjoyed for a long time the unique position of being the only thermally stable bivalent organotin species known. The substituted stannocenes **LXVIIIc–n** and also the stannocenophanes **LXVIIIo–r** were subsequently synthesized following the same salt elimination method (*192,215–217,221,222*) or other synthetic strategies (*218–220*), as indicated in Eqs. (30)–(37) and in Table VIII. Finally, a $(C_5H_5)_2Sn$ unit is present in $[(\mu\text{-}\eta^5\text{-}C_5H_5)_2Sn(\mu\text{-}\eta^5\text{-}C_5H_5Sn)]^+ \cdot THF[BF_4]^-$, the reaction product of stannocene (**LXVIIIa**) with boron trifluoride etherate.

$$2\ R_5C_5M \xrightarrow{\ SnCl_2\ } Sn(C_5R_5)_2\ (\textbf{LXVIIIa–g}) \tag{30}$$
$$M = Li,\ Na,\ K$$

$$X(C_5R_4Li)_2 \xrightarrow{\ SnCl_2\ } Sn(C_5R_4)_2X\ (\textbf{LXVIIIo–r}) \tag{31}$$

$$(C_5R_nH_{5-n})_2Sn \xrightarrow{\ 2\ BuLi\ } (C_5R_nH_{4-n}Li)_2Sn$$

$$(C_5R_nH_{4-n}Li)_2Sn \xrightarrow{\ RCl\ } [C_5R_{n+1}H_{4-n}]_2Sn\ (\textbf{LXVIIId,e,f})$$

$$(C_5R_nH_{4-n}Li)_2Sn \xrightarrow{\ (R'_2N)_2PCl\ } \{C_5R_nH_{4-n}[P(NR'_2)_2]\}_2Sn\ (\textbf{LXVIIIh}) \tag{32}$$

$$(C_5H_5)_2Sn + 2\ Me_3SnNMe_2 \longrightarrow [(Me_3Sn)C_5H_4]_2Sn\ (\textbf{LXVIIIi}) \tag{33}$$

$$(Me_5C_5)_2SnCl_2 \xrightarrow[\text{or Li}_2COT]{\text{Li Naphth.}} (Me_5C_5)_2Sn\ (\textbf{LXVIIIc}) \tag{34}$$

$$H_5C_5SnCl + Ph_5C_5Li \longrightarrow H_5C_5(Ph_5C_5)Sn\ (\textbf{LXVIIIk}) \tag{35}$$

$$H_5C_5SnCl + (RCO)H_4C_5Na \longrightarrow H_5C_5[(RCO)H_4C_5]Sn\ (\textbf{LXVIIIl–n}) \tag{36}$$

$$(H_5C_5)_2Sn + (R_2N)_2P^+AlCl_4^- \longrightarrow \{(R_2N)_2PH{-}C_5H_4\}(H_5C_5)Sn^+AlCl_4^-\ (\textbf{LXVIIIj}) \tag{37}$$

The stannocenes **LXVIIIa–r** are colorless to deep yellow compounds of varying sensitivity to air, soluble in the common aprotic organic solvents (with the exception of **LXVIIIl–n** and the ionic species **LXVIIIj**). It can be assumed that nearly all the neutral stannocenes so far known are monomeric species in the solid state, in the gas phase, and in solution. X-Ray crystal structure and gas phase diffraction data for stannocene species are collected in Table IX (see also Fig. 14). They reveal interesting information about the steric effects of substituents at the cyclopentadienyl ring and show trends similar to those already observed for germanocenes. Stannocene (**LXVIIIa**) and most of its

TABLE VIII

STANNOCENES LXVIIa–r SYNTHESIZED ACCORDING TO EQS. (30)–(37)

CpCp'Sn	LXVIIIa (213)	b (214)	c (192,219)	d (215,216)	e (215)	f (215)	g (217)	h (216)
Cp	H_5C_5	MeH_4C_5	Me_5C_5	$(Me_3Si)H_4C_5$	$(Me_3Si)_2H_3C_5$	$(Me_3Si)_3H_2C_5$	Ph_5C_5	$\{[(i\text{-Pr})_2N]_2P\}H_4C_5$
Cp'	H_5C_5	MeH_4C_5	Me_5C_5	$(Me_3Si)H_4C_5$	$(Me_3Si)_2H_3C_5$	$(Me_3Si)_3H_2C_5$	Ph_5C_5	$\{[(i\text{-Pr})_2N]_2P\}H_4C_5$

CpCp'Sn	i (218)	j (220)	k (217)	l (222)	m (222)	n (222)	o (221)
Cp	$(Me_3Sn)H_4C_5$	$[(i\text{-Pr})_2N]_2PH\!-\!H_4C_5$	Ph_5C_5	$(MeCO)H_4C_5$	$(MeOCO)H_4C_5$	$(EtOCO)H_4C_5$	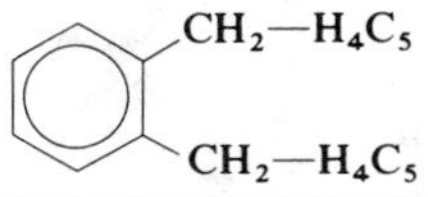 benzene ring, ortho $CH_2\!-\!H_4C_5$ / $CH_2\!-\!H_4C_5$
Cp'	$(Me_3Sn)H_4C_5$	$H_5C_5^+\,AlCl_4^-$	H_5C_5	H_5C_5	H_5C_5	H_5C_5	

CpCp'Sn	p (221)	q (221)	r (267)
Cp	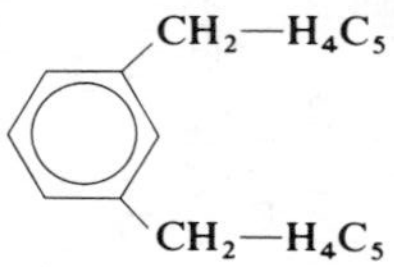 benzene ring, meta $CH_2\!-\!H_4C_5$ / $CH_2\!-\!H_4C_5$	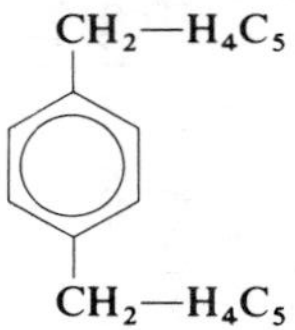 benzene ring, para $CH_2\!-\!H_4C_5$ / $CH_2\!-\!H_4C_5$	$Me_2Si\!-\!C_5Me$
Cp'			$(CH_2)_2$; $Me_2Si\!-\!C_5Me_4$

TABLE IX

Bond Angles and Bond Lengths in Stannocenes

Compound	LXVIIIa[a] (224)	a[b] (224)	b (195)	c[a] (224)	f (215)	g (217)	h (216)
Method	X ray	GED	GED	X ray	X ray	X ray	X ray
α (°)	44.00	—	—	36.4	—	—	—
	46.60	55	50(6)	35.4	—	0	43
β (°)	148.00	—	—	—	—	—	—
	143.70	125	—	—	162	180	150.2
a (Å)	2.40	—	—	2.39	—	—	—
	2.42	2.42	2.40	2.38	—	2.40	—
d (Å)	0.21	—	—	0.23	—	—	—
	0.36	—	—	0.22	—	—	—
Sn—C (Å)	2.58 –	—	—	2.58 –	—	2.68 –	2.53 –
	2.85	2.71	—	2.77	—	2.70	2.81

[a] Two independent molecules.
[b] Poor quality on account of extraneous scattering.

derivatives are bent sandwich compounds with a stereochemically active lone pair. As expected the angle α between the cyclopentadienyl ring planes becomes smaller going from the unsubstituted (**LXVIIIa**) to the permethylated (**LXVIIIc**) stannocene. Very recently it has been shown by Zuckerman et al. (217) that in the case of the perphenylated stannocene (**LXVIIIg**) the angle has even become 0; this complex is the first ferrocene-like sandwich with a group IV element as central atom (see Fig. 14). The perpendicular distance from tin to the cyclopentadienyl ring planes (a in Table IX) is very similar in all stannocene structures. Generally the tin atom slips away from the ring centroid position (d in Table IX); as a consequence very different Sn—C distances of between 2.53 and 2.85 Å are observed. A further consequence is that the angle between the ring normals ($180° - \alpha$) and the ring centroid–tin–ring centroid angle (β in Table IX) are different. To avoid confusion note should be taken of this in discussions of stannocene structures.

The distorted pentahapto attachment of the cyclopentadienyl rings in bent stannocenes is not reflected in their low-temperature ^{1}H- or ^{13}C-NMR spectra, where only averaged resonance signals are observed. The barrier for

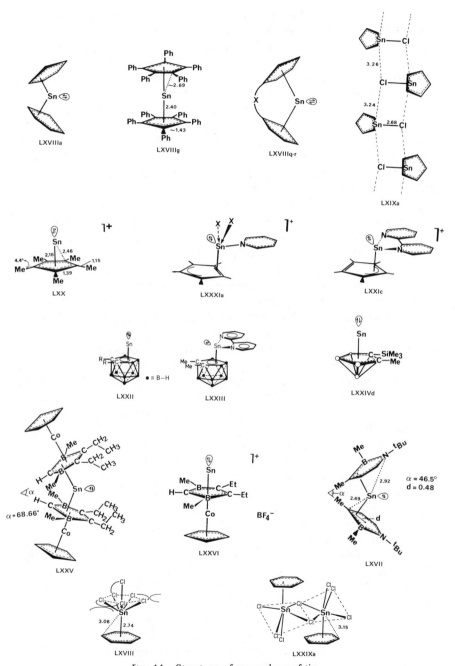

FIG. 14. Structures of π complexes of tin.

rotation must therefore be very low. Intermolecular exchange processes have been observed for stannocene **LXVIIIb** (*191*); satellite lines, which are present at $-40°C$ due to coupling with ^{117}Sn and ^{119}Sn nuclei, collapse on raising the temperature. On the other hand, in the NMR spectra of the stannocene **LXVIIIc** the coupling is still present at room temperature, but vanishes after addition of small amounts of $Me_5C_5Sn^+$ BF_4^- owing to fast dissociative processes (*225*). ^{119}Sn-NMR data for stannocenes reveal a very pronounced upfield chemical shift of more than 2100 ppm in comparison to organotin(IV) compounds [**LXVIIIa**, -2199 (*225,226*); **b**, -2171 (*191*); **c**, -2129 (*225*)]. This corresponds to a large increase in shielding for the tin(II) species and is attributable to substantially increased s electron density at the tin atom. The ^{119}Sn-NMR spectrum of **LXVIIIc** consists of a 31-line multiplet, confirming the equivalence on the NMR time scale between the methyl protons of the cyclopentadienyl rings; only 17 lines are intense enough to be observed (*225*).

The tin-119m Mössbauer spectra of some stannocenes were recorded by Zuckerman's group (*221,222*). The observed isomer shifts are compatible with the presence of divalent tin, while the dependence of the quadrupole splitting on the interplanar angle in these complexes is still unknown.

A qualitative MO bonding scheme for linear and bent group IV metallocenes has already been discussed in Section XIII,A. For the simplest bent tin sandwich **LXVIIIa**, a semiempirical MO study of the extended Hückel type (*192*) and a SCF–X–SW calculation (*227*) have been performed. The X_α–SW calculation revealed that the highest occupied MO's ($3a_1$ and $1a_2$) are π type and highly localized on the cyclopentadienyl rings. In order of decreasing energy, the MO's associated with the π bonding of the cyclopentadienyl systems to the tin atom are $2b_1, 1b_2, 1b_1, 1a_1$; the one exhibiting the largest tin lone pair character is the orbital $2a_1$. The $2b_1$ orbital provides the weakest ring–tin bonding; the primary interaction occurs between the ring MO's and the Sn $5p_z$ AO. Somewhat stronger ring–tin interaction is evident in the $1b_2$ MO, which contains a substantial amount of Sn p_x character. The strongest bonding between a tin $5p$ orbital and the cyclopentadienyl rings occurs in the $1b_1$ MO. The lowest energy MO involved in ring–tin bonding is $1a_1$, resulting from interaction of the Sn $5s$ orbital with an a_1 combination of cyclopentadienyl ring π MO's.

The UV and PE spectral data available for group IV metallocenes have been interpreted with the aid of the theoretical ionization energies computed for **LXVIIIa** (*227*); the assignments are presented in Table X. As expected the energies of the HOMO's in the unsubstituted and permethylated metallocenes are nearly the same and independent of the central atom, indicating the nonbonding character of these orbitals. Furthermore it is interesting to note that the energies of the "lone-pair" MO's are lower for the lead compounds than for the comparable tin analogs. The reverse would have been expected

TABLE X

IONIZATION ENERGIES AND ORBITAL ASSIGNMENTS FOR GROUP IV METALLOCENES[a]

Orbital	Calculated $(H_5C_5)_2Sn$	Experimental				
		$(Me_5C_5)_2Ge$	$(H_5C_5)_2Sn$	$(Me_5C_5)_2Sn$	$(H_5C_5)_2Pb$	$(Me_5C_5)_2Pb$
3a₁ / 1a₂	6.60 / 6.61	6.60 / 6.75	7.57 / 7.91	6.60 / 6.60	7.55 / 7.85	6.33 / 6.88
2b₁ / 1b₂	7.31 / 7.64	7.91 / 8.05	8.85 / 8.85	7.64 / 7.64	8.54 / 8.88	7.38 / 7.38
2a₁	8.74	8.36	9.58	8.40	10.10	8.93
1b₁	11.25	—	10.5	9.4	10.6	9.38
1a₁	~13.40	—	—	—	—	—

[a] All values in eV (197,227).

assuming equivalent bonding and structure for the two series of compounds. The observed effects can be explained on the basis of the larger ring–metal–ring angles in the lead complexes than in the tin analogs thus implying that in the former the lone pair possesses more s character (227).

Another type of divalent organotin π complex has been reported by Noltes et al. (228,229). The cyclopentadienyltin halides **LXIXa,b** precipitate upon mixing concentrated THF solutions of stannocene with the tin dihalides [Eq. (38)]. The complex **LXIXa** can also be obtained by the reaction of **LXVIIIa** with hydrogen chloride in THF [Eq. (39)] (228). Similarly, the pentamethylcyclopentadienyltin compounds **LXIXc,d** have been prepared from decamethylstannocene **LXVIIIc** and the appropriate carboxylic acid [Eq. (40)] (230).

$$(H_5C_5)_2Sn + SnX_2 \xrightarrow{\text{THF}} 2\,H_5C_5SnX\,(\textbf{LXIXa,b}) \qquad (38)$$

$$(H_5C_5)_2Sn + HCl \xrightarrow[-C_5H_6]{} H_5C_5SnCl\,(\textbf{LXIXa}) \qquad (39)$$

$$(Me_5C_5)_2Sn + HX \xrightarrow[-Me_5C_5H]{} Me_5C_5SnX\,(\textbf{LXIXc,d}) \qquad (40)$$

LXIXa	b	c	d
X Cl	Br	CF$_3$COO	CCl$_3$COO

The tin complexes **LXIXa–d** are colorless, crystalline, and very air-sensitive compounds, monomeric in solution and in the gas phase. The species **LXIXc,d** tend to disproportionate in solution. Averaged resonance signals with no (117,119) tin coupling are observed in the ^{1}H- and ^{13}C-NMR spectra of these compounds (228,230).

An X-ray crystal structure analysis has shown that in the solid state **LXIXa** is a coordination polymer in which the tin atom is surrounded by one near chlorine atom, by two rather weakly bonded next-nearest-neighbor chlorine atoms, and by an asymmetrically pentahapto-bonded cyclopentadienyl ring, as indicated in Fig. 14. The perpendicular tin to ring distance (2.30 Å) is significantly shorter than in stannocene; the nearest tin–chlorine distance is even shorter in **LXIXa** (2.68 Å) than in the solid state structure of SnCl$_2$ (229).

Cyclopentadienyltin cations, the representatives of a third class of π-cyclopentadienyltin compounds, were first observed in the mass spectra of cyclopentadienyl-substituted organotin(II) and organotin(IV) species. The first such species to be isolated was the Me$_5$C$_5$Sn$^+$ cation, in the form of the tetrafluoroborate salt (**LXXa**) (Scheme 6) (192,231). Subsequently, other electrophiles have been employed leading to the compounds **LXXb–d** (Scheme 6) (203,230).

The compound H$_5$C$_5$Sn$^+$ BF$_4^-$ (**LXXe**), containing the parent ion in this series, was synthesized in an analogous way (230). Furthermore, the H$_5$C$_5$Sn$^+$

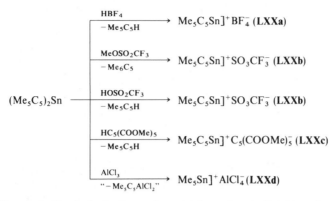

SCHEME 6. Synthesis of compounds containing cyclopentadienyltin cations.

unit is present in the complex structure resulting from the reaction of stannocene with boron trifluoride (222). Electrophilic attack at the π system of a stannocene molecule is common to all synthetic routes. Low nucleophilicity of the counteranion is important for the stabilization of these salt like species; neutral covalent species are otherwise formed, as already described in this chapter. The compounds **LXXa–e** are colorless, crystalline species rather sensitive to air and moisture, and readily soluble in polar aprotic solvents. A coupling to the tin isotopes has been observed in the ^{1}H- and ^{13}C-NMR spectra. The ^{119}Sn resonances appear at rather high field (-2100 ppm). The isomer shifts in the tin-119m Mössbauer spectra are typical for tin(II) species (221). Mass spectra show that these salts exist as covalent species in the gas phase.

The X-ray crystal structure of **LXXa** confirms the presence of isolated cations and anions in the solid state. In the cation, the pentamethylcyclopentadienyl π system is symmetrically pentahapto bonded to the tin atom (see **LXX** in Fig. 14). The tin–ring centroid distance is considerably shorter than in decamethylstannocene. The methyl groups are bent away from the plane of the cyclopentadienyl ring.

The bonding in the cation **LXX** can be qualitatively explained in the same way as the bonding in the corresponding germanium species. The cationic pentamethylcyclopentadienyltin unit acts as a Lewis acid in its reactions with pyridine, pyrazine, and 2,2′-bipyridine, in which the adducts **LXXXIa–d** are formed [Eq. (41)] (205). X-Ray crystal structure data of adducts **LXXIa** and **c**

$$Me_5C_5Sn]^+ \ X^- + base \ \rightarrow \ Me_5C_5Sn \cdot base]^+ \ X^-$$

$$\text{(LXXa,b)} \qquad\qquad \text{(LXXIa–d)}$$

(41)

reveal a distortion of the pentahapto bonding towards di- and trihapto structures (see Fig. 14). Interestingly, trifluoromethylsulfonate anions are

LXXI	a	b	c	d
base	(pyridine)	(pyridine, N)	(bipyridine)	(bipyridine)
X⁻	F_3CSO_3	F_3CSO_3	F_3CSO_3	BF_4

coordinated to the tin atom in **LXXIa** but not in **LXXIc**. These differences in structure can be understood by considering the interactions of the frontier orbitals in these compounds (205). The orange–red color of the complexes **LXXIc,d** is caused by charge transfer from the tin "lone-pair" into the π^x orbital of the bipyridine system. According to NMR measurements the adducts are highly fluxional in solution.

B. Dicarbollyl Complexes

Tin atom insertion into a vacant site of an icosahedral carborane fragment was first reported by Rudolph and co-workers (232,209). Treatment of the 7,8-$C_2H_2B_9H_9^{2-}$ dianion with tin dichloride afforded the icosahedral 1-stanna-2,3,-dicarba-*closo*-dodecaborane (**LXXIIa**), here formulated as a π complex of a tin center. The methyl derivative **LXXIIb** has been obtained similarly (210). Surprisingly, **LXXIIa** can also be prepared starting from diorganotin dichlorides instead of tin dichloride (232). The isomeric 1-stanna-2,4-dicarba-*closo*-dodecaborane is thermally unstable and decomposes easily to form *closo*-$C_2H_2B_9H_9$ and tin metal (211). The complexes **LXXIIa,b** are colorless, sublimable, rather air-stable solids, monomeric in solution, in the gas phase, and presumably also in the solid state. The proposed structures (see Fig. 14) are consistent with the ^{11}B-NMR data and in accord with other known metallacarborane structures. A crystalline, orange–red bipyridyl adduct **LXXIII** of **LXXIb** has been recently described (145). The X-ray crystal structure of this complex shows a distorted icosahedral arrangement with the tin atom displaced toward the boron atoms in the open face of the π ligand. Trihapto bonding similar to that observed for the bipyridyl adduct of $Me_5C_5Sn^+$ (**LXXIc**) can be inferred (see Fig. 14).

The insertion of a tin atom into a nonicosahedral carborane framework was first reported by Grimes and co-workers (233), following approaches similar to those which had previously been successful with transition metal reagents. Reduction of *closo*-2,4-$C_2B_5H_7$ with sodium naphthalenide generates the $C_2B_5H_7^{2-}$ dianion, which on treatment with tin dichloride yields primarily the neutral carborane and traces of a species corresponding to $SnC_2B_5H_7$. The anions produced by deprotonation of $2,3\text{-}C_2B_4H_8$ and its dimethyl derivative proved more amenable to insertion of tin than the $C_2B_5H_7^{2-}$ species. Addition of tin dichloride to a THF solution of these anions produced the stannacarboranes **LXXIVa,b** as colorless, sublimable crystalline solids [equation (42)] in rather low yields. Very recently high yield syntheses of the trimethylsilylated stannacarboranes **LXXIVc–e** have been described by Hosmane *et al.* (*234*). The spectroscopic data are all consistent with a pentagonal–bipyramidal structure for these complexes. In the solid state this was confirmed by a recent single-crystal X-ray diffraction study of **LXXIVd** see Fig. 14) (*145*).

$$2,3\text{-}C_2R_2B_4H_6 \xrightarrow[\text{THF}]{\text{NaH}} Na[2,3\text{-}C_2R_2B_4H_5] \xrightarrow[-\text{NaCl}]{\frac{1}{2}\text{SnCl}_2}$$

$$\tfrac{1}{2}1,2,3SnC_2R_2B_4H_4(\textbf{LXXIVa–e}) + \tfrac{1}{2}C_2R_2B_4H_6 \quad (42)$$

	LXXIVa	b	c	d	e
R^1	H	Me	Me_3Si	Me_3Si	Me_3Si
R^2	H	Me	Me_3Si	Me	H

C. Complexes with Other Anionic 6π Ligands

An interesting tin complex has recently been reported by Siebert and co-workers (*235*). Reaction of the sandwich anion $[(H_5C_5)Co\text{-}(C_2Et_2B_2Me_2CH)]^-$ with tin dichloride results in the formation of **LXXV**, a dark orange, crystalline compound, readily soluble in aprotic, nonpolar solvents. The NMR data for **LXXV** are in accord with a tetradecker sandwich structure that is highly symmetrical on the NMR timescale. A single-crystal X-ray diffraction study revealed that two virtually identical $(H_5C_5)Co\text{-}(C_2B_2C)$ sandwich units are bound to the central tin atom in an asymmetrical pentahapto fashion via the 1,3-diborolenyl ligands (see Fig. 14). The complex exhibits a bent structure; the angle between the planes of the diborolenyl rings is 68 and 66° in the two independent molecules of the unit cell. Using the isolobal relationship (*236*) between $H_5C_5^-$ and $(H_5C_5)Co\text{-}(C_2B_2C)^-$ the bonding in this π complex can be described in the same way as that in stannocene. Similarities in chemical behavior are also found: Just as ionic species of the type $R_5C_5Sn]^+ X^-$ are formed by the reaction of stannocenes with tetrafluoroboric acid, an ionic compound **LXXVI** (see Fig. 14) containing the cationic dinuclear $(H_5C_5)Co(C_2B_2C)Sn]^+$ unit arises from the analogous reaction of **LXXV** (*235*).

Another interesting π complex, bis(1-*tert*-butyl-2,3-dimethyl-1,2-azaborolinyl)tin (**LXXVII**), has been very recently reported by Schmid *et al.*

(*239*). Reaction of tin(II) dichloride with the azaborolinyllithium leads to the orange–yellow complex, which decomposes above $-20°C$ with the formation of tin. Both diastereomers are obtained; the X-ray crystal structure of the isomer with C_2 symmetry shows a stannocene-like molecule (see Fig. 14) with an interplane angle of $46.5°$. A substantial displacement of the tin atom away from the nitrogen atoms and in the direction of the BBC ring fragment is observed, implying trihapto bonding of the heterocyclic π ligands. A low-field shift in the ^{11}B-NMR spectrum of **LXXVII** may reflect the fact that effective back-bonding is not possible in π complexes with main-group elements as central atoms.

D. Arene Complexes

The interaction of acceptor metal ions with donor aromatic molecules has been studied by Amma's group for a number of years. A number of arene complexes with tin as central atom have been investigated and characterized. Two different types of benzene complexes have been isolated from a reaction mixture consisting of tin dichloride, aluminum trichloride, and benzene, namely the compounds π-$C_6H_6Sn(AlCl_4)_2 \cdot C_6H_6$ (**LXXVIII**) (*237*) and π-$C_6H_6SnCl(AlCl_4)$ (**LXXIXa**) (*238*). The arene complex π-(p-$Me_2C_6H_4$)$SnCl(AlCl_4)$ (**LXXIXb**) has been similarly synthesized (*238*). These complexes are obtained as colorless, air-sensitive crystals, which readily decompose.

X-Ray crystal structure analyses have been performed for all three compounds. The structure of **LXXVIII** may be described as consisting of a π-benzene tin moiety weakly bonded to two chlorine atoms from each of three different tetrachloroaluminate ions (see Fig. 14). This corresponds to tetrahedral $AlCl_4^-$ ions sharing edges to form a linear chain structure. Two chlorine atoms on one $AlCl_4^-$ unit are not bound to any tin atom, whereas all the chlorine atoms of the others are involved in weak tin–chlorine interactions. The tin to carbon distances in the π-benzene tin unit seem to be comparatively long, but they are still shorter than the sum of the van der Waals radii of 4.0 Å (*237*). The second benzene molecule lies in the cleft between the chains without any contact to a tin center. The structure of **LXXIXa** contains the $Sn_2Cl_2^{2+}$ unit as an integral feature. Each tin atom in this dimer is chelated by two chlorine atoms of an $AlCl_4^-$ moiety and in addition bound to one chlorine of another $AlCl_4^-$ unit to produce an overall chain structure. The coordination polyhedron of tin is completed by a slightly asymmetric π interaction with the benzene molecule, as portrayed in Fig. 14. The complex **LXXIXb** can be considered as isostructural with the complex **LXXIXa**. A possible description of the bonding in terms of a qualitative molecular orbital treatment has been given (237). In principle, a similar type of bonding can be assumed in a $H_5C_5Sn^+$ cation and a $H_6C_6Sn]^{2+}$ dication, since they are isoelectronic species possessing isolobal $H_5C_5^-$ and H_6C_6 fragments.

XV

π COMPLEXES OF LEAD

A. Cyclopentadienyl Complexes

Dicyclopentadienyllead(II) (**LXXXa**, plumbocene) was prepared in 1956 by the reaction of sodium cyclopentadienide with lead(II) nitrate (240). The substituted plumbocenes **LXXXb–e** have subsequently been obtained in similar fashion [Eq. (43)] (223,241). A colorless, salt-like species (**LXXXf**) containing the cationic $(H_5C_5)H_4(HPNMe_2)C_5Pb^+$ unit has recently been reported (220). The plumbocene **LXXXa** has also been obtained by the thermal decomposition of bis(cyclopentadienyl)diphenyllead (271). The plumbocenes **LXXXa–e** are colored yellow to deep red. Their sensitivity to oxidation depends on the substituents at the cyclopentadienyl ring; compound **LXXXa** is rather air sensitive, whereas compound **LXXXe** can be handled for minutes in air without decomposition.

$$2\,CpM + PbX_2 \xrightarrow[-2MX]{} Cp_2Pb \qquad (43)$$

Cp_2Pb	**LXXXa** (240)	**b** (223)	**c** (241)	**d** (241)	**e** (241)
Cp	H_5C_5	Me_5C_5	$(Me_3Si)H_4C_5$	$(Me_3Si)_2H_3C_5$	$(Me_3Si)_3H_2C_5$

According to the result of an X-ray structure analysis, the orthorhombic modification (242) of the parent plumbocene has a polymeric structure in the solid state, consisting of zigzag chains of lead atoms separated by bridging cyclopentadienyl ligands lying perpendicular to the vector joining adjacent lead centers. An additional cyclopentadienyl ligand is bound to each lead atom in a pentahapto fashion, as indicated in Fig. 15. In contrast to the polymeric solid-state structure, a monomeric bent-sandwich structure has been confirmed in the gas phase by an electron diffraction study (see Fig. 15) (224). Interestingly, solutions of **LXXXa** in diethyl ether or tetrahydrofuran behave as 1:1 electrolytes (243) presumably due to ionization as $H_5C_5Pb^+$ cations and $H_5C_5^-$ anions. In contrast to **LXXXa**, the permethylated plumbocene **LXXXb** is also monomeric in the solid state (see Fig. 15) (223). A

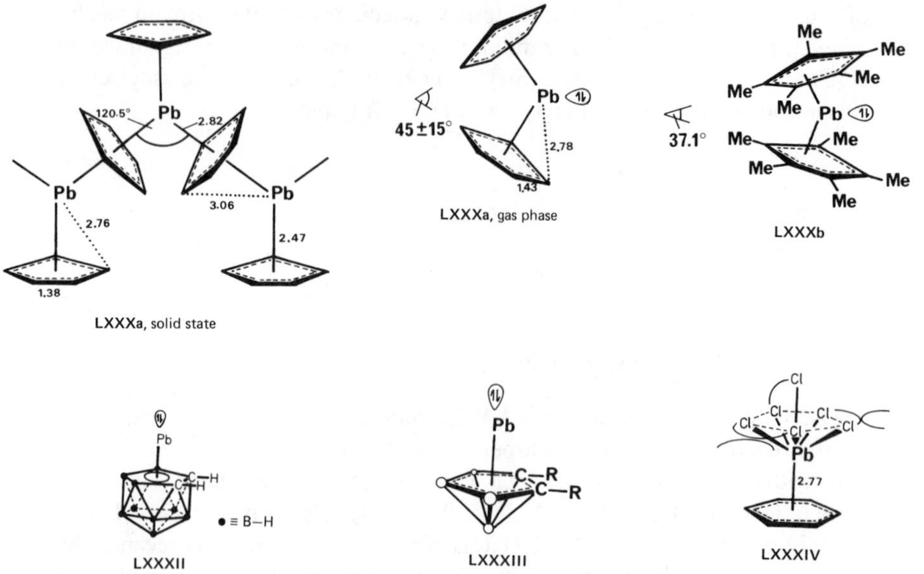

FIG. 15.   Structures of π complexes of lead.

cyclopentadienyl ring slippage similar to that observed for germanocenes and stannocenes leads to lead–carbon distances ranging from 2.69 to 2.90 Å.

The electronic and structural properties of the plumbocene polymer $[(H_5C_5)_2Pb]_n$ have been studied using the tight-bending (LCAO) method of band structure calculation (244). The observed polymeric structure minimizes the antibonding interaction between occupied lead and bridging cyclopentadienyl orbitals. The bonding in monomeric plumbocene moieties can be explained in the same way as that in other monomeric group IV metallocenes. This view is supported by the successful correlation of PE spectroscopic data and a calculated molecular orbital scheme for **LXXXb** (227) (see Section XIV,A).

Another class of cyclopentadienyl lead complexes has been prepared by Puddephatt's group (245). Monocyclopentadienyllead(II) compounds, H_5C_5PbX, can be synthesized by the reaction of plumbocene with various electrophiles as shown in Scheme 7. The complexes **LXXXIa–d** are colorless or pale yellow solids. The high melting points and low solubilities suggest that these compounds have a polymeric structure with bridging halide or acetate groups. π Bonding of the cyclopentadienyl ligands was inferred from IR spectroscopic data. In the ^1H- and ^{13}C-NMR spectra of the lead complexes **LXXX** and **LXXXI** no coupling to the ^{207}Pb nucleus could be observed thus indicating the existence of very fast exchange processes.

$$(H_5C_5)_2Pb \xrightarrow{\begin{array}{c} HX \\ \overline{-H_6C_5} \end{array}} H_5C_5PbX \qquad (LXXXIa-d)$$

(LXXXa)

$$\xrightarrow{\begin{array}{c} a)\,I_2 \\ b)\,MeI \end{array}} H_5C_5PbI \qquad (LXXXIc)$$

$$\xrightarrow{RCOCl} H_5C_5PbCl \qquad (LXXXIa)$$

LXXXIa	b	c	d
X Cl	Br	I	MeCOO

SCHEME 7. Syntheses of monocyclopentadienyllead compounds.

B. *Dicarbollyl Complexes*

Complexes with carbollide ligands have been prepared according to Eqs. (44) and (45) (*232,246*). These complexes are light yellow sublimable solids, monomeric in solution, in the gas phase, and presumably also in the solid state. Their stability is lower than that of the corresponding germanium and tin species. Their structures (see Fig. 15) have been deduced from spectroscopic data and by analogy with other metallacarborane species.

$$Na_2[7,8\text{-}C_2H_2B_9H_9] + Pb(ac)_2 \xrightarrow{\text{THF}} 1,2,3\text{-}PbC_2H_2B_9H_9 \ (LXXXII) \qquad (44)$$

$$Na[2,3\text{-}C_2R_2B_4H_5] + PbBr_2 \rightarrow 1,2,3\text{-}PbC_2R_2B_4H_4 \ (LXXXIIIa,b) \qquad (45)$$

LXXXIII	a	b
R	H	Me

C. *Arene Complexes*

The synthesis, structure, and bonding of the lead complex π-$C_6H_6Pb(AlCl_4)_2 \cdot C_6H_6$ (**LXXXIV**) have been reported by Amma and coworkers (*247*). Colorless, air-sensitive, and presumably thermolabile crystals of **LXXXIV** can be obtained from the reaction of lead(II) chloride and aluminum trichloride in benzene [Eq. (46)]. An X-ray crystallographic study has revealed the polymeric chain structure of this complex. The description of the structure is similar to that of the isomorphous tin(II) compound. The distance from lead to the center of the weakly bonded benzene ring is longer than the lead–cyclopentadienyl ring centroid distance in the plumbocenes **LXXX**. The lead–chlorine distances are rather long when compared to the sum of the covalent radii (2.50 Å), but still shorter than the sum of the van der Waals radii (4.0 Å), thus indicating the existence of weak bonding interactions (see Fig. 15).

$$PbCl_2 + 2\,AlCl_3 \xrightarrow{\text{benzene}} C_6H_6Pb(AlCl_4)_2 \cdot C_6H_6 \ (LXXXIV) \qquad (46)$$

XVI

π COMPLEXES OF PHOSPHORUS, ARSENIC, AND ANTIMONY

A. Cyclopentadienyl Complexes

The first cyclopentadienyl π complexes of phosphorus, arsenic, and antimony have been characterized only recently. The salt-like species **LXXXV–LXXXVII** containing a cationic π complex are prepared via halide ion abstraction from pentamethylcyclopentadienyl element halides according to Eq. (47) (248–250).

$$Me_5C_5(R)ElHal \xrightarrow{El'Hal_3} Me_5C_5(R)El^+El'Hal_4^- \qquad (47)$$

El = P, As, Sb; El' = B, Al
(**LXXXVa–c, LXXXVIa–c, LXXXVII**

	LXXXVa	b	c	LXXXVIa	b	c	LXXXVII
El	P	P	P	As	As	As	Sb
R	NMe_2	CMe_3	$CH(SiMe_3)_2$	NMe_2	H_5C_5	Me_5C_5	Me_5C_5
El'Hal₄	$AlCl_4$	$AlCl_4$	$AlCl_4$	$AlCl_4$	$AlCl_4$	BF_4	BF_4

The complexes **LXXXVa–c** and **LXXXVIa,b** are thermally unstable (278), their structure (see Fig. 16) has been deduced mainly from NMR spectroscopic data and from the analogy to neutral IVB element complexes of the type R_5C_5ElR (248,249). MNDO calculations for the phosphorus cations reveal a dihapto structure as a global minimum and a very low barrier to circumannular migration of the PR moiety (248). The only isolable compounds are the crystalline, yellow to orange bis(pentamethylcyclopentadienyl)arsenium and -stibenium tetrafluoroborates (**LXXXVIc, LXXXVII**) (250). An X-ray crystal structure analysis of **LXXXVIc** shows the molecule to have a bent-sandwich structure; the metal–ring bonding is distorted in the direction of dihapto bonding to the one and trihapto bonding to the other cyclopentadienyl ring (see Fig. 16) (250). In solution, the cations in **LXXXV–LXXXVII** are all highly fluxional. Once more, a low nucleophilicity of the counter anion is important for the stabilization of these cationic π complexes.

B. Carbollyl Complexes

Carbollyl complexes of group VB elements have been mainly investigated by Todd's group (121). The heterogeneous reaction of the $CHB_{10}H_{10}^{3-}$ anion

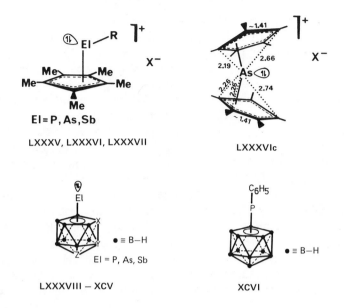

FIG. 16.   Structures of π complexes of phosphorus, arsenic, and antimony.

with VB element trihalides produces the compounds **LXXXVIII, XCI**, and **XCIV** [Eq. (48)] (see Fig. 16) (*251,252*).

$$Na_3CHB_{10}H_{10}(THF)_2 + ElHal_3 \xrightarrow{-3\,NaHal} 1,2\text{-}ElCHB_{10}H_{10} \qquad (48)$$

$$\textbf{(LXXXVIII, XCI, XCIV)}$$

	LXXXVIII	**LXXXIX**	**XC**	**XCI**	**XCII**	**XCIII**	**XCIV**	**XCV**
El	P	P	P	As	As	As	Sb	Sb
X	CH	BH	BH	CH	BH	BH	CH	BH
Y	BH	CH	BH	BH	CH	BH	BH	CH
Z	BH	BH	CH	BH	BH	CH	BH	BH

A low yield synthesis of **XCI** starts from $Na_3CHB_{10}H_{10}$ and solid As_2O_3 in aqueous KOH–heptane (253). The complexes described rearrange thermally to the 1,7- (**LXXXIX, XCII, XCV**) and the 1,12-isomers (**XC, XCIII**) (252,254). All these compounds are sublimable, air-stable colorless solids. Their structures (see Fig. 16) have been deduced from spectroscopic data. Halogen substituents at the boron atoms may be introduced by halogenation catalyzed by aluminum chloride as well as by photochlorination reactions (254–256), and substitutions at the carbon atom can be performed after metallation with organolithium compounds (251,257,258). Reaction of the complexes **LXXXVIII, LXXXIX, XCI, XCII**, and **XCIV** with amines affords anionic species of the type $ElCHB_9H_{10}^-$ via boron extrusion from the carbollide system (257,206). In these anions the lone pair at the group VB element shows a pronounced nucleophilicity, in sharp contrast to its inertness in the neutral complexes (121). Finally a synthesis which formally involves the insertion of a PR^{4+} fragment into the open face of a $B_{11}H_{11}^{4-}$ anion has been described. Reaction of phenylphosphorus dichloride with $Na_2B_{11}H_{13}$ yields the complex $P(C_6H_5)B_{11}H_{11}$ (**XCVI**) in low yield (251). The ^{11}B-NMR spectrum is consistent with an icosahedral structure, as indicated in Fig. 16.

C. Arene Complexes

Crystalline adducts between antimony trichloride and aromatic hydrocarbons have been known since the nineteenth century (272,273) and have been investigated in detail since the 1970s (274). Since such adducts have been postulated as intermediates in Friedel–Crafts reactions, their structure is of interest.

A crystalline 2:1 complex (**CI**) of antimony trichloride and naphthalene was obtained from a hot petroleum ether solution on cooling. An X-ray analysis (275) revealed an interaction between the antimony atom and the arene π system. The structure consists of layers of antimony trichloride molecules alternating with layers of naphthalene molecules. The coordination sphere around the antimony atom is portrayed in Fig. 16. Two antimony–chlorine distances are equal, while the third is significantly longer. The antimony atom is 3.2 Å away from the plane of the naphthalene molecule, thus indicating a weak π interaction.

The structure of a 2:1 complex between antimony trichloride and phenanthrene consists of alternating double layers of $SbCl_3$ and arene molecules (276). Unlike complex **CI** and $SbCl_3$ itself, Sb—Cl contacts between neighboring $SbCl_3$ units are observed in this complex. In both structures mentioned here, vacant coordination sites are observed, which might be occupied by a stereochemically active lone pair at the antimony center.

XVII

π COMPLEXES OF GROUP VI AND GROUP VII ELEMENTS

It is difficult for the elements of group VI and VII to act as central atoms in π complexes. Vacant orbitals or coordination sites are generally necessary for the stabilization of a π complex. This would require high oxidation states of the group VI or VII elements, as illustrated by the structures **V**, **W**, and **X** for a series of cyclopentadienylsulfur complexes. It is therefore not surprising that attempts to generate complexes of this type so far have been unsuccessful (259).

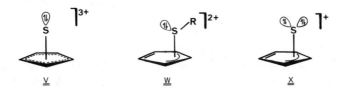

The use of more highly negatively charged π systems as ligands increases the chance of π complex formation by group VI elements. This is nicely confirmed by the synthesis of the carbollyl complexes **XCVII**, **XCVIII**, and **XCIX**, which are formally composed of a $B_{11}H_{11}^{4-}$ π ligand and a Group VI element in the $+4$ oxidation state. The sublimable, colorless compounds are synthesized according to Eq. (49) (253,260). Their structure (see Fig. 17) has been deduced mainly from ^{11}B-NMR spectroscopic data. These and a few other complexes,

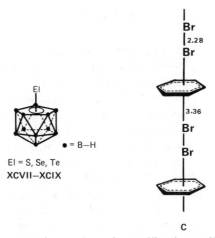

FIG. 17.   Structures of π complexes of group VI and group VII elements.

which can also be regarded as π complexes, have recently been described by Todd (*121*).

$$NaB_{11}H_{14} + \quad ElO_2 \quad \rightarrow \quad ElB_{11}H_{11} \ (\textbf{XCVII–XCIX}) \tag{49}$$
$$(NaHElO_3)$$

XCVII	**XCVIII**	**XCIX**
El S	Se	Te

π Complexes between some group VI or VII molecules (Hal_2, SO_2) and alkenes or arenes have been known for a long time (*261,262*). Most of these electron donor–acceptor complexes are stable only in solution; some of them are important intermediates in electrophilic addition and substitution reactions (*263*). They are not discussed in detail here.

Crystalline 1:1 complexes of a few arenes with halogen molecules have been isolated and structurally characterized. As an example, the structure of the π complex **C** between benzene and bromine is given in Fig. 17 (*264*). In chains of alternate benzene and Br_2 species the bromine atoms are localized on the sixfold symmetry axis of the benzene ring. The bromine–bromine distance of 2.28 Å is nearly the same as that observed in the free molecule. The distance from each bromine atom to the adjacent benzene plane is 3.36 Å (~ 0.30 Å less than the sum of the van der Waals radii). The extent of charge transfer is estimated to be 0.06 electrons in the electronic ground state (*265*), indicating the weakness of π bonding in this and other comparable complexes.

XVIII

CONCLUSION

The collection of so many examples in this article clearly demonstrates that π complexes of main-group elements are now a well established class of compounds. The bonding in these complexes differs importantly from that in most transition-metal π species. Main group fragments do not possess empty or partially filled d subshells. As a consequence, π-electron density from the ligand can be transferred to a smaller extent and only into s- or p-type orbitals. In many situations, π electrons therefore remain in nonbonding orbitals centered at the ligands; even purely electrostatic interactions are sometimes used to explain the bonding. Furthermore, the energy difference between filled d subshells—if present at all—and π^* orbitals at the ligand is so high that electron transfer (back donation) cannot be significant. Thus, synergistic effects—typical in transition metal π complexes—are not observed in π complexes of main-group elements.

This review has dealt with the synthesis, structure, and bonding of main-group π complexes, but has covered their chemistry only incidentally. Recent results promise that main-group π complexes will simplify synthesis of known products and open paths to new ones. It is hoped that this review will stimulate and support research in this direction.

IXX

ADDENDUM

Some very recent progress is summarized here.

Microwave dielectric loss measurements in the region 1–9 GHz have been reported for beryllocene in different solvents (cyclohexane, decalin, benzene, and 1,4-dioxane) (*279*). The dipolar nature of beryllocene was confirmed. The dielectric absorption was interpreted to suggest rocking movements of the cyclopentadienyl groups synchronous with oscillation of the beryllium atom between two equivalent positions in an η^5, σ structure, as indicated in **CII**.

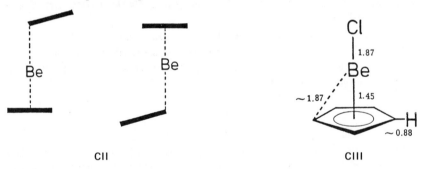

| CII | CIII |

The crystal structure of cyclopentadienylberyllium chloride has been determined by X-ray diffraction (*280*). As portrayed in **CIII**, the molecule has the same structure in the solid state as in the gas phase. These results rule out the possibility suggested by the mass spectrum (*281*) that H_5C_5BeCl is associated. Although not precisely located, the hydrogen atoms of the cyclopentadienyl ring show a slight tendency to bend toward the beryllium atom, in agreement with theoretical predictions (*63*).

π Complexes of monovalent gallium, indium, and thallium with arene ligands (see Sections IX,D, X,D, and XI,D) have been reviewed recently by Schmidbaur (*282*), who puts special emphasis on the practical application of these compounds, on ^{69,71}Ga NMR spectra, on X-ray crystal structure data, and on bonding concepts.

The first π complex with silicon as the central atom meanwhile has been synthesized in our group. *Bis*(pentamethylcyclopentadienyl)dichlorosilane reacts with alkali metal naphthalenides to give decamethylsilicocene, a thermally stable, but extremely air-sensitive, colorless compound. Surprisingly, an X-ray diffraction analysis showed that two conformers (see **CIV** and **CV**) are

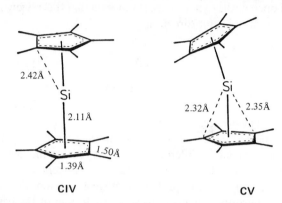

CIV   CV

present in the unit cell in the ratio 1:2. Decamethylsilicocene is the first molecular compound of divalent silicon that is stable under normal conditions (*283*).

Synthesis and structure of decabenzylgermanocene have been recently described (*284*). In this π complex, the lone pair at the germanium atom seems to be effectively shielded by benzyl groups, thus explaining the stability of the compound against air oxidation.

ACKNOWLEDGMENTS

The work of the author's group described here was done by a team of active co-workers, whose names are given in the references and to whom I wish to express my appreciation. Generous support for our studies by the Deutsche Forschungsgemeinschaft and the Fonds der Chemischen Industrie is gratefully acknowledged as well. I am thankful to Dr. Ch. Morley and Dr. J. Broad for improving the English, and to Mrs. A. Zingler for preparing the manuscript for publication.

REFERENCES

1. G. Stucky, *Adv. Chem. Ser.* **130,** Chap. 3 (Am. Chem. Soc., Washington, D.C.) (1974).
2. J. L. Wardell, *in* "Comprehensive Organometallic Chemistry" (G. Wilkinson, F. G. A. Stone, and E. W. Abel, eds.), Chap. 2, Pergamon, Oxford, 1982.
3. W. N. Setzer and P. V. R. Schleyer, *Adv. Organomet. Chem.* **24,** 354 (1985).
4. J. B. Smart, R. Hogan, P. A. Scherr, M. T. Emerson, and J. P. Oliver, *J. Organomet. Chem.* **64,** 1 (1974).
5. K. Jonas and C. Krüger, *Angew. Chem.* **92,** 513 (1980).
6. K. Jonas, *Adv. Organomet. Chem.* **19,** 97 (1981).
7. H. Klein, H. Witty, and U. Schubert, *J. Chem. Soc., Chem. Commun.,* 231 (1983).

8. G. Boche, H. Dietrich, W. Mahdi, H. Etzrodt, M. Marsch, W. Massa, and G. Baum, *Angew. Chem.* **98**, 84 (1986).
9. H. Köster and E. Weiss, *Chem. Ber.* **115**, 3422 (1982).
10. S. D. Patterman, J. L. Karle, and G. D. Stucky, *J. Am. Chem. Soc.* **92**, 1150 (1970).
11. W. E. Rhine and G. D. Stucky, *J. Am. Chem. Soc.* **97**, 737 (1975).
12. J. J. Brooks, W. E. Rhine, and G. D. Stucky, *J. Am. Chem. Soc.* **94**, 7339 (1972).
13. S. K. Arora, R. B. Bates, W. A. Beavers, and R. S. Cutler, *J. Am. Chem. Soc.* **97**, 6271 (1975).
14. L. Reichardt, *in* "Monographs in Modern Chemistry," Vol. 3, Verlag Chemie, Weinheim, 1979.
15. M. Schlosser, "Struktur und Reaktivität Polarer Organometalle." Springer-Verlag, Berlin and New York, 1973.
16. M. Schlosser and M. Staehle, *Angew. Chem.* **92**, 494 (1980).
17. S. Brownstein, S. Bywater, and D. J. Worsfold, *J. Organomet. Chem.* **199**, 1 (1980).
18. W. Neugebauer and P. V. R. Schleyer, *J. Organomet. Chem.* **198**, C1 (1981).
19. M. Staehle and M. Schlosser, *J. Organomet. Chem.* **220**, 277 (1981).
20. P. West, J. J. Purmont, and S. V. Mc Kinley, *J. Am. Chem. Soc.* **90**, 797 (1968).
21. G. Boche, K. Buckl, and D. R. Schneider, *Liebigs Ann. Chem.* 1135 (1980).
22. H. Albrecht, K. Zimmermann, G. Boche, and G. Decher, *J. Organomet. Chem.* **262**, 1 (1984).
23. T. Clark, C. Rhode, and P. V. R. Schleyer, *Organometallics* **2**, 1344 (1983).
24. G. Decher and G. Boche, *J. Organomet. Chem.* **259**, 31 (1983).
25. P. Jutzi, E. Schlüter, S. Pohl, and W. Saak, *Chem. Ber.* **118**, 1959 (1985).
26. H. Preuss and G. Diercksen, *Int. J. Quantum Chem.* **1**, 349 (1967).
27. M. J. S. Dewar and R. C. Haddon, *J. Am. Chem. Soc.* **95**, 5836 (1973).
28. N. T. Anh, M. Elian, and R. Hoffmann, *J. Am. Chem. Soc.* **100**, 110 (1978).
29. S. Alexandratos, A. Streitwieser, Jr., H. F. Schaefer, III, *J. Am. Chem. Soc.* **98**, 7959 (1976).
30. E. D. Jemmis and P. V. R. Schleyer, *J. Am. Chem. Soc.* **104**, 4781 (1982).
31. M. Lattman and A. H. Cowley, *Inorg. Chem.* **23**, 241 (1984).
32. K. C. Waterman and A. Streitwieser, Jr., *J. Am. Chem. Soc.* **106**, 3138 (1984).
33. M.F. Lappert, A. Singh, L. M. Engelhardt, and A. H. White, *J. Organomet. Chem.* **262**, 271 (1984).
34. D. Bladauski, W. Broser, H. J. Hecht, D. Rewicki, and H. Dietrich, *Chem. Ber.* **112**, 1380 (1979).
35. M. Walczak and G. D. Stucky, *J. Am. Chem. Soc.* **98**, 5531 (1976).
36. G. Boche, H. Etzrodt, W. Massa, and G. Baum, *Angew. Chem.* **97**, 858 (1985).
37. W. E. Rhine, J. H. Davis, and G. D. Stucky, *J. Organomet. Chem.* **134**, 139 (1977).
38. J. J. Brooks, W. Rhine, and G. D. Stucky, *J. Am. Chem. Soc.* **94**, 7346 (1972).
39. W. E. Rhine, J. Davis, and G. D. Stucky, *J. Am. Chem. Soc.* **97**, 2079 (1975).
40. D. Wilhelm, H. Dietrich, T. Clark, W. Mahdi, I. Kos, and P. V. R. Schleyer, *J. Am. Chem. Soc.* **106**, 7279 (1984).
41. H. Köster and E. Weiss, *J. Organomet. Chem.* **168**, 273 (1979).
42. J. J. Brooks and G. D. Stucky, *J. Am. Chem. Soc.* **94**, 7333 (1972).
43. R. D. Rogers, J. L. Atwood, M. D. Rausch, D. W. Macomber, and W. P. Heist, *J. Organomet. Chem.* **238**, 79 (1982).
44. T. Aoyagi, H. M. M. Shearer, K. Wade, and G. Whitehead, *J. Organomet. Chem.* **175**, 21 (1979).
45. R. Zerger, W. Rhine, and G. D. Stucky, *J. Am. Chem. Soc.* **96**, 5441 (1974).
46. D. J. Brauer and G. D. Stucky, *J. Am. Chem. Soc.* **92**, 3956 (1970); *idem, J. Organomet. Chem.* **37**, 217 (1972).
47. Values from the literature or calculated from simple trigonometric functions.
48. N. A. Bell, J. W. Nowell, and H. M. M. Shearer, *J. Chem. Soc., Chem. Commun.*, 147 (1982).

49. E. O. Fischer and H. P. Hofmann, *Chem. Ber.* **92**, 482 (1959).
50. C. Wong, T. Y. Lee, and S. Lee, *Acta Crystallogr.* **B28**, 1662 (1972).
51. D. A. Drew and A. Haaland, *Acta Crystallogr.* **B28**, 3671 (1972).
52. J. Lusztyk and K. B. Starowieyski, *J. Organomet. Chem.* **170**, 293 (1979).
53. C. Wong, T. Y. Lee, T. J. Lee, T. W. Chang, and C. S. Lin, *Inorg. Nucl. Chem. Lett.* **9**, 667 (1973).
54. A. Haaland, *Acta Chem. Scand.* **22**, 3030 (1968).
55. A. Almenningen, A. Haaland, and J. Lusztyk, *J. Organomet. Chem.* **170**, 271 (1979).
56. R. Gleiter, M. C. Böhm, A. Haaland, R. Johansen, and J. Lusztyk, *J. Organomet. Chem.* **170**, 285 (1979).
57. C. Wong and S. Wang, *Inorg. Nucl. Chem. Lett.* **11**, 677 (1975).
58. R. Blom, A. Haaland, and J. Weidlein, *J. Chem. Soc., Chem. Commun.*, 266 (1985).
59. M. Sundbom, *Acta Chem. Scand.* **20**, 1608 (1966)
60. O. Y. Lopatko, N. M. Klimenko, and M. E. Dyatkina, *Zh. Strukt. Khim.* **13**, 1128 (1972).
61. D. S. Marynick, *J. Am. Chem. Soc.* **99**, 1436 (1977).
62. N. S. Chin and L. Schäfer, *J. Am. Chem. Soc.* **100**, 2604 (1978).
63. E. D. Jemmis, S. Alexandratos, P. v. R. Schleyer, and A. Streitwieser, Jr., III, H. F. Schaefer, *J. Am. Chem. Soc.* **100**, 5695 (1978), and references cited therein.
64. M. J. S. Dewar and H. S. Rzepa, *J. Am. Chem. Soc.* **100**, 777 (1978).
65. J. Demuynk and M. M. Rohmer, *Chem. Phys. Lett.* **54**, 567 (1978).
66. T. C. Bartke, A. Bjorseth, A. Haaland, K. M. Marstokk, and H. Møllendal, *J. Organomet. Chem.* **85**, 271 (1975).
67. D. A. Drew and A. Haaland, *Acta Chem. Scand.* **26**, 3351 (1972).
68. A. Haaland and D. P. Novak, *Acta Chem. Scand. Ser. A* **28**, 153 (1974).
69. D. A. Drew and G. L. Morgan, *Inorg. Chem.* **16**, 1704 (1977).
70. A. Haaland and D. A. Drew, *Acta Chem. Scand.* **26**, 3079 (1972).
71. G. E. Coates, D. L. Smith, and R. C. Srivastava, *J. Chem. Soc., Dalton Trans.*, 618 (1973).
72. W. E. Stewart and T. H. Siddal, *Chem. Rev.* **70**, 517 (1970).
73. M. C. Böhm, R. Gleiter, G. L. Morgan, J. Lusztyk, and K. B. Starowieyski, *J. Organomet. Chem.* **194**, 257 (1980).
74. D. A. Coe, J. W. Nibler, T. H. Cook, D. A. Drew, and G. L. Morgan, *J. Chem. Phys.* **63**, 4842 (1975).
75. D. F. Gaines, K. M. Coleson, and J. C. Calabrese, *J. Am. Chem. Soc.* **101**, 3979 (1979).
76. D. F. Gaines, J. L. Walsh, J. H. Morris, and D. F. Hillenbrand, *Inorg. Chem.* **17**, 1516 (1978).
77. D. F. Gaines and J. L. Walsh, *Inorg. Chem.* **17**, 1238 (1978).
78. D. A. Drew, G. Gundersen, and A. Haaland, *Acta Chem. Scand.* **26**, 2147 (1972).
79. *D. A. Drew and A. Haaland, J. Chem. Soc. Chem. Commun.*, 1551 (1971).
80. E. M. Marlett and R. N. Saunders, U.S. Patent 3,783,054 (1974), *Chem. Abstr.* **80**, 122953 (1974).
81. J. Bicerano and W. N. Lipscomb, *Inorg. Chem.* **18**, 1565 (1979).
82. P. Fischer, J. Stadelhofer, and J. Weidlein, *J. Organomet. Chem.* **116**, 65 (1976)
83. K. Wade, *Adv. Inorg. Radiochem.* **18**, 1 (1976).
84. G. Popp and M. F. Hawthorn, *Inorg. Chem.* **10**, 391 (1971).
85. G. Popp and M. F. Hawthorn, *J. Am. Chem. Soc.* **90**, 6553 (1968).
86. W. E. Lindsell, *in* "Comprehensive Organometallic Chemistry" (G. Wilkinson, F. G. A. Stone, and E. W. Abel, eds.), Chap. 5.0, Pergamon, Oxford, 1982.
87. F. A. Cotton and G. Wilkinson, *Chem. Ind.*, 307 (1954).
88. F. A. Cotton, G. Wilkinson, and J. M. Birmingham, *J. Inorg. Nucl. Chem.* **2**, 95 (1956).
89. E. O. Fischer and W. Hafner, *Z. Naturforsch.* **B9**, 503 (1954).
90. W. A. Barber, *J. Inorg. Nucl. Chem.* **4**, 373 (1957).

91. W. Bünder and E. Weiss, *J. Organomet. Chem.* **92**, 1 (1975).
92. A. Haaland, J. Lusztyk, J. Brunvoll, K. B. Starowieyski, *J. Organomet. Chem.* **85**, 279 (1975).
93. J. L. Robbins, N. Edelstein, B. Spencer, and J. C. Smart, *J. Am. Chem. Soc.* **104**, 1882 (1982).
94. C. P. Morley, P. Jutzi, C. Krüger, and J. M. Wallis, *Organometallics* (in press).
95. C. Morley, P. Jutzi, and C. Krüger, in preparation.
96. H. O. House, R. A. Latham, and G. M. Whitesides, *J. Org. Chem.* **32**, 2481 (1967).
97. G. E. Parris and E. C. Ashby, *J. Organomet. Chem.* **72**, 1 (1974).
98. C. Johnson, J. Toney, and G. D. Stucky, *J. Organomet. Chem.* **40**, C11 (1972).
99. W. Ford, *J. Organomet. Chem.* **32**, 27 (1971).
100. K. Faegri, Jr., J. Almlöf, and H. P. Lüthi, *J. Organomet. Chem.* **249**, 303 (1983).
101. S. Evans, M. L. H. Green, B. Jewitt, A. F. Orchard, and C. F. Pygall, *J. Chem. Soc. Faraday Trans. 2* **68**, 1847 (1972).
102. C. Cauletti, J. C. Green, M. R. Kelly, J. van Tilborg, J. Robbins, and J. Smart, *J. Electron. Spectr. Rel. Phenom.* **19**, 327 (1980).
103. R. Benn, H. Lehmkuhl, K. Mehler, and A. Rufinska, *Angew. Chem.* **96**, 521 (1984).
104. K. D. Smith and J. L. Atwood, *Inorg. Synth.* **16**, 137 (1976).
105. K. D. Smith and J. L. Atwood, *J. Am. Chem. Soc.* **96**, 994 (1974).
106. F. Fujita, Y. Ohnuna, H. Yasuda, and H. Tani, *J. Organomet. Chem.* **113**, 201 (1976).
107. Y. Kai, N. Kanehisa, K. Miki, N. Kasai, K. Mashima, H. Yasuda, and A. Nakamura, *Chem. L ett.*, 1277 (1982).
108. B. G. Gowenlock and W. E. Lindsell, *J. Organomet. Chem. Library*, **3**, 1 (1977).
109. R. Zerger and G. Stucky, *J. Organomet. Chem.* **80**, 7 (1974).
110. B. Bogdanović, S. Liao, R. Mynott, K. Schlichte, and U. Westeppe, *Chem. Ber.* **117**, 1378 (1984).
111. W. Strohmeier, H. Landsfeld, F. Germert, and W. Langhanser, *Z. Anorg. Allg. Chem.* **307**, 120 (1960).
112. M. E. O'Neill and K. Wade, *in* "Comprehensive Organometallic Chemistry" (G. Wilkinson, F. G. A. Stone, and E. W. Abel, eds.), Chap. 1. Pergamon, Oxford, 1982.
113. R. Köster, Organoboron heterocycles, *in* "Progress Boron Chemistry," Vol. 1, p. 314, Pergamon, Oxford, 1964.
114. P. Jutzi, A. Seufert, and W. Buchner, *Chem. Ber.* **112**, 2488 (1979).
115. P. Jutzi and A. Seufert, *J. Organomet. Chem.* **161**, C5 (1978).
116. P. Jutzi and B. Krato, unpublished results.
117. R. E. Williams, *Adv. Inorg. Radiochem.* **18**, 67 (1976).
118. R. W. Rudolph, *Accounts Chem. Res.*, 446 (1976).
119. T. Onak, *in* "Comprehensive Organometallic Chemistry" (G. Wilkinson, F. G. A. Stone, and E. W. Abel, eds.), Chap. 5.4. Pergamon, Oxford, 1982.
120. R. N. Grimes, *in* "Comprehensive Organometallic Chemistry" (G. Wilkinson, F. G. A. Stone, and E. W. Abel, eds.), Chap. 5.5. Pergamon, Oxford, 1982.
121. C. J. Todd, *in* "Comprehensive Organometallic Chemistry" (G. Wilkinson, F. G. A. Stone, and E. W. Abel, eds.), Chap. 5.6. Pergamon, Oxford, 1982.
122. M. F. Hawthorne and P. A. Wegner, *J. Am. Chem. Soc.* **90**, 896 (1968).
123. F. X. Kohl and P. Jutzi, *Angew. Chem.* **95**, 55 (1983).
124. J. J. Eisch, *in* "Comprehensive Organometallic Chemistry" (G. Wilkinson, F. G. A. Stone, and E. W. Abel, eds.), Chap. 6. Pergamon, Oxford, 1982.
125. G. Hata, *J. Chem. Soc., Chem. Commun.*, 7 (1968).
126. T. W. Dolzine, J. St. Denis, and J. P. Oliver, *J. Am. Chem. Soc.* **94**, 8260 (1972).
127. J. P. Oliver, T. W. Dolzine, *J. Am. Chem. Soc.* **96**, 1737 (1974).
128. T. Mole and J. R. Surtees, *Aust. J. Chem.* **17**, 1229 (1964).

129. G. D. Stucky, M. M. Mc Pherson, W. E. Rhine, J. J. Eisch, and J. L. Considine, *J. Am. Chem. Soc.* **96**, 1941 (1974).
130. A. Almenningen, L. Fernholt, and A. Haaland, *J. Organomet. Chem.* **155**, 245 (1978).
131. H. Lehmkuhl, W. Haaf, and W. Storm, unpublished.
132. J. P. Oliver, *Adv. Organomet. Chem.* **15**, 235 (1977).
133. P. Jutzi, *Chem. Rev.* (in press).
134. W. R. Kroll and B. E. Hudson, Jr., *J. Organomet. Chem.* **28**, 205 (1971).
135. W. R. Kroll and W. Naegele, *J. Chem. Soc., Chem. Commun.*, 246 (1969).
136. A. Haaland and J. Weidlein, *J. Organomet. Chem.* **40**, 29 (1972).
137. B. Teclé, W. R. Corfield, and J. P. Oliver, *Inorg. Chem.* **21**, 458 (1982).
138. D. Drew and A. Haaland, *Acta Chem. Scand.* **27**, 3735 (1973).
139. O. Gropen and A. Haaland, *J. Organomet. Chem.* **92**, 157 (1975).
140. P. R. Schonberg, R. T. Paine, and C. F. Campana, *J. Am. Chem. Soc.* **101**, 7726 (1979).
141. D. A. T. Young, R. J. Wiersema, and M. F. Hawthorne, *J. Am. Chem. Soc.* **93**, 5687 (1971).
142. M. R. Churchill, A. H. Reis, D. A. T. Young, G. R. Willey, and M. F. Hawthorne, *J. Chem. Soc., Chem. Commun.*, 298 (1971).
143. D. A. T. Young, G. R. Willey, M. F. Hawthorne, A. H. Reis, and M. R. Churchill, *J. Am. Chem. Soc.* **92**, 6663 (1970).
144. B. M. Mikhailov and T. V. Potapova, *Izv. Akad. Nauk SSSR, Ser. Khim.* **5**, 1153 (1968).
145. A. H. Cowley, P. Galow, N. S. Hosmane, P. Jutzi, and N. C. Norman, *J. Chem. Soc., Chem. Commun.*, 1564 (1984).
146. E. A. Jeffrey and T. Mole, *J. Organomet. Chem.* **11**, 393 (1968).
147. B. Teclé, W. H. Ilsley, and J. P. Oliver, *Inorg. Chem.* **20**, 2335 (1981).
148. W. Fries, W. Schwarz, H. D. Hausen, and J. Weidlein, *J. Organomet. Chem.* **159**, 373 (1978).
149. T. Fjedberg, A. Haaland, R. Seip, and J. Weidlein, *Acta Chem. Scand.* **A35**, 437 (1981).
150. J. Stadelhofer, J. Weidlein, and A. Haaland, *J. Organomet. Chem.* **84**, C1 (1975); *idem, ibid.* **116**, 55 (1976).
151. K. Mertz, F. Zettler, H. D. Hausen, and J. Weidlein, *J. Organomet. Chem.* **122**, 159 (1976).
152. R. N. Grimes, W. J. Rademaker, M. L. Denniston, R. F. Bryaw, and P. T. Greene, *J. Am. Chem. Soc.* **94**, 1865 (1972).
153. E. Canadell, O. Eisenstein, and J. Rubio, *Organometallics* **3**, 759 (1984).
154. E. Kinsella, J. Chadwick, and J. Coward, *J. Chem. Soc. A*, 968 (1968).
155. R. E. Rundle and J. D. Corbett, *J. Am. Chem. Soc.* **79**, 752 (1957).
156. J. G. Oliver and J. Worrall, *J. Inorg. Nucl. Chem. Lett.* **3**, 575 (1967).
157. H. Schmidbaur, U. Thewalt, and Th. Zafiopoulos, *Organometallics* **2**, 1550 (1983).
158. H. Schmidbaur, U. Thewalt, and Th. Zafiopoulos, *Chem. Ber.* **117**, 3381 (1984).
159. H. Schmidbaur, U. Thewalt, and Th. Zafiopoulos, *Angew. Chem.* **96**, 60 (1984).
160. E. O. Fischer and H. P. Hofmann, *Angew. Chem.* **69**, 639 (1957).
161. J. S. Poland and D. G. Tuck, *J. Organomet. Chem.* **42**, 307 (1972).
162. F. Einstein, M. M. Gilbert, and D. G. Tuck, *Inorg. Chem.* **11**, 2832 (1972).
163. S. Shibata, L. S. Bartell, and R. M. Gavin, Jr., *J. Chem. Phys.* **41**, 717 (1964).
164. E. Frasson, F. Menegus, and C. Panattoni, *Nature (London)* **199**, 1087 (1963).
165. S. Cradock and W. Duncan, *J. Chem. Soc. Faraday Trans. 2* **74**, 194 (1978).
166. R. G. Edgell, J. Fragala, and A. F. Orchard, *J. Electron. Spectr. Relat. Phenom.* **14**, 467 (1978).
167. C. Lin and D. G. Tuck, *Can. J. Chem.* **60**, 699 (1982).
168. J. Ebenhöch, G. Müller, S. Riede, and H. Schmidbaur, *Angew. Chem.* **96**, 367 (1984).
169. G. M. Kuz'yants, *Izv. Akad. Nauk SSSR, Ser. Khim.*, 1785 (1976).
170. R. Nast and K. Käb, *J. Organomet. Chem.* **6**, 456 (1966).
171. H. Kurosowa, Thallium *in* "Comprehensive Organometallic Chemistry" (G. Wilkinson, F. G. A. Stone, and E. W. Abel, eds.). Vol. 1, p. 749. Pergamon, Oxford, 1982.

172. H. P. Fritz and F. H. Köhler, *J. Organomet. Chem.* **30**, 177 (1971).
173. A. A. Koridze, N. A. Ogorodnikova, and P. V. Petrovsky, *J. Organomet. Chem.* **157**, 145 (1978).
174. T. J. Katz and J. J. Mrowka, *J. Am. Chem. Soc.* **89**, 1105 (1967).
175a. G. A. Rupprecht, L. W. Messerle, J. D. Fellmann, and R. R. Schrock, *J. Am. Chem. Soc.* **102**, 6236 (1980).
175b. P. Jutzi and W. Leffers, *Chem. Soc., Chem. Commun.*, 1735 (1985).
176. M. B. Freeman, L. G. Sneddon, and J. C. Huffmann, *J. Am. Chem. Soc.* **99**, 5194 (1977).
177. G. Wulfsberg and R. West, *J. Am. Chem. Soc.* **94**, 6069 (1972).
178. J. K. Tyler, A. P. Cox, and J. Sheridan, *Nature (London)* **183**, 1182 (1959).
179. C. S. Ewig, R. Osman, and J. R. van Wazer, *J. Am. Chem. Soc.* **100**, 5017 (1978).
180. J. L. Spencer, M. Green, and F. G. A. Stone, *J. Chem. Soc., Chem. Commun.*, 1178 (1972).
181. H. M. Colquhoum, T. J. Greenhough, and M. G. H. Wallbridge, *J. Chem. Soc., Chem. Commun.*, 737 (1977).
182. K. Stumpf, H. Pritzkow, and W. Siebert, *Angew. Chem.* **97**, 64 (1985).
183. G. E. Herberich, H. J. Becker, and C. Engelke, *J. Organomet. Chem.* **153**, 265 (1978).
184. T. Auel and L. Amma, *J. Am. Chem. Soc.* **90**, 5941 (1968).
185. H. Schmidbaur, W. Bublak, J. Riede, and G. Müller, *Angew. Chem.* **97**, 402 (1985).
186. H. Hogeveen and P. Kwant, *Acc. Chem. Res.* **8**, 413 (1975).
187. M. J. S. Dewar and C. H. Reynolds, *J. Am. Chem. Soc.* **106**, 1744 (1984).
188. M. J. S. Dewar and M. K. Halloway, *J. Am. Chem. Soc.* **106**, 6619 (1984).
189. J. V. Scibelli and M. D. Curtis, *J. Am. Chem. Soc.* **95**, 924 (1973).
190. M. Grenz, E. Hahn, W. W. du Mont, and J. Pickart, *Angew. Chem.* **96**, 68 (1984).
191. A. Bonny, A. D. Mc Master, and S. R. Stobart, *Inorg. Chem.* **17**, 935 (1978).
192. P. Jutzi, F. Kohl, P. Hofmann, C. Krüger, Y. H. Tsay, *Chem. Ber.* **113**, 757 (1980).
193. P. Jutzi, E. Schlüter, M. B. Hursthouse, A. M. Arif, and R. L. Short, *J. Organomet. Chem.* **299**, 19 (1986).
194. P. Jutzi, B. Hampel, M. B. Hursthouse, and A. J. Howes, *Organometallics*, in press.
195. J. Almlöf, L. Fernholt, K. Faegri, Jr., A. Haaland, B. E. R. Schilling, R. Seip, and K. Taugbøl, *Acta Chem. Scand.* **A37**, 131 (1983).
196. L. Fernholt, A. Haaland, P. Jutzi, F. X. Kohl, and R. Seip, *Acta Chem. Scand.* **A38**, 211 (1984).
197. G. Bruno, E. Ciliberto, J. L. Fragala, and P. Jutzi, *J. Organomet. Chem.* **289**, 263 (1985).
198. F. X. Kohl and P. Jutzi, *J. Organomet. Chem.* **243**, 31 (1983).
199. C. F. Childs and S. Winstein, *J. Am. Chem. Soc.* **90**, 7146 (1968).
200. A. Haaland and B. E. R. Schilling, *Acta Chem. Scand.* **A38**, 217 (1984).
201. P. Jutzi, B. Hampel, K. Stroppel, C. Krüger, K. Angermund, and P. Hofmann, *Chem. Ber.* **118**, 2789 (1985).
202. P. Jutzi, B. Hampel, M. B. Hursthouse, and A. J. Howes, *Organometallics* **299**, 19 (1986).
203. P. Jutzi, F. X. Kohl, E. Schlüter, M. B. Hursthouse, and N. P. C. Walker, *J. Organomet. Chem.* **271**, 393 (1984).
204. P. Jutzi and B. Hampel, unpublished.
205. F. X. Kohl, E. Schlüter, P. Jutzi, C. Krüger, G. Wolmershäuser, P. Hofmann, and P. Stauffert, *Chem. Ber.* **117**, 1178 (1984).
206. L. J. Todd, A. R. Burke, H. T. Silverstein, J. L. Little, and G. S. Wikholm, *J. Am. Chem. Soc.* **91**, 3376 (1969).
207. L. J. Todd and G. S. Wikholm, *J. Organomet. Chem.* **71**, 219 (1974).
208. T. Yamamoto and L. E. Todd, *J. Organomet. Chem.* **67**, 75 (1974).
209. R. W. Rudolph, R. L. Voorhees, and R. E. Cochy, *J. Am. Chem. Soc.* **92**, 3351 (1970).
210. P. Jutzi and P. Galow, to be published.
211. V. Chowdry, W. R. Pretzer, D. N. Rai, and R. W. Rudolph, *J. Am. Chem. Soc.* **95**, 4560 (1973).

212. D. C. Beer and L. E. Todd, *J. Organomet. Chem.* **50**, 93 (1973).
213. E. O. Fischer and H. Grubert, *Z. Naturforsch.* **B11**, 423 (1956).
214. L. D. Dave, D. F. Evans, and G. Wilkinson, *J. Chem. Soc.*, 3684 (1959).
215. A. H. Cowley, P. Jutzi, F. X. Kohl, J. G. Lasch, N. C. Norman, and E. Schlüter, *Angew. Chem.* **96**, 603 (1984).
216. A. H. Cowley, J. G. Lasch, N. C. Norman, C. A. Stewart, and T. C. Wright, *Organometallics* **2**, 1691 (1983).
217. M. J. Heeg, C. Janiak, and J. J. Zuckerman, *J. Am. Chem. Soc.* **106**, 4259 (1984).
218. E. J. Bulten and H. A. Budding, *J. Organomet. Chem.* **157**, C3 (1978).
219. P. Jutzi and B. Hielscher, *J. Organomet. Chem.* **219**, C25 (1985).
220. A. H. Cowley, R. A. Kemp, and C. A. Stewart, *J. Am. Chem. Soc.* **104**, 3239 (1982).
221. T. S. Dory and J. J. Zuckerman, *J. Organomet. Chem.* **264**, 295 (1984).
222. T. S. Dory, J. J. Zuckerman, and M. D. Rausch, *J. Organomet. Chem.* **281**, C8 (1985).
223. J. L. Atwood, W. E. Hunter, A. H. Cowley, R. A. Jones, and C. A. Stewart, *J. Chem. Soc., Chem. Commun.*, 925 (1981).
224. A. Almenningen, A. Haaland, and T. Motzfeld, *J. Organomet. Chem.* **7**, 97 (1967).
225. P. Jutzi and B. Hielscher, in preparation.
226. R. K. Harris and B. E. Mann, "NMR and the Periodic Table," p. 351. Academic Press, New York, 1978.
227. S. G. Baxter, A. H. Cowley, J. G. Lasch, M. Lattman, W. P. Sharum, and C. A. Steart, *J. Am. Chem. Soc.* **104**, 4064 (1982).
228. K. D. Bos, E. J. Bulten, and J. G. Noltes, *J. Organomet. Chem.* **39**, C52 (1972).
229. K. D. Bos, E. J. Bulten, and J. G. Noltes, *J. Organomet. Chem.* **99**, 71 (1975).
230. F. X. Kohl and P. Jutzi, *Chem. Ber.* **114**, 488 (1981).
231. P. Jutzi, F. X. Kohl, and C. Krüger, *Angew Chem.* **91**, 81 (1979).
232. R. L. Vorhees and R. W. Rudolph, *J. Am. Chem. Soc.* **91**, 2173 (1969).
233. K. S. Wong and R. N. Grimes, *Inorg. Chem.* **16**, 2053 (1977).
234. N. S. Hosmane, N. N. Sirmokadam, and R. H. Herber, *Organometallics* **3**, 1665 (1984).
235. H. Wadepohl, H. Pritzkow, and W. Siebert, *Organometallics* **2**, 1899 (1983).
236. J. Edwin, M. C. Böhm, N. Chester, D. Hoffmann, R. Hoffmann, H. Pritzkow, W. Siebert, K. Stumpf, and H. Wadepohl, *Organometallics* **2**, 1666 (1983).
237. P. F. Rodesiler, Th. Auel, and E. L. Amma, *J. Am. Chem. Soc.* **97**, 7405 (1975).
238. M. S. Weininger, P. F. Rodesiler, and E. L. Amma, *Inorg. Chem.* **18**, 751 (1979).
239. G. Schmid, D. Zaika, and R. Boese, *Angew Chem.* **97**, 581 (1985).
240. E. O. Fischer and H. Grubert, *Z Anorg. Allg. Chem.* **286**, 237 (1956).
241. P. Jutzi and E. Schlüter, *J. Organomet. Chem.* **253**, 313 (1983).
242. C. Panattoni, G. Bombieri, and U. Croatto, *Acta Crystallogr.* **21**, 823 (1966).
243. W. Strohmeier, H. Landsfeld, and F. Gernert, *Z. Electrochem.* **66**, 823 (1962).
244. E. Canadell, O. Eisenstein, and T. Hughbanks, *Inorg. Chem.* **23**, 2435 (1984).
245. A. K. Holliday, P. H. Makin, and R. J. Puddephatt, *J. Chem. Soc. Dalton Trans.*, 435 (1976).
246. R. W. Rudolph, R. L. Vorhees, and R. Cochroy, *J. Am. Chem. Soc.* **92**, 351 (1970).
247. A. G. Gash, P. F. Rodesiler, and E. L. Amma, *Inorg. Chem.* **13**, 2429 (1974).
248. S. G. Baxter, A. H. Cowley, and S. K. Mehrotra, *J. Am. Chem. Soc.* **103**, 5572 (1981).
249. A. H. Cowley and S. K. Mehrotra, *J. Am. Chem. Soc.* **105**, 2074 (1983).
250. P. Jutzi and Th. Wippermann, *Angew. Chem.* **95**, 244 (1983).
251. J. L. Little, T. J. Moran, and L. J. Todd, *J. Am. Chem. Soc.* **89**, 5495 (1967).
252. L. J. Todd, A. R. Burke, A. R. Garber, H. T. Silverstein, and B. N. Stochoff, *Inorg. Chem.* **9**, 2175 (1970).
253. G. D. Friesen and L. J. Todd, *J. Chem. Soc., Chem. Commun.*, 349 (1978).
254. L. J. Todd, J. L. Little, and H. T. Silverstein, *Inorg. Chem.* **8**, 1698 (1969).

255. L. J. Zakharkin and V. J. Kyskin, *Zh. Obshch. Khim.* **40**, 2234 (1970).
256. H. S. Wong and W. N. Lipscomb, *Inorg. Chem.* **14**, 1350 (1975).
257. L. J. Zakharkin and V. J. Kyskin, *Zh. Obshch. Khim.* **40**, 2241 (1970).
258. L. A. Fedorov, V. J. Kyskin, and L. J. Zakharkin, *Zh. Obshch. Khim.* **42**, 536 (1972).
259. A. J. Bard, A. H. Cowley, J. K. Leland, P. Jutzi, Ch. P. Morley, and E. Schlüter, *J. Chem. Soc., Dalton Trans.*, 1303 (1985).
260. J. Plasek, S. Hermanek, and Z. Janousek, *Coll. Czech. Chem. Commun.* **42**, 785 (1977).
261. L. J. Andrews, *Chem. Rev.*, 713 (1954).
262. R. Foster, "Organic Charge Transfer Complexes." Academic Press, London, 1969.
263. S. Fukuzumi and J. K. Kochi, *J. Am. Chem. Soc.* **104**, 7599 (1982).
264. O. Hassel and K. O. Stromme, *Acta Chem. Scand.* **12**, 1146 (1958).
265. H. Sakai, H. Matsuyama, H. Yamaoka, and Y. Maeda, *Bull. Chem. Jpn.* **56**, 1016 (1983).
266. R. Goddard, J. Akhtar, and K. Starowieyski, *J. Organomet. Chem.* **282**, 149 (1985).
267. P. Jutzi and R. Dickbreder, *Chem. Ber.* **119**, 1750 (1986).
268. B. G. Conway and M. D. Rausch, *Organometallics* **4**, 688 (1985).
269. S. Z. Goldberg, K. N. Raymond, C. A. Harmon, and D. H. Templeton, *J. Am. Chem. Soc.* **96**, 1348 (1974).
270. M. D. Rausch, B. H. Edwards, R. D. Rogers, and J. L. Atwood, *J. Am. Chem. Soc.* **105**, 3882 (1983).
271. Chr. Gaffney and P. G. Harrison, *J. Chem. Soc., Dalton Trans.*, 1055 (1982).
272. W. Smith, *J. Chem. Soc.* **35**, 309 (1879).
273. W. Smith and G. W. Davies, *J. Chem. Soc.* **41**, 411 (1882).
274. H. H. Perkampus and E. Schonberger, *Z. Naturforsch.* **31b**, 73 (1976).
275. R. Hulme and J. T. Szymansiki, *Acta Cryst.* **B25**, 753 (1969).
276. A. Demaldé, A. Mangia, M. Nardelli, G. Pelizzi, and M. E. Vidoni Tani, *Acta Cryst.* **B28**, 147 (1972).
277. O. T. Beachley, Jr., L. Victoriano, J. C. Pazik, and M. R. Churchill, J. C. Fettinger, 30 IUPAC Conference, Manchester 1985.
278. The crystalline, thermally stable complex $Me_5C_5PNH^tBu^+AlCl_4^-$ was recently described; E. Niecke and D. Gudat, in preparation.
279. S. J. Pratten, M. K. Cooper, M. J. Aroney, and S. W. Filipczvk, *J. Chem. Soc., Dalton Trans.*, 1761 (1985).
280. R. Goddard, J. Akhtar, and K. B. Starowieyski, *J. Organomet. Chem.* **282**, 149 (1985).
281. K. B. Starowieyski and J. Lusztyk, *J. Organomet. Chem.* **133**, 281 (1977).
282. H. Schmidbaur, *Angew. Chem.* **97**, 893 (1985).
283. P. Jutzi, D. Kanne, and C. Krüger, *Angew. Chem.* **98**, Int. Ed. **25**, 164 (1986).
284. H. Schumann, C. Janiak, E. Hahn, J. Loebel, and J. Zuckerman, *Angew. Chem.* **97**, 765 (1985).

Dienes, Versatile Reactants in the Photochemistry of Group 6 and 7 Metal Carbonyl Complexes

CORNELIUS G. KREITER

Fachbereich Chemie der Universität Kaiserslautern
D-6750 Kaiserslautern, Federal Republic of Germany

I

INTRODUCTION

Photochemical activation of transition metal carbonyls has been used as a preparative tool for substitution of carbonyl ligands by donor molecules or unsaturated hydrocarbons for many years (*1–6*). The advantage of photochemical activation in comparison with thermal activation is the possibility of conducting reactions at fairly low temperatures. Hence even thermolabile products can be prepared and isolated by appropriate treatment of the reaction mixtures. However, due to the various activation modes of transition metal carbonyls by UV light, often more than one product is obtained, and chromatographic separation is necessary. Limitations are set primarily by the amount of substance which can be irradiated in solution at one time.

Photochemical reactions have been used for the preparation of various olefin, and acetylene complexes (*7*). Application to the coordination of dienes as ligands has not been used extensively, so far. In this article the preparative aspects of the photochemistry of carbonyls of the group 6 and group 7 elements and some key derivatives, with the exception of technetium, with conjugated and cumulated dienes will be described. Not only carbonyl substitution reactions by the dienes, but also C—C bond formation, C—H activation, C—H cleavage, and isomerizations due to H shifts, have been observed, thereby leading to various types of complexes.

A. *General Considerations*

With mononuclear transition metal carbonyls, photochemically induced substitutions of one or two carbonyl ligands by a diene can be expected to be the only reactions. Dinuclear transition metal carbonyls or carbonyl complexes with metal–carbon σ or π bonds offer additional possibilities. The most simple ones are shown in the generalized reaction scheme of Eqs. (1)–(6).

$$[XM(CO)_n] + \text{diene} \rightarrow [XM(CO)_{n-1}(\eta^2\text{-diene})] + CO \tag{1}$$

$$[XM(CO)_n] + \text{diene} \rightarrow [M(CO)_{n-1}(COX)(\eta^2\text{-diene})] \tag{2}$$

$$[XM(CO)_n] + \text{diene} \rightarrow [XM(CO)_{n-2}(\eta^4\text{-diene})] + 2\,CO \tag{3}$$

$$[XM(CO)_n] + \text{diene} \rightarrow [M(CO)_{n-1}(COX)(\eta^4\text{-diene})] + CO \tag{4}$$

$$[XM(CO)_n] + \text{diene} \rightarrow [M(CO)_{n-1}(\eta^3\text{-endiyl}{-}X)] + CO \tag{5}$$

$$[XM(CO)_n] + \text{diene} \rightarrow [M(CO)_{n-1}(\eta^3\text{-endiyl}{-}COX)] \tag{6}$$

$$X = M(CO)_n, H, CH_3$$

II

PHOTOREACTIONS OF THE COMPLEXES $[M(CO)_{6-n}L_n]$ (M = Cr, Mo, W; L = donor ligand; n = 0, 1, 2) WITH CONJUGATED DIENES

A. Syntheses of $[M(CO)_{6-2m}(\eta^4\text{-diene})_m]$ Complexes (m = 1, 2)

Upon UV irradiation in hydrocarbon solution, the hexacarbonyls of chromium, molybdenum, and tungsten react differently with conjugated dienes like 1,3-butadiene (**1a**), (*E*)-1,3-pentadiene (**1b**), 2-methyl-1,3-butadiene (**1c**), (*E,E*)-2,4-hexadiene (**1d**), (*E*)-2-methyl-1,3-pentadiene (**1e**), 2-ethyl-1,3-butadiene (**1f**), or 1,3-cyclohexadiene (**1g**). Chromium hexacarbonyl (**2**) yields, with the acyclic dienes **1a–1f**, tetracarbonyl-η^2-dienechromium(0) complexes (**3a–3f**) in a smooth reaction (*8–10*). With 1,3-cyclohexadiene, in addition to **3g**, dicarbonylbis(η^4-1,3-cyclohexadiene)chromium(0) (**4g**) is obtained [Eqs. (7) and (8)]. During chromatography on silica gel, the 1,3-cyclohexadiene complex **3g** dismutates readily to [Cr(CO)$_6$] and **4g** [Eq. (9)]. Under the same conditions with **2** 1,3-cyclopentadiene (**1h**) yields, in a hydrogen-transfer reaction, the stable dicarbonyl-η^5-cyclopentadienyl-η^3-cyclopent-enylchromium (**5**) (*11–13*) [Eq. (10)].

$$[Cr(CO)_6] + \text{diene} \rightarrow [Cr(CO)_4(\eta^4\text{-diene})] + 2\,CO \tag{7}$$

$$\quad\; \mathbf{2} \qquad\quad \mathbf{1a\text{--}1g} \qquad\qquad\quad \mathbf{3a\text{--}3g}$$

$$[Cr(CO)_6] + 2\,C_6H_8 \rightarrow [Cr(CO)_2(\eta^4\text{-}C_6H_8)_2] + 4\,CO \tag{8}$$

$$\quad\; \mathbf{2} \qquad\quad \mathbf{1g} \qquad\qquad\quad \mathbf{4g}$$

$$2\,[Cr(CO)_4(\eta^4\text{-}C_6H_8)] \rightarrow [Cr(CO)_6] + [Cr(CO)_2(\eta^4\text{-}C_6H_8)_2] \tag{9}$$

$$\qquad \mathbf{3g} \qquad\qquad\qquad\quad \mathbf{2} \qquad\qquad\quad \mathbf{4g}$$

$$[Cr(CO)_6] + C_5H_6 \rightarrow [Cr(CO)_2(\eta^5\text{-}C_5H_5)(\eta^3\text{-}C_5H_7) + 4\,CO \qquad (10)$$

$$\quad 2 \qquad\qquad 1h \qquad\qquad\qquad\qquad 5$$

In contrast to $[Cr(CO)_6]$, the homolog $[Mo(CO)_6]$ (6) does not form the corresponding tetracarbonyl complexes with 1a–1g, but thermolabile species $[Mo(CO)_5(\eta^2\text{-diene})]$ (7) (14). Using the more reactive tricarbonyl-η^6-mesitylenemolybdenum(0) (8) with the dienes 1a, 1c, and 1g as starting materials, dicarbonylbis(η^4-diene)molybdenum complexes (9a, 9c, and 9g) can be isolated (15,16) [Eq. (11)].

$$[Mo(CO)_3(\eta^6\text{-}C_6H_3(CH_3)_3)] + 2\ \text{diene} \rightarrow$$

$$\quad 8 \qquad\qquad\qquad\qquad 1a, 1c, 1g$$

$$[Mo(CO)_2(\eta^4\text{-diene})_2] + CO + C_6H_3(CH_3)_3 \quad (11)$$

$$9a, 9c, 9g$$

$[W(CO)_6]$ (10) reacts with 1a to give dicarbonylbis(η^4-1,3-butadiene)-tungsten(0) (11a) in good yield (8) [Eq. (12)]. Surprisingly, 2-methyl-1,3-butadiene (1c) with 10 forms $[W(CO)_5(\eta^2\text{-}C_5H_8)]$ (12) as the only product [Eq. (13)], as was also reported for 1a as diene (17). However, dicarbonylbis(η^4-2-methyl-1,3-butadiene)tungsten(0) (11c) is accessible from $[W(CO)_5(CH_3CN)]$ and 1c [Eq. (14)], although only in low yield (18).

$$[W(CO)_6] + 2\,C_4H_6 \rightarrow [W(CO)_2(\eta^4\text{-}C_4H_6)_2] + 4\,CO \qquad (12)$$

$$\quad 10 \qquad\qquad 1a \qquad\qquad\qquad 11a$$

$$[W(CO)_6] + C_5H_8 \rightarrow [W(CO)_5(\eta^2\text{-}C_5H_8)] + CO \qquad (13)$$

$$\quad 10 \qquad\qquad 1c \qquad\qquad\qquad 12$$

$$[W(CO)_5(CH_3CN)] + 2\,C_5H_8 \rightarrow [W(CO)_2(\eta^4\text{-}C_5H_8)_2] + 3\,CO + CH_3CN \quad (14)$$

$$\quad 13 \qquad\qquad\qquad 1c \qquad\qquad\qquad\qquad 11c$$

The dicarbonylbis(η^4-1,3-cyclohexadiene)molybdenum(0) (9g) and tungsten (11g) complexes are formed only in moderate yields, photochemically. They are readily accessible, however, by thermal reactions of the respective tricarbonyltris(acetonitrile)metal(0) complexes (19) and 1g (20).

B. Syntheses of [M(CO)$_{4-n}$L$_n$(η^4-diene)] Complexes (n = 1, 2)

The mono- and disubstituted complexes $[M(CO)_5\{P(CH_3)_3\}]$ [M = Cr (14) (21), W (16) (22)], $[M(CO)_5\{P(OCH_3)_3\}]$ [M = Cr (18) (21,23), W (21) (24)], cis- or trans-$[Cr(CO_4\{P(CH_3)_3\}_2]$ (23) (25), cis- or trans-$[M(CO)_4\{P(OCH_3)_3\}_2]$ [M = Cr (25) (25), Mo (27) (26), W (29) (26)],

TABLE I

$[M(CO)_{4-n}L_n(\eta^4\text{-diene})]$ Complexes Synthesized According to Eq. (15)

	Diene							
	1a	1b	1c	1d	1e	1f	1g	1i
$[Cr(CO)_3\{P(CH_3)_3\}(\eta^4\text{-diene})]$	15a	15b	15c	15d	15e	15f	15g	
$[W(CO)_3\{P(CH_3)_3\}(\eta^4\text{-diene})]$		17b		17d				
$[Cr(CO)_3\{P(OCH_3)_3\}(\eta^4\text{-diene})]$	19a	19b	19c	19d	19e	19f		
$[W(CO)_3\{P(OCH_3)_3\}(\eta^4\text{-diene})]$	22a	22b	22c	22d	22e		22g	22i
$[Cr(CO)_2\{P(CH_3)_3\}_2(\eta^4\text{-diene})]$	24a	24b	24c	24d	24e	24f		
$[Cr(CO)_2\{P(OCH_3)_3\}_2(\eta^4\text{-diene})]$	26a	26b	26c	26d	26e	26f		
$[Mo(CO)_2\{P(OCH_3)_3\}_2(\eta^4\text{-diene})]$	28a		28c					
$[W(CO)_2\{P(OCH_3)_3\}_2(\eta^4\text{-diene})]$	30a	30b	30c	30d	30e		30g	
$[Cr(CO)_2\{(CH_3)_2PC_2H_4P(CH_3)_2\}(\eta^4\text{-diene})]$	32a	32b	32c	32d	32e	32f		
$[W(CO)_2\{(CH_3)_2PC_2H_4P(CH_3)_2\}(\eta^4\text{-diene})]$	34a	34b		34d	34e			
$[Cr(CO)_2\{P(CH_3)_3\}\{P(OCH_3)_3\}(\eta^4\text{-diene})]$	36a			36d				

$[M(CO)_4\{(CH_3)_2PC_2H_4P(CH_3)_2\}]$ [M = Cr (**31**) (*27*), W (**33**) (*28*)], and *cis*- or *trans*-$[Cr(CO)_4\{P(CH_3)_3\}\{P(OCH_3)_3\}]$ (**35**) (*29*) react analogously, upon UV irradiation, with the dienes **1a–1g** and **1i**, with substitution of two carbonyl ligands to afford $[M(CO)_{4-n}L_n(\eta^4\text{-diene})]$ complexes [Eq. (15)] (Table I). With **18** 1,3-cyclopentadiene forms $[Cr(CO)\{P(OCH_3)_3\}-(\eta^5\text{-}C_5H_5)(\eta^3\text{-}C_5H_7)]$ (**20**) (*8*).

$$[M(CO)_{6-n}L_n] + \text{diene} \rightarrow [M(CO)_{4-n}L_n(\eta^4\text{-diene})] + 2\,CO \qquad (15)$$

C. Stereochemistry of $[M(CO)_{4-n}L_n(\eta^4\text{-diene})]$ Complexes

The coordination sphere of $[M(CO)_3L(\eta^4\text{-diene})]$ complexes can be regarded in good approximation as pseudooctahedral. Two cis positions are occupied by the formally bidentate η^4-diene ligand. For an unambiguous assignment of the conceivable stereoisomers, the positions in the octahedron are marked according to IUPAC rules (*30*) with the letters a–f. Arbitrarily, the η^4-diene ligand is set into positions b and c, the outer carbon atoms C-1 and C-4 adjacent to a (o = oben), the inner ones, C-2 and C-3 to f (u = unten) (*23*). So, the four stereoisomers of $[M(CO)_3L(\eta^4\text{-diene})]$ complexes, with η^4-diene ligands of C_s symmetry, as well as the corresponding four isotopomers, $[Cr(CO)_3(^{13}CO)(\eta^4\text{-diene})]$ complexes, can be characterized by only one letter consequent upon the position of the donor ligand L or the ^{13}CO ligand. Two of these isomers, a and f, have C_s symmetry, the other two, d and e, are enantiomers. η^4-Diene ligands with C_1 symmetry produce a doubling of the

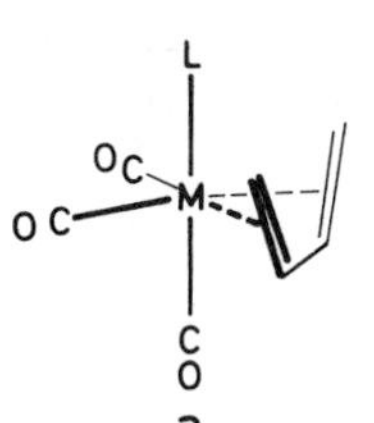

a 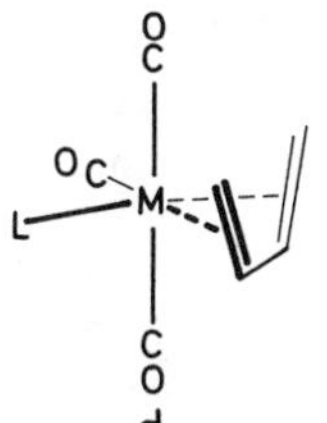   d 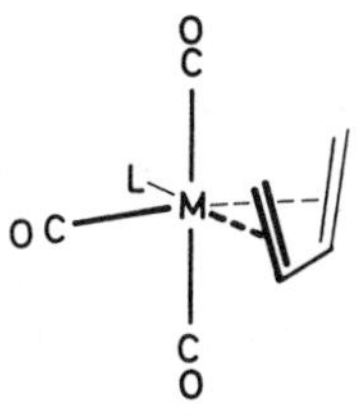   e 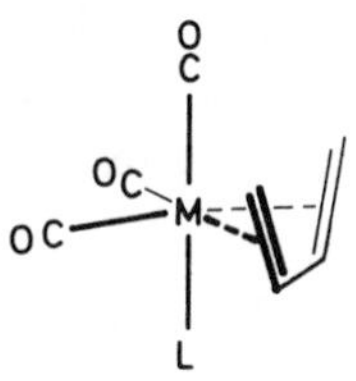   f

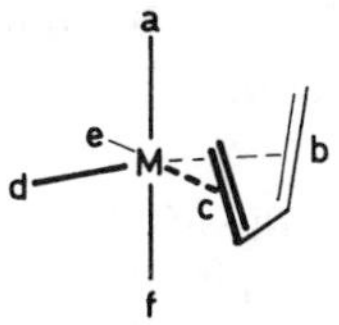

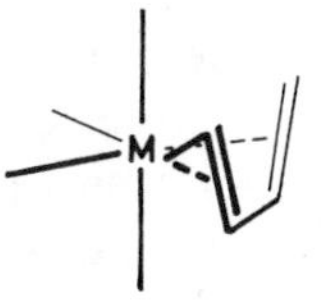

S 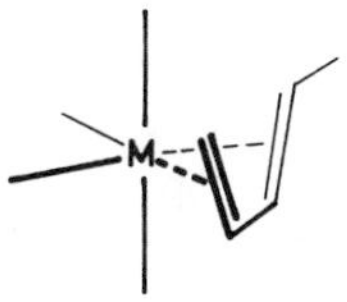   R

stereoisomers of the $[M(CO)_3L(\eta^4\text{-diene})]$ complexes, due to the two enantiomeric coordination modes R and S.

Using the same principles, the stereoisomers of $[M(CO)_2L_2(\eta^4\text{-diene})]$ complexes can also be assigned unambiguously. For complexes with diene ligands of C_s symmetry six stereoisomers have to be considered, namely, two with C_s symmetry, af and de, and four with C_1 symmetry, ad, ae, df, and ef. Unsymmetrical η^4-diene ligands or two different donor ligands double the number of possible stereoisomers. When $L_2 = (CH_3)_2PC_2H_4P(CH_3)_2$, the

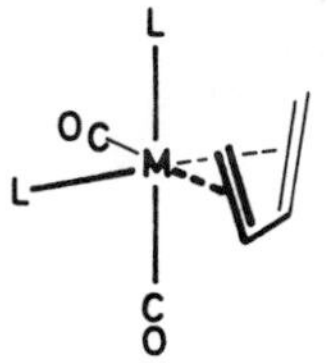

ad 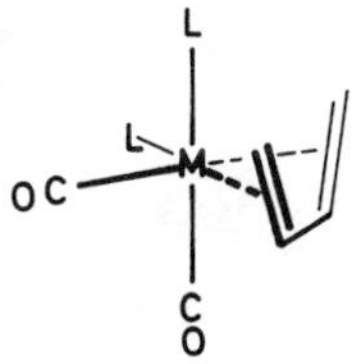   ae 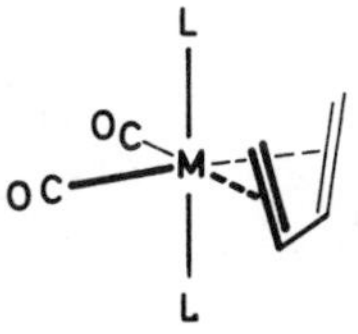   af

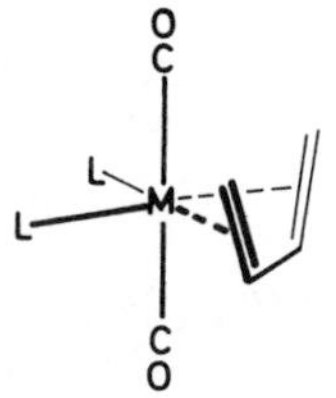

de 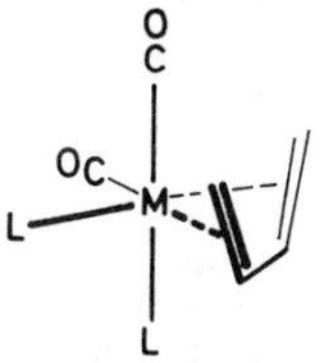   df 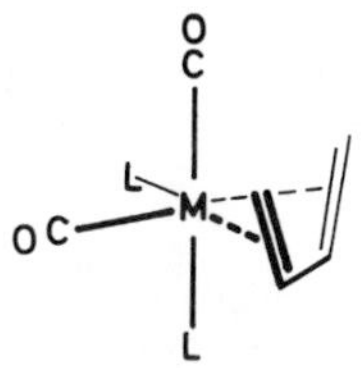   ef number of isomers is reduced to five. Due to steric reasons, the *af* isomer cannot exist. For complexes with unsymmetrical diene ligands, 12 or 10 isomers, respectively, have to be taken into account.

D. *Molecular Structures of [M(CO)$_{4-n}$L$_n$(η^4-diene)] Complexes*

Crystal and molecular structures of tricarbonyl-η^4-tricyclo[6.3.0.0^{2,7}]-undeca-3,5-dienetrimethylphosphitetungsten(0) (**22i**) (*31*) (Fig. 1), dicarbonyl-η^4-(*E*)-1,3-pentadienebis(trimethylphosphine)chromium(0) (**24b**) (*25*), and dicarbonyl-η^4-(E,E)-2,4-hexadienebis(trimethylphosphine)chromium(0) (**24d**) (*25*) (Fig. 2) have been determined by X-ray structure analyses. The coordination spheres of the metal atoms in the three complexes are approximately octahedral. There are distinct distortions within the M(CO)$_{4-n}$L$_n$ fragments, which are associated with a decreased angle of the axial *a* and *f* ligands from 180 to 160° or less and an increased angle of the equatorial ligands *d* and *e* from 90 to 100° or more. The distortions of the complexes **22i**, **24b**, and **24d** tend from octahedral toward bicapped tetrahedral coordination spheres. The diene and the ligands in *a* and *f* positions mark the corners of the tetrahedron, the ligands in *d* and *e* positions represent the two caps (*25*). Nevertheless, the binding angles at the chromium atoms are closer to those

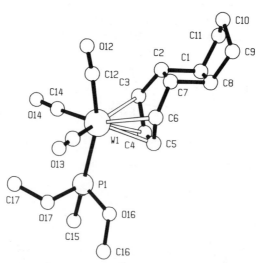

FIG. 1. Molecular structure of tricarbonyl-η^4-tricyclo[6.3.0.0^{2,7}]undeca-3,5-dienetrimethylphosphitetungsten(0) (**22i**) (*31*): W-1—C-3 241.5, W-1—C-4 228.1, W-1—C-5 227.7, W-1—C-6 244.1, C-3—C-4 139.5, C-4—C5 142.1, C-5—C-6 139.4 pm; C-3—C-4—C-5 115.5°, C-4—C-5—C-6 119.7°, P-1—W-1—C-12 160.0°, C-13—W-1—C-14 101.2°.

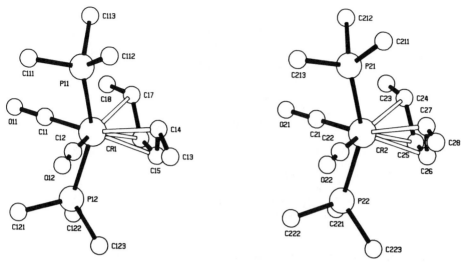

FIG. 2. Molecular structure of dicarbonyl-η^4-(E,E)-2,4-hexadienebis(trimethylphosphine)-chromium(0) (**24d**) (*25*). Reprinted with permission from *Z. Naturforsch.* **39B**, 1553 (1984). Copyright by Verlag der Zeitschrift für Naturforschung.

in an octahedron than in a bicapped tetrahedron. The donor ligands are found only in *a* or/and *f* positions. Dihedral angles of 84.6° (**22i**) and 80.9°/82.3° (**24d**) were found between the planes of the four coordinated diene carbon atoms, and the planes of the central metal and the ligands in the *d* and *e* positions. The η^4-diene ligands are slightly tilted toward position *a*. In **22i** the tricyclic hydrocarbon adopts the same conformation as in the corresponding complexes $[Fe(CO)_3(\eta^4\text{-}C_{11}H_{14})]$ (*32*) and $[Mo(CO)_2(\eta^4\text{-}C_{11}H_{14})_2]$ (*33*).

E. Dynamic Behavior of $[M(CO)_{4-n}L_n(\eta^4\text{-diene})]$ Complexes

Temperature-dependent ^{13}C-NMR carbonyl signals of the $[Cr(CO)_4$-$(\eta^4\text{-diene})]$ complexes **3a–3f**, indicate an easy mutual conversion of the four $[Cr(CO)_3(^{13}CO)(\eta^4\text{-diene})]$ isotopomers *a, d, e,* and *f* (*10,23*). For this interconversion the barriers of activation were found to be between 39 and 47 kJ/mol for complexes with acyclic diene ligands (Table II). Only in the case of tetracarbonyl-η^4-1,3-cyclohexadienechromium(0) (**3g**) is the carbonyl scrambling fast, even at 190 K.

Although the complexity of the dynamic stereochemistry of **3a–3e** does not allow a conclusive elucidation of the mechanism of the hindered ligand movement on the basis of the NMR data, a rotation of the diene ligands

TABLE II

ACTIVATION PARAMETERS FOR
CARBONYL SCRAMBLING IN
$[M(CO)_4(\eta^4\text{-diene})]$ COMPLEXES

Complex	$\Delta G_{200}{}^*$ (kJ/mol)
3a $[Cr(CO)_4(\eta^4\text{-}C_4H_6)]$	42.0
3b $[Cr(CO)_4(\eta^4\text{-}C_5H_8)]$	40.7
3c $[Cr(CO)_4(\eta^4\text{-}C_5H_8)]$	44.4
3d $[Cr(CO)_4(\eta^4\text{-}C_6H_{10})]$	46.2
3e $[Cr(CO)_4(\eta^4\text{-}C_6H_{10})]$	43.4
3f $[Cr(CO)_4(\eta^4\text{-}C_6H_{10})]$	40.5
3g $[Cr(CO)_4(\eta^4\text{-}C_6H_8)]$	< 30

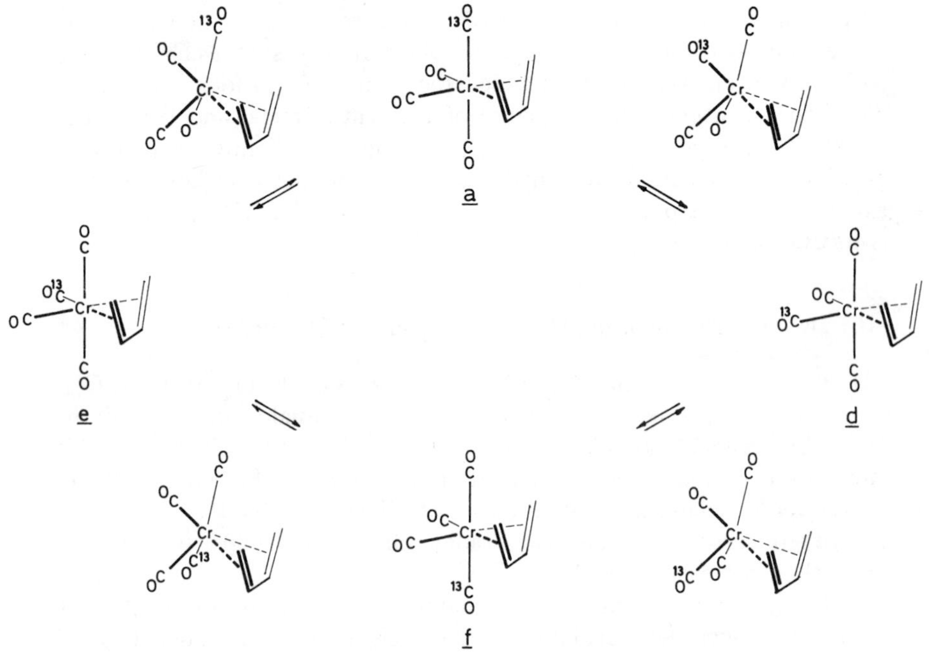

around the axis through the chromium and the center of gravity of the four-coordinated carbon atoms reasonably explains the spectroscopic results. During rotation, the diene ligands always turn their same side toward the central metal. This is proved convincingly by the ^{1}H-NMR spectra of tetracarbonyl-η^4-2-ethyl-1,3-butadienechromium(0) (**3f**) (*10*), which is formed as a pair of enantiomers due to the substitution pattern of **1f** and the two modes of coordination. The chirality of **3f**, which can be monitored, via the diastereotopic methylene protons of the 2-ethyl group, by ^{1}H-NMR spectroscopy, is maintained even during fast carbonyl scrambling. Only a turnover of the diene ligand, which means $R-S$ isomerization, would destroy the chirality. If such a process is occurring, its rate must be much smaller than the rate of diene rotation.

A $\pm 90°$ rotation of the diene ligands against the Cr(CO)$_4$ fragments in **3a–3g**, has to be accompanied by an appropriate change of the C—Cr—C angles of the four carbonyl ligands, according to an $a-d-e-f$ isotopomerization. The latter process may be inferred from the temperature-dependent line shapes of the carbonyl signals. Without change in the geometry of the Cr(CO)$_4$ fragment, a $\pm 90°$ diene rotation would lead to a bicapped tetrahedral transition state, which corresponds only to a mutual exchange between the a and f and the d and e isotopomers, respectively. The proposed mechanism for the carbonyl scrambling can be related to the Berry pseudo-rotation (*34*) for trigonal bipyramidal molecules. The rotational axis of the diene ligand corresponds to the pivot atom, the displacements of the other ligands are similar to those occurring during the pseudo-rotation. The transition states between the four isotopomers are reached when the diene ligand is rotated by an angle of $\pm 45°$ against the Cr(CO)$_4$ moiety. A trigonal prismatic geometry for these transition states is reasonable (*10*).

According to their temperature-dependent NMR spectra, the [M(CO)$_3$L-(η^4-diene)] complexes **15a–15f** (*21*), **17b**, **17d** (*22*), **19a–19f** (*21,23*), and **22a–22i** (*24*) show hindered ligand movements. With very few exceptions, a preference for the a isomers is found. For compounds **15b**, **15d**, **19b**, **19d**, and **22g** f isomers can be detected in addition to a isomers. The bulky tricyclo[6.3.0.0^{2,7}]undeca-3,5-triene ligand in **22i** causes the f isomer to be preferred, in agreement with its molecular structure.

For complexes **15b**, **15d**, **19b**, **19d**, and **22g**, hindered $a-f$ isomerizations are observed with barriers of activation between 40 and 50 kJ/mol (Table III). The mechanism involved in these movements corresponds to the mutual interchange of the isotopomers in the [Cr(CO)$_4$(η^4-diene)] complexes. However, in contrast to the energetically equal isotopomers a, d, e, and f of the [Cr(CO)$_4$(η^4-diene)] complexes, the isomers a, d, e, and f of **15b**, **15d**, **19b**, **19d**, and **22g** differ in their energies, according to their populations. Only a and f isomers can be detected for **15b**, **15d**, **19b**, **19d**, and **22g** in substantial amounts. So far, there is no evidence for the d and e isomers. However, the d and e isomers have to be taken into account as intermediates in the $a-f$ isomerization.

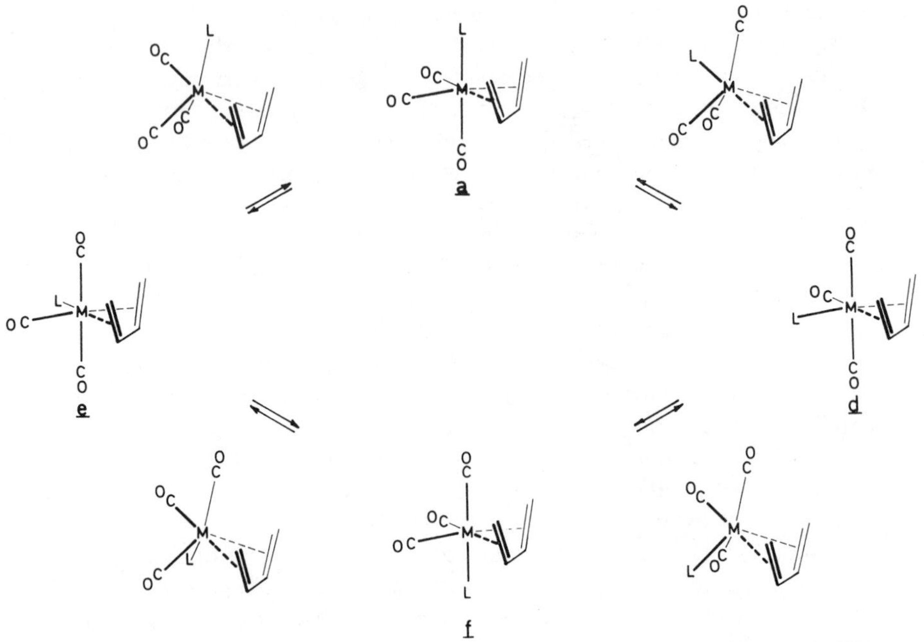

All the $[M(CO)_3L(\eta^4\text{-diene})]$ complexes exhibit a second ligand movement with a higher barrier of activation. At low temperatures, complexes with unsymmetrically substituted diene ligands show three different carbonyl signals in their ^{13}C-NMR spectra, according to the different $[M(CO)_2(^{13}CO)L\text{-}(\eta^4\text{-diene})]$ isotopomers d, e, and f. At higher temperatures, the coalescence

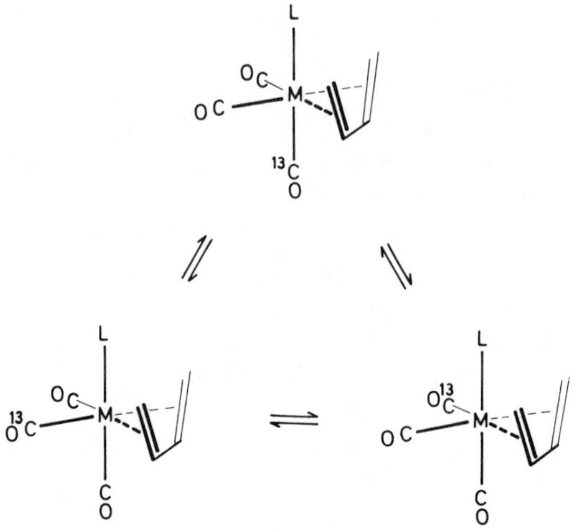

TABLE III

ACTIVATION PARAMETERS FOR $a-f$ ISOMERIZATION AND CARBONYL SCRAMBLING IN
$[M(CO)_3L(\eta^4\text{-diene})]$ COMPLEXES

Complex	ΔH^* (kJ/mol)	ΔS^* (J/mol · K)	ΔG_T^* (kJ/mol)
15a $[Cr(CO)_3\{P(CH_3)_3\}(\eta^4\text{-}C_4H_6)]$	41.6	−49	51.4_{200}
15b $[Cr(CO)_3\{P(CH_3)_3\}(\eta^4\text{-}C_5H_8)]$	57.0	−2	57.4_{200}
15c $[Cr(CO)_3\{P(CH_3)_3\}(\eta^4\text{-}C_5H_8)]$	58.7	−2	59.0_{200}
15d $[Cr(CO)_3\{P(CH_3)_3\}(\eta^4\text{-}C_6H_{10})]$	75.7	54	64.8_{200}
15e $[Cr(CO)_3\{P(CH_3)_3\}(\eta^4\text{-}C_6H_{10})]$	40.0	−63	52.6_{200}
15f $[Cr(CO)_3\{P(CH_3)_3\}(\eta^4\text{-}C_6H_{10})]$	45.3	−37	55.1_{200}
19a $[Cr(CO)_3\{P(OCH_3)_3\}(\eta^4\text{-}C_4H_6)]$	51.7	−2	52.0_{200}
19b $[Cr(CO)_3\{P(OCH_3)_3\}(\eta^4\text{-}C_5H_8)]$	32.1	−29	37.9_{200}
	59.5	−16	62.8_{200}
19c $[Cr(CO)_3\{P(OCH_3)_3\}(\eta^4\text{-}C_5H_8)]$	48.5	−24	53.2_{200}
19d $[Cr(CO)_3\{P(OCH_3)_3\}(\eta^4\text{-}C_6H_{10})]$	51.1	26	45.9_{200}
	78.1	−20	74.1_{200}
19e $[Cr(CO)_3\{P(OCH_3)_3\}(\eta^4\text{-}C_6H_{10})]$	41.6	−49	51.4_{200}
19f $[Cr(CO)_3\{P(OCH_3)_3\}(\eta^4\text{-}C_6H_{10})]$	45.6	−37	53.0_{200}
22a $[W(CO)_3\{P(OCH_3)_3\}(\eta^4\text{-}C_4H_6)]$	—	—	56.0_{250}
22b $[W(CO)_3\{P(OCH_3)_3\}(\eta^4\text{-}C_5H_8)]$	—	—	51.1_{250}
22c $[W(CO)_3\{P(OCH_3)_3\}(\eta^4\text{-}C_5H_8)]$	—	—	57.4_{250}
22d $[W(CO)_3\{P(OCH_3)_3\}(\eta^4\text{-}C_6H_{10})]$	—	—	54.2_{250}
22e $[W(CO)_3\{P(OCH_3)_3\}(\eta^4\text{-}C_6H_{10})]$	—	—	57.9_{250}
22g $[W(CO)_3\{P(OCH_3)_3\}(\eta^4\text{-}C_6H_8)]$	—	—	36.2_{200}
			50.6_{250}

of the carbonyl signals of these isotopomers establish their mutual interconversion. This means the carbonyl ligands change their positions with respect to the donor ligand in the $Cr(CO)_3L$ fragment. Some type of turnstile rearrangement of the $M(CO)_3$ unit within the $M(CO)_3L$ moiety is suggested as the mechanism for this exchange. The barriers of activation are distinctly higher than those for $a-f$ isomerization in **15b**, **15d**, **19b**, **19d**, and **22g** (Table III). Owing to the much faster rotation of the diene ligands, it is not necessary to consider a definite orientation of the diene ligands, with respect to the $Cr(CO)_3L$ fragements, for this ligand movement. In the time average process, the diene has to be taken into account only as a ligand of rotational symmetry sited on the apex of a "tetragonal pyramid."

In the $[M(CO)_2L_2(\eta^4\text{-diene})]$ complexes, the monodentate donor ligands are found almost exclusively at the a and f positions. There is evidence for the ad and ae stereoisomers only for $[Cr(CO)_2\{P(CH_3)_3\}\{P(OCH_3)_3\}\text{-}(\eta^4\text{-butadiene})]$ (**36a**) (29). A hindered ligand movement, which causes temperature-dependent NMR signals for the donor ligands, is explained by a transformation of these ligands from the a to the f position and vice versa. A mechanism similar to carbonyl scrambling in the $[Cr(CO)_4(\eta^4\text{-diene})]$ complexes, or to $a-f$ isomerization in the $[M(CO)_3L(\eta^4\text{-diene})]$ complexes, is quite reasonable (Table IV). The rotation of the η^4-diene ligand against the $Cr(CO)_2L_2$ fragment affords intermediates with the donor ligands at the d and e sites.

Donor atoms at the d and e sites are also found in the chelate complexes $[M(CO)_2\{(CH_3)_2PC_2H_4P(CH_3)_2\}(\eta^4\text{-diene})]$ [**32a–32f** (27), **34a**, **34b**, **34d**, and **34e** (28)]. From the ^{1}H-NMR spectra of **32a** and **34a**, one can infer a preference for the ad and ae enantiomers (Table V). They are interconverted via an achiral intermediate. For complexes with unsymmetrically substituted diene ligands (**32b**, **32c**, **32e**, **32f**, **34b**, and **34e**), two diastereomeric pairs of enantiomers are detected. Here, too, a mutual transformation of the diastereomers is observed, but the complexes remain chiral, even during fast isomerization. Again, rotation of the diene ligands satisfactorily explains the experimental results. Isomers with the chelate ligand at

TABLE IV

Activation Parameters for Diene Rotation in $[M(CO)_2L_2(\eta^4\text{-diene})]$ Complexes

Complex	ΔH^* (kJ/mol)	ΔS^* (J/mol · K)	ΔG_{293}^* (kJ/mol)
24a $[Cr(CO)_2\{P(CH_3)_3\}_2(\eta^4\text{-}C_4H_6)]$	30.6	6	32.3
24b $[Cr(CO)_2\{P(CH_3)_3\}_2(\eta^4\text{-}C_5H_8)]$	49.7	−2	49.2
24c $[Cr(CO)_2\{P(CH_3)_3\}_2(\eta^4\text{-}C_5H_8)]$	—	—	< 30
24d $[Cr(CO)_2\{P(CH_3)_3\}_2(\eta^4\text{-}C_6H_{10})]$	55.1	24	62.2
24e $[Cr(CO)_2\{P(CH_3)_3\}_2(\eta^4\text{-}C_6H_{10})]$	49.4	−13	45.5
24f $[Cr(CO)_2\{P(CH_3)_3\}_2(\eta^4\text{-}C_6H_{10})]$	—	—	< 30
26a $[Cr(CO)_2\{P(OCH_3)_3\}_2(\eta^4\text{-}C_4H_6)]$	55.4	−72	34.2
26b $[Cr(CO)_2\{P(OCH_3)_3\}_2(\eta^4\text{-}C_5H_8)]$	45.0	19	50.5
26c $[Cr(CO)_2\{P(OCH_3)_3\}_2(\eta^4\text{-}C_5H_8)]$	—	—	< 30
26d $[Cr(CO)_2\{P(OCH_3)_3\}_2(\eta^4\text{-}C_6H_{10})]$	49.7	−2	49.2
26e $[Cr(CO)_2\{P(OCH_3)_3\}_2(\eta^4\text{-}C_6H_{10})]$	56.5	−36	45.9
26f $[Cr(CO)_2\{P(OCH_3)_3\}_2(\eta^4\text{-}C_6H_{10})]$	—	—	< 30
28a $[Mo(CO)_2\{P(OCH_3)_3\}_2(\eta^4\text{-}C_4H_6)]$	—	—	52.0
28c $[Mo(CO)_2\{P(OCH_3)_3\}_2(\eta^4\text{-}C_5H_8)]$	—	—	48.4
30a $[W(CO)_2\{P(OCH_3)_3\}_2(\eta^4\text{-}C_4H_6)]$	—	—	47.4
30b $[W(CO)_2\{P(OCH_3)_3\}_2(\eta^4\text{-}C_5H_8)]$	—	—	58.8
30c $[W(CO)_2\{P(OCH_3)_3\}_2(\eta^4\text{-}C_5H_8)]$	—	—	44.8
22d $[W(CO)_2\{P(OCH_3)_3\}_2(\eta^4\text{-}C_6H_{10})]$	—	—	64.2
22e $[W(CO)_2\{P(OCH_3)_3\}_2(\eta^4\text{-}C_6H_{10})]$	—	—	55.5
22g $[W(CO)_2\{P(OCH_3)_3\}_2(\eta^4\text{-}C_6H_8)]$	—	—	42.0

TABLE V

ACTIVATION PARAMETERS FOR DIENE ROTATION IN $[M(CO)_2L_2(\eta^4\text{-diene})]$ COMPLEXES

	Complex	ΔH^* (kJ/mol)	ΔS^* (J/mol·K)	ΔG_{293}^* (kJ/mol)
32a	$[Cr(CO)_2\{(CH_3)_2PC_2H_4P(CH_3)_2\}(\eta^4\text{-}C_4H_6)]$	46.7	2	46.2
32c	$[Cr(CO)_2\{(CH_3)_2PC_2H_4P(CH_3)_2\}(\eta^4\text{-}C_5H_8)]$	40.4	−29	48.9
32d	$[Cr(CO)_2\{(CH_3)_2PC_2H_4P(CH_3)_2\}(\eta^4\text{-}C_6H_{10})]$	56.5	−8	58.8
32e	$[Cr(CO)_2\{(CH_3)_2PC_2H_4P(CH_3)_2\}(\eta^4\text{-}C_6H_{10})]$	40.8	−26	48.4
34a	$[W(CO)_2\{(CH_3)_2PC_2H_4P(CH_3)_2\}(\eta^4\text{-}C_4H_6)]$	—	—	50.0
34d	$[W(CO)_2\{(CH_3)_2PC_2H_4P(CH_3)_2\}(\eta^4\text{-}C_6H_{10})]$	—	—	59.0

the *df* and *ef* sites are not detectable, but may act as intermediates during the *ad–ae* isomerization.

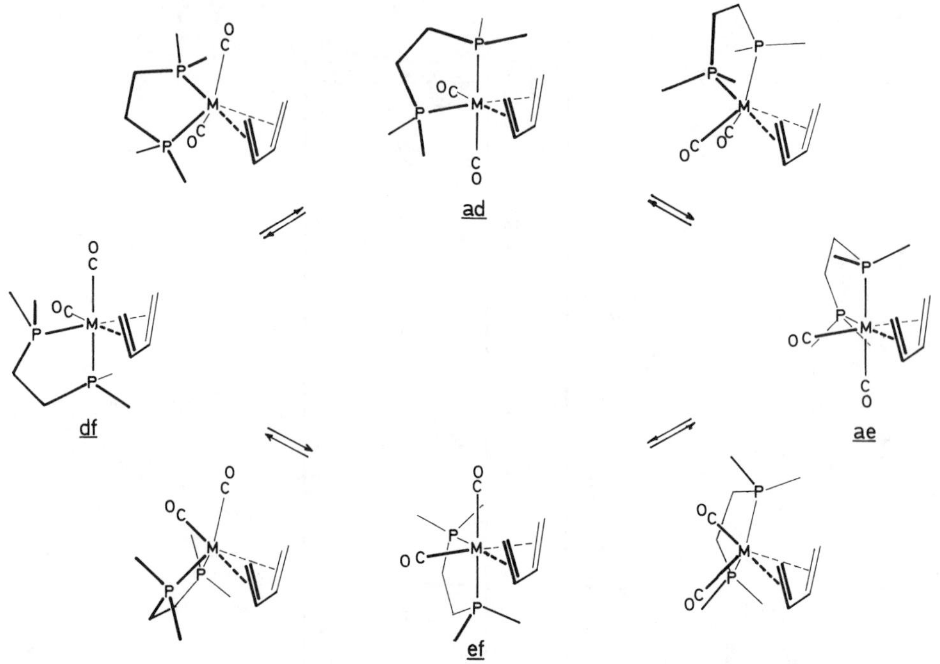

F. Ligand Exchange in [M(CO)₄₋ₙLₙ(η⁴-diene)] Complexes

When equimolar amounts of $[Cr(CO)_3\{P(CH_3)_3\}(\eta^4\text{-}1,3\text{-butadiene})]$ (**15a**) and $[Cr(CO)_3\{P(OCH_3)_3\}(\eta^4\text{-}(E,E)\text{-}2,4\text{-hexadiene})]$ (**19d**) are heated in benzene, a ligand exchange takes place (*22*) [Eq. (16)]. The activation parameters for this ligand exchange were established as $\Delta G_{298}^* = 98$ kJ/mol, $\Delta H^* = 84.5$ kJ/mol, and $\Delta S^* = -45.5$ J/mol · K.

$$[Cr(CO)_3\{P(CH_3)_3\}(\eta^4\text{-}C_4H_6)] + [Cr(CO)_3\{P(OCH_3)_3\}(\eta^4\text{-}C_6H_{10})] \rightleftharpoons$$

$$\quad\quad\quad\quad \textbf{15a} \quad\quad\quad\quad\quad\quad\quad\quad\quad\quad \textbf{19d}$$

$$[Cr(CO)_3\{P(CH_3)_3\}(\eta^4\text{-}C_6H_{10})] + [Cr(CO)_3\{P(OCH_3)_3\}(\eta^4\text{-}C_4H_6)] \quad (16)$$

$$\quad\quad\quad\quad \textbf{15d} \quad\quad\quad\quad\quad\quad\quad\quad\quad\quad \textbf{19a}$$

In order to investigate whether the donor or the diene ligands migrate, the behavior of a solution of $[W(CO)_3\{P(CH_3)_3\}(\eta^4\text{-}(E)\text{-}1,3\text{-pentadiene})]$ (**17b**) and $[Cr(CO)_3\{P(OCH_3)_3\}(\eta^4\text{-}(E,E)\text{-}2,4\text{-hexadiene})]$ **(19d)** was studied (22). Only the diene ligands exchanged between the two different metals in this case [Eq. (17)].

$$[W(CO)_3\{P(CH_3)_3\}(\eta^4\text{-}C_5H_8)] + [Cr(CO)_3\{P(OCH_3)_3\}(\eta^4\text{-}C_6H_{10})] \rightleftharpoons$$

$$\text{17b} \qquad\qquad\qquad\qquad \text{19d}$$

$$[W(CO)_3\{P(CH_3)_3\}(\eta^4\text{-}C_6H_{10})] + [Cr(CO)_3\{P(OCH_3)_3\}(\eta^4\text{-}C_5H_8)] \quad (17)$$

$$\text{17d} \qquad\qquad\qquad\qquad \text{19b}$$

For $[Cr(CO)_2L_2(\eta^4\text{-diene})]$ complexes migration of the donor ligands was also observed. Solutions of $[Cr(CO)_2\{P(CH_3)_3\}\{P(OCH_3)_3\}(\eta^4\text{-diene})]$ **(36a** and **36d)** are instable upon standing at room temperature and form an equilibrium mixture with $[Cr(CO)_2\{P(CH_3)_3\}_2(\eta^4\text{-diene})]$ **(24a** and **24d)** and $[Cr(CO)_2\{P(OCH_3)_3\}_2(\eta^4\text{-diene})]$ **(26a** and **26d)** (29). An identical mixture is obtained starting from **24a** or **24d** and **26a** or **26d**. For a mixture of **24d** and **26d**, $\Delta G_{298}^* = 95.3 \pm 1$ kJ/mol was determined for the consumption of **26d** in a first-order reaction during establishment of the equilibrium with **36d** [Eq. (18)].

$$[Cr(CO)_2\{P(CH_3)_3\}_2(\eta^4\text{-diene})] + [Cr(CO)_2\{P(OCH_3)_3\}_2(\eta^4\text{-diene})] \rightleftharpoons$$

$$\begin{array}{ll}\text{24a (12.1\%)} & \text{26a (12.1\%)}\\ \text{24d (6.3\%)} & \text{26d (6.3\%)}\end{array}$$

$$2\,[Cr(CO)_2\{P(CH_3)_3\}\{P(OCH_3)_3\}(\eta^4\text{-diene})] \quad (18)$$

$$\text{36a (75.8\%)}$$
$$\text{36d (87.4\%)}$$

These results show clearly the possibility of intramolecular ligand exchange in $[M(CO)_{4-n}L_n(\eta^4\text{-diene})]$ complexes. Donor ligands as well as diene ligands migrate from one metal center to the other. However, the activation barriers are definitely higher than those for the intramolecular diene ligand rotations and carbonyl scrambling.

G. Stereochemistry of [M(CO)₂(η⁴-diene)₂] Complexes

The stereochemistry of the compounds $[M(CO)_2(\eta^4\text{-diene})_2]$ [M = Cr **(4)**, Mo **(9)**, W **(11)** (16,18,20)] corresponds to that of octahedral $[ML_2(LL)_2]$ complexes containing two mono- and two bidentate ligands. For such complexes, cis and trans isomers are conceivable, with the cis isomers existing as a pair of enantiomers (Λ, Δ) (30). Due to the local C_s symmetry of 1,3-butadiene or 1,3-cyclohexadiene ligands, two trans and three diastereomeric pairs of enantiomers of the cis isomers are possible. In analogy with the $[M(CO)_{4-n}L_n(\eta^4\text{-diene})]$ complexes, one CO ligand is set at the *a*, and one diene ligand in the *b* and *c* sites. The positions *d*–*f* are occupied by the two remaining ligands. In order to define the stereochemistry of the conceivable isomers, the orientation of the diene ligand with respect to the carbonyl, with which three facial positions are shared, is classified as *o* (oben) when C-1 and C-4 are turned toward that CO and as *u* (unten) when C-2 and C-3 are in the neighborhood of that CO. For the trans isomers, two choices of CO ligand sites are possible. In this situation the orientation of both dienes with respect to the same CO ligand has to be defined in order to characterize the stereochemistry.

oo

ou

Λoo

Λou

Λuu

Δoo

Δou

Δuu

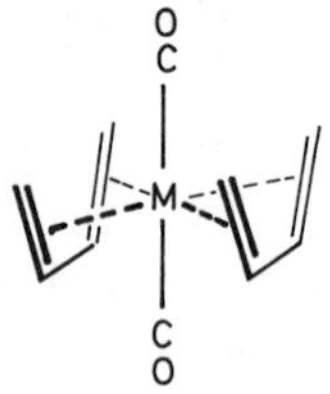

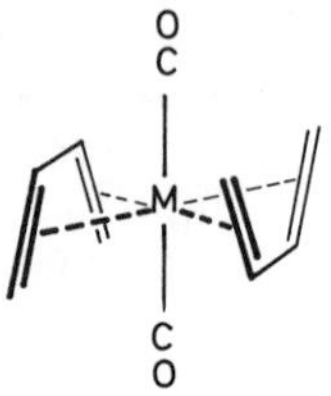

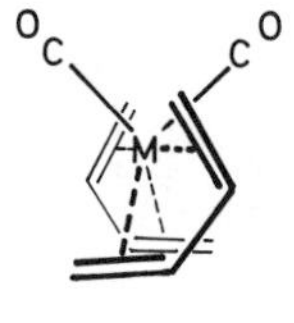

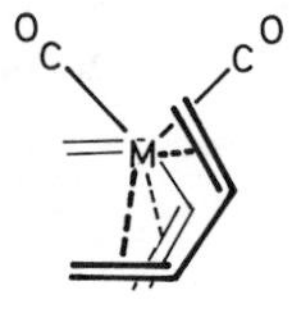

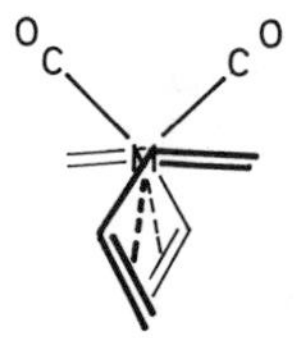

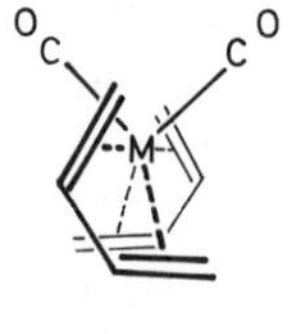

The complexes **9c** and **11c** with 2-methyl-1,3-butadiene as ligands form three diasteromers, according the coordinations modes of the diene ligands, Λ*RR*,Δ*SS*, Λ*SS*,Δ*RR*, and Λ*RS*,Δ*SR*. The enantiomers Λ*RR*,Δ*SS* and Λ*SS*,Δ*RR* have C_2 symmetry and contain two chemically equivalent diene ligands. The Λ*RS* ≡ Λ*SR* and Δ*SR* ≡ Δ*RS* forms are asymmetric, containing diene ligands of opposite chirality which are chemically different.

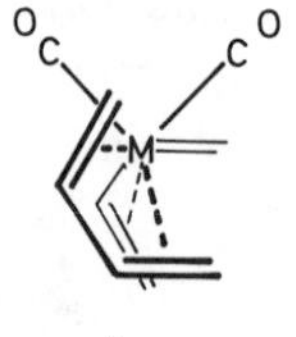

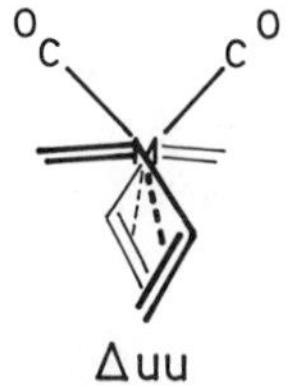

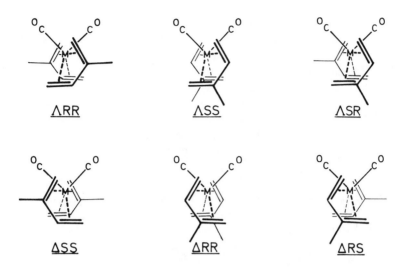

ΛRR   ΛSS   ΛSR

ΔSS   ΔRR   ΔRS

H. *Molecular Structure of cis-[W(CO)$_2$(η^4-1,3-butadiene)$_2$]*

The molecular structures of *cis*-[M(CO)$_2$(η^4-diene)$_2$] [M, diene = Mo, tricyclo[6.3.0.0^{2,7}]undeca-3,5-diene (*33*); W, 1,3-butadiene (**11a**) (*31*) (Fig. 3)] show a cis configuration and an *o,o* orientation of the diene ligands. These

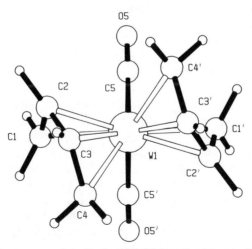

FIG. 3.   Molecular structure of *cis*-dicarbonylbis(1,3-butadiene)tungsten(0) (**11a**) (*31*): W-1—C-1 229.6, W-1—C-2 225.8, W-1—C-3 227.4, W-1—C-4 237.2, C-1—C-2 146.0, C-2—C-3 139.2, C-3—C-4 144.1 pm; C-5—W-1—C-5′ 99.8°.

structures correspond well with the observed ^{1}H-NMR spectra of the complexes at low temperatures. Similar structural features were established for cis-[Mo{P(CH$_3$)$_3$}$_2$(η^4-butadiene)$_2$] (35). The coordination spheres are distorted octahedral, in contrast to [Mo(η^4-butadiene)$_3$] (36–38), which has an idealized trigonal prismatic geometry.

cis-Dicarbonylbis(1,3-butadiene)tungsten(0) (**11a**), adopts C$_2$ symmetry (31). As in other η^4-butadiene complexes (35,39), the terminal methylene groups of the η^4-butadiene ligands are tilted out of the plane of C-1—C-4, due to a rehybridization of the sp^2 carbon atoms.

I. Dynamic Behavior of cis-[M(CO)$_2$(η^4-diene)$_2$] Complexes

At 200 K the ^{1}H- and ^{13}C-NMR spectra of **9a**, **9g**, **11a**, and **11g** agree with the molecular structures, determined for **11a** in the solid state. The ^{1}H and ^{13}C signals show one kind of diene ligands in a chiral environment, as can be expected for these complexes, due to the inherent C$_2$ symmetry. At room temperature, the resonance signals simulate local C$_s$ symmetry for the diene ligands. This corresponds to a hindered ligand movement with an achiral transition state. For **4g**, even at 147 K only signal broadening is observed, which corresponds to a fast ligand movement even at this temperature. As pointed out for [M(CO)$_{4-n}$L$_n$(η^4-diene)] complexes, a rotation of the η^4-diene ligands around their coordination axis is quite likely and explains the D (Dynamic) NMR results. The temperature-dependent NMR spectra of **4g**, **9a**, **9g**, **11a**, and **11g** can be similarly understood. The activation barriers (ΔG_{250}*) were determined: 26.1 kJ/mol for **4g**, 50.3 kJ/mol for **9a**, 43.1 kJ/mol for **9g**, 44.9 kJ/mol for **11a**, and 40.4 kJ/mol for **11g**.

More insight into the molecular dynamics of [M(CO)$_2$(η^4-diene)$_2$] complexes is provided by the ^{1}H-NMR spectra of the 2-methyl-1,3-butadiene complexes (**9c** and especially **11c**) (18). Two species are detected in comparable amounts, corresponding to $\Lambda RR,\Delta SS$ with C$_2$ symmetry and $\Lambda RS \equiv \Lambda SR,\Delta RS \equiv \Delta SR$ with C$_1$ symmetry. While the species with C$_2$ symmetry shows no distinctly temperature-dependent signals, so also the ^{1}H-NMR signals of the C$_1$ species simulate C$_2$ symmetry at room temperature.

The mutual transformation of the $\Lambda RS \equiv \Lambda SR$ and $\Delta SR \equiv \Delta RS$ enantiomers, can be explained by a fast rotation of the diene ligands around their coordination axis, with ΔG_{250}* = 47 kJ/mol. In the time average, the complex corresponds to a tetrahedron, formed by the two CO ligands and the rotational axis of the diene ligands.

Two rotamers are possible for the C$_2$ form with o,o orientation of the diene ligands. In the $\Lambda SS,\Delta RR$ forms, steric interactions between the methyl groups can be expected, which are a minimum in the $\Delta SS,\Lambda RR$ forms. Therefore the populations of the former are very small and cause the slight temperature dependence of the NMR signals.

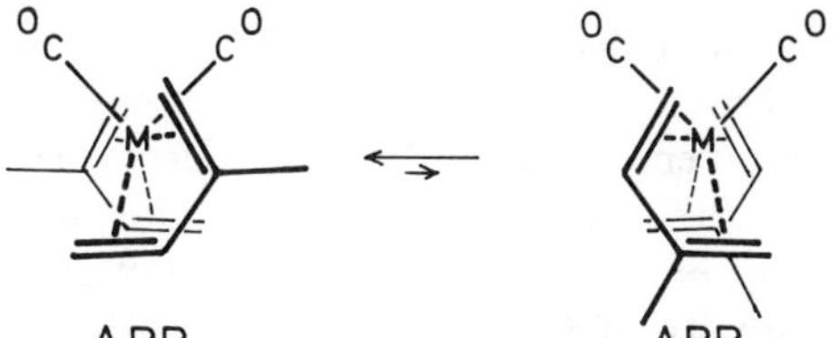

III

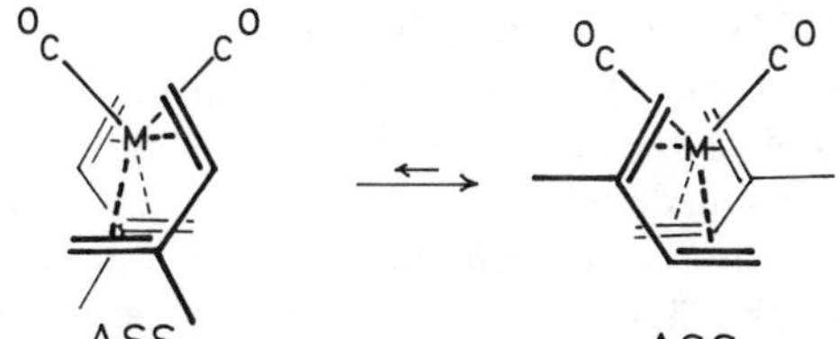

ELECTRON DEFICIENT $\eta^{4:CH}$-DIENE CHROMIUM(0) COMPLEXES

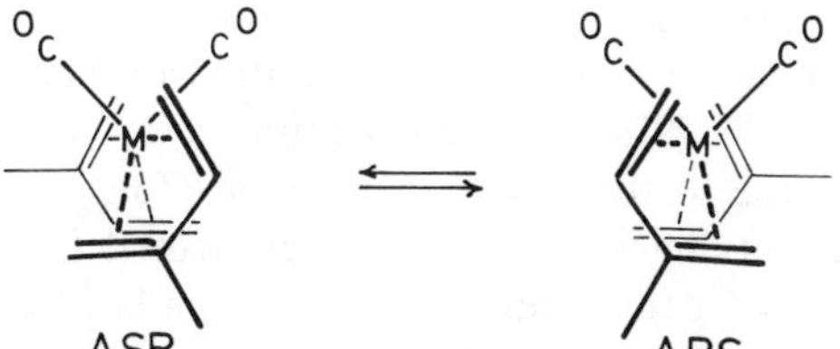

Although complexes with C—H—metal three-center, two-electron bonds were first observed several years ago (40–42), they have received increasing attention recently as model systems for C—H activation by transition metal complexes (43). A general route to such compounds involves the protonation of diene (35,44–51) or olefin complexes (52–56). The resulting 16-electron species are stabilized by the formation of C—H—metal bridges. Irradiation of the complexes [Cr(CO)$_5$L] [L = CO, P(CH$_3$)$_3$, P(OCH$_3$)$_3$] in presence of conjugated dienes having certain substituents provides a photochemical route to electron-deficient $\eta^{4:CH}$-diene complexes.

A. Photochemical Syntheses of [Cr(CO)$_2$L($\eta^{4:CH}$-diene)] Complexes [L = CO, P(CH$_3$)$_3$, P(OCH$_3$)$_3$]

Dienes with methyl or methylene groups in 1Z and/or 4Z position like (Z)-1,3-pentadiene (**1j**), (E,Z)-2,3-hexadiene (**1k**), 4-methyl-1,3-pentadiene (**1l**), (Z)-3-methyl-1,3-pentadiene (**1m**), 2,4-dimethyl-1,3-pentadiene (**1n**) (44), 1,3-cycloheptadiene (**1o**) (45), 1,5-dimethyl-1,3-cyclohexadiene (**1p**), 5,5-dimethyl-1,3-cyclohexadiene (**1q**), or 2-methyl-5-isopropyl-1,3-cyclohexadiene (**1r**) react photochemically with [Cr(CO)$_6$], [Cr(CO)$_5${P(CH$_3$)$_3$}], or [Cr(CO)$_5$-{P(OCH$_3$)$_3$}] at approximately 200 K to give deep red [Cr(CO)$_2$L-($\eta^{4:CH}$-diene)] complexes [L = CO (**37g**, **37n**, **37o**, and **37r**), P(CH$_3$)$_3$ (**38g**, **38n**, and **38o**), P(OCH$_3$)$_3$ (**39g**, and **39j–39r**)] [see Eq. (19)]. The yields are markedly increased when a vigorous stream of nitrogen is bubbled through the pentane solution of the reactants during irradiation, to sweep out the carbon monoxide. Immediately after irradiation, low-temperature column chromatography of the reaction mixtures and recrystallization of **37g**, **37n**, **37o**, **37r**, **38g**, **38n**, **38o**, **39g**, and **39j–39r** is necessary in order to prevent spontaneous decomposition of the electron deficient metal–diene complexes. Surprisingly, the complexes are thermally stable in the pure state up to 390 K in toluene. Prolonged heating causes substitution of the diene ligands by toluene, and tricarbonyl-η^6-toluenechromium(0) is formed.

The formulas of **37–39** show the substitution of three carbonyls by a conjugated diene. Due to the fact that dienes are four-electron donors, the complexes are electron-deficient species. The electron deficiency is reduced by formation of C—H—Cr three-center, two-electron bonds, previously found for a number of other complexes (43). When 2-methyl-5-isopropyl-1,3-cyclohexadiene is used as a potential ligand, the expected complexes **37r** and **39r** are not obtained, but instead complexes with the isomeric 1-methyl-4-isopropyl-1,3-cyclohexadiene group (**37r′**, and **39r′**) are formed [Eq. (19)]. Similarly, **1k** and **1l** form not only the corresponding complexes **39k** and **39l**, but also the isomeric complexes with (Z)-1,3-hexadiene (**39k′**) and with (Z)-2-methyl-1,3-pentadiene (**39l′**).

$$[Cr(CO)_5L] \ + \ diene \ \rightarrow \ [Cr(CO)_2L(\eta^{4:CH}\text{-diene})] + 3\ CO$$

2	L = CO	1g,1j–1r	37g, 37n, 37o, 37r′
14	P(CH$_3$)$_3$		38g, 38n, 38o
18	P(OCH$_3$)$_3$		39g, 39j–39r′, 39k′, 39l′

(19)

B. Stereochemistry of [Cr(CO)$_2$L($\eta^{4:CH}$-diene)] Complexes [L = CO, P(CH$_3$)$_3$, P(OCH$_3$)$_3$]

Formally, the $\eta^{4:CH}$-diene ligands are tridentate and occupy three facial positions in the octahedral [Cr(CO)$_2$L($\eta^{4:CH}$-diene)] complexes [L = CO,

$P(CH_3)_3$, $P(OCH_3)_3$]. For each of the [Cr(CO)$_3$($\eta^{4:CH}$-diene)] complexes **39g**, **37n**, and **37o** two enantiomers are expected. As a consequence of the C—H—Cr bridge from C-5 or C-6 of the 1,3-cyclohexadiene, and from C-5 or C-7 of the 1,3-cycloheptadiene, both ligands lose their inherent mirror planes. The situation is somewhat different with the 2,4-dimethyl-1,3-pentadiene complex (**37n**). Here only the 4Z methyl group is able to form the C—H—Cr bridge. Due to the two coordination modes of the diene, R or S, a pair of enantiomers is present for **37n**. Finally, **37r'** forms two diastereomeric pairs of enantiomers, because the C—H—Cr bridge can be formed from C-5 or C-6 and the diene can be coordinated in the R or S mode.

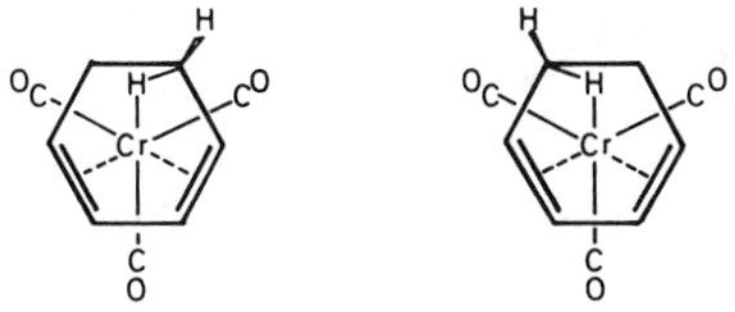

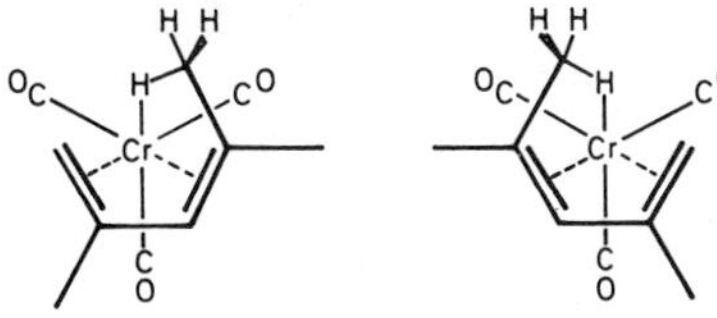

37g   37n

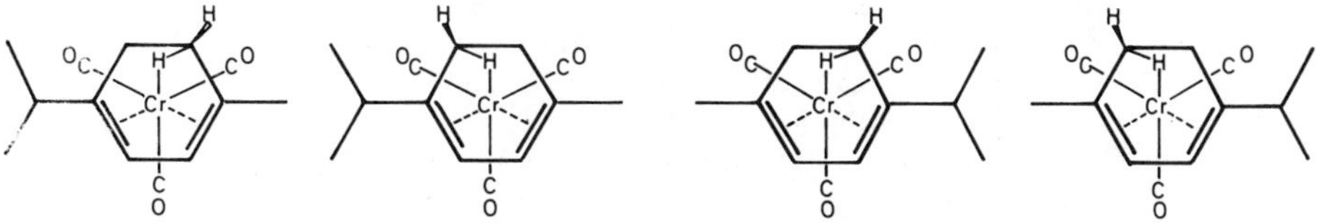

37r'

As in the [M(CO)$_{4-n}$L$_n$(η^4-diene)] complexes (*10,23*), the diene unit is set arbitrarily into the *b* and *c* sites, and thus the bridging C—H bond is in site *a*. In the complexes **38** and **39**, the donor ligand can be coordinated to three different sites for each enantiomer. The steric properties of the $\eta^{4:CH}$-diene ligands **1j**, **1k**, **1k'**, **1l**, **1l'**, **1m**, and **1p** correspond to those of **1n**, and the properties of **1q** resemble those of **1r'**.

C. *Molecular Structures of [Cr(CO)$_2$L($\eta^{4:CH}$-diene)] Complexes*

The molecular structures of dicarbonyl-$\eta^{4:CH}$-1,3-cycloheptadienetrimethylphosphinechromium(0) (**38o**) (Fig. 4) (*57*) and dicarbonyl-$\eta^{4:CH}$-2,4-dimethyl-1,3-pentadienetrimethylphosphitechromium(0) (**39n**) (Fig. 5) (*58*) have been determined by X-ray structure analysis. In good approximation, the coordination spheres of the metal atoms of **38o** and **39n** are octahedral. Two carbonyls and one donor ligand occupy three facial coordination sites.

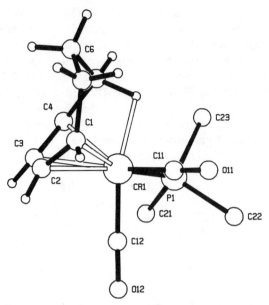

FIG. 4. Molecular structure of dicarbonyl-$\eta^{4:CH}$-1,3-cycloheptadienetrimethylphosphine-chromium(0) (**38o**) (*57*): Cr-1—C-1 220.8, Cr-1—C-2 214.7, Cr-1—C-3 216.9, Cr-1—C-4 213.9, Cr-1—C-5 235.5, Cr-1—H-51 185, C-5—H-51 116, C-1—C-2 139.2, C-2—C-3 140.2, C-3—C-4 137.5, C-4—C-5 147.8 pm; H-51—Cr-1—C-12 165.7°, P-1—Cr-1—C-11 92.4°. Reprinted with permission from *Chem. Ber.* **118**, 3944 (1985). Copyright by VCH Verlagsgesellschaft mbH.

In the remaining positions, the diene entity and the bridging hydrogen are located. The binding angles of the three monodentate ligands are close to 90°. The bridging hydrogen at site *a* of **38o** and **39n** is bent somewhat toward the neighbor ligands at the *d* and *e* positions.

All the chromium–carbon distances from C-1 to C-4 range between 214 and 221 pm. Remarkably, the carbon atom, which is connected only via the H bridge with the central metal, also has a short distance of 235.5(2), or 239.9(4) pm to Cr, respectively, much shorter than the van der Waals radii of both atoms. The bridging hydrogen shows elongated bond length to C, in comparison with the other C—H bonds. The Cr—H—C angle is about 100°.

Normal C—C bond lengths were found within the $\eta^{4:CH}$-diene ligands. The C—C distances of the diene unit are quite similar, with a slight tendency toward alternating short–long–short. There is some, but not too significant, shortening of the C-4—C-5 bond. The atoms C-1 to C-5 are essentially coplanar.

In **38o**, the $\eta^{4:CH}$-1,3-cycloheptadiene group adopts the *R* configuration, and the trimethylphosphine ligand is found at the *d* position. In **39n**, the trimethylphosphite ligand occupies the *d* site also, but the $\eta^{4:CH}$-2, 4-dimethyl-1, 3-pentadiene ligand shows an *S* configuration. The bonding in the $\eta^{4:CH}$-

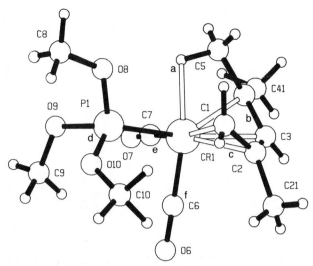

FIG. 5. Molecular structure of dicarbonyl-$\eta^{4:CH}$-2,4-dimethyl-1,3-pentadienetrimethyl-phosphitechromium(0) **(39n)** *(58)*: Cr-1—C-1 224.6, Cr-1—C-2 219.9, Cr-1—C-3 216.0, Cr-1—C-4 214.0, Cr-1—C-5 239.9, Cr-1—H-51 194, C-5—H-51 108, C-1—C-2 138.7, C-2—C-3 141.4, C-3—C-4 137.3, C-4—C-5 148.8 pm; H-51—Cr-1—C-6 169.9°, P-1—Cr-1—C-7 92.1°. Reprinted with permission from *Angew. Chem.* **97**, 503 (1985). Copyright by VCH Verlagsgesellschaft mbH.

diene complexes **38o** and **39n** is closely related to those in $\eta^{3:CH}$-enyl complexes of iron *(44,45)* and manganese *(47,48)*. Similar binding properties to those in **38o** or **39n** probably have to be considered for $[Cr(PF_3)_3(H)(\eta^5\text{-}C_8H_{11})]$ *(59)* and $[Cr(PF_3)_3(H)(\eta^5\text{-}C_6H_7)]$ *(60)*, owing to the more or less identical Cr—C bond lengths and similar ^{1}H-NMR spectroscopic behavior of these compounds.

D. *Dynamic Behavior of [Cr(CO)₂L(η⁴:CH-diene)] Complexes*

By D-NMR spectroscopy, up to three hindered ligand movements were detected in $[Cr(CO)_2L(\eta^{4:CH}\text{-diene})]$ complexes, a situation comparable to $[Mn(CO)_3(\eta^{3:CH}\text{-cyclohexenyl})]$ complexes *(46–51)*. Only small activation barriers are found for exchange processes of the C—H—Cr bridge. Higher barriers are observed for the site exchange of the three monodentate ligands. Similar energies are necessary to activate 1,5-H shifts (Table VI).

It is necessary to distinguish between two different H-bridge exchange processes. Complexes with cyclic, symmetrical diene ligands like **37g**, **37o**, **38g**, **38o**, **39g**, and **39o** show an exchange between *endo*-H-5 and *endo*-H-6 or *endo*-H-7, respectively. By this process, the *R* and *S* configurations are interconverted via an achiral 16-electron transition state. The relative positions of the monodentate ligands and the diene unit are unaffected by this

TABLE VI

ACTIVATION PARAMETERS FOR HINDERED LIGAND MOVEMENTS IN $[M(CO)_2L(\eta^{4:CH}\text{-diene})]$ COMPLEXES[a]

	Complex	Sd / Re	Se / Rd	T(K)	ΔG_{1T}^{*}	ΔG_{2T}^{*}	ΔG_{3T}^{*}
37m	$[Cr(CO)_3(\eta^{4:CH}\text{-}C_6H_{10})]$				31_{180}	—	63_{350}
37n	$[Cr(CO)_3(\eta^{4:CH}\text{-}C_7H_{12})]$				33_{165}	—	72_{298}
37g	$[Cr(CO)_3(\eta^{4:CH}\text{-}C_6H_8)]$				27_{165}	—	39_{210}
37o	$[Cr(CO)_3(\eta^{4:CH}\text{-}C_7H_{10})]$				28_{165}	—	67_{298}
38m	$[Cr(CO)_2\{P(CH_3)_3\}(\eta^{4:CH}\text{-}C_6H_{10})]$	82	18	240	25_{150}	66_{320}	76_{380}
					$>25_{200}$		—
38n	$[Cr(CO)_2\{P(CH_3)_3\}(\eta^{4:CH}\text{-}C_6H_{10})]$	100		240	25_{150}	—	73_{400}
38o	$[Cr(CO)_2\{P(CH_3)_3\}(\eta^{4:CH}\text{-}C_7H_{10})]$	80	20	125	28_{165}	—	64_{350}
39k	$[Cr(CO)_2\{P(OCH_3)_3\}(\eta^{4:CH}\text{-}C_6H_{10})]$	16	72	240	29_{170}	55_{300}	—
					31_{180}		—
39k'	$[Cr(CO)_2\{P(OCH_3)_3\}(\eta^{4:CH}\text{-}C_6H_{10})]$	12		240	$\approx 30_{180}$	—	$>60_{300}$
39l		40				70_{350}	—
39l'		50	10			—	61_{310}
39m	$[Cr(CO)_2\{P(OCH_3)_3\}(\eta^{4:CH}\text{-}C_6H_{10})]$	70	30	240	29_{170}	62_{300}	73_{360}
					$>30_{190}$		—
39n	$[Cr(CO)_2\{P(OCH_3)_3\}(\eta^{4:CH}\text{-}C_7H_{12})]$	100		240	29_{165}	—	70_{350}
39g	$[Cr(CO)_2\{P(OCH_3)_3\}(\eta^{4:CH}\text{-}C_6H_8)]$	89	11	120	27_{165}	—	32_{185}
39o	$[Cr(CO)_2\{P(OCH_3)_3\}(\eta^{4:CH}\text{-}C_7H_{10})]$	80	20	125	28_{165}	—	61_{350}
39p	$[Cr(CO)_2\{P(OCH_3)_3\}(\eta^{4:CH}\text{-}C_8H_{12})]$	92	8	140	—	31_{160}	54_{280}

[a] L = CO, $P(CH_3)_3$, $P(OCH_3)_3$. ΔG_{1T}^{*}, fluctuation of C—H—Cr bridges; ΔG_{2T}^{*}, rotation of the $\eta^{4:CH}$-diene ligands; ΔG_{3T}^{*}, 1,5-H shift within the $\eta^{4:CH}$-diene ligands in kJ/mol for $T(K)$.

movement. Hence, in compounds with donor ligands at the *d* or *e* positions, the two diastereomers are transformed into each other by the fluctuations of the C—H—Cr bridges.

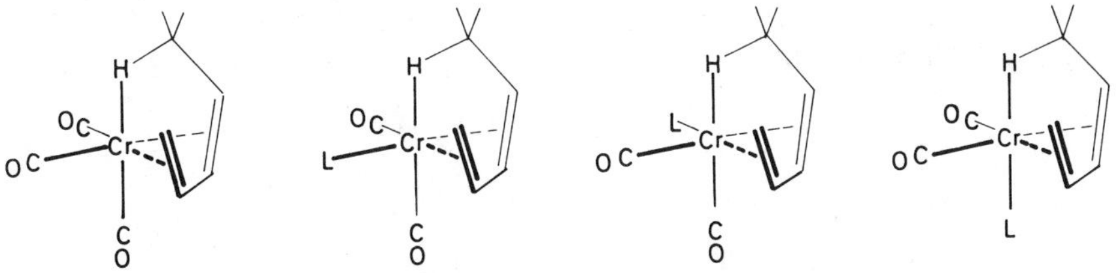

Complexes with C—H—Cr bridges involving methyl groups, for example, **37n, 38n, 39j, 39k, 39k′, 39l, 39l′, 39m**, and **39n**, exhibit a hindered rotation of this substituent. In $[Fe\{P(OCH_3)_3\}_3(\eta^{3:CH}\text{-butenyl})]^+$, a hindered rotation of the bridging methyl group was also observed (*44*).

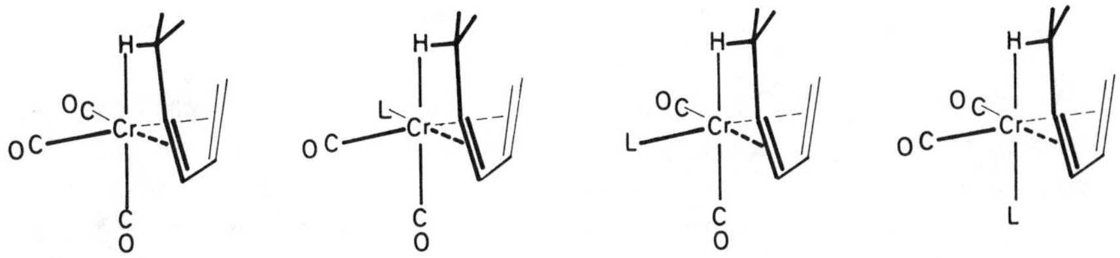

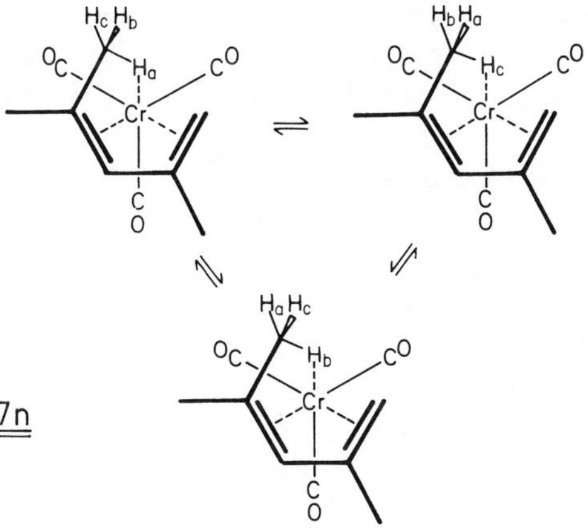

For various reasons, some diene ligands do not show exchange processes of the C—H—Cr bridges, for instance, in complexes **39k′, 39p**, and **39q**. In **39k′**, an exchange occurs involving the two 5-methylene protons. The two conceivable rotamers obviously differ so much in their energies due to steric effects of the 6-methyl group that only one form can be detected by NMR spectroscopy. Similar reasons have to be proposed for the occurrence of only one of the diastereomers of **39p**, the other being energetically unfavored. Finally, in **39q** no suitable CH unit is present for an exchange to occur.

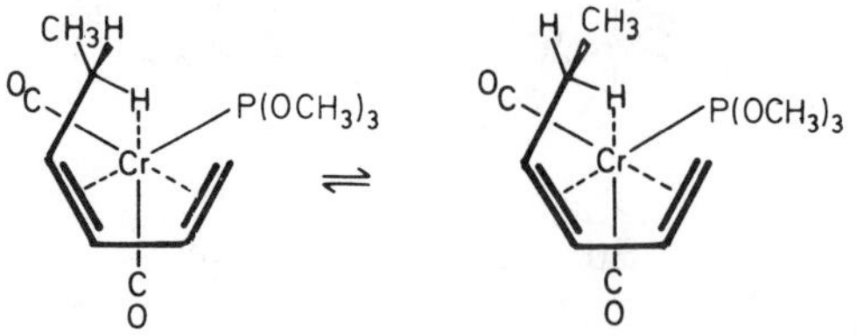

39k′

At ambient temperatures, fast rotation of the formally tridentate $\eta^{4:CH}$-diene ligands occurs around an axis through the chromium and the approximate center of the diene unit and the C—H—Cr bridge, relative to the $Cr(CO)_2L$ fragment. This process represents a $d-e-f$ isomerization and eliminates the influence of the $Cr(CO)_2L$ moiety on the symmetry of the $\eta^{4:CH}$-diene ligands.

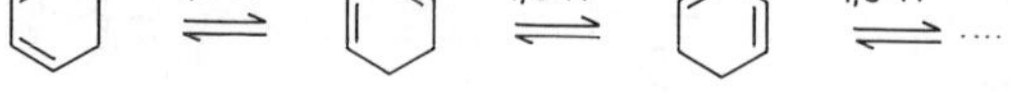

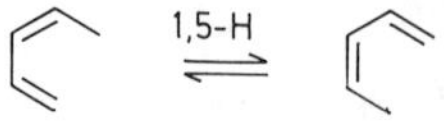

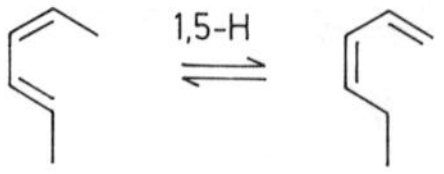

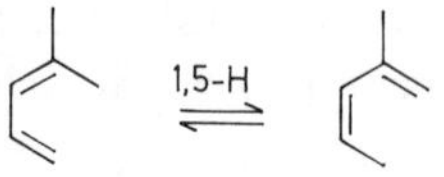

At still higher temperatures 1,5-H shifts become active. Three different consequences of these hydrogen migrations have to be considered. The diene ligands **1g** and **1o** show in their complexes only degenerate transformations. In contrast, the diene ligands **1j**, **1m**, **1n**, **1p**, and **1q** can rearrange by 1,5-H shifts to their enantiomers. Moreover, complexes of the $\eta^{4:CH}$-diene ligands **1k** and **1l** are transformed into different isomers (**1k′** and **1l′**) which are in equilibrium with the complexes of the parent dienes. Complexes of **1r** transform by 1,5-H shifts into three isomers, from which only one, with the 1-methyl-4-isopropyl ligand, has been detected. Obviously the others have only very small populations, according to their higher energies.

The observed much lower activation energies for the 1,5-H shifts in the complexes **37g**–**39r′** than in the free hydrocarbons (61–63) provide good evidence for an activation of the C—H bonds by the central metal. Short-lived $[Cr(H)(CO)_{2-n}L(\eta^5\text{-dienyl})]$ complexes have to be considered as intermediates during the 1,5-H shift processes.

E. Synthesis and Structure of
$[Cr(CO)_2\{P(OCH_3)_3\}(\eta^{4:2}\text{-5H-benzocycloheptene})]$

5*H*-Benzocycloheptene (**1s**) can be classified as a derivative of 1,3-cycloheptadiene (**1o**) or 1,3,5-cycloheptatriene (**68a**) with an annelated benzene ring. The thermal reactions of **1s** with $[Cr(CO)_6]$ (64) and $[Fe(CO)_5]$ (65) have been described earlier. The chromium complex contains the hydrocarbon coordinated with C-1 through C-4, C-10, and C-11 to the metal, and so it belongs to the large family of η^6-benzenetricarbonylchromium(0) compounds. In the corresponding $Fe(CO)_3$ complex, the diene unit within the seven-membered ring, encompassing C-6 through C-9, is bound to the metal. In both compounds the aromatic character of the annelated benzene ring is maintained.

The photoreaction of **18** with **1s** yields a deep green complex (**40**) [Eq. (20)], having the $Cr(CO)_2\{P(OCH_3)_3\}$ fragment coordinated to the seven-membered ring. Hence compound **40** is structurally related to tricarbonyl-η^6-1,3,5-cycloheptatrienechromium(0). The X-ray structure analysis of **40** proved that coordination of the 5*H*-benzocycloheptene occurs via the atoms C-6 through C-11 (Fig. 6) (66). The coordination sphere of the metal is approximately octahedral, with 5*H*-benzocycloheptene as a tridentate ligand at positions *a*–*c*. Two carbonyl ligands and the trimethylphosphite ligand occupy the remaining three facial positions. Trimethylphosphite is found in the *e* site. There are noticeable differences concerning the Cr—C bond lengths (Fig. 6). The four carbon atoms C-6 through C-9 have normal distances, but C-10 and in particular C-11 show distinctly increased Cr—C bond lengths, which suggest only a weak interaction of C-10, and even a weaker one of C-11 with the chromium. The C—C bonds of the 5*H*-benzocycloheptene exhibit values within the normal range. There is no significant effect on the bond length of C-10—C-11, in comparison with the other C—C bonds of the annelated benzene ring. In contrast, $Fe(CO)_3$ fragments coordinated to two diene moieties within 3α-dimethylstyrene show normal Fe—C distances, and cause different C—C bond lengths in the benzene ring (67,68).

$$[Cr(CO)_5\{P(OCH_3)_3\}] + C_{11}H_{10} \rightarrow [Cr(CO)_2\{P(OCH_3)_3\}(\eta^{4:2}\text{-}C_{11}H_{10})] + 3\,CO \quad (20)$$

$$\qquad\quad \textbf{18} \qquad\qquad\qquad \textbf{1s} \qquad\qquad\qquad\qquad \textbf{40}$$

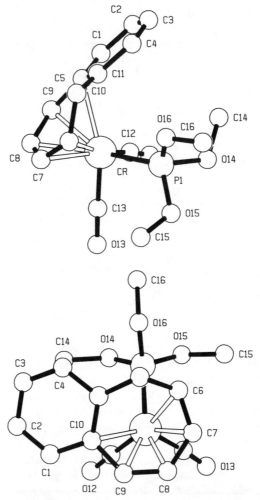

FIG. 6.   Molecular structure of η^{6-11}-5*H*-benzocycloheptenedicarbonyltrimethylphosphite-chromium(0) (**40**) (*66*): Cr-1—C-6 227.7, Cr-1—C-7 217.7, Cr-1—C-8 219.2, Cr-1—C-9 218.1, Cr-1—C-10 237.5, Cr-1—C-11 258.9, C-6—C-7 137.6, C-7—C-8 142.5, C-8—C-9 138.0, C-9—C-10 143.2, C-10—C-11 140.2 pm; P-1—Cr-1—C-12 91.1°, P-1—Cr-1—C-13 91.2°, C-12—Cr-1—C-13 79.2°.

The binding properties of the $\eta^{4:2}$-5*H*-benzocycloheptene ligand resembles that of $\eta^{4:CH}$-1,3-cycloheptadiene ligand in **39n**. In place of the C—H—Cr bridge, a weak interaction with C-10 and C-11 is recognized. A comparison of the Cr—C bond lengths of **40** with those of **38o** shows a surprising similarity for the coordinated diene unit, and for C-5 in **38o** with C-10 in **40**.

D-NMR spectroscopy reveals a facile 1,5-H shift for the *endo*-H-5 in **40**, with $\Delta G_{320}^* = 62.0 \pm 2$ kJ/mol. The low activation barrier is in accord with a metal-assisted H migration like that found in **37o**, **38o**, and **39o**. Although **40** contains no C—H—Cr bridge in the solid state, in solution a small population of the $[Cr(CO)_2\{P(OCH_3)_3\}(\eta^{4:CH}\text{-}C_{11}H_{10})]$ species (**40′**) has to be assumed from the ^{1}H-NMR spectra. The population of **40′** increases with elevation of temperature and is responsible for the 1,5 shift in **40**. Even faster is rotation of the $Cr(CO)_2\{P(OCH_3)_3\}$ unit relative to the 5*H*-benzocycloheptene ligand.

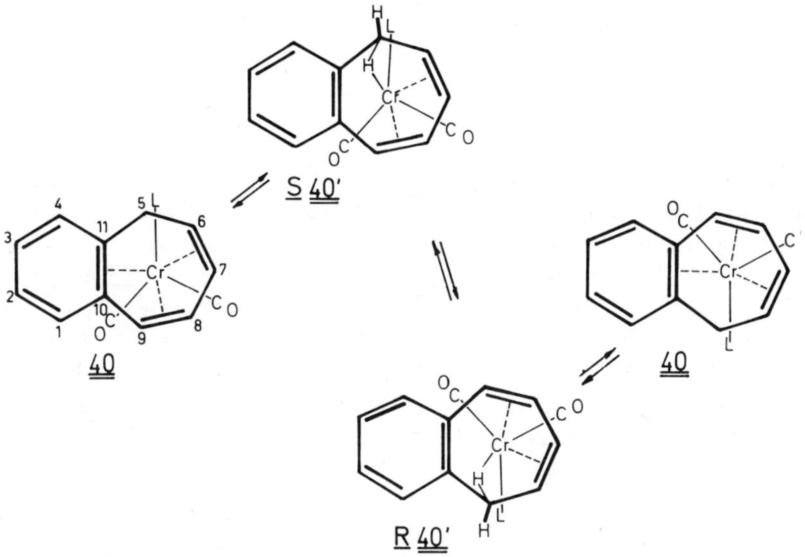

IV

PHOTOREACTIONS OF TRICARBONYL-η^6-1,3,5-CYCLOHEPTATRIENECHROMIUM(0) AND ITS DERIVATIVES WITH CONJUGATED DIENES

A. *Cycloaddition Reactions of [Cr(CO)$_3$(η^6-1,3,5-cycloheptatriene)] and [Cr(CO)$_3$(η^6-8,8-dimethylheptafulvene)] with Dienes*

Ultraviolet irradiation of $[Cr(CO)_3(\eta^6\text{-}1,3,5\text{-cycloheptatriene})]$ (**41**) in the presence of conjugated dienes causes no CO evolution. A photochemical [6 + 4] cycloaddition takes place with the dienes **1a–1d**, **1i**, and 2,3-dimethyl-1,3-butadiene (**1t**) (*69–71*) [Eq. (21)]. Within the coordination sphere of chromium, two C—C bonds are formed between C-1 and C-6 of the η^6-1,3,5-cycloheptatriene ligand and C-1 and C-4 of 1,3-butadiene

$$[Cr(CO)_3(\eta^6\text{-}C_7H_8)] + C_4H_{6-n}(CH_3)_n \;\rightarrow\; [Cr(CO)_3(\eta^{4:2}\text{-}C_{11}H_{14-n}(CH_3)_n)] \quad (21)$$

41   **1a–1d, 1i, 1t**   **42a–42d, 42i, 42t**

and its derivatives. Orange $\eta^{4:2}$-bicyclo[4.4.1]undeca-2,4,8-triene complexes
(**42a–42d, 42i,** and **42t**) are obtained in high yields.

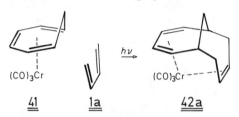

    41   **1a**   **42a**

An extension of this reaction principle to other complexes related to **41** and other dienes is limited. However, 1,3-cyclopentadiene (**1h**) or 1,3,5,7-cyclooctatetraene substitute the η^6-1,3,5-cycloheptatriene ligand, and $[Cr(CO)_2(\eta^5\text{-}C_5H_5)(\eta^3\text{-}C_5H_7)]$ (**5**) and $[Cr(CO)_3(\eta^6\text{-}C_8H_8)]$ (**72**), are formed, respectively, in high yield (*73*). In order to prevent the formation of **5**, spiro[4.2]hepta-1,3-diene and spiro[4.4]nona-1,3-diene (**1u**) (*74*) were used instead of **1h**. The diene **1u** should not readily form derivatives of η^5-cyclopentadienyl ligands. While spiro[4.2]hepta-1,3-diene does not form well-defined products, **1u** adds to **41** in a smooth reaction (*73*) [Eq. (22)]. The constitution of **43** was

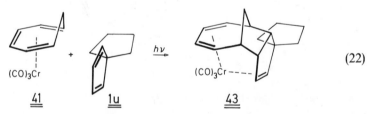

    41   **1u**   **43**   (22)

elucidated from its ^{1}H-NMR spectrum measured at 500 MHz. Obviously, on steric grounds the expected [6 + 4] cycloadduct is not formed, but a [6 + 2] cycloadduct is produced with an unaffected spiro system.

  With 1,3-cyclohexadiene (**1g**), a 1:1 adduct, tricarbonyl-$\eta^{4:2}$-tricyclo-[6.3.2.0^{2,7}]trideca-3,5,9-trienechromium(0) (**44**) is also isolated [Eq. (23)].

$$[Cr(CO)_3(\eta^6\text{-}C_7H_8)] + C_6H_8 \;\rightarrow\; [Cr(CO)_3(\eta^{4:2}\text{-}C_{13}H_{16})] \quad (23)$$

41   **1g**   **44**

Its constitution was determined by the 500-MHz ^{1}H-NMR spectrum. In the first step of the formation of **44**, a [4 + 2] cycloaddition with the η^6-1,3,5-cycloheptatriene ligand acting as diene and **1g** acting as ene components occurs. Two diastereomeric forms have to be considered for this intermediate. Probably, assisted by the metal, 1,5-H shifts from C-5 and C-6 to positions 12 and 13, respectively, complete the reaction.

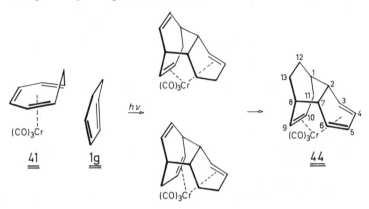

The reaction path **41** + **1g** → **44** was proved by the use of tricarbonyl-η^6-(1-6-d_6-1,3,5-cycloheptatriene)chromium(0) (**41d**). In the reaction product of **41d** with **1g** only one CH_2 but two CHD groups are present. This shows unequivocally that H migration from the 1,3-cyclohexadiene moiety to the 1,3,5-cycloheptatriene ligand has taken place. Photoreactions of the heavier homologs of **41**, $[Mo(CO)_3(\eta^6\text{-}C_7H_8)]$ and $[W(CO)_3(\eta^6\text{-}C_7H_8)]$ with conjugated dienes end up with substitution of the 1,3,5-cycloheptatriene ligand by the dienes *(70)*.

Tricarbonyl-η^6-1,3,5-cyclooctadienechromium(0)   and   tricarbonyl-η^6-1,3,5,7-cyclooctatetraenechromium(0) do not react with dienes photochemically. However, a similar reactivity to that for **41** is observed for tricarbonyl-η^6-8,8-dimethylheptafulvenechromium(0) (**45**) *(75)*. [6 + 4] Cycloaddition is observed with **1a**, **1b**, **1d**, and **1g** *(76)* [Eq. (24)]. So, interestingly, 1,3-cyclohexadiene (**1g**) behaves in this instance like the other dienes. 2-Methyl-1,3-butadiene (**1c**) and (*E*)-2-methyl-1,3-pentadiene (**1e**) form, under the same reaction conditions, complexes **46c** and **46e** in addition to the [6 + 4] cycloadducts $\eta^{3:5}$-[1-(3-butene-1,2-diyl)-7-isopropylidenecycloheptadienyl]-dicarbonylchromium(0) (**47c** and **47e**) *(76)* [Eq. (25)]. When 2,3-dimethyl-1,3-butadiene (**1t**) is used in the photoreaction with **45**, only the dicarbonyl complex **47t** can be detected *(76, 77)* [Eq. (26)]. Under ambient conditions, the complexes **47c**, **47e**, and **47t** add carbon monoxide, and [4 + 6] cycloaddition to the tricarbonyl compounds **46c**, **46e**, and **46t** is completed instantaneously [Eq. (27)].

$$[Cr(CO)_3(\eta^6\text{-}C_{10}H_{12})] + C_4H_{6-n}(CH_3)_n \rightarrow [Cr(CO)_3(\eta^{4:2}\text{-}C_{14}H_{18-n}(CH_3)_n)] \quad (24)$$

45   **1a, 1b, 1d**   **46a, 46b, 46d**

$$[Cr(CO)_3(\eta^6\text{-}C_{10}H_{12})] + C_4H_{6-n}(CH_3)_n \;\rightarrow$$

45   **1c, 1e**

$$[Cr(CO)_3(\eta^6\text{-}C_{14}H_{18-n}(CH_3)_n)] + [Cr(CO)_2(\eta^{3:5}\text{-}C_{14}H_{18-n}(CH_3)_n)] \quad (25)$$

46c, 46e   **47c, 47e**

$$[Cr(CO)_3(\eta^6\text{-}C_{10}H_{12})] + C_4H_4(CH_3)_2 \;\rightarrow\; [Cr(CO)_2(\eta^{3:5}\text{-}C_{14}H_{16}(CH_3)_2)] + CO \quad (26)$$

45   **1t**   **47t**

$$[Cr(CO)_2(\eta^{3:5}\text{-}C_{14}H_{18-n}(CH_3)_n)] + CO \;\rightarrow\; [Cr(CO)_3(\eta^6\text{-}C_{14}H_{18-n}(CH_3)_n)] \quad (27)$$

47c, 47e, 47t   **46c, 46e, 46t**

Formation of the complexes **47c**, **47e**, and **47t** and their smooth reaction with carbon monoxide sheds some light on the mechanism of [6 + 4] cyclo-addition in the coordination sphere of chromium. Obviously, the photo-reaction proceeds stepwise. First, the diene is coordinated to an activated complex (**45**) and forms tricarbonyl-η^4-8,8-dimethylheptafulvene-η^2-

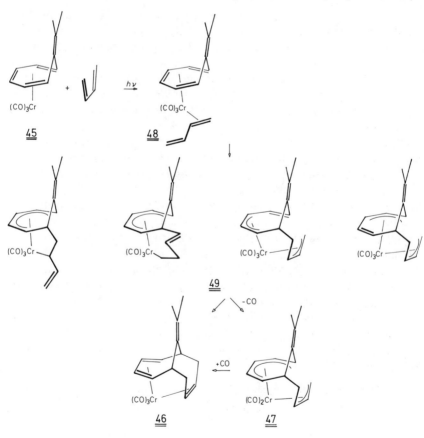

dienechromium(0) (**48**). The C—C bond formation between C-1 of the η^4-8,8-dimethylheptafulvene and C-1 of the η^2-diene ligands produces the $\eta^{3:3}$-1-(butene-1,2-diyl)-7-isopropylidenecycloheptadienyl ligand. Four different modes of coordination have to be considered for these ligands in the intermediates **49**. The intermediates formed from dienes without methyl substituents at the 2 and/or 3 position, immediately form a second C—C bond and yield the [6 + 4] cycloadducts (**46**). Intermediates of the type **49** from the dienes **1c**, **1e**, and **1t** obviously do not establish the second C—C bond as readily as those of **1a**, **1b**, **1d**, and **1g**. Under photochemical conditions CO is lost, and the complexes **47c**, **47e**, and **47t** are formed. Similar intermediates as discussed for the reaction of **45** with dienes have to be taken into account for the [6 + 4] cycloaddition of dienes to **41**.

B. *Molecular Structures of Tricarbonyl-$\eta^{4:2}$-bicyclo[4.4.1]-undeca-2,4,8-trienechromium(0) Complexes*

X-Ray structure analyses have been performed on the compounds tricarbonyl-$\eta^{4:2}$-7,10-dimethylbicyclo[4.4.1]undeca-2,4,8-trienechromium(0) (**42d**) (*69*) and tricarbonyl[$\eta^{3-6:9,10}$-(13-isopropylidenetricyclo[6.2.2.1^{2,7}]-trideca-3,5,9-triene)]chromium(0) (**46g**) (*76*) (Fig. 7). In a formal sense, **42d**

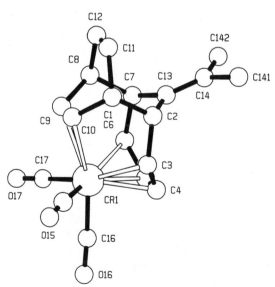

FIG. 7. Molecular structure of tricarbonyl[$\eta^{3-6:9,10}$-(13-isopropylidenetricyclo[6.2.2.1^{2,7}]-trideca-3,5,9-triene)]chromium(0) (**46g**) (*76*): Cr-1—C-3 227.9, Cr-1—C-4 216.8, Cr-1—C-5 218.0, Cr-1—C-6 230.6, Cr-1—C-9 243.8, Cr-1—C-10 244.6 pm; C-17—Cr-1—C-18 80.5°, C-17—Cr-1—C-19 102.2°, C-18—Cr-1—C-19 81.9°. Reprinted with permission from *Chem. Ber* **118**, 964 (1985). Copyright by VCH Verlagsgesellschaft mbH.

and **46g** have octahedral coordination for the metal atoms, with the bicyclo-[4.4.1]undeca-2,4,8-triene units functioning as tridentate ligands. The bond angles of the three carbonyl ligands deviate by about $\pm 10°$ from 90°. The Cr—C distances vary between 220 and 230 pm for the coordinated diene unit, encompassing C-2 to C-5. For the isolated C=C double bond a separation of 244 pm is found. Taking bond distances as a rough measure of bonding character and strength, one has to conclude that the diene unit is more tightly coordinated to chromium, than the isolated C—C bond. This weaker interaction is caused by the steric nature of the hydrocarbon frame. A comparison of the stereochemistry of **46d** and **46g** with that in the $[Cr(CO)_2 L$-$(\eta^{4:CH}$-diene)] complexes **38g**, **38o**, **39n**, and **40** shows close relations between the three groups of complexes. Coordination of six sp^2 carbon atoms of the tricyclic hydrocarbon ligand in **46g** causes a fixation of the 1,3-cyclohexadiene portion in a boat conformation with energetically unfavorable eclipsed methylene groups.

C. *Cleavage of the Hydrocarbon Ligands of*
   $[Cr(CO)_3\{\eta^6\text{-}C_{11}H_{14-n}(CH_3)_n\}]$ *and*
   $[Cr(CO)_3\{\eta^6\text{-}C_{14}H_{18-n}(CH_3)_n\}]$ *Complexes*

Cleavage of the Cr—C bonds in $[Cr(CO)_3(\eta^6\text{-}C_{11}H_{14-n}(CH_3)_n)]$ (**42**) and $[Cr(CO)_3(\eta^6\text{-}C_{14}H_{18-n}(CH_3)_n)]$ (**46**) is achieved in a smooth reaction with an excess of trimethylphosphine (*70*) or trimethylphosphite (*76*) in *n*-hexane at room temperature [Eqs. (28) and (29)]. Due to the low solubility of **51** in *n*-

$[Cr(CO)_3(\eta^6\text{-}C_{11}H_{14-n}(CH_3)_n)] + 3\ P(OCH_3)_3 \rightarrow$

   42a–42d, 42i, 42t

$$C_{11}H_{14-n}(CH_3)_n + [Cr(CO)_3\{P(OCH_3)_3\}_3] \quad (28)$$

   50a–50d, 50i, 50t   **51**

$[Cr(CO)_3(\eta^6\text{-}C_{14}H_{18-n}(CH_3)_n)] + 3\ P(OCH_3)_3 \rightarrow$

   46a–46e, 46g

$$C_{14}H_{18-n}(CH_3)_n + [Cr(CO)_3\{P(OCH_3)_3\}_3] \quad (29)$$

   52a–52e, 52g   **51**

hexane, separation of the reaction products is easy. By cooling the reaction mixture to 253 K, **51** can be almost completely removed. The hydrocarbons **50** and **52** are purified by column chromatography and/or distillation. Photochemical [6 + 4] cycloaddition of dienes to coordinated 1,3,5-cycloheptatriene (*69*, *70*) or 8,8-dimethylheptafulvene (*76*), followed by cleavage of the hydrocarbons formed in the coordination sphere of chromium, represents an interesting preparative route to di- and tricyclic hydrocarbons, not easily accessible by other methods.

D. *Molybdenum and Chromium Complexes of*
$C_{11}H_{14-n}(CH_3)_n$ *and* $C_{14}H_{18-n}(CH_3)_n$ *Hydrocarbons*

As already mentioned, no [6 + 4] cycloaddition is observed when tri-carbonyl-η^6-1,3,5-cycloheptatrienemolybdenum(0) is irradiated in presence of conjugated dienes. So an explanation of these experimental results has to be provided. One reason might be an unfavorable fit of the $C_{11}H_{14-n}(CH_3)_n$ and $C_{14}H_{18-n}(CH_3)_n$ hydrocarbons into the coordination sphere of molybdenum, arising from the increased Mo—C bond lengths. The other reason concerns the possible different stabilities of the intermediates of the photochemical [6 + 4] cycloaddition with molybdenum as the central metal.

When tricarbonyl(diglyme)molybdenum(0) (**53**) (*78*) is heated with the hydrocarbons **50a–50d** and **50t** in *n*-hexane, the analogous molybdenum complexes of the chromium species **42** are readily formed (*70*) [Eq. (30)]. These results prove the second explanation to be the correct one as to why tricarbonyl-η^6-1,3,5-cycloheptatrienemolybdenum shows no [4 + 6] cyclo-addition with conjugated dienes.

$$[Mo(CO)_3(diglyme)] + C_{11}H_{14-n}(CH_3)_n \rightarrow$$

53   **50a–50d, 50t**

$$[Mo(CO)_3(\eta^{4:2}\text{-}C_{11}H_{14-n}(CH_3)_n)] + diglyme \quad (30)$$

54a–54d, 54t

When the hydrocarbon $C_{13}H_{16}$ (**50g**), obtained from **44** by displacement with trimethylphosphite, is coordinated to molybdenum [Eq. (31)], the resulting complex contains a rearranged hydrocarbon ligand, whose structure conforms with the [6 + 4] cycloadduct of 1,3-cyclohexadiene to 1,3,5-cycloheptatriene. Obviously, during the complexation reaction, C—C bonds are cleaved and otherwise reconstituted in the coordination sphere of the molybdenum (*73*).

$$[Mo(CO)_3(diglyme)] \quad + \qquad\qquad \xrightarrow{\Delta} \qquad\qquad + \quad diglyme \quad (31)$$

$\underset{\displaystyle \underline{53}}{}$ $\underset{\displaystyle \underline{50g'}}{}$ $\underset{\displaystyle \underline{54g}}{}$

The hydrocarbons **52a–52e** and **52g** contain four C=C double bonds. In addition to the conjugated diene unit, two isolated bonds are present. The exocyclic one is parallel and the endocyclic one is perpendicular to the local plane of symmetry of the diene portion. When the hydrocarbons **52a–52e** and **52g** are recoordinated to chromium by the reaction with tris(acetonitrile)-tricarbonylchromium(0) (**55**) [Eq. (32)], the complexes **46** are not obtained but instead isomeric chromium compounds, in which the diene moiety and the exocyclic C=C double bond are coordinated to the chromium, are produced (79).

$$[Cr(CO)_3(CH_3CN)_3] \ + \quad \underset{\underline{\underline{52a\text{-}e,g}}}{} \quad \xrightarrow{\Delta} \quad \underset{\underline{\underline{56a\text{-}e,g}}}{(CO)_3Cr} \quad + \quad 3\ CH_3CN \quad (32)$$

<u><u>55</u></u>

The structures of **56a–56e** and **56g** have been elucidated by ^{1}H- and ^{13}C-NMR spectroscopy, and for **56a** also by X-ray structure analysis (Fig. 8). The structural properties of **56a** have similarities to those of the isomeric **46a**. The diene portion is normally bound to the metal with Cr—C distances of 220 and 230 pm. For the isolated C=C double bond, Cr—C bond lengths of 241.8 and 271.6 pm were found.

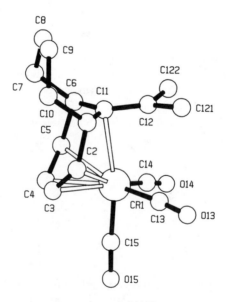

Fig. 8. Molecular structure of tricarbonyl-$\eta^{2-5:11,12}$-11-isopropylidenebicyclo[4.4.1]undeca-2,4,8-trienechromium(0) (**56a**) (79): Cr-1—C-2 233.9, Cr-1—C-3 219.4, Cr-1—C-4 218.2, Cr-1—C-5 229.9, Cr-1—C-11 241.8, Cr-1—C-12 271.6 pm; C-15—Cr-1—C-16 92.8°, C-15—Cr-1—C-17 84.3°, C-16—Cr-1—C-17 84.1°.

A preference for the coordination of an exocyclic C=C double bond in place of an endocyclic one was also observed for 9-methylene-bicyclo[4.2.1]nona-2,4,7-triene. With **55** this hydrocarbon forms [Cr(CO)$_3$-($\eta^{4:2}$-C$_{10}$H$_{10}$)], in which comparable stereochemistry with **56a** is found. However, the diene portion shows Cr—C distances of 220 and 240 pm, as does the exocyclic C=C-double bond (*80,81*). Hence the differences of bond strengths are not as pronounced as in **56a**.

The reactions of **53** with the tetraenes **52a–52e** and **52g** yield, in the case of **52a, 52b, 52d**, and **52g**, mixtures of isomeric complexes (**57** and **58**) [Eq. (33)], which are homologs of **46** and **56**. Two of the hydrocarbons (**52c** and **52e**), however, form only molybdenum complexes corresponding to **56**.

[Mo(CO)$_3$(diglyme)] + C$_{14}$H$_{18-n}$(CH$_3$)$_n$ →

   53   **52a–52e, 52g, 52t**

$$[Mo(CO)_3(\eta^{4:2}\text{-}C_{14}H_{18-n}(CH_3)_n)] + \text{diglyme} \quad (33)$$

                        57a–57e, 57g, 57t
                        58a–58e, 58g

E. Reactions of [Cr(CO)$_3$(η^6-C$_7$H$_8$)], [Cr(CO)$_3$(η^6-C$_8$H$_8$)], and [Cr(CO)$_3$(η^6-C$_{10}$H$_{12}$)] with Pentafulvenes

Photoreactions of [Cr(CO)$_3$(η^6-C$_7$H$_8$)] (**41**) with 6-mono- and 6,6-disubstituted pentafulvenes (**59a–59f**) preferentially yield dicarbonyl complexes with substituted $\eta^{3:5}$-2-cyloheptadienylene-2-cyclopentadienylidenemethane chelate ligands (*82, 83*). In the course of the reaction, C-6 of the fulvene forms a C—C bond to C-1 of the 1,3,5-cycloheptatriene ligand, and one CO ligand is displaced. This reaction is of the same type as the formation of the $\eta^{3:5}$-[1-(3-butene-1,2-diyl)-7-isopropylidenecycloheptadienyl] complexes **47c**, **47e** and **47t**. The fulvene unit is transformed into a monosubstituted cyclopentadienyl entity, η^5-coordinated to the chromium, with the 1,3,5-cyclo-

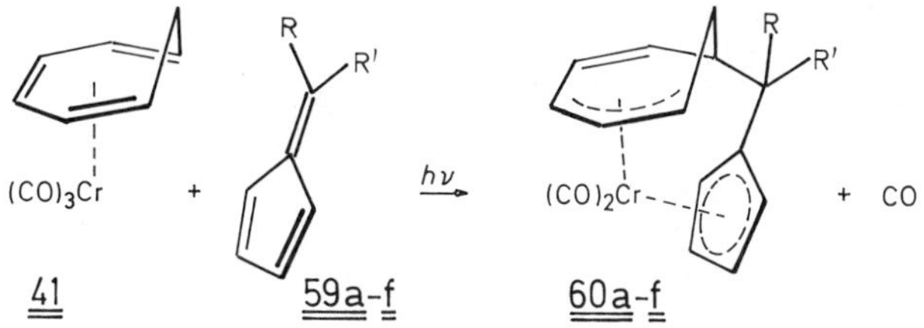

$$ (34) $$

  41   **59a–f**   **60a–f**

	59a	59b	59c	59d	59e	59f
R	CH$_3$	CH$_3$	C$_6$H$_5$	OCH$_3$	H	H
R'	CH$_3$	C$_6$H$_5$	C$_6$H$_5$	OCH$_3$	OCOCH$_3$	N(CH$_3$)$_2$

heptatriene converted to an η^3-coordinated, substituted cycloheptadienyl unit.

6,6-Diphenylpentafulvene (59c) shows another reaction with 41. By H transfer from 1,3,5-cycloheptatriene to C-6 of 59c, dicarbonyl-η^3-cyloheptatrienyl-η^5-diphenylmethylcyclopentadienylchromium (61) is formed [Eq. (35)]. In a similar manner, $[Cr(CO)_3(\eta^6\text{-}C_{10}H_{12})]$ (45) reacts with pentafulvenes 59a–59c and 59f to give dicarbonyl complexes (84) [Eq. (36)]. Independent of the substituents, three different types of complexes with $\eta^{3:5}$-coordinated $C_{16}H_{16}RR'$ ligands are isolated.

$$[Cr(CO)_3(\eta^6\text{-}C_7H_8)] + C_6H_4(C_6H_5)_2 \rightarrow$$

$$\quad\quad\quad 41 \quad\quad\quad\quad\quad\quad 59c$$

$$[Cr(CO)_2\{\eta^5\text{-}C_5H_4CH(C_6H_5)_2\}(\eta^3\text{-}C_7H_7)] + CO \quad (35)$$

$$\quad\quad\quad\quad\quad\quad\quad\quad\quad\quad 61$$

$$[Cr(CO)_3(\eta^6\text{-}C_{10}H_{12})] + C_6H_4RR' \rightarrow [Cr(CO)_2(\eta^{3:5}\text{-}C_{16}H_{16}RR')] + CO \quad (36)$$

$$\quad\quad 45 \quad\quad\quad\quad 59a\text{--}59c, 59f \quad\quad 62a, 62f, 63a, 63b, 63f, 64b, 64c$$

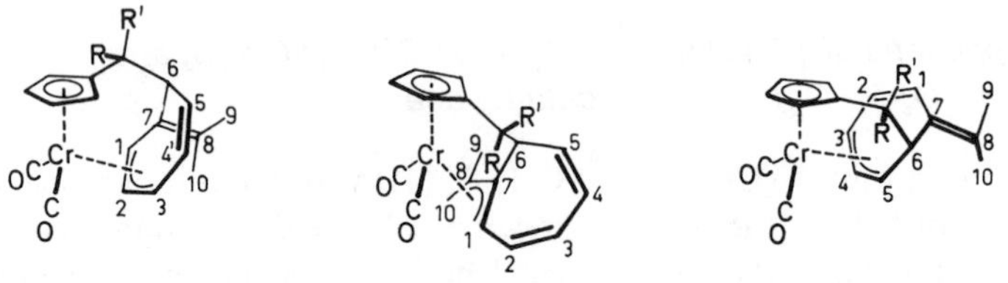

62a,f   63a,b,f   64b,c

In contrast to the [6 + 4] cycloaddition, which can be applied only to 41 and 45, pentafulvenes also react with tricarbonyl-η^6-1,3,5,7-cyclooctatetraenechromium(0) (65) to form $\eta^{3:5}$-coordinated hydrocarbon chelate

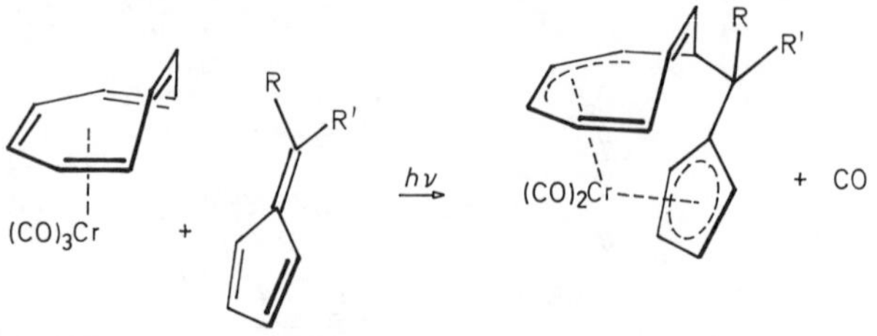

$$\quad\quad 65 \quad\quad\quad\quad 59a\text{--}f \quad\quad\quad 66a\text{--}f$$

ligands (85) [Eq. (37)]. With 6-dimethylaminofulvene (59f), complex 65 reacts further in an unexpected ring-contraction reaction to afford dicarbonyl-

$\eta^{3:5}$-2-cycloheptatrienylidene-1-cyclopentadienylidene-1-dimethylaminoethyl-chromium (**67**) [Eq. (38)].

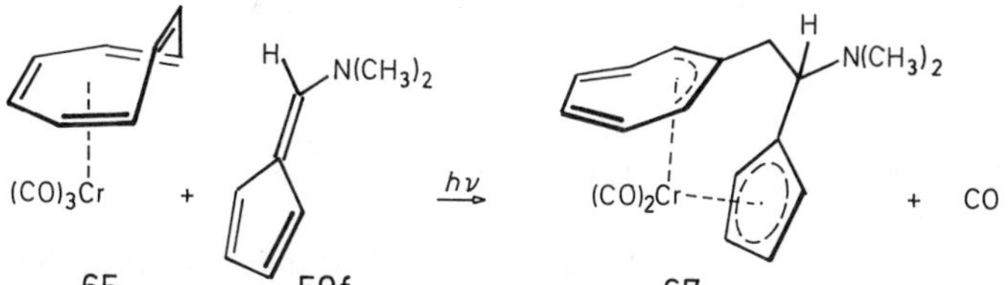

$$65 \qquad\qquad 59f \qquad\qquad\qquad\qquad 67 \qquad\qquad\qquad (38)$$

Not only η^6-1,3,5-cycloheptatriene and η^6-1,3,5,7-cyclooctatetraene ligands react with pentafulvenes in the coordination sphere of chromium. Reversion of this reaction principle is demonstrated by the photoreaction of tricarbonyl-η^6-6,6-dimethylfulvene (**68**) with 1,3,5-cycloheptatriene (**69a**), 7-methyl-1,3,5-cycloheptatriene (**69b**), or 7-methoxy-1,3,5-cycloheptatriene (**69c**) (*86*) [Eq. (39)]. The trienes add to **68**, displace a CO ligand, and form correspondingly substituted dicarbonyl-$\eta^{3:5}$-2-cycloheptadienylen-2-cyclopentadienylidenepropanechromium complexes. Compound **69b** forms only one isomer, with the methyl group at the *endo*-7 position (**70b**). In **70b** the η^3-enyl system encompasses C-1 to C-3 of the seven-membered ring. Complex **69c** yields three isomers, which were separated by column chromatography. One (**70c**) has the same constitution as **70b**. The others (**71** and **72**) bear the methoxy group at the 7-*exo* position. The η^3-enyl system of **71** is located at C-1 to C-3, that of **72** at C-3 to C-5. Compound **72** rearranges readily above room temperature to give **71**. This demonstrates a higher stability of the η^3-enyl system at C-1 to C-3 than at C-3 to C-5.

$$[Cr(CO)_3(\eta^6\text{-}C_8H_{10})] + C_7H_7R \rightarrow [Cr(CO)_2(\eta^{3:5}\text{-}C_{15}H_{17}R)] + CO \qquad (39)$$

$$\quad\; 68 \qquad\qquad 69a\text{-}69c \qquad\qquad 60a, 70b, 70c, 71, 72$$

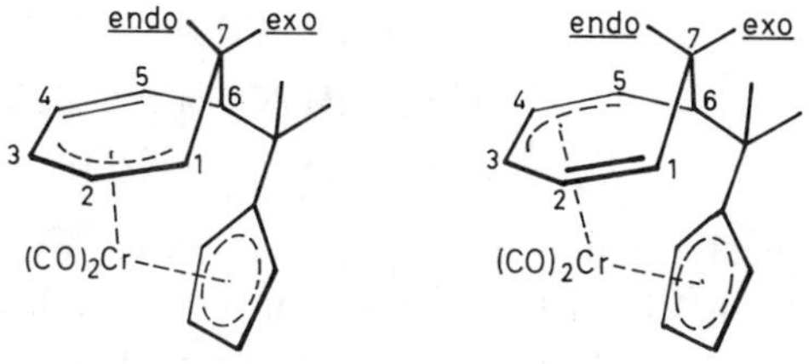

$$\underline{70b,c,} \quad \underline{71} \qquad\qquad\qquad \underline{72}$$

	70b	70c	71, 72
endo	CH_3	OCH_3	H
exo	H	H	OCH_3

V

PHOTOREACTIONS OF $[(\eta^5\text{-}C_5H_5)M(CO)_3CH_3]$ (M = Mo, W) WITH CONJUGATED DIENES

A. Reactions of $[(\eta^5\text{-}C_5H_5)Mo(CO)_3CH_3]$ with 1,3-Butadiene and 1,3-Cyclopentadiene

No simple η^2- or η^4-diene complexes could be detected from the reaction of tricarbonyl-η^5-cyclopentadienylmethylmolybdenum (73) (87,88) with 1,3-butadiene (1a). The reaction mixture contains four enyl complexes (74–77). The enyl ligands are formed from 1,3-butadiene by acetyl (74–76) or methyl

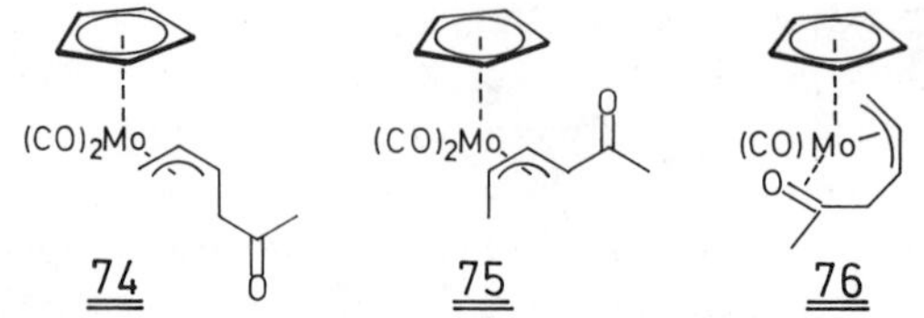

$\underline{74}$ $\qquad$ $\underline{75}$ $\qquad$ $\underline{76}$

group transfer to C-1 (77) (89) [Eq. (40)]. The major product of the reaction is 76, a complex with the $\eta^{3:2}$-(Z)-2-hexen-1-yl-5-one chelate ligand. An η^3-enyl system encompasses C-1 to C-3, and the ketonic C=O double bond is η^2-coordinated to molybdenum. In 74, the same ligand is present, but instead of the keto function, a second CO ligand is coordinated to the chromium. The η^3-(Z, E)-3-hexen-2-yl-5-one complex (75) is formed from 74 by a 1,4-H shift. The η^3-(E)-2-penten-1-yl complex (76) is obtained as a minor by-product.

$[(\eta^5\text{-}C_5H_5)Mo(CO)_3CH_3] + C_4H_6 \rightarrow$

$\qquad$ 73 $\qquad\qquad\qquad$ 1a

$\qquad\qquad$ $[(\eta^5\text{-}C_5H_5)Mo(CO)_2(\eta^3\text{-}C_4H_6COCH_3]$ 74, 75

$\qquad\qquad$ $[(\eta^5\text{-}C_5H_5)Mo(CO)(\eta^{3:2}\text{-}C_4H_6COCH_3]$ 76

$\qquad\qquad$ $[(\eta^5\text{-}C_5H_5)Mo(CO)_2(\eta^3\text{-}C_4H_6CH_3]$ $\qquad$ 77 $\qquad$ (40)

The photochemical reaction corresponds to facile insertion of 1a into $[(\eta^5\text{-indenyl})Mo(CO)_3(CH_3)]$ with formation of $[(\eta^5\text{-indenyl})Mo(CO)_2(\eta^3\text{-}(Z)\text{-2-hexen-1-yl-5-one})]$ (90). Cationic complexes of the type $[(\eta^5\text{-}C_5H_5)\text{-}M(\eta^{3:n}\text{-enylketone})]$ $[PF_6]$ (M = Rh, Ir) have been prepared recently. An η^3-coordination of the enyl moiety and of the carbonyl function by a free pair of electrons has been discussed (91).

Only with 1,3-cyclopentadiene (1h) does 73 give two η^4-1,3-cyclopentadiene complexes [Eq. (41)]. The two products are the thermo- and photolabile acetylcarbonyl-η^4-1,3-cyclopentadiene-η^5-cyclopentadienylmolybdenum **(78)**, and the moderately stable carbonyl-η^4-1,3-cyclopentadiene-η^5-cyclopentadienyl-molybdenum **(79)** *(92,93)*.

η^4-Cyclopentadiene complexes are well known, for example, $[M(\eta^5\text{-}C_5H_5)(\eta^4\text{-}C_5H_6)]$ [M = Co *(94)*, Rh, Ir *(95)*] or $[M(CO)_3(\eta^4\text{-}C_5H_6)]$ [M = Fe *(96–98)*, Ru *(99)*]. The latter readily release CO, and hydride is transferred from the η^4-C_5H_6 ligand to the metal. The $[(\eta^5\text{-}C_5H_5)M(CO)_2(H)]$ complexes evolve hydrogen and form $[(\eta^5\text{-}C_5H_5)_2M_2(CO)_4]$. Similarly, compound **78** loses CO by consecutive photoreaction and hydride transfer, with the molybdenocene derivative acetylbis(η^5-cyclopentadienyl)hydridomolybdenum **(80)** being formed [Eq. (42)]. Prolonged UV irradiation substantially increases the yield of **80**. This complex is also accessible from $[(\eta^5\text{-}C_5H_5)_2Mo(H)MgBr(THF)_2]$ by treatment with acetyl chloride *(100,101)*.

$$[(\eta^5\text{-}C_5H_5)Mo(CO)_3CH_3] + C_5H_6 \rightarrow [(\eta^5\text{-}C_5H_5)Mo(CO)(COCH_3)(\eta^4\text{-}C_5H_6)]\ \textbf{78}$$

$$\textbf{73} \qquad\qquad \textbf{1h} \qquad\qquad [(\eta^5\text{-}C_5H_5)Mo(CO)(CH_3)(\eta^4\text{-}C_5H_6)]\quad \textbf{79}\quad (41)$$

$$[(\eta^5\text{-}C_5H_5)Mo(CO)(COCH_3)(\eta^4\text{-}C_5H_6)] \rightarrow [(\eta^5\text{-}C_5H_5)_2MoH(COCH_3)] + CO\quad (42)$$

$$\textbf{78} \qquad\qquad\qquad\qquad \textbf{80}$$

When $[\{\eta^5\text{-}C_5(CH_3)_5\}Mo(CO)_3(CH_3)]$ **(81)** is reacted photochemically with **1a**, only the substitution product $[\{\eta^5\text{-}(CH_3)_5C_5\}Mo(CO)(CH_3)(\eta^4\text{-}C_4H_6)]$ **(82)** is obtained *(102)* [Eq. (43)]. The molecular structure of **82** was determined by X-ray diffraction (Fig. 9). The arrangement of the local fivefold axis of the η^5-$C_5(CH_3)_5$ ligand and of the three remaining ligands

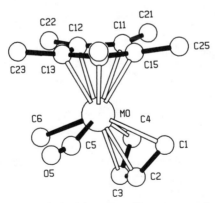

Fig. 9. Molecular structure of η^4-1,3-butadienecarbonylmethyl-η^5-pentamethylcyclopentadienylmolybdenum **(82)** *(102)*: Mo-1—C-1 243.3, Mo-1—C-2 228.6, Mo-1—C-3 229.1, Mo-1—C-4 221.9, Mo-1—C-5 210.2, Mo-1—C-6 214.5, C-1—C-2 145.3, C-2—C-3 143.2, C-3—C-4 138.9 pm.

is square pyramidal, if η^4-butadiene is regarded as a bidentate ligand. The η^5-$C_5(CH_3)_5$ and η^4-butadiene ligands are oriented in the o form. Similar molybdenum coordination spheres are present in $[(\eta^5$-$C_5H_5)Mo(Cl)\{P(C_2H_5)_3\}(\eta^4$-$endo$-$C_5H_5C_2H_5)]$ *(103)* and in $[(\eta^5$-$C_5H_5)Mo\{(C_6H_5)_2$-$PC_2H_4P(C_6H_5)_2\}(\eta^4$-diene)][PF$_6$] *(104–106)*.

$$[\{\eta^5\text{-}(CH_3)_5C_5\}Mo(CO)_3(CH_3)] + C_4H_6 \;\rightarrow$$

$$\qquad\qquad\textbf{81}\qquad\qquad\qquad\textbf{1}$$

$$[\{\eta^5\text{-}(CH_3)_5C_5\}Mo(CO)(CH_3)(\eta^4\text{-}C_4H_6)] + 2\,CO \quad (43)$$

$$\textbf{82}$$

B. Reactions of $[(\eta^{5:1}$-$C_5H_4C_2H_4)Mo(CO)_3]$ with Dienes

In contrast to **73**, the related tricarbonyl[2-(η^5-cyclopentadienyl)ethyl]-molybdenum (**83**) *(107)* forms fairly stable diene complexes with **1a–1c** in a stepwise reaction, first η^2- and, after prolongated irradiation, η^4-coordinated [Eq. (44)]. Only to a minor extent is insertion of the dienes **1a** and **1b** into the Mo—C σ bond observed in the complexes **86a** and **86b** *(108)*.

$$[(\eta^{5:1}\text{-}C_5H_4C_2H_4)Mo(CO)_3] + \text{diene} \;\rightarrow$$

$$\qquad\textbf{83}\qquad\qquad\qquad\textbf{1a–1c}$$

$$[(\eta^{5:1}\text{-}C_5H_4C_2H_4)Mo(CO)_2(\eta^2\text{-diene})] \quad \textbf{84a–84c}$$

$$[(\eta^{5:1}\text{-}C_5H_4C_2H_4)Mo(CO)(\eta^4\text{-diene})] \quad \textbf{85a–85c}$$

$$[(\eta^{5:3}\text{-}C_5H_4C_6H_{10})Mo(CO)_2] \qquad\qquad \textbf{86a}$$

$$[(\eta^{5:3}\text{-}C_5H_4C_7H_{12})Mo(CO)_2] \qquad\qquad \textbf{86b} \quad (44)$$

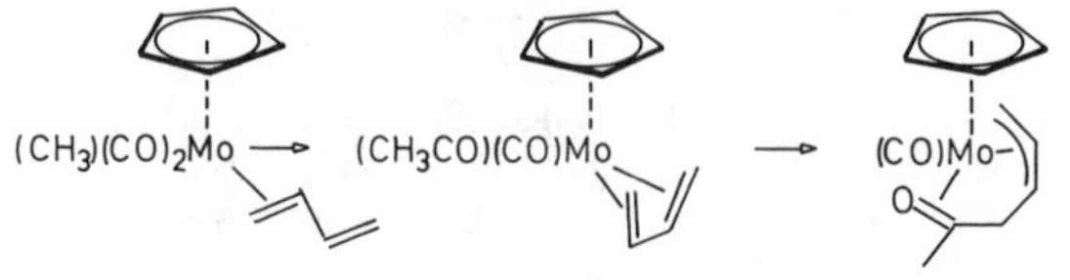

This reaction elucidates the mechanism of the photoreaction of **73** with dienes. In the first step, **73** loses one carbonyl ligand with formation of the reactive 16-electron species $[(\eta^5$-$C_5H_5)Mo(CO)_2CH_3]$ (**87**) *(109–113)*, which adds a diene molecule. η^2-Diene complexes $[(\eta^5$-$C_5H_5)Mo(CO)_2CH_3$-$(\eta^2$-diene)] (**88**) are quite likely as intermediates. Coordination of the free C=C double bond of the η^2-diene ligand causes insertion of CO into the Mo—C σ bond and leads, with **1h** as diene, to **78**. Photochemical loss of CO would explain the formation of **79**. With **1a** as the diene component, complexes comparable with **78** and **79** are obviously unstable, and they rearrange by insertion of the diene into the Mo—C σ bond of the acetyl group to give **76** or by simultaneous addition of CO to yield **77**.

C. Stereochemistry and Dynamic Behavior of [($\eta^{5:1}$-C₅H₄C₂H₄)Mo(CO)₂(η^2-diene)] Complexes

The coordination sphere of molybdenum in $[(\eta^{5:1}\text{-C}_5\text{H}_4\text{C}_2\text{H}_4)\text{Mo(CO)}_2\text{-}(\eta^2\text{-diene})]$ complexes can be considered to be quasi-square pyramidal like that of **82**. A strong distortion is to be expected from the chelating $\eta^{5:1}$-C₅H₄C₂H₄ ligand, which pulls together the apex and one basal position of the coordination pyramid. In a countermove, the space for ligands in the basal position trans to the Mo—C σ bond is substantially enlarged. Thus not only dienes can be coordinated to the molybdenum in **80** instead of the *trans*-CO ligand, as already mentioned, but also bulky olefins like *cis*-2-butene (*114*).

In spite of the enlarged space for η^2-diene ligands in **84**, there is a distinct preference of coordination of the least substituted C=C double bond. With (*E*)-1,3-pentadiene (**1b**) or 2-methyl-1,3-butadiene (**1c**), only one η^2-diene complex for each species is obtained. This seems to be a general feature of the ligand behavior of substituted conjugated dienes, as demonstrated also for $[(\eta^5\text{-C}_5\text{H}_5)\text{Ni(CH}_3)(\eta^2\text{-diene})]$ complexes (*115*).

84b   **84c**

At low temperatures, two rotamers can be detected for compounds **84a–84c** by NMR spectroscopy, with the free double bond turned toward the five-membered ring (syn) or away from it (anti). The hindered syn–anti isomerization becomes fast at room temperature, and is reasonably explained by rotations of the η^2-diene ligands around their coordination axis. It is important to note, that there is no interchange between the free and the coordinated C=C double bonds during this isomerization. Furthermore, the chirality, caused by the unsymmetrically substituted coordinated C=C double bonds of the η^2-diene ligands, is preserved in the complexes. This is convincingly monitored by the signals of the diastereotopic protons of the $\eta^{5:1}$-C₅H₄C₂H₄ ligands. This proves a hindered rotation of the η^2-diene ligands around the axis of coordination without a rotation around the C=C axis at a comparable rate. The barriers of activation were $\Delta G_{300}^* = 62.0 \pm 0.2$ kJ/mol for **84a**, $\Delta G_{280}{}^* = 66.2 \pm 0.2$ kJ/mol for **84b**, and $\Delta G_{298}{}^* = 61.0 \pm 0.2$ kJ/mol for **84c**.

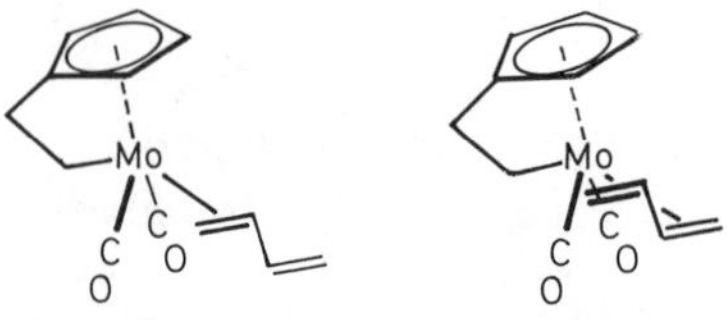

anti   syn

84a

D. Stereochemistry of [($\eta^{5:1}$-$C_5H_4C_2H_4$)Mo(CO)(η^4-diene)] Complexes

Stereochemically, the [($\eta^{5:1}$-$C_5H_4C_2H_4$)Mo(CO)(η^4-diene)] complexes **85a–85c** are closely related to the η^2-diene complexes **84a–84c**, with the free C=C double bond in the former coordinated to the molybdenum instead of a CO ligand. Two rotamers can be expected, with C-1 and C-4 of the dienes pointing toward the five-membered ring (*o*) or away from it (*u*).

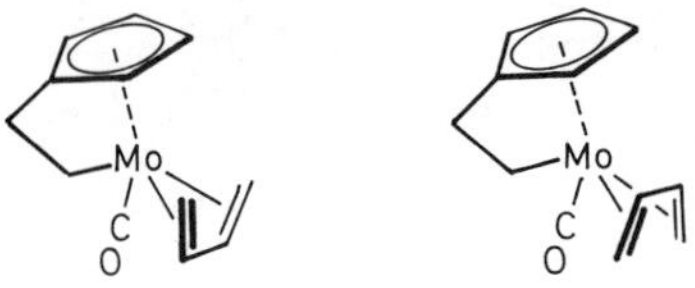

o   u

85a

The unsymmetrical diene ligands in **85b** and **85c** may form two diastereomers with the less substituted double bond in the cis or trans configuration with respect to the Mo—C σ bond. For **85b**, only the isomer with the less substituted double bond cis to this bond is found. Compound **85c** forms both isomers, but the cis isomer is distinctly preferred. In contrast with the fluxional η^2-diene complexes **84a–84c**, compounds **85a–85c** do not show temperature-dependent ^{1}H-NMR spectra for the η^4-diene ligands. However, this does not disprove a hindered rotation of the η^4-diene ligands when one of the two conformers (*o* or *u*) is energetically more favored than the other.

E. Reactions of [(η^5-C_5H_5)W(CO)$_3CH_3$] with Dienes

With tricarbonyl-η^5-cyclopentadienylmethyltungsten (**89**) (*87*), the tungsten homolog of **73**, the photoreactions with dienes proceed quite normally.

Primarily, acetylcarbonyl-η^5-cyclopentadienyl-η^4-dienetungsten complexes (**90**) are obtained, together with minor amounts of the corresponding methyl complexes (**91**) (*89,116*) [Eq. (45)].

$[(\eta^5\text{-}C_5H_5)W(CO)_3CH_3]$ +   diene   →

89   1a–1c, 1g, 1h, 1t

$[(\eta^5\text{-}C_5H_5)W(CO)(COCH_3)(\eta^4\text{-diene})]$ **90a–90c, 90g, 90h, 90t**

$[(\eta^5\text{-}C_5H_5)W(CO)(CH_3)(\eta^4\text{-diene})]$ **91a–91c, 91g, 91h, 91t**   (45)

F. Hindered Diene Rotation in [(η^5-C₅H₅)W(CO)(COCH₃)-(η^4-diene)] Complexes

The quasi-square pyramidal η^4-diene complexes $[(\eta^5\text{-}C_5H_5)W(CO)(COCH_3)(\eta^4\text{-diene})]$ (**90a–90c, 90g, 90h,** and **90t**) form two different isomers with the η^4-diene ligands in an *o* or *u* orientation with respect to the apical η^5-C₅H₅ ring, which are detected in the NMR spectra at 200 K. For **87a**, both rotamers have approximately the same populations. Thus the combined steric and electronic effects on the η^4-diene ligands must be very similar in both isomers, in sharp contrast to the related complexes **85a–85c**. Diene ligands like **1h**, with a methylene group between C-1 and C-4, show a preference for the *u* isomer on steric grounds. The complexes **90a–90c, 90g, 90h, 90t** contain four different ligands in a nonplanar arrangement and are therefore chiral.

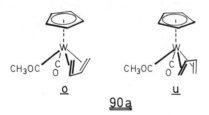

At room temperature the *o–u* isomerizations of the η^4-diene complexes **90** are fast (Table VII). As can be seen for the ^{1}H-NMR signals in **90a**, the η^4-1,3-butadiene ligand retains six different proton signals, proving the maintainance of the chirality. Furthermore, there is no evidence for an exchange of protons between the *E* and *Z* positions on C-1 and C-4. A hindered rotation of the diene ligands around the coordination axis, showing always the same side to the tungsten atom, is in agreement with the experimental results.

For the cationic complexes $[(\eta^5\text{-}C_5H_5)Mo(CO)_2(\eta^4\text{-diene})]^+$, an alternative movement for the diene ligands has been considered, with a metallacyclopentene transition state (*117*). Such transition states or intermediates are well-known for η^4-diene complexes of the early transition metals (*118*). In the

TABLE VII

ACTIVATION PARAMETERS FOR η^4-DIENE LIGAND ROTATION IN $[(\eta^5$-$C_5H_5)W(CO)(COCH_3)(\eta^4$-diene)] COMPLEXES

	Complex	Population (%)		ΔG_{300}^* (kJ/mol)
		o	u	
90a	$[(\eta^5$-$C_5H_5)W(CO)(COCH_3)(\eta^4$-$C_4H_6)]$	54	46	60.0
90b	$[(\eta^5$-$C_5H_5)W(CO)(COCH_3)(\eta^4$-$C_5H_8)]$	87	13	59.7
90c	$[(\eta^5$-$C_5H_5)W(CO)(COCH_3)(\eta^4$-$C_5H_8)]$	95	5	57.8
90t	$[(\eta^5$-$C_5H_5)W(CO)(COCH_3)(\eta^4$-$C_6H_{10})]$	79	21	61.0
90g	$[(\eta^5$-$C_5H_5)W(CO)(COCH_3)(\eta^4$-$C_6H_8)]$	—	—	< 30.0
90h	$[(\eta^5$-$C_5H_5)W(CO)(COCH_3)(\eta^4$-$C_5H_6)]$	91	9	58.5

case of **90a–90h**, however, this kind of movement can be excluded with certainty.

G. Thermal Rearrangements of $[(\eta^5$-$C_5H_5)Mo(CO)(COCH_3)(\eta^4$-$C_5H_6)]$

In hydrocarbon solution, $[(\eta^5$-$C_5H_5)Mo(CO)(COCH_3)(\eta^4$-$C_5H_6)]$ (**78**) is unstable at room temperature and rearranges to the molybdenocene derivatives η^5-acetylcyclopentadienyl-η^5-cyclopentadienyldihydridomolybdenum (**92**) and η^5-acetylcyclopentadienylcarbonyl-η^5-cyclopentadienylmolybdenum (**93**) (92,93) [Eq. (46)]. Small amounts of the green dicarbonyl-η^5-cyclopentadienyl-η^3-exo-4-acetylcyclopentenylmolybdenum (**94**) are also present in the reaction mixture.

$[(\eta^5$-$C_5H_5)Mo(CO)(COCH_3)(\eta^4$-$C_5H_6)]$ →

78

$[(\eta^5$-$C_5H_4COCH_3)(\eta^5$-$C_5H_5)MoH_2]$ **92**

$[(\eta^5$-$C_5H_4COCH_3)(\eta^5$-$C_5H_5)Mo(CO)]$ **93**

$[(\eta^5$-$C_5H_5)Mo(CO)_2(\eta^3$-exo-$C_5H_6COCH_3)]$ **94**   (46)

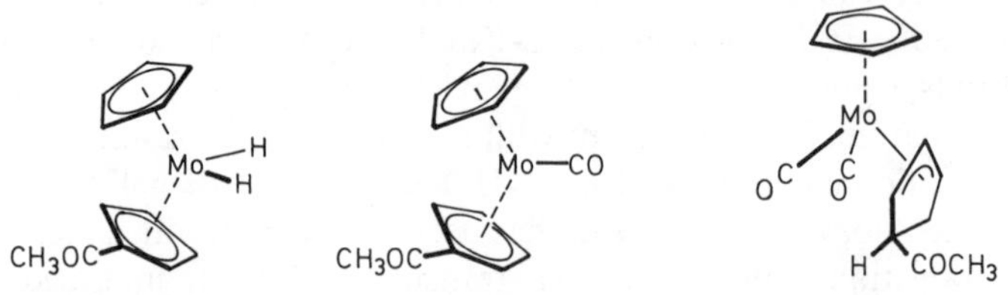

A conceivable intermediate of the same type as **76a**, $[(\eta^5\text{-}C_5H_5)Mo(CO)\text{-}(\eta^{3:2}\text{-}C_5H_6COCH_3)]$ (**76h**), in equilibrium with **78** [Eq. (47)] could explain the formation of the products **92–94**. By two successive hydride transfers from the $\eta^{3:2}\text{-}C_5H_6COCH_3$ ligand to molybdenum, via $[(\eta^5\text{-}C_5H_5)Mo(CO)\text{-}H(\eta^4\text{-}C_5H_5COCH_3)]$ (**95**) and substitution of the CO ligand, **92** is produced [Eqs. (48) and (49)]. Complex **92** is susceptible to substitution of the two hydride ligands by CO, present in the solution, as is well-known for $[(\eta^5\text{-}C_5H_5)_2MoH_2]$ (*119–121*), and so yields the carbonyl complex **93** [Eq. (50)].

$$[(\eta^5\text{-}C_5H_5)Mo(CO)(COCH_3)(\eta^4\text{-}C_5H_6)] \;\rightarrow\; [(\eta^5\text{-}C_5H_5)Mo(CO)(\eta^{3:2}\text{-}C_5H_6COCH_3)] \quad (47)$$

$$\qquad\quad \textbf{78} \qquad\qquad\qquad\qquad\qquad\qquad\qquad\qquad \textbf{76h}$$

$$[(\eta^5\text{-}C_5H_5)Mo(CO)(\eta^{3:2}\text{-}C_5H_6COCH_3)] \;\rightarrow\; [(\eta^5\text{-}C_5H_5)Mo(CO)H(\eta^4\text{-}C_5H_5COCH_3)] \quad (48)$$

$$\qquad\quad \textbf{76h} \qquad\qquad\qquad\qquad\qquad\qquad\qquad \textbf{95}$$

$$[(\eta^5\text{-}C_5H_5)Mo(CO)H(\eta^4\text{-}C_5H_5COCH_3)] \;\rightarrow$$

$$\textbf{95}$$

$$[(\eta^5\text{-}C_5H_4COCH_3)(\eta^5\text{-}C_5H_5)MoH_2] + CO \quad (49)$$

$$\textbf{92}$$

$$[(\eta^5\text{-}C_5H_4COCH_3)(\eta^5\text{-}C_5H_5)MoH_2] + CO \;\rightarrow$$

$$\textbf{92}$$

$$[(\eta^5\text{-}C_5H_4COCH_3)(\eta^5\text{-}C_5H_5)Mo(CO)] + H_2 \quad (50)$$

$$\textbf{93}$$

By D-NMR spectroscopy, hindered rotation of the acetyl substituents in **92** ($\Delta G_{208}{}^* = 40.3 \pm 0.2$ kJ/mol) and in **93** ($\Delta G_{208}{}^* = 43.1 \pm 0.2$ kJ/mol) has been detected, thus indicating bond order higher than unity for the acetyl–ring C—C bond, due to a "fulvenolate" character of the acetylcyclopentadienyl ligand.

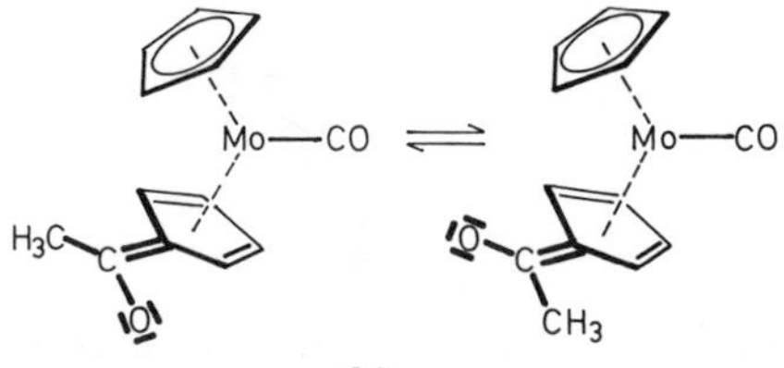

$$\underline{\underline{93}}$$

The formation of **94** can be understood by reaction of **76h** with CO, present in the solution [Eq. (51)]. The *exo* position of the acetyl group suggests an intermolecular transfer of the latter or of the 4-acetylcyclopentenyl ligand.

Unsubstituted $[(\eta^5\text{-}C_5H_5)Mo(CO)_2(\eta^3\text{-}C_5H_7)]$ has been prepared by different routes, photochemically (13) or by reduction of molybdenocene derivatives (122,123).

$[(\eta^5\text{-}C_5H_5)Mo(CO)(\eta^{3:2}\text{-}C_5H_6COCH_3)] + CO \rightarrow$

76h

$$[(\eta^5\text{-}C_5H_5)Mo(CO)_2(\eta^{3:2}\text{-}exo\text{-}C_5H_6COCH_3)] \quad (51)$$

94

H. Thermal Reactions of $[(\eta^5\text{-}C_5H_5)Mo(CO)(COCH_3)$ $(\eta^4\text{-}C_5H_6)]$ with Carbon Monoxide, Trimethylphosphite, and Conjugated Dienes

Acetylcarbonyl-η^4-1,3-cyclopentadiene-η^5-cyclopentadienylmolybdenum (**78**) is thermo- and photolabile, so one can also expect a high reactivity toward potential ligands. The reactions of carbon monoxide, trimethylphosphite, and several conjugated dienes have been studied in this respect (124).

With CO, a fast addition reaction and transfer of the acetyl group from the molybdenum to the η^4-1,3-cyclopentadiene ligand is observed, and the red endo isomer of **94**, $[(\eta^5\text{-}C_5H_5)Mo(CO)_2(\eta^4\text{-}endo\text{-}C_5H_6COCH_3)]$ (**96**), is formed in high yields [Eq. (52)]. Trimethylphosphite releases the η^4-1,3-

$[(\eta^5\text{-}C_5H_5)Mo(CO)(COCH_3)(\eta^4\text{-}C_5H_6)] + CO \rightarrow$

78

$$[(\eta^5\text{-}C_5H_5)Mo(CO)_2(\eta^3\text{-}endo\text{-}C_5H_6COCH_3)] \quad (52)$$

96

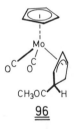

<u>96</u>

cyclopentadiene ligand in **78**. In a retro insertion reaction, a mixture of the cis and trans isomers of $[(\eta^5\text{-}C_5H_5)Mo(CO)_2CH_3\{P(OCH_3)_3\}]$ (**97** and **98**) is formed [Eq. (53)]. Although $[(\eta^5\text{-}C_5H_5)Mo(CO)_3(COCH_3)]$ is unstable (125–127), substituted complexes of the form $[(\eta^5\text{-}C_5H_5)Mo(CO)_2\text{-}L(COCH_3)]$ are well-known and have been thoroughly studied (128–134). Formation of **97** and **98** proceeds owing to the absence of carbon monoxide.

$[(\eta^5\text{-}C_5H_5)Mo(CO)(COCH_3)(\eta^4\text{-}C_5H_6)]$ + $P(OCH_3)_3$ →

78

$$[(\eta^5\text{-}C_5H_5)Mo(CO)_2CH_3\{P(OCH_3)_3\}] + C_5H_6 \quad (53)$$

cis **97**
trans **98**

Substitution reactions are also observed with the dienes **1a–1c**, **1g** and **1t** [Eq. (54)]. The η^4-1,3-cyclopentadiene ligand is displaced by the dienes, and the acetyl group immediately transfers to C-1 or C-4 of the η^4-diene ligands in the intermediates **78a–78c**, **78g**, and **78t**. Carbonyl-η^5-cyclopentadienyl-$\eta^{3:2}$-(Z)-2-hexen-1-yl-5-onemolybdenum (**76a**) and the corresponding derivatives are obtained in high yield (*124*). With the unsymmetrically substituted dienes **1b** and **1c**, two different isomers each, **76b**, **76b′**, **76c** and **76c′** are formed

$[(\eta^5\text{-}C_5H_5)Mo(CO)(COCH_3)(\eta^4\text{-}C_5H_6)]$ + $C_4H_{6-n}(CH_3)_n$ →

78   **1a–1c, 1g, 1t**

$$[(\eta^5\text{-}C_5H_5)Mo(CO)\{\eta^{3:2}\text{-}C_4H_{6-n}(CH_3)_nCOCH_3\}] + C_5H_6 \quad (54)$$

76a–76c, 76g, 76t, 76b′, 76c′   **1h**

and can be separated by column chromatography. Addition of the acetyl group shows no exclusive preference for attack at C-1 or C-4 of the η^4-diene ligands in the intermediates **78b** and **78c**.

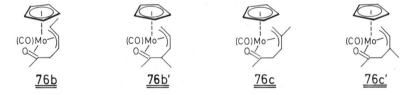

76b   **76b′**   **76c**   **76c′**

1,3-Cyclopentadiene (**1h**) has a very different reactivity toward **78** than have the dienes **1a–1c**, **1g** and **1t**. The three isomeric complexes **99–101**

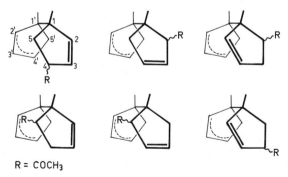

R = COCH₃

are produced (*124*) [Eq. (55)]. In addition, the acetyl group is transferred from molybdenum to the $C_{10}H_{12}$ hydrocarbon ligand. For the $\eta^{3:2}$-$C_{10}H_{12}COCH_3$ ligands formed, several isomeric forms have to be con-

$$[(\eta^5\text{-}C_5H_5)Mo(CO)(COCH_3)(\eta^4\text{-}C_5H_6)] + C_5H_6 \rightarrow$$

78   **1h**

$$[(\eta^5\text{-}C_5H_5)Mo(CO)(\eta^{3:2}\text{-}C_{10}H_{12}COCH_3)] \quad (55)$$

99, 100, 101

sidered. Molecular models show, however, that only two of them fit properly into the coordination sphere of the molybdenum. The nature of the three different isomers isolated (**99–101**) were determined by ^{1}H-NMR spectroscopy and, in the case of **99**, also by X-ray structure analysis (Fig. 10). The molecular structure of **99** shows an almost total coverage of the molybdenum by the centrosymmetrically coordinated η^5-C_5H_5 and the chelating $\eta^{3:2}$-$C_{10}H_{12}COCH_3$ ligand.

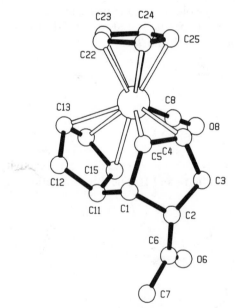

FIG. 10. Molecular structure of $\eta^{3'-5':4,5}$-*endo*-acetyl-2,2',3-trihydrobis(1,1'-cyclopentadienyl)carbonyl-η^5-cyclopentadienylmolybdenum (**99**) (*124*): Mo-1—C-4 231.1, Mo-1—C-5 227.2, Mo-1—C-13 231.3, Mo-1—C-14 217.8, Mo-1—C-15 228.6, C-4—C-5 142.4, C-13—C-14 141.6, C-14—C-15 143.1 pm.

VI

PHOTOREACTIONS OF $[(\eta^5\text{-}CH_3C_5H_4)M(CO)_3]_2$ (M = Mo, W) WITH CONJUGATED DIENES

Ultraviolet activation of the complexes hexacarbonylbis(η^5-cyclopentadienyl)dimolybdenum and -ditungsten (135) has been studied in detail (136–138). In addition to homolytic cleavage of the metal–metal bonds, loss of carbon monoxide has also been observed. The products of the photochemical reactions of $[(\eta^5\text{-}CH_3C_5H_4)M(CO)_3]_2$ (M = Mo, W) with the dienes 1a–1c, 1g and 1t and 1,3,5-cycloheptatriene (68a) differ markedly from those obtained from the thermal reaction of $[(\eta^5\text{-}C_5H_5)Mo(CO)_n]_2$ (n = 2,3) with dienes (139,140).

A. Reaction of $[(\eta^5\text{-}CH_3C_5H_4)Mo(CO)_3]_2$ with Conjugated Dienes

Due to low solubility in hydrocarbon solvents at low temperatures, $[(\eta^5\text{-}C_5H_5)Mo(CO)_3]_2$ is not well suited for photoreactions. In order to increase the solubility, the methylcyclopentadienyl derivative, $[(\eta^5\text{-}CH_3C_5H_4)\text{-}Mo(CO)_3]_2$ (102), has been used. After photoreaction with 1a–1c and 1t in n-pentane solution, the reaction mixtures contain two products each, dinuclear tricarbonylbis(η^5-methylcyclopentadienyl)(η^4-diene)molybdenum (103) and dicarbonyl-η^3-enyl-η^5-methylcyclopentadienylmolybdenum complexes (104) (141) [Eq. (56)].

$[(\eta^5\text{-}CH_3C_5H_4)Mo(CO)_3]_2$ +   diene   →

       102   1a–1c, 1t

           $[(\eta^5\text{-}CH_3C_5H_4)_2Mo_2(CO)_3(\eta^4\text{-diene})]$ 103a–103c, 103t

           $[(\eta^5\text{-}CH_3C_5H_4)Mo(CO)_2(\eta^3\text{-enyl})]$ 104a–104c, 104t   (56)

When solutions of 103a–103c and 103t are heated [Eq. (57)], the complexes release one carbonyl ligand and rearrange to $[(\eta^5\text{-}CH_3C_5H_4)_2Mo_2(CO)_2\text{-}(\mu\text{-}\eta^4\text{-diene})]$ (105a–105c and 105t), the η^5-methylcyclopentadienyl derivatives of the already known complex, $[(\eta^5\text{-}C_5H_5)_2Mo_2(CO)_2(\mu\text{-}\eta^4\text{-}s\text{-}cis\text{-}(Z)\text{-}1,3\text{-pentadiene})]$ (140). A similar bonding situation was also found for $[(\eta^5\text{-}C_5H_5)_2Co_2(\mu\text{-}\eta^4\text{-}s\text{-}cis\text{-butadiene})]$ (142).

$[(\eta^5\text{-}CH_3C_5H_4)_2Mo_2(CO)_3(\eta^4\text{-diene})]$ →

    103a–103c, 103t

        $[(\eta^5\text{-}CH_3C_5H_4)_2Mo_2(CO)_2(\mu\text{-}\eta^4\text{-diene})]$ + CO   (57)

        105a–105c, 105t

1,3-Cycloheptadiene (**1o**) reacts somewhat differently with complex **102** [Eq. (58)]. Although the reaction product has a formula corresponding with **103a**–**103c**, **103t**, it contains a μ-$\eta^{4:1}$-1,3-cycloheptadiene-1-yl group and a terminal hydrido ligand. 1,3,5-Cycloheptatriene (**69a**) with **102** forms under

$$[(\eta^5\text{-}CH_3C_5H_4)Mo(CO)_3]_2 + C_7H_{10} \rightarrow$$

$$\quad\quad\quad\quad 102 \quad\quad\quad\quad\quad\quad\quad 1o$$

$$[(\eta^5\text{-}CH_3C_5H_4)_2Mo_2(CO)_3(H)(\mu\text{-}\eta^{4:1}\text{-}C_7H_9)] + 3\,CO \quad (58)$$

$$\quad\quad\quad\quad\quad\quad\quad 106$$

photochemical conditions the mononuclear complex dicarbonyl-η^5-cyclopentadienyl-η^3-2,4-cycloheptadien-1-ylmolybdenum (**107**) [Eq. (59)].

$$[(\eta^5\text{-}CH_3C_5H_4)Mo(CO)_3]_2 + 2\,C_7H_8 \rightarrow$$

$$\quad\quad\quad\quad 102 \quad\quad\quad\quad\quad\quad\quad 69a$$

$$[(\eta^5\text{-}CH_3C_5H_4)Mo(CO)_2(\eta^3\text{-}C_7H_9)] + 2\,CO \quad (59)$$

$$\quad\quad\quad\quad\quad\quad\quad 107$$

B. Molecular Structure and Dynamics of Tricarbonyl-η^4-2,3-dimethyl-1,3-butadiene-bis(η^5-methylcyclopentadienyl)dimolybdenum

According to X-ray analysis, the molecular structure of $[(\eta^5\text{-}CH_3C_5H_4)_2\text{-}Mo_2(CO)_3(\eta^4\text{-}C_4H_4(CH_3)_2)]$ (**103t**) can be rationalized as consisting of a tricarbonyl-η^5-methylcyclopentadienylmolybdenum and an η^4-2,3-dimethyl-1,3-butadiene-η^5-methylcyclopentadienylmolybdenum moiety connected by a Mo—Mo σ bond (Fig. 11). The bond length is in the same range as for the parent complex $[(\eta^5\text{-}C_5H_5)_2Mo_2(CO)_6]$ [323.5(1) pm] (*143*). Formally, the former is a 17-, the latter a 15-electron species. The electron deficiency of the 15-electron entity is reduced by the bridging carbonyl ligand (at position 7), σ-bonded to Mo-1, and π-coordinated by its CO multiple bond to Mo-2. A second carbonyl ligand (at position 8) is bent toward Mo-2. The η^4-2,3-dimethyl-1,3-butadiene ligand is oriented in the o form with respect to the η^5-methylcyclopentadienyl ring. Compounds **103a**–**103c** are the first examples of homonuclear Mo complexes with μ-$\eta^{1:2}$-CO, four-electron ligand. This type of coordination for carbon monoxide was first detected for $[Mn_2(CO)_5\{(C_6H_5)_2PCH_2P(C_6H_5)_2\}]$ (*144a*). Meanwhile a limied number of heteronuclear complexes with μ-$\eta^{1:2}$-CO ligands have been prepared (*145*).

The NMR spectra of the complexes $[(\eta^5\text{-}CH_3C_5H_4)_2Mo_2(CO)_3(\eta^4\text{-}$ diene)] (**103a**–**103c** and **103t**) are temperature dependent. At low temperatures the spectra of **103a** and **103t** demonstrate chirality for the complexes,

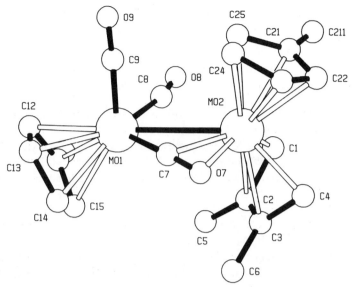

FIG. 11. Molecular structure of tricarbonyl-η^4-2,3-dimethyl-1,3-butadienebis(η^5-methyl-cyclopentadienyl)dimolybdenum (**103t**) (*141*): Mo-1—Mo-2 318.9, Mo-2—C-1 224.8, Mo-2—C-2 239.6, Mo-2—C-3 235.4, Mo-2—C-4 223.2, Mo-1—C-7 192.6, Mo-1—C-8 197.6, Mo-1—C-9 193.4, Mo-2—C-7 217.7, Mo-2—O-7 224.9, Mo-2—C-8 284.7 pm.

although the η^4-diene ligands are achiral. The chirality can be well understood on the basis of the molecular structure of **103t**. Opposite to the η^4-diene ligand, a four-electron π-CO bridge and a semibridging CO ligand are found, which cause C_1 symmetry for the complexes.

<u>103a</u>

By raising the temperature, the NMR signals of all diastereotopic protons collapse, according to a hindered ligand movement, by which an achiral transition state is passed. A reasonable explanation for the spectroscopic results can be found in an interchange of the μ-$\eta^{1:2}$-CO bridge and the CO semibridge. There are further examples for complexes with four-electron μ-$\eta^{1:2}$-CO bridges, in which a hindered exchange between the μ-$\eta^{1:2}$-CO bridge and an other carbonyl ligand is observed (*144b*). Activation barriers of $\Delta G_{257}{}^* = 52.2 \pm 2$ kJ/mol (**103a**), $\Delta G_{235}{}^* = 51.9 \pm 2$ kJ/mol (**103c**), and $\Delta G_{280}{}^* = 56.3 \pm 2$ kJ/mol (**103t**) were determined.

C. *Molecular Structure of Tricarbonyl(μ-$\eta^{4:1}$-1,3-cycloheptadien-1-yl)-hydridobis(η^5-methylcyclopentadienyl)dimolybdenum*

Stabilization of the tricarbonylbis(η^5-methylcyclopentadienyl)dimolybdenum unit by a conjugated diene can be very different from that found for **103t**, as is shown by the X-ray structure analysis of tricarbonyl(μ-$\eta^{4:1}$-1,3-cycloheptadien-1-yl)hydriodobis(η^5-methylcyclopentadienyl)dimolybdenum (**106**) (Fig. 12) (*141*). In complex **106**, the 1,3-cycloheptadiene has added oxidatively with C-1 bonded to Mo-1 and shows a normal η^4-coordination toward Mo-2. The exact position of the hydrido ligand could not be determined by the X-ray structure analysis, but its most probable location is near the bisection of the Mo-2—Mo-1—C-8 angle, based on the ligand positions on Mo-1. Another complex with a μ-$\eta^{4:1}$-1,3-butadiene-1-yl bridge was recently obtained from the reaction of $[Rh_2\{(i\text{-}C_3H_7)_2PC_2H_4P(C_3H_7)_2\}_2\text{-}(\mu\text{-}H)_2]$ with butadiene (*146*).

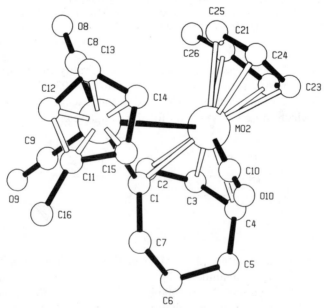

FIG. 12. Molecular structure of tricarbonyl-μ-$\eta^{4:1}$-1,3-cycloheptadien-1-ylhydridobis-(η^5-methylclopentadienyl)dimolybdenum (**106**) (*141*): Mo-1—Mo-2 313.3, Mo-1—C-1 220.8, Mo-2—C-1 224.7, Mo-2—C-2 221.0, Mo-2—C-3 223.3, Mo-2—C-4 234.6 pm; Mo-2—Mo-1—C-1 45.8°, C-1—Mo-1—C-9 70.7°, C-9—Mo-1—C-8 79.9°, C-8—Mo-1—Mo-2 92.6°.

D. Reactions of $[(\eta^5\text{-}CH_3C_5H_4)W(CO)_3]_2$ with Dienes

The photoreactions of hexacarbonylbis(η^5-methylcyclopentadienyl)tungsten (**108**) with the dienes **1b**, **1c**, **1g** and **1t** yield the tungsten complexes **109b**, **109c**, **109g**, and **109t** [Eq. (60)], which are homologs of **103**. In contrast,

$$(\eta^5\text{-}CH_3C_5H_4)W(CO)_3]_2 + \quad \text{diene} \quad \rightarrow$$

$$\quad\quad \textbf{108} \quad\quad\quad\quad \textbf{1b, 1c, 1g, 1t}$$

$$[(\eta^5\text{-}CH_3C_5H_4)_2W_2(CO)_3(\eta^4\text{-diene})] + 3\,CO \quad (60)$$

$$\textbf{109b, 109c, 109g, 109t}$$

$$\textbf{(1a)}$$

with 1,3-butadiene only a tetracarbonylbis(η^5-methylcyclopentadienyl)-ditungsten complex (**110**) was isolated [Eq. (61)]. Naively, one could ex-

$$[(\eta^5\text{-}CH_3C_5H_4)W(CO)_3]_2 + C_4H_6 \rightarrow$$

$$\quad \textbf{108} \quad\quad\quad\quad \textbf{1a}$$

$$[(\eta^5\text{-}CH_3C_5H_4)_2W_2(CO)_4(\mu\text{-}\eta^{3:1}\text{-}C_4H_6)] + 2\,CO \quad (61)$$

$$\textbf{110}$$

pect two different modes for the substitution of two carbonyl ligands in $[(\eta^5\text{-}CH_3C_5H_4)W(CO)_3]_2$ (**108**) by one diene molecule, in analogy with the isomeric dinuclear $[Re_2(CO)_8(\eta^4\text{-}1,3\text{-butadiene})]$ complexes (147,148). One of the rhenium complexes contains s-trans-1,3-butadiene as a bridging ligand between two Re atoms of the $(CO)_4Re—Re(CO)_4$ fragment, the other a s-cis-η^4-1,3-butadiene ligand coordinated to one Re atom of the $(CO)_5Re—Re(CO)_3$ fragment. In μ-$\eta^{4:1}$-2-buten-1,1-diyltetracarbonylbis(η^5-methylcyclopentadienyl)ditungsten (**110**), the 1,3-butadiene molecule is found in a further coordination mode (Fig. 13).

By an 1,4-H shift, 1,3-butadiene is transformed into a 2-buten-1,1-diyl ligand, which is σ-bonded by C-1 to W-1 and π-bonded by C-1 through C-3 to W-2. In a formal sense, the coordination sphere of W-2 can be compared with those in η^4-1,3-butadienecarbonylmethyl-η^5-pentamethylcyclopentadienylmolybdenum (**82**). The W-1—C-1 bond length is distinctly shorter than a normal W—C single bond. This and the bond angles on the 2-buten-1,1-diyl bridge can be explained satisfactorily by involving a W-1=C-1 double bond. The bonding properties in **107** are best described as a metalladiene, encompassing W-1 and C-1 through C3, which is π-bonded to W-2. By taking into account the carbonyl bridge between W-1 and W-2, the 18-electron rule is fulfilled for W-1 and W-2. A π-character for the W-1—W-2 bond therefore has to be proposed.

The binding properties of $[Re_2(CO)_8\{\mu\text{-}\eta^{1:3}\text{-}CH{=}CH—C(CH_3)_2\}]$ (149), prepared from $[Re_2(CO)_8(H)(\mu\text{-}\eta^{1:2}\text{-}C_2H_3)]$ (150) and 3,3-dimethylcyclo-

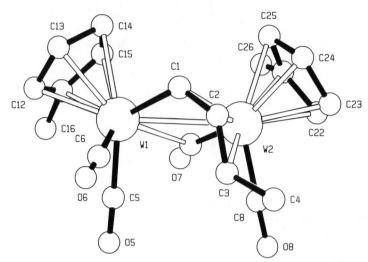

FIG. 13. Molecular structure of μ-$\eta^{4:1}$-2-butene-1,1-diyltetracarbonylbis(η^5-methylcyclo-pentadienyl)ditungsten (110) (141): W-1—W-2 311.0, W-1—C-1 212.2, W-2—C-1 226.7, W-2—C-2 226.2, W-2—C-3 238.3, W-2—C-7 201.1, W-1—C-7 251.5, C-1—C-2 142.0, C-2—C-3 143.2, C-3—C-4 155.1 pm; W-2—C-7—O-7 153.9°.

propene, are closely related to those of 107; however, there is no semibridg-ing CO ligand present. 1,3,5-Cycloheptatriene (69a) with 108 yields dicar-bonyl-η^5-methylcyclopentadienyl-η^3-2,4-cycloheptadien-1-yltungsten (111) [Eq. (62)].

$$[(\eta^5\text{-}CH_3C_5H_4)W(CO)_3]_2 + C_7H_8 \rightleftharpoons$$

   109   69a

$$2\,[(\eta^5\text{-}CH_3C_5H_4)W(CO)_2(\eta^3\text{-}C_7H_9)] + 2\,CO \quad (62)$$

                                 111

VII

PHOTOREACTIONS OF $M_2(CO)_{10}$ (M = Mn, Re) WITH CONJUGATED DIENES

Photochemical reactions of the homoleptic carbonyls of manganese and rhenium with donor ligands have been thoroughly studied in recent years (2–4). The nature of the primary photoproducts of the complexes $M_2(CO)_{10}$ (M = Mn, Re) has been investigated by different techniques (151–165). The photochemical formation of two primary products has been unambiguously proved. Homolytic fission of the M—M bonds by UV irradiation yields the well-established $M(CO)_5$ radicals. On the other hand, by CO elimination the coordinatively unsaturated species $M_2(CO)_9$ are formed.

The preparative potential of the photochemical reactions of $Mn_2(CO)_{10}$ and $Re_2(CO)_{10}$ with conjugated dienes has hitherto been used only rarely (*51,147,166–168*). The reason for this might be that the reactions do not proceed in a simple fashion. Normally, mixtures of several products are obtained. Powerful separation techniques such as low-temperature column chromatography or high-performance liquid (HPL) chromatography are necessary to isolate the products, and this has thrown some light on the nature of the reactions.

A. *Reaction of* $Mn_2(CO)_{10}$ *with 1,3-Butadiene*

When decacarbonyldimanganese(0) (**112**) is irradiated photochemically at 253 K with 1,3-butadiene (**1a**) in *n*-hexane solution [Eq. (63)], four dinuclear reaction products are observed by analytical HPL chromatography (*169,170*). After a short period of irradiation, μ-$\eta^{3:1}$-2-butene-1-diylenneacarbonyldimanganese (**113a**) is the predominate reaction product. Prolonged

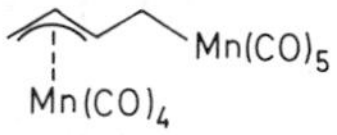

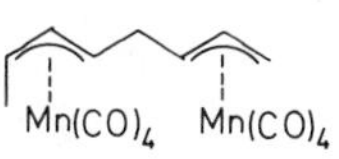

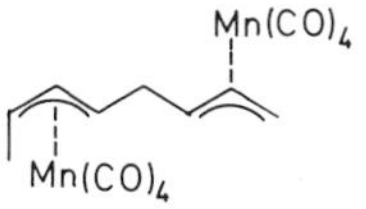

photoreaction yields two diastereomeric octacarbonyl-μ-$\eta^{3:3}$-(*E,E,Z*)-2,6-octadiene-1,5-diyldimanganese complexes (**114** and **115**). They differ with respect to the coordination of the two $Mn(CO)_4$ units to the (*E,E,Z*)-2,6-octadiene-1,5-diyl hydrocarbon bridge. A complete separation of **114** and **115** has not been possible as yet, owing to very similar retention times. A minor by-product of the reaction is the previously reported and structurally well-characterized complex μ-$\eta^{2:2}$-*s-trans*-1,3-butadieneoctacarbonyldimanganese(0) (**116**) (*166–168*).

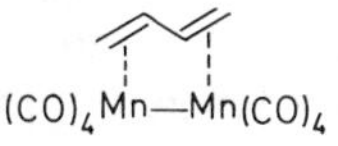

$$Mn_2(CO)_{10} + C_4H_6 \rightarrow [Mn_2(CO)_9(\eta^{3:1}\text{-}C_4H_6)] \quad \textbf{113a} \qquad (63)$$

$$\textbf{112} \qquad \textbf{1a} \qquad [Mn_2(CO)_8(\eta^{3:3}\text{-}C_8H_{12})] \quad \textbf{114,115}$$

$$[Mn_2(CO)_8(\eta^{2:2}\text{-}C_4H_6)] \quad \textbf{116}$$

Upon standing at room temperature, **114** and **115** rearrange in *n*-hexane solution into three different isomers. First, there is a $Z-E$ isomerization to give two diastereomeric octacarbonyl-μ-$\eta^{3:3}$-(E,E,E)-2,6-octadiene-1,5-diyl-dimanganese complexes (**117** and **118**). To a minor extent **114** and **115** transform, by an 1,4-H shift, into octacarbonyl-μ-$\eta^{3:3}$-(E,E,E)-2,5-octadiene-1,4-diyldimanganese (**119**). Complex **119** is the sole product when **114** and **115** are heated to reflux in *n*-hexane solution for 4 hours. In contrast with the (E,E,Z)- and (E,E,E)-2,6-octadiene-1,5-diyl complexes (**114** and **115**, and **117** and **118**), which are obtained as diastereomeric pairs of enantiomers, only one pair of enantiomers of **119**, with an (E,E,E)-2,5-octadiene-1,4-diyl bridge, was detected. Obviously, there is only one favorable possibility for the coordination of two $Mn(CO)_4$ fragments toward two linearly conjugated enyl units.

The formation of **113a** can be rationalized by an attack of the $[Mn(CO)_5 \cdot]$ radical (**120**) at C-1 of **1a** to give $[(CO)_5MnC_4H_6 \cdot]$. The latter then reacts with a second **120** molecule to yield an intermediate, $[(CO)_5MnC_4H_6Mn(CO)_5]$, which, like other carbonyl-η^1-enyl complexes, loses CO (*171–174*) and yields **113a**. A related species $[(CO)_5ReC_4H_6Re(CO)_5]$ was recently prepared (*175*). By photochemical displacement of CO in **113a** by **1a** and insertion into the Mn—C σ bond, compounds **114** and **115** are formed. The minor by-product **116** may come from the addition of **1a** to the coordinatively unsaturated fragment $[Mn_2(CO)_9]$ (**121**), with $[Mn_2(CO)_9(\eta^2\text{-}C_4H_6)]$ as an intermediate.

B. *Reactions of Mn$_2$(CO)$_{10}$ with 1,3-Pentadiene and Its Derivatives*

When the photoreactions of decacarbonyldimanganese(0) (**112**) are conducted with conjugated dienes containing chains of five or more carbon atoms, the reaction products differ substantially from those of the reaction with 1,3-butadiene [Eq. (64)]. With (E)- and (Z)-1,3-pentadiene (**1b** and **1j**), (E,E)-, and (E,Z)-2,4-hexadiene (**1d** and **1k**), (E)-2-methyl-1,3-pentadiene (**1e**), 4-methyl-1,3-pentadiene (**1l**), and 2,4-dimethyl-1,3-pentadiene (**1n**), the main reaction is the formation of mononuclear tetracarbonylmanganese complexes with η^3-(E)-dienyl and η^3-enyl ligands (*176–178*). The formation of these ligands is the result of a hydrogen transfer between two diene molecules, and it may be classified as a disproportionation. In these reactions, the configurations of the dienes show no influence on the constitutions of the products, in contrast to the results with $Cr(CO)_{6-n}L_n$ complexes. The complexes **123b**, **123d** and **123n** are obtained as mixtures together with their diastereomers **123b′**, **123d′**, and **123n′**.

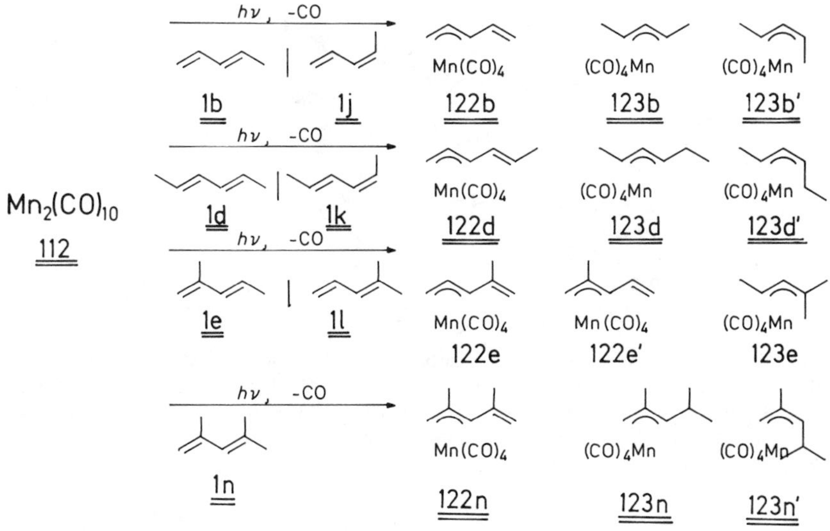

(64)

Interesting by-products are isolated from the reaction of **112** with (E,E)- or (E,Z)-2,4-hexadiene (**1d** and **1k**) (177). These are a heptacarbonyl-μ-$(\eta^{2-5:2,5}$-2,4-hexadiene-2,5-diyl)dimanganese complex (**124**) (Fig. 14), a dinuclear manganacyclopentadiene complex, which is a link between the corresponding hexacarbonyldiiron (179), and octacarbonylditungsten complexes (180). In small amounts, octacarbonyl-μ-$\eta^{3:3}$-(E,E)-2,5-hexadiene-1,4-diyldimanganese (**125**) is isolated as well. The latter can also be prepared together with its EZ-isomer photochemically from **112** and 1,3,5-hexatriene (181).

Somewhat surprising is the occurrence of the tricarbonyl-$\eta^{3:2}$-2,6-cyclo-nonadiene-1-yl-manganese complexes **126d** and **126n**, with methyl groups in 4, 5, and 8 and 1, 3, 5, 5, and 7 positions, respectively, during the photoreaction

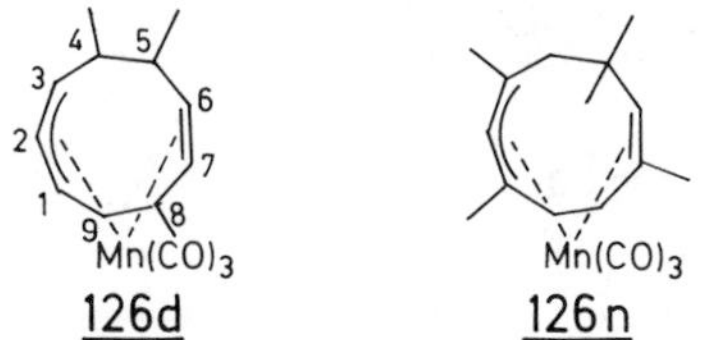

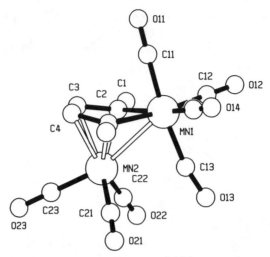

FIG. 14.   Molecular structure of heptacarbonyl-μ-$\eta^{2-5:2,5}$-2,4-hexadiene-2,5-diyldimanganese (124) (177): Mn-1—Mn-2 265.4, Mn-1—C-2 207.1, Mn-1—C-5 207.4, Mn-2—C-2 212.6, Mn-2—C-3 213.1, Mn-2—C-4 213.4, Mn-2—C-5 211.6, Mn-1—C-13 185.0, Mn-2—C-13 265.1 pm; Mn-1—C-13—O-13 168.3°. Reprinted with permission from *J. Organomet. Chem.* **270**, C37 (1984). Copyright by Elsevier Sequoia S. A.

of **112** with **1d** and **1n** (*182*). The molecular structure of **126n** (Fig. 15) (*182*) shows similarities with the structure of **46g**. The enyl unit, C-1 through C-3, is closely bonded to Mn-1, like the diene moiety in **46g**. The Mn—C distances (C-6 and C-7) indicate a weaker interaction of the isolated C=C double bond to the metal.

The reactions of **112** with 1,3-pentadiene and its derivatives are easily understood, when intermediates corresponding to **113a** are assumed. In contrast with **113a**, an enneacarbonyl-μ-$\eta^{3:1}$-2-pentene-1,4-diyldimanganese complex (**113b**) should be unstable with respect to β-elimination, by which **122b** and pentacarbonylhydridomanganese (**127**) are formed [Eqs. (65) and

$$\text{Mn}_2(\text{CO})_{10} \ + \ \underset{\textbf{1b}}{\diagdown\!\!\!\diagup\!\!\!\diagdown} \ \xrightarrow{\ h\nu\ } \ \underset{\underset{\textbf{113b}}{\text{Mn(CO)}_4}}{\diagup\!\!\!\diagdown}\text{Mn(CO)}_5 \ + \ \text{CO} \qquad (65)$$

$$\underset{\textbf{112}}{} \qquad \underset{\textbf{1b}}{}$$

$$\underset{\underset{\textbf{113b}}{\text{Mn(CO)}_4}}{\diagup\!\!\!\diagdown}\text{Mn(CO)}_5 \qquad \underset{\underset{\textbf{122b}}{\text{Mn(CO)}_4}}{\diagdown\!\!\!\diagup\!\!\!\diagdown} \ + \ \underset{\textbf{127}}{\text{Mn(CO)}_5\text{H}} \qquad (66)$$

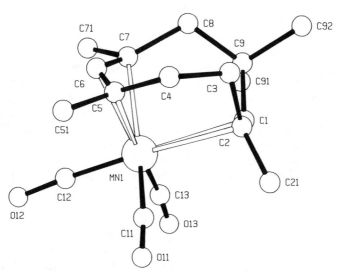

Fig. 15. Molecular structure of tricarbonyl-$\eta^{1-3:6,7}$-1,3,5,5,7-pentamethyl-2,6-cyclo-nonadiene-1-ylmanganese (**126n**) (*182*): Mn-1—C-1 243.9, Mn-1—C-2 250.3, Mn-1—C-5 220.2, Mn-1—C-6 211.6, Mn-1—C-7 224.6 pm; C-11—Mn-1—C-12 86.6°, C-11—Mn-1—C-13 98.3°, C-12—Mn-1—C-13 85.7°.

(66)]. In a further step, **127** reacts photochemically with free diene with loss of CO to give **123b, 123b′** [Eq. (67)].

$$[Mn(CO)_5H] + C_5H_8 \rightleftharpoons [Mn(CO)_4(\eta^3\text{-}C_5H_9)] + CO \qquad (67)$$

127   **1b**   **123b, 123b′**

Heating *n*-hexane solutions of the tetracarbonyl-η^3-(*E*)-dienylmanganese complexes **122b, 122d, 122e, 122k** and **122n** causes loss of CO. Nearly quantitatively the tricarbonyl-η^5-dienylmanganese complexes **128b, 128d, 128e** and **128n** are formed as yellow crystals (*168*) [Eq. (68)]. In the course of this reaction an *E–Z* isomerization of the dienyl ligands takes place. To some

$$[Mn(CO)_4(\eta^3\text{-}C_5H_{7-n}(CH_3)_n)] \rightleftharpoons [Mn(CO)_3(\eta^5\text{-}C_5H_{7-n}(CH_3)_n)] + CO \qquad (68)$$

122b, 122d, 122e, 122k, 122n   **128b, 128d, 128e, 128n**

Mn(CO)₄
122b   →   Mn(CO)₃ + CO
128b

extent, the formation of **128d** and **128n** has already been observed during the photochemical reaction of **112** with **1d** and **1n**. In reaction mixtures of these dienes the $\eta^{3:2}$-2,6-cyclononadiene-1-yl complexes **126d** and **126n** are also present. We propose, therefore, that **128d**, and **128n** react photochemically with **1d**, or **1n** in a [5 + 4] cycloaddition to yield **126d**, or **126n**.

Although η^5-pentadienyl complexes have been known for a long time, the first examples being [Fe(CO)$_3$(η^5-hexadienyl)][ClO$_4$] (*183*) or [Cr(η^5-pentadienyl)$_2$] and [Cr(η^5-pentadienyl)$_2$(CO)] (*184*), increasing interest in these ligands has arisen in the last few years. Homoleptic and mixed η^5-dienyl complexes of Ti (*185*), V, Cr (*186*), Mo (*187*), Mn (*188–190*), and Fe (*191*) have been prepared by metathetical reactions. Photochemistry provides an easy access to Mn derivatives. Quite recent is the development of syntheses for η^3-dienyl complexes. Examples are known for Mn (*192,193*), Fe (*193*), Co (*194*), and Ni (*195,196*).

C. Reactions of [Mn₂(CO)₁₀] with Cumulated Dienes

The behavior of allene (1,2-propadiene) and its derivatives as ligands in transition metal complexes has been investigated for more than 20 years (*197,198*). Allene may be coordinated by only one C=C double bond to a metal (*197,199*), or it may link to two metal centers as a 2:2 or 3:1 electron donor. Dinuclear μ-$\eta^{2:2}$-allene complexes are known for Mo (*200,201*), W (*201*), Mn (*202–203*), and Rh (*199,204*). In [Fe$_2$(CO)$_7$(μ-$\eta^{3:1}$-C$_3$H$_4$)] a σ and an allylic π-interaction characterize the bonding mode (*205,206*). In the course of the reaction, often two or more allene molecules become connected with transition metal complexes. Two allenes linked by C-2 yield tetramethyleneethane (*207,208*), which is stabilized by coordination on Fe (*205,208*), Ni (*210*), or Pd (*210,211*).

During the course of these investigations, the thermal and photochemical reactions of [Mn$_2$(CO)$_{10}$] with allene and some substituted allenes have been studied, but with no positive results. However, when the photochemical reaction is conducted with allene (**129a**) at reduced temperatures, three products can be isolated from the reaction mixture [Eq. (69)]. The yellow complex μ-$\eta^{2:2}$-alleneoctacarbonyldimanganese(0) (**130a**) predominates, and in addition the by-products octacarbonyl-μ-$\eta^{3:3}$-tetramethyleneethanedi-

$$(CO)_4Mn\!-\!Mn(CO)_4$$
130a

131

132

manganese (131) and enneacarbonyl-μ-$\eta^{3:1}$-2-methylene-4-pentene-1,4-diyl-dimanganese (132) are obtained (212,213).

$$[Mn_2(CO)_{10}] + C_3H_4 \rightleftharpoons [Mn_2(CO)_8(C_3H_4)] \ \mathbf{130a} \qquad (69)$$

$$\mathbf{112} \qquad \mathbf{129a} \qquad [Mn_2(CO)_8(C_6H_8)] \ \mathbf{131}$$

$$[Mn_2(CO)_9(C_6H_8)] \ \mathbf{132}$$

The formation of the products of photoreaction of 112 with allene can be rationalized as reactions of the two primary photochemical products of 112 [Mn(CO)$_5$ (120) and Mn$_2$(CO)$_9$ (121)] on the cumulated hydrocarbon. Complexation of allene with 121 should yield an intermediate, [Mn$_2$(CO)$_9$(η^2-C$_3$H$_4$)] (133), which loses CO thermally or photochemically and forms 130a. Two different intermediate radicals are to be expected from 120 and allene, with the C—Mn bond to C-1 (134) or to C-2 (135). Dimerization of 134 leads to a tetramethyleneethane complex with two η^1-coordinated Mn(CO)$_5$ groups (136). Complex 131 is readily formed from 136 by photochemical loss of two CO ligands, as is usual for [Mn(CO)$_5$(η^1-enyl)] complexes (171,172). Combination of the radical intermediates 134 and 135 gives a reasonable precursor (137) for 132. In 137, only one η^1-bonded Mn(CO)$_5$ group can release one CO without hydrogen migration and form an [Mn(CO)$_4$(η^3-enyl)] unit.

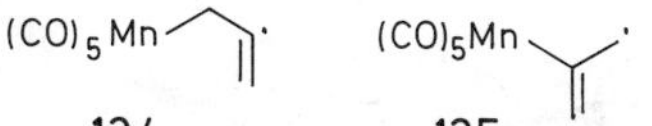

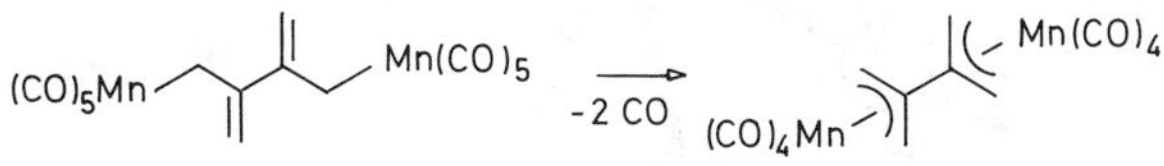

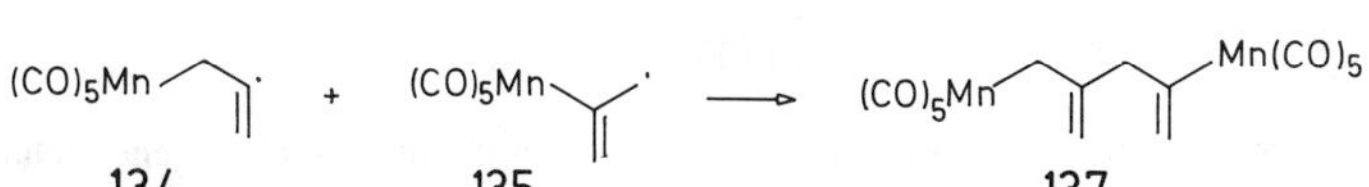

Interesting structural features are found for the complexes 130a (Fig. 16) and 131 (Fig. 17). Compound 130a is the first structurally characterized complex with an allene bridge between two metal centers with octahedral coordination and a metal–metal bond. Thus further insight into the specific bonding properties of the cumulated hydrocarbon ligand in a bridging function is possible.

In contrast with the related μ-$\eta^{2:2}$-s-trans-1,3-butadiene complexes of manganese and rhenium, [M$_2$(CO)$_8$(C$_4$H$_6$)] [M = Mn (116) (167), Re

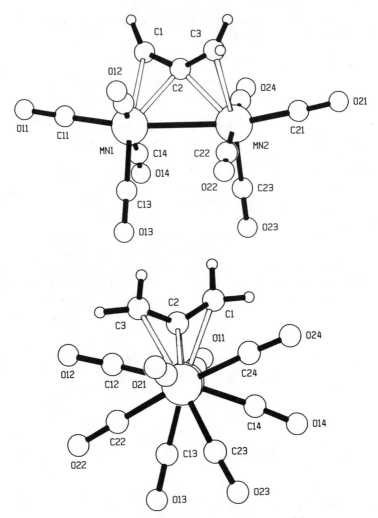

FIG. 16. Two views of the molecular structure of μ-($\eta^{2:2}$-allene)-octacarbonyl-dimanganese(0) (**130a**) (*213*): Mn-1—Mn-2 279.9, Mn-1—C-1 218.4, Mn-1—C-2 202.7, Mn-2—C-2 202.5, Mn-2—C-3 219.3, C-1—C-2 140.1, C-2—C-3 137.5 pm; C-1—C-2—C-3 140.7°. Reprinted with permission from *J. Organomet. Chem.* **302**, 35 (1986). Copyright by Elsevier Sequoia S. A.

(**158a**) (*147*)], which retain the basic features of the parent $[M_2(CO)_{10}]$ complexes, with a linear arrangement of the axial CO ligands and the metal–metal bonds, the torsional angle of the axial CO ligands in **130a** is 90.7°. Moreover, in the complexes **138** and **139** the lengths of the metal–metal bonds are increased [301.2(2) and 311.4(1) pm] in comparison with those of

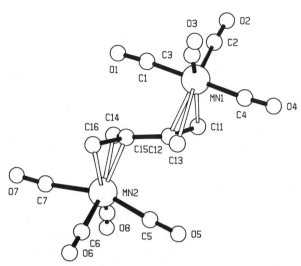

Fig. 17. Molecular structure of octacarbonyl-μ-$\eta^{3:3}$-tetramethylene-ethanedimanganese (131) (213). Reprinted with permission from *J. Organomet. Chem.* **302**, 35 (1986). Copyright by Elsevier Sequoia S. A.

the [$M_2(CO)_{10}$] complexes [M = Mn, 290.38(6); Re, 304.1(1) pm] (214). In **130a**, the Mn—Mn distance [279.9(1) pm] is distinctly shorter than in **112**. The geometry of **130a** resembles in some respects that of the [M(CO)$_4$-(η^3-enyl)] complexes [M = Mn (215), Re]. Formally, the portion C-1, C-2, and Mn-2 can be considered as an η^3-manganaallyl unit toward Mn-1, and vice versa for C-3, C-2, and Mn-1 toward Mn-2. From this picture, a π character for the Mn—Mn bond must be deduced, which could explain the torsional angle of the axial CO ligands in **130a**. The C-1—C-2—C-3 angle of 140.7(3)° corresponds well with those in other μ-$\eta^{2:2}$-allene complexes (200,202,203).

The crystal lattice of **131** contains two disordered molecules, which are related by an crystallographic center of symmetry. Therefore, bond lengths and angles could not be determined precisely. The overall structure shows the planar tetramethyleneethane ligand, with two Mn(CO)$_4$ groups on opposite sides (Fig. 17).

1,1-Dimethylallene (**129b**) shows some parallels in its photoreaction with allene with **112** [Eq. (70)]. Octacarbonyl-μ-$\eta^{2:2}$-1,1-dimethylallenedimanganese(0) (**130b**) is obtained in high yield as the main product. A mixture of the isomeric complexes tetracarbonyl-η^2-2-methyl-2-buten-1-ylmanganese (**123c**) and tetracarbonyl-η^3-3-methyl-2-buten-1-ylmanganese (**123c′**), as well as the electron-deficient compounds hexacarbonyl-μ-$\eta^{3:CH:3:CH}$-2,3,4,5-tetramethyl-2,5-hexadien-1,4-diyldimanganese (**138**) and tricarbonyl-$\eta^{3:CH}$-2-methyl-(E)-3-{3(8),6-m-menthadien-6-yl}-2-buten-1-yl-manganese (**139**), are

$$[Mn_2(CO)_{10}] + C_3H_2(CH_3)_2 \rightarrow [Mn(CO)_8(\mu\text{-}\eta^{2:2}\text{-}C_5H_8)] \qquad \textbf{130b} \qquad (70)$$

$$\textbf{112} \qquad\qquad \textbf{129b} \qquad\qquad [Mn(CO)_4(\eta^3\text{-}C_5H_9)] \qquad\qquad \textbf{123c, 123c}'$$

$$[Mn_2(CO)_6(\mu\text{-}\eta^{3:CH:3:CH}\text{-}C_{10}H_{16})] \; \textbf{138}$$

$$[Mn(CO)_3(\eta^{3:CH}\text{-}C_{15}H_{23})] \qquad\qquad \textbf{139}$$

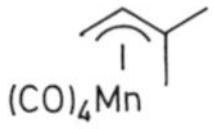

| $\underline{\underline{123c}}$ | $\underline{\underline{123c'}}$ | $\underline{\underline{138}}$ | $\underline{\underline{139}}$ |

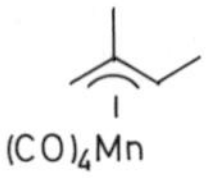

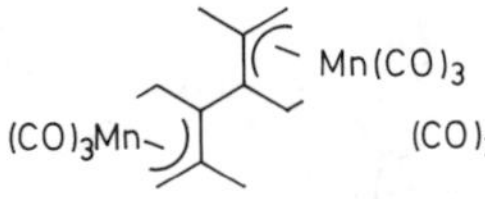

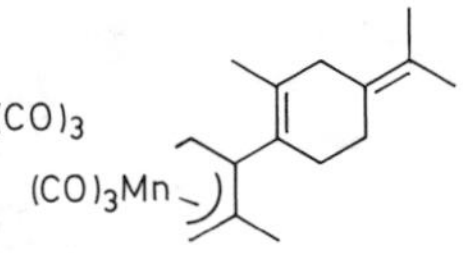

isolated in substantial amounts (*216*). Under ambient conditions, **138** and **139** add carbon monoxide reversibly and form instable hepta- and octacarbonyl complexes **140** and **141**, or, respectively, the moderatly stable tetracarbonyl **142** [Eqs. (71)–(73)].

$$[Mn_2(CO)_6(\mu\text{-}\eta^{3:CH:3:CH}\text{-}C_{10}H_{16})] + CO \rightarrow [Mn_2(CO)_7(\mu\text{-}\eta^{3:CH:3}\text{-}C_{10}H_{16})] \; \textbf{140} \quad (71)$$

$$\textbf{138}$$

$$[Mn_2(CO)_7(\mu\text{-}\eta^{3:CH:3}\text{-}C_{10}H_{16})] + CO \rightarrow [Mn_2(CO)_8(\mu\text{-}\eta^{3:3}\text{-}C_{10}H_{16})] \; \textbf{141} \qquad (72)$$

$$\textbf{140}$$

$$[Mn(CO)_3(\eta^{3:CH}\text{-}C_{15}H_{23})] + CO \rightarrow [Mn(CO)_4(\eta^3\text{-}C_{15}H_{23})] \; \textbf{142} \qquad (73)$$

$$\textbf{139}$$

Three radical intermediates are conceivable from $Mn(CO)_5$ (**120**) and **129b** by formation of a manganese–carbon bond to C-1, C-2, or C-3 (**143, 144** and **145**). Because, none of the isolated products contains 1,1-dimethylallene units connected via C-1 or C-3, the intermediate **144** does not need to be considered further.

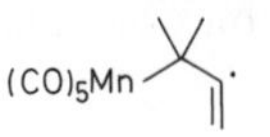

| $\underline{\underline{143}}$ | $\underline{\underline{144}}$ | $\underline{\underline{145}}$ |

| $\underline{\underline{146}}$ | $\underline{\underline{147}}$ | $\underline{\underline{148}}$ |

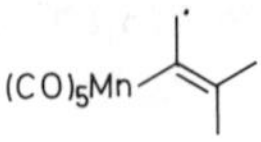

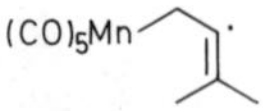

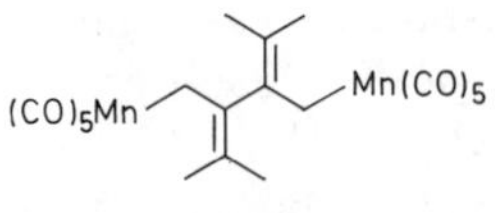

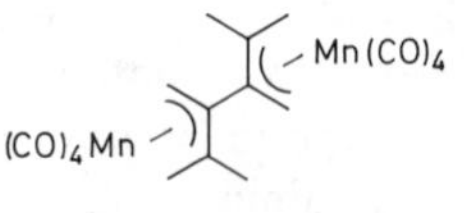

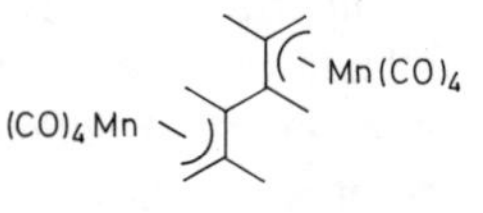

Dimerization of **145** opens a reasonable route to **138**. In the dimer **146**, two $Mn(CO)_5(\eta^1$-enyl) groups are present, these being photochemically unstable due to loss of CO. The resulting tetramethyltetramethyleneethane complex (**147**) rearranges by two 1,4-H shifts, often observed with $Mn(CO)_4$ (η^3-enyl) complexes (*171*), to a tetramethyl-2,5-hexadiene-1,4-diyl compound (**148**). Obviously, two linearly conjugated enyl units are more stable than crosswise conjugated ones. Both enyl systems bear methyl groups at the Z position, which interact sterically with the a site of the coordination octahedron; so, carbonyl ligands are displaced by two Mn—H—C bridges, and compound **138** results.

There are three possibilities for the location of the enyl units in the 2,3,4,5-tetramethyl-2,5-hexadiene-1,4-diyl frame (**149**), E,E, E,Z, and Z,Z. Moreover, the enyl units may adopt s-trans or s-cis configurations. On steric grounds,

EE   EZ   ZZ

s-trans   s-cis

the latter should not be energetically favored. The coordination of two $Mn(CO)_n$ groups ($n = 3,4$) is conceivable to each of these forms in four different ways, R,R, S,S, and meso forms. A definite decision between these numerous structural possibilities cannot be based on NMR results. Therefore, an X-ray structure analysis was performed for **138**.

The dinuclear complex **138** shows C_2 symmetry and consists of two identical tricarbonyl-$\eta^{3:CH}$-2-methyl-2-buten-1,3-diyl moieties, which are linked in the $3E$ position (Fig. 18). The coordination spheres of the Mn atoms are quasi-octahedral. Three facial positions are occupied by the $\eta^{3:CH}$-2-methyl-2-buten-1,3-diyl ligand, which is coordinated by an enyl unit (C-1 to C-3) and by a C—H—Mn bridge, formed from the 3Z-methyl group, to the metal. In the other sites three CO ligands are found. The bond angles between the CO ligands and the bridging hydrogen deviate only by 3.7° from 90° or 180°, respectively, and correspond well with the angles for octahedral coordination. The plane of the enyl unit is tilted toward the Mn-1 H-41 axis at 41.6°.

Small differences are found for the Mn—C bond lengths of C-1 to C-3 [215.4(1), 213.5(1), and 207.9(1) pm], in contrast to η^3-enyl complexes, in which the central carbon atom is closer to the metal than the terminal ones (*171,215*). However, there is good agreement with the bond lengths found in $[Mn(CO)_3(\eta^{3:CH}\text{-}C_6H_8CH_3)]$ (*47,48*). The distance of C-4 from Mn-1

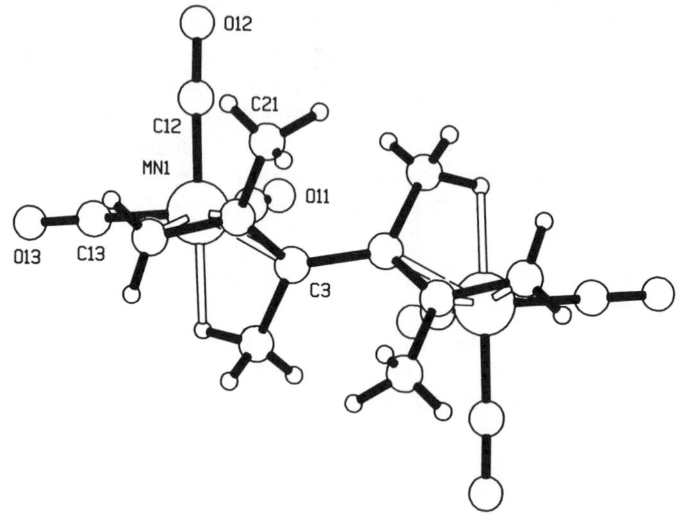

Fig. 18. Molecular structure of hexacarbonyl-μ-$\eta^{3:\text{CH}:3:\text{CH}}$-2,3,4,5-tetramethyl-2,5-hexadiene-1,4-diyldimanganese (**138**) (*216*): Mn-1—C-1 215.4, Mn-1—C-2 213.5, Mn-1—C-3 207.9, Mn-1—C-4 229.1, Mn-1—H-41 182, C-4—H-41 96 pm. Reprinted with permission from *J. Organomet. Chem.* **302**, 217 (1986). Copyright by Elsevier Sequoia S. A.

[229.1(1) pm] is just in the expected range for Mn—C bonds, although H-41 is part of a three-center, two-electron Mn—H—C bond. The bond lengths of H-41 to C-4 and Mn-1 are stretched in comparison to the other C—H distances, and to the Mn—H bond length in [Mn(CO)$_5$(H)] (160.2(16) pm) (*217*).

All five carbon atoms of the $\eta^{3:\text{CH}}$-2-methyl-2-buten-1,3-diyl ligand deviate only slightly from a planar arrangement. The planes of both moieties are oriented orthogonal to each other. The coordination of the $\eta^{3:\text{CH}}$-2-methyl-(*E*)-3-[3(8),6-*m*-menthadien-6-yl]-2-buten-1-yl ligand to the Mn(CO)$_3$ fragment in **139** (Fig. 19) is very similar to that in **138**. There are only minor differences in bond lengths and angles of the coordinated moiety.

Bond formation between the radicals **143** and **145** with **150** offers an explanation for the formation of the 2-methyl-(*E*)-3-[3(8),6-*m*-menthadien-6-yl]-2-buten-1-yl ligands in **139** and **142**. Loss of CO and β-elimination in **150** leads to **151**, which forms the η^3-2-methyl-(*E*)-3-[3(8),6-*m*-menthadien-6-yl]-

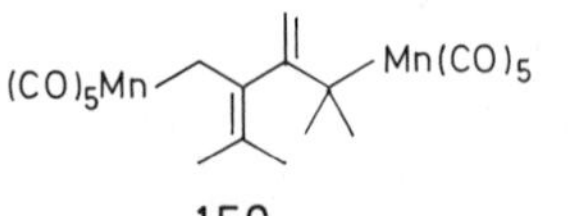

150

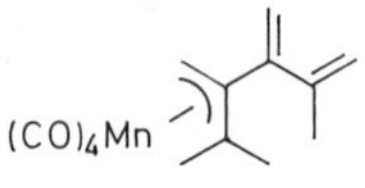

151

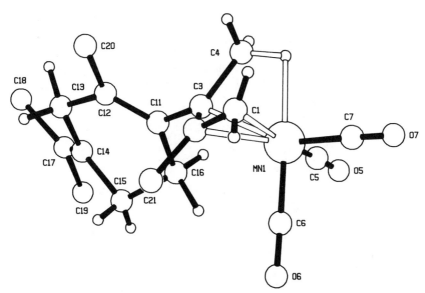

Fig. 19. Molecular structure of tricarbonyl-$\eta^{3:CH}$-2-methyl-(E)-3-[3(8),6-m-menthadiene-6-yl]-2-butene-1-ylmanganese (**139**): Mn-1—C-1 212.7, Mn-1—C-2 211.3, Mn-1—C-3 208.1, Mn-1—C-4 229.6, Mn-1—H-41 190, C-4—H-41 96 pm. Reprinted with permission from *J. Organomet. Chem.* **302**, 217 (1986). Copyright by Elsevier Sequoia S. A.

2-buten-1-yl complex (**142**) by a [4 + 2] cycloaddition of free 1,1-dimethyl-allene onto the conjugated diene unit in **151**.

By ^{1}H-NMR spectroscopy, a hindered rotation of the (4Z)-methyl groups has been detected, with activation barriers of $\Delta G_{298}^* = 41.7 \pm 1$ kJ/mol for **138** and $\Delta G_{298}^* = 43.5$ kJ/mol for **139**. Similar rotations of methyl groups forming part of a C—H metal bridge were found in the [Cr(CO)$_2$L($\eta^{4:CH}$-diene)] complexes (**37n**, **38n**, and **39j–39n**) (*57*), and in [Fe{P(OCH$_3$)$_3$}-($\eta^{3:CH}$-enyl)]$^+$ (*44,45*). For [Ir{P(C$_6$H$_5$)$_3$}$_2$(H)($\eta^{3:CH}$-enyl)]$^+$ a C—H—Ir interaction is inferred from the NMR spectra. However, the rotation of the methyl group involved in the bridge is fast down to 153 K (*218*). A close structural and electronic relationship exists between **138**, **139**, and the recently prepared, complex [Mn(CO)$_3$($\eta^{3:CH}$-3-methyl-2-buten-1-yl)] (*219*). Corresponding intermediates [RuL$_3$($\eta^{3:CH}$-enyl)][PF$_6$] are assumed for the isomerization of 1,5- to 1,3-cyclooctadiene (*220*).

D. Reactions of [Re$_2$(CO)$_{10}$] with Dienes

The photoreaction of [Re$_2$(CO)$_{10}$] (**152**) with 1,3-butadiene (**1a**) [Eq. (74)] yields a mixture of six colorless or pale yellow complexes, which were separated by HPLC. The main product of the reaction is μ-$\eta^{2:1}$-1,3-butadien-1-yloctacarbonyl-μ-hydridodirhenium (**153a**). A mixture of minor amounts of the *E*- and *Z*-isomers of tetracarbonyl-η^3-2-buten-1-ylrhenium (**154**), η^4-1,3-butadieneoctacarbonyldirhenium (**155a**), and two diastereomeric octacarbonyl-μ-$\eta^{3:3}$-*EE*-2,7-octadien-1,6-diyldirhenium (**156** and **157**) is isolated (*148*).

$$[Re_2(CO)_{10}] + C_4H_6 \rightleftharpoons [Re_2(CO)_8(H)(\mu\text{-}\eta^{2:1}\text{-}C_4H_5)] \quad \textbf{153a} \qquad (74)$$

$$\textbf{152} \qquad \textbf{1a} \qquad [Re(CO)_4(\eta^3\text{-}C_4H_7)] \; E, Z \quad \textbf{154}$$

$$[Re_2(CO)_8(\eta^4\text{-}C_4H_6)] \qquad \textbf{155a}$$

$$[Re_2(CO)_8(\mu\text{-}\eta^{3:3}\text{-}C_8H_{12})] \qquad \textbf{156, 157}$$

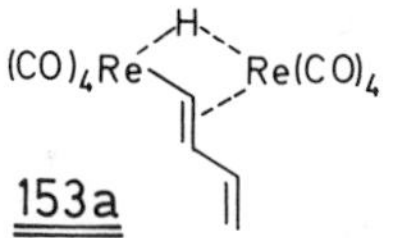

(CO)₄Re, H, Re(CO)₄ **153a**

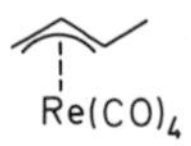

Re(CO)₄ **E-154**

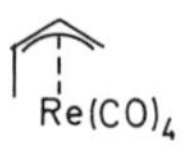

Re(CO)₄ **Z-154**

(CO)₃Re—Re(CO)₅ **155a**

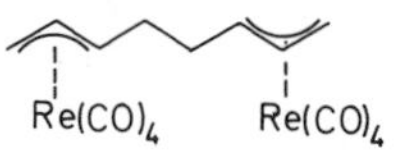

Re(CO)₄ Re(CO)₄ **156**

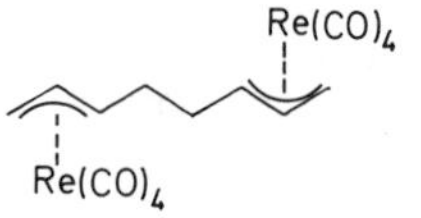

Re(CO)₄ Re(CO)₄ **157**

Complex **153a** rearranges fast in polar solvents to the Re homolog of **116**, μ-$\eta^{2:2}$-butadieneoctacarbonyldirhenium (**158a**) [Eq. (75)]. The hydrido ligand is retransferred, probably in an intramolecular reductive rearrangement onto the 1,3-butadien-1-yl-bridge. The homologs of **113** and **158a**, $[Re_2(CO)_9(\mu\text{-}\eta^{3:1}\text{-}C_4H_6)]$, are also reported to be the photoproducts of **152** and **1a** (*147*).

$$[Re_2(CO)_8(H)(\mu\text{-}\eta^{2:1}\text{-}C_4H_5)] \rightleftharpoons [Re_2(CO)_8(\mu\text{-}\eta^{2:2}\text{-}C_4H_6)] \qquad (75)$$

$$\textbf{153a} \qquad\qquad\qquad \textbf{158a}$$

While **154–157** are specific products for **1a**, complexes of the type of **153a** are the predominate products when **152** is reacted photochemically with acetylene (*221*), olefins (*150,222,223*), dienes, trienes, and tetraenes (*224–226*). In contrast to a previous report (*227*), 1,3,5-cycloheptatriene reacts with **152** like the other unsaturated hydrocarbons. Similar reaction products are obtained with **1c** as the diene component [Eq. (76)]. Although complexes comparable with **154, 156,** and **157** are missing, a trinuclear species is formed in addition (*225*).

$$[Re_2(CO)_{10}] + C_5H_8 \rightarrow [Re_2(CO)_8(H)(\mu\text{-}\eta^{2:1}\text{-}C_5H_7)] \quad \mathbf{153c} \quad\quad (76)$$

$$\mathbf{52} \quad\quad \mathbf{1c} \quad\quad [Re_2(CO)_8(\eta^4\text{-}C_5H_8)] \quad\quad\quad \mathbf{155c}$$

$$[Re_2(CO)_8(\mu\text{-}\eta^{2:2}\text{-}C_5H_8)] \quad\quad \mathbf{158c}$$

$$[Re_3(CO)_{13}(\mu\text{-}\eta^{2:1}\text{-}C_5H_7)] \quad\quad \mathbf{159}$$

The cyclic dienes **1g** and **1h** yield isomeric $[Re_2(CO)_8(H)(\mu\text{-}\eta^{2:1}\text{-dienyl})]$ complexes (**153g, 153g′, 153h** and **153h′**) [Eqs. (77) and (78)], which differ in the position of the Re—C bond. Both dienes form $[Re(CO)_3(\eta^5\text{-dienyl})]$ **160g**, **160h** as by-products. Finally, a derivative of **160h**, $[Re_2(CO)_8(\mu\text{-}\eta^{5:1}\text{-}C_5H_4)]$ (**161**), was isolated (*224*).

$$[Re_2(CO)_{10}] + C_6H_8 \rightarrow [Re_2(CO)_8(H)(\mu\text{-}\eta^{2:1}\text{-}C_6H_7)] \quad \mathbf{153g, 153g'} \quad (77)$$

$$\mathbf{152} \quad\quad \mathbf{1g} \quad\quad [Re(CO)_3(\eta^5\text{-}C_6H_7)] \quad\quad\quad \mathbf{160g}$$

$$[Re_2(CO)_{10}] + C_5H_6 \rightarrow [Re_2(CO)_8(H)(\mu\text{-}\eta^{2:1}\text{-}C_5H_5)] \quad \mathbf{153h, 153h'} \quad (78)$$

$$\mathbf{152} \quad\quad \mathbf{1h} \quad\quad [Re(CO)_3(\eta^5\text{-}C_5H_5)] \quad\quad\quad \mathbf{160h}$$

$$[Re_2(CO)_8(\mu\text{-}\eta^{5:1}\text{-}C_5H_4)] \quad\quad \mathbf{161h}$$

The smooth formation of the $[Re_2(CO)_8(H)(\mu\text{-}\eta^{2:1}\text{-dienyl})]$ complexes **153a, 153c, 153g, 153g′, 153h** and **153h′** demonstrates the ease of C—H bond cleavage by Re under photochemical conditions. From linear dienes only H-1 *E* and from the cyclic dienes **1g** and **1h** only H-1 or H-2 are substituted by a Re—C bond.

The complexes of type **153** show temperature-dependent carbonyl signals in their ^{13}C-NMR spectra due to hindered movements of the $\mu\text{-}\eta^{2:1}$-dienyl ligands. This movement can be described as a windshield wiper type oscillation, by which the σ and π bonds are interchanged between the two rhenium atoms. For octacarbonyl-$\mu\text{-}\eta^{2:1}$-1,3,5-cycloheptatrien-2-yl-μ-hydridodirhenium (**162**) and for octacarbonyl-$\mu\text{-}\eta^{2:1}$-1,3,5-cycloheptatrien-1-yl-μ-hydridodirhenium (**162′**) activation barriers of $\Delta G_{255}^* = 48.0 \pm 2$ kJ/mol and $\Delta G_{207}^* = 39.9 \pm 2$ kJ/mol (*218*) have been determined.

An X-ray structure analysis was carried out on **162** (Fig. 20). Two coordination octahedra with a common corner occupied by the μ-H ligand and

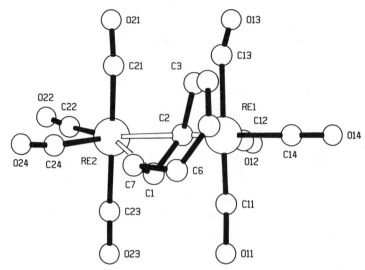

FIG. 20. Molecular structure of octacarbonyl-μ-$\eta^{2:1}$-1,3,5-cycloheptatrien-2-yl-μ-hydrido-dirhenium (**162**) (*226*): Re-1—Re-2 304.2, Re-1—C-2 230, Re-2—C-1 245, Re-2—C-2 239 pm. Reprinted with permission from *Z. Naturforsch.* **40B**, 1188 (1985). Copyright by Verlag der Zeitschrift für Naturforschung.

two additional corners, linked by the μ-$\eta^{2:1}$-trienyl ligand, properly describe the structure. The atom C-2 is σ-bonded to Re-1, and C-1 and C-2 are η^2-coordinated to Re-2. Because the σ-bond axis to Re-1 cuts through C-2, and the π-bond axis to Re-2 through the center of C-1 and C-2, the coordination octahedra are tilted against each other by approximately 15°. In spite of the short Re—Re distance of 304.2(3) pm, a Re—Re bond is excluded on basis of the mutual orientation of the two coordination octahedra.

Several dinuclear complexes with μ-$\eta^{2:1}$-olefinyl bridges have already been prepared. Examples are $[(\eta^5\text{-}C_5H_5)_2Mo_2(CO)_4(\mu\text{-}\eta^{2:1}\text{-}RC{=}CHR)]^+$ (*228*), $[Mn_2(CO)_7\{\mu\text{-}P(C_6H_5)_2\}(\mu\text{-}\eta^{2:1}\text{-}C_2H_2R)]$ (*229*), $[Fe_2(CO)_6(\mu\text{-}SR)\text{-}(\mu\text{-}\eta^{2:1}\text{-}C_2H_3)]$ (*230*), $[Rh_2\{(i\text{-}C_3H_7)_2PC_2H_4P(i\text{-}C_3H_7)_2\}_2(\mu\text{-}H)(\mu\text{-}\eta^{2:1}\text{-}C_2H_3)]$ (*231*), $[(\eta^5\text{-}C_9H_7)_2Rh_2(CO)_2(\mu\text{-}\eta^{2:1}\text{-}i\text{-}C_3H_5)][BF_4]$ (*204*), and $[(\eta^5\text{-}C_9H_7)_2Rh_2(\mu\text{-}\eta^{2:1}\text{-}C_2H_3)(\mu\text{-}\eta^{2:1}\text{-}i\text{-}C_4H_7)]$ (*232*). The trinuclear Os complexes of the type $[Os_3(CO)_{10}(\mu\text{-}H)(\mu\text{-}\eta^{2:1}\text{-}olefinyl)]$ (*233–236*) also have structural patterns closely related to **153**.

The molecular structure of $[Re_3(CO)_{13}(\mu\text{-}\eta^{2:1}\text{-}C_5H_7)]$ (**159**) shows an angular arrangement of the three metal centers (Fig. 21). Re-1 and Re-2 bear four and Re-3 five CO ligands. The atoms Re-1 and Re-2 are linked by a μ-$\eta^{2:1}$-(E)-3-methyl-1,3-butadien-1-yl bridge and by a semibridging CO ligand. The overall structure corresponds to three coordination octahedra, two of which share an edge, and the third is bonded to an equatorial

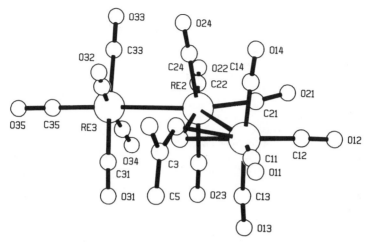

FIG. 21. Molecular structure of tridecacarbonyl-μ-$\eta^{2:1}$-(E)-3-methyl-1,3-butadiene-1-yl-trirhenium (**159**) (*225*): Re-1—Re-2 300.5, Re-2—Re-3 306.4, Re-1—Cl 223.4, Re-1—C-2 241.0, Re-2—C-1 215.7, C-1—C-2 131.2 pm; Re-1—Re-2—Re-3 132.5°. Reprinted with permission from *Z. Naturforsch.* (in press). Copyright by Verlag der Zeitschrift für Naturforschung.

corner. Compound **159** is related to $[Re_3(CO)_{14}(\mu\text{-H})]$ (*237,238*), with one CO and the hydrido ligand formally replaced by the μ-$\eta^{2:1}$-olefinyl bridge.

ACKNOWLEDGEMENTS

The author gratefully acknowledges the cooperation of Prof. W. S. Sheldrick and his co-worker J. Kaub, who have performed most of the X-ray structure analyses. I would also like to thank the Deutsche Forschungsgemeinschaft and the Fond der Chemischen Industrie for financial support.

REFERENCES

1. W. Strohmeier, *Angew. Chem.* **76**, 873 (1964); idem, *Angew. Chem., Int. Ed. Engl.* **3**, 730 (1964).
2. E. Koerner von Gustorf and F.-W. Grevels, *Fortschr. Chem. Forsch.* **13**, 366 (1969).
3. M. Wrighton, *Chem. Rev.* **74**, 401 (1974).
4. G. L. Geoffroy and M. S. Wrighton, "Organometallic Photochemistry." Academic Press, New York, 1979.
5. H. G. Alt, *Angew. Chem.* **96**, 752 (1984).
6. D. B. Pourreau and G. L. Geoffroy, *Adv. Organomet. Chem.* **24**, 249 (1985).
7. M. Herberhold, "Metal-π-Complexes, Vol. II, Complexes with Monoolefinic Ligands," Part I. Elsevier, Amsterdam, London, New York, 1972.
8. S. Özkar, Thesis, Technische Univ. München, 1976.
9. I. Fischler, M. Budzwait, and E. A. Koerner v. Gustorf, *J. Organomet. Chem.* **105**, 325 (1976).
10. M. Kotzian, C. G. Kreiter, and S. Özkar, *J. Organomet. Chem.* **229**, 29 (1982).
11. E. O. Fischer and K. Ulm, *Z. Naturforsch.* **15B**, 59 (1960).

12. H. P. Fritz, H. Keller, and E. O. Fischer, *Naturwissenschaften* **48**, 518 (1961).
13. W. C. Mills and M. S. Wrighton, *J. Am. Chem. Soc.* **101**, 5830 (1979).
14. I. W. Stolz, G. R. Dobson, and R. K. Sheline, *Inorg. Chem.* **2**, 1264 (1963).
15. M. Herberhold, Thesis, Univ. München, 1961.
16. E. O. Fischer, H. P. Kögler, and P. Kuzel, *Chem. Ber.* **93**, 3006 (1960).
17. M. Wrighton, G. S. Hammond, and H. B. Gray, *J. Am. Chem. Soc.* **92**, 6068 (1970).
18. C. G. Kreiter and S. Özkar, *Z. Naturforsch.* **38B**, 1424 (1983).
19. D. P. Tate, W. R. Knipple, and J. M. Augl, *Inorg. Chem.* **1**, 433 (1962).
20. R. B. King and A. Fronzaglia, *Inorg. Chem.* **5**, 1837 (1966).
21. M. Kotzian, C. G. Kreiter, G. Michael, and S. Özkar, *Chem. Ber.* **116**, 3637 (1983).
22. G. Michael and C. G. Kreiter, unpublished.
23. C. G. Kreiter and S. Özkar, *J. Organomet. Chem.* **152**, C13 (1978).
24. S. Özkar and C. G. Kreiter, *J. Organomet. Chem.* **303**, 367 (1986).
25. C. G. Kreiter, M. Kotzian, U. Schubert, R. Bau, and M. A. Bruck, *Z. Naturforsch.* **39B**, 1553 (1984).
26. S. Özkar and C. G. Kreiter, *J. Organomet. Chem.* **256**, 57 (1983).
27. M. Kotzian and C. G. Kreiter, *J. Organomet. Chem.* **289**, 295 (1985).
28. S. Özkar and C. G. Kreiter, unpublished results.
29. G. Michael and C. G. Kreiter, *Z. Naturforsch.* **39B**, 1738 (1984).
30. IUPAC Nomenclature of Inorganic Chemistry, *Pure Appl. Chem.* **28**, 1 (1971).
31. C. G. Kreiter, M. Kotzian, and S. Özkar, unpublished results.
32. F. A. Cotton, V. W. Day, B. A. Frenz, K. I. Hardcastle, and J. M. Troup, *J. Am. Chem. Soc.* **95**, 4522 (1973).
33. F. A. Cotton and B. A. Frenz, *Acta Cryst.* **B30**, 1772 (1974).
34. R. S. Berry, *J. Chem. Phys.* **32**, 933 (1960).
35. M. Brookhart, K. Cox, G. N. Cloke, J. C. Green, M. L. H. Green, P. M. Hare, J. Bashkin, A. E. Derome, and P. D. Grebenik, *J. Chem. Soc., Dalton Trans.*, 423 (1985).
36. P. S. Skell, E. M. Van Dam, and P. M. Silvon, *J. Am. Chem. Soc.* **96**, 626 (1974).
37. P. S. Skell and M. J. McGlinchey, *Angew. Chem.* **87**, 215 (1975); idem, *Angew. Chem., Int. Ed. Engl.* **14**, 195 (1975).
38. W. Gausing and G. Wilke, *Angew. Chem.* **93**, 201 (1981); idem, *Angew. Chem., Int. Ed. Engl.* **20**, 186 (1981).
39. G. Huttner, D. Neugebauer, and A. Razavi, *Angew. Chem.* **87**, 353 (1975); idem, *Angew. Chem., Int. Ed. Engl.* **14**, 352 (1975).
40. F. A. Cotton and V. W. Day, *J. Chem. Soc., Chem. Commun.*, 415 (1974).
41. F. A. Cotton, T. La Cour, and A. G. Stanislowski, *J. Am. Chem. Soc.* **96**, 754 (1974).
42. F. A. Cotton and A. G. Stanislowski, *J. Am. Chem. Soc.* **96**, 5074 (1974).
43. M. Brookhart and M. L. H. Green, *J. Organomet. Chem.* **250**, 395 (1983).
44. S. D. Ittel, F. A. Van-Catledge, and J. P. Jesson, *J. Am. Chem. Soc.* **101**, 6905 (1979).
45. R. K. Brown, J. M. Williams, A. J. Schultz, G. D. Stucky, S. D. Ittel, and R. L. Harlow, *J. Am. Chem. Soc.* **102**, 981 (1980).
46. W. Lamanna and M. Brookhart, *J. Am. Chem. Soc.* **103**, 989 (1981).
47. M. Brookhart, W. Lamanna, and M. B. Humphrey, *J. Am. Chem. Soc.* **104**, 2117 (1982).
48. A. J. Schultz, R. G. Teller, M. A. Beno, J. M. Williams, M. Brookhart, W. Lamanna, and M. B. Humphrey, *Science* **220**, 197 (1983).
49. M. Brookhart, W. Lamanna, and A. R. Pinkas, *Organometallics* **2**, 638 (1983).
50. M. Brookhart and A. Lukacs, *J. Am. Chem. Soc.* **106**, 4161 (1984).
51. M. Zöller, H.-E. Sasse, and M. Ziegler, *Z. Anorg. Allg. Chem.* **425**, 257 (1976).
52. Z. Dawoodi, M. L. H. Green, V. S. B. Mtetwa, and K. Prout, *J. Chem. Soc., Chem. Commun.*, 802 (1982).

53. Z. Dawoodi, M. L. H. Green, V. S. B. Mtetwa, and K. Prout, *J. Chem. Soc., Chem. Commun.*, 1410 (1982).
54. M. Brookhart, M. L. H. Green, and R. B. Pardy, *J. Chem. Soc., Chem. Commun.*, 691 (1983).
55. R. B. Cracknell, A. G. Orpen, and J. L. Spencer, *J. Chem. Soc., Chem. Commun.*, 326 (1984).
56. G. F. Schmidt and M. Brookhart, *J. Am. Chem. Soc.* **107**, 1443 (1985).
57. G. Michael, J. Kaub, and C. G. Kreiter, *Angew. Chem.* **97**, 503 (1985); *idem, Angew. Chem., Int. Ed. Engl.* **24**, 502 (1985).
58. G. Michael, J. Kaub, and C. G. Kreiter, *Chem. Ber.* **118**, 3944 (1985).
59. J. R. Blackborow, C. R. Eady, F.-W. Grevels, E. A. Koerner von Gustorf, A. Scrivanti, O. S. Wolfbeis, R. Benn, D. J. Brauer, C. Krüger, P. J. Roberts, and Y.-H. Tsay, *J. Chem. Soc., Dalton Trans.*, 661 (1981).
60. J. R. Blackborow, R. H. Grubbs, A. M. Ashita, A. Scrivanti, and E. A. Koerner von Gustorf, *J. Organomet. Chem.* **122**, C6 (1976).
61. W. R. Roth, *Chimia (Zürich)* **20**, 229 (1966).
62. W. R. Roth, J. König, and K. Stein, *Chem. Ber.* **103**, 426 (1970).
63. D. J. McLennan and P. M. W. Gill, *J. Am. Chem. Soc.* **107**, 2971 (1985).
64. T. W. Beall and L. W. Houk, *Inorg. Chem.* **13**, 2280 (1974).
65. D. J. Bertelli and J. M. Viebrock, *Inorg. Chem.* **7**, 1240 (1968).
66. C. G. Kreiter, G. Michael, and J. Kaub, unpublished results.
67. H. Herbstein and G. Reisner, *J. Chem. Soc., Chem. Commun.*, 1077 (1972).
68. H. Herbstein and G. Reisner, *Acta Cryst.* **33B**, 3304 (1977).
69. S. Özkar, H. Kurz, D. Neugebauer, and C. G. Kreiter, *J. Organomet. Chem.* **160**, 115 (1978).
70. C. G. Kreiter and H. Kurz, *Chem. Ber.* **116**, 1494 (1983).
71. S. Özkar and C. G. Kreiter, *J. Organomet. Chem.* **293**, 229 (1985).
72. C. G. Kreiter, A. Maasböl, F. A. L. Anet, H. D. Kaesz, and S. Winstein, *J. Am. Chem. Soc.* **88**, 3444 (1966).
73. C. G. Kreiter, E. Michels, and H. Kurz, *J. Organomet. Chem.* **232**, 249 (1982).
74. C. F. Wilcox and R. Craig, *J. Am. Chem. Soc.* **93**, 3866 (1961).
75. J. A. S. Howell, B. F. G. Johnson, and J. Lewis, *J. Chem. Soc., Dalton Trans.*, 293 (1974).
76. E. Michels, W. S. Sheldrick, and C. G. Kreiter, *Chem. Ber.* **118**, 964 (1985).
77. E. Michels and C. G. Kreiter, *J. Organomet. Chem.* **252**, C1 (1983).
78. R. P. M. Werner and T. H. Coffield, *Chem. Ind. (London)*, 936 (1960).
79. C. G. Kreiter, E. Michels, and J. Kaub, *Z. Naturforsch.* **41B**, 722 (1986).
80. A. Salzer and W. von Philipsborn, *J. Organomet. Chem.* **170**, 63 (1979).
81. G. B. Jameson and A. Salzer, *Organometallics* **1**, 689 (1982).
82. C. G. Kreiter and H. Kurz, *Z. Naturforsch.* **33B**, 1285 (1978).
83. C. G. Kreiter and H. Kurz, *J. Organomet. Chem.* **214**, 339 (1981).
84. C. G. Kreiter and E. Michels, *Chem. Ber.* **117**, 344 (1984).
85. C. G. Kreiter and H. Kurz, *Z. Naturforsch.* **38B**, 841 (1983).
86. C. G. Kreiter and H. Kurz, *Z. Naturforsch.* **37B**, 1322 (1982).
87. T. S. Piper and G. Wilkinson, *J. Inorg. Nucl. Chem.* **3**, 104 (1956).
88. R. B. King, *Organomet. Synth.* **1**, 145 (1965).
89. K. Nist, Thesis, Univ. Kaiserslautern, 1984.
90. M. Bottrill and M. Green, *J. Chem. Soc., Dalton Trans.*, 820 (1979).
91. P. Powell, *J. Organomet. Chem.* **243**, 205 (1983).
92. J. Kögler, Thesis, Univ. Kaiserslautern, 1985.
93. C. G. Kreiter, J. Kögler, and K. Nist, *J. Organomet. Chem.* (in press).
94. M. L. H. Green, L. Pratt, and G. Wilkinson, *J. Chem. Soc.*, 3753 (1959).
95. E. O. Fischer and U. Zahn, *Chem. Ber.* **92**, 1624 (1959).
96. A. Davison, M. L. H. Green, and G. Wilkinson, *J. Chem. Soc.*, 3172 (1961).

97. R. K. Kochhar and R. Pettit, *J. Organomet. Chem.* **6,** 272 (1966).
98. T. H. Whitesides and J. Shelly, *J. Organomet. Chem.* **92,** 215 (1975).
99. A. P. Humphries and S. A. R. Knox, *J. Chem. Soc., Dalton Trans.,* 1710 (1975).
100. S. G. Davies, M. L. H. Green, K. Prout, A. Coda, and V. Tazzoli, *J. Chem. Soc., Chem. Commun.,* 135 (1977).
101. S. G. Davies and M. L. H. Green, *J. Chem. Soc., Dalton Trans.,* 1510 (1978).
102. C. G. Kreiter, G. Wendt, and W. S. Sheldrick, unpublished results.
103. E. Cannillo and K. Prout, *Acta Cryst.* **B33,** 3916 (1977).
104. J. A. Segal, M. L. H. Green, J.-C. Daran, and K. Prout, *J. Chem. Soc., Chem. Commun.,* 766 (1976).
105. K. Prout and J.-C. Daran, *Acta Cryst.* **B33,** 2303 (1977).
106. M. L. H. Green, J. Knight, and J. A. Segal, *J. Chem. Soc., Dalton Trans.,* 2189 (1977).
107. P. Eilbracht, *Chem. Ber.* **109,** 1429 (1976).
108. C. G. Kreiter, W. Michels, and M. Wenz, *Chem. Ber.* **119,** 1994 (1986).
109. R. G. Severson and A. Wojcicki, *J. Organomet. Chem.* **157,** 173 (1978).
110. E. Samuel, R. D. Rausch, T. E. Gismondi, E. A. Mintz, and C. Giannotti, *J. Organomet. Chem.* **172,** 309 (1979).
111. R. J. Kazlauskas and M. S. Wrighton, *J. Am. Chem. Soc.* **102,** 1727 (1980).
112. R. J. Kazlauskas and M. S. Wrighton, *J. Am. Chem. Soc.* **104,** 6005 (1982).
113. K. A. Mahmoud, A. J. Rest, H. G. Alt, M. E. Eichner, and B. M. Jansen, *J. Chem. Soc., Dalton Trans.,* 175, 187 (1984).
114. C. G. Kreiter and M. Wenz, unpublished results.
115. H. Lehmkuhl, C. Naydowski, F. Danowski, M. Bellenbaum, R. Benn, A. Rufinka, G. Schroth, R. Mynott, and S. Pasynkiewicz, *Chem. Ber.* **117,** 3231 (1984).
116. C. G. Kreiter, K. Nist, and J. Kögler, *Z. Naturforsch.* **41B,** 599 (1986).
117. F. W. Faller and A. M. Rosan, *J. Am. Chem. Soc.* **99,** 4858 (1977).
118. R. Benn and G. Schroth, *J. Organomet. Chem.* **228,** 71 (1982).
119. E. O. Fischer and Y. Hristidu, *Z. Naturforsch.* **15B,** 135 (1960).
120. M. A. Adams, K. Folting, J. C. Huffman, and K. G. Caulton, *J. Organomet. Chem.* **164,** C29 (1979).
121. M. A. Adams, K. Folting, J. C. Huffman, and K. G. Caulton, *Inorg. Chem.* **18,** 3020 (1979).
122. F. W. S. Benfield, R. A. Forder, M. L. H. Green, G. A. Moser, and K. Prout, *J. Chem. Soc., Chem. Commun.,* 759 (1973).
123. K. L. T. Wong and H. H. Brintzinger, *J. Am. Chem. Soc.* **97,** 5143 (1975).
124. C. G. Kreiter, J. Kögler, W. S. Sheldrick, and K. Nist, *J. Organomet. Chem.* (in press).
125. K. W. Barnett and D. W. Slocum, *J, Organomet. Chem.* **44,** 1 (1972).
126. R. B. King, A. D. King, Jr., M. Z. Iqbal, and C. C. Frazier, *J. Am. Chem. Soc.* **100,** 1687 (1978).
127. S. B. Butts, S. H. Strauss, E. M. Holt, R. E. Stimson, N. W. Alcock, and D. F. Shrifer, *J. Am. Chem. Soc.* **102,** 5093 (1980).
128. G. Capron-Cotigny and R. Poilblanc, *C. R. Acad. Sci. Paris, Ser. C* **263,** 885 (1966).
129. I. S. Butler, F. Basolo, and R. G. Pearson. *Inorg. Chem.* **6,** 2074 (1967).
130. K. W. Barnett and P. M. Treichel, *Inorg. Chem.* **6,** 294 (1967).
131. M. R. Churchill and J. P. Fennessy, *Inorg. Chem.* **7,** 953 (1968).
132. P. J. Craig and M. Green, *J. Chem. Soc., A,* 1978 (1968).
133. P. J. Craig and M. Green, *J. Chem. Soc., A,* 157 (1969).
134. K. W. Barnett, *Inorg. Chem.* **8,** 2009 (1969).
135. G. Wilkinson, *J. Am. Chem. Soc.* **76,** 209 (1954).
136. M. S. Wrighton and D. S. Ginley, *J. Am. Chem. Soc.* **97,** 4246 (1975).
137. J. L. Hughey, IV, C. R. Bock, and T. J. Meyer, *J. Am. Chem. Soc.* **97,** 4440 (1975).
138. A. S. Goldman and D. R. Tyler, *J. Am. Chem. Soc.* **106,** 4066 (1984).

139. R. Goddard, S. A. R. Knox, R. F. D. Stansfield, F. G. A. Stone, M. J. Winter, and P. Woodward, *J. Chem. Soc., Dalton Trans.*, 147 (1982).

140. M. Griffiths, S. A. R. Knox, R. F. D. Stansfield, F. G. A. Stone, M. J. Winter, and P. Woodward, *J. Chem. Soc., Dalton Trans.*, 159 (1982).

141. C. G. Kreiter, G. Wendt, and J. Kaub, unpublished results.

142. J. A. King, Jr., and K. P. C. Vollhardt, *Organometallics* **2**, 684 (1983).

143. R. D. Adams, D. E. Collins, and F. A. Cotton, *Inorg. Chem.* **13**, 1086 (1974).

144a C. J. Commons and B. Hoskins, *Aust. J. Chem.* **28**, 1663 (1975); R. Colton and C. J. Commons, *ibid.* **28**, 1673 (1975).

144b J. A. Marsella and K. G. Caulton, *Organometallics* **1**, 274 (1982).

145. C. P. Horwitz and D. F. Shriver, *Adv. Organomet. Chem.* **23**, 219 (1984).

146. M. D. Fryzuk, T. Jones, and F. W. B. Einstein, private communication.

147. E. Guggolz, F. Oberdorfer, and M. L. Ziegler, *Z. Naturforsch.* **36B**, 1060 (1981).

148. K. H. Franzreb and C. G. Kreiter, *Z. Naturforsch.* **37B**, 1058 (1982).

149. M. Green, A. G. Orpen, C. J. Schaverien, and I. D. Williams, *J. Chem. Soc., Chem. Commun.*, 1399 (1983).

150. P. O. Nubel and T. L. Brown, *J. Am. Chem. Soc.* **104**, 4955 (1982).

151. M. S. Wrighton and D. S. Ginley, *J. Am. Chem. Soc.* **97**, 2065 (1975).

152. A. F. Hepp and M. S. Wrighton, *J. Am. Chem. Soc.* **103**, 1258 (1981).

153. A. F. Hepp and M. S. Wrighton, *J. Am. Chem. Soc.* **105**, 5934 (1983).

154. U. L. Waltz, O. Hackelberg, L. M. Dorfman, and A. J. Wojcicki, *J. Am. Chem. Soc.* **100**, 7259 (1978).

155. R. W. Wegman, R. J. Olsen, D. R. Gard, L. R. Faulkner, and T. L. Brown, *J. Am. Chem. Soc.* **103**, 6089 (1981).

156. L. J. Rothberg, N. J. Cooper, K. S. Peters, and V. Vaida, *J. Am. Chem. Soc.* **104**, 3536 (1982).

157. A. M. Stolzenberg and E. L. Muetterties, *J. Am. Chem. Soc.* **105**, 822 (1983).

158. N. J. Coville, A. M. Stolzenberg, and E. L. Muetterties, *J. Am. Chem. Soc.* **105**, 2499 (1983).

159. D. G. Leopold and V. Vaida, *J. Am. Chem. Soc.* **106**, 3720 (1984).

160. H. Yesaka, T. Kobayashi, K. Yasufuku, and S. Nagakura, *J. Am. Chem. Soc.* **105**, 6249 (1983).

161. K. Yasufuku, H. Noda, J. Iwai, H. Ohtani, M. Hoshino, and T. Kobayashi, *Organometallics* **4**, 2174 (1985).

162. T. Kobayashi, H. Ohtani, H. Noda, S. Teratani, H. Yamazaki, and K. Yasufuku, *Organometallics* **5**, 110 (1986).

163. A. Fox and A. Poë, *J. Am. Chem. Soc.* **102**, 2497 (1980).

164. I. R. Dunkin, P. Härter, and C. J. Shields, *J. Am. Chem. Soc.* **106**, 7248 (1984).

165. S. P. Church, H. Hermann, F.-W. Grevels, and K. Schaffner, *J. Chem. Soc., Chem. Commun.*, 785 (1984).

166. M. L. Ziegler, H. Haas, and R. K. Sheline, *Chem. Ber.* **98**, 2454 (1965).

167. H. E. Sasse and M. L. Ziegler, *Z. Anorg. Allg. Chem.* **392**, 167 (1972).

168. M. Zöller and M. L. Ziegler, *Z. Anorg. Allg. Chem.* **425**, 265 (1976).

169. C. G. Kreiter and W. Lipps, *Angew. Chem.* **93**, 191 (1981); *idem, Angew. Chem., Int. Ed. Engl.* **20**, 201 (1981).

170. C. G. Kreiter and W. Lipps, *Chem. Ber.* **115**, 973 (1982).

171. H. L. Clarke, *J. Organomet. Chem.* **80**, 155 (1974).

172. H. D. Kaesz, R. B. King, and F. G. A. Stone, *Z. Naturforsch.* **15B**, 682 (1960).

173. M. L. H. Green and P. L. I. Nagy, *J. Chem. Soc.*, 189 (1963).

174. M. Cousins and M. L. H. Green, *J. Organomet. Chem.* **3**, 421 (1965).

175. W. Beck, K. Raab, U. Nagel, and W. Sacher, *Angew. Chem.* **97**, 498 (1985).

176. M. Leyendecker and C. G. Kreiter, *J. Organomet. Chem.* **249**, C31 (1983).

177. M. Leyendecker, W. S. Sheldrick, and C. G. Kreiter, *J. Organomet. Chem.* **270**, C37 (1984).

178. C. G. Kreiter and M. Leyendecker, *J. Organomet. Chem.* **280,** 225 (1985).
179. G. Dettlaf and E. Weiss, *J. Organomet. Chem.* **108,** 213 (1978).
180. J. Levisalles, F. Rose-Munch, and H. Rudler, *J. Chem. Soc., Chem. Commun.,* 1057 (1981).
181. C. G. Kreiter and M. Leyendecker, *J. Organomet. Chem.* **292,** C18 (1985).
182. C. G. Kreiter, M. Leyendecker, and W. S. Sheldrick, unpublished results.
183. J. E. Mahler and R. Pettit, *J. Am. Chem. Soc.* **84,** 1511 (1962).
184. U. Giannini, E. Pellino, and M. P. Lachi, *J. Organomet. Chem.* **12,** 551 (1968).
185. J.-Z. Liu and R. D. Ernst, *J. Am. Chem. Soc.* **104,** 3737 (1982).
186. C. F. Campana, R. D. Ernst, D. R. Wilson, and J.-Z. Liu, *Inorg. Chem.* **23,** 2732 (1984).
187. L. Stahl, J. P. Hutchinson, D. R. Wilson, and R. D. Ernst, *J. Am. Chem. Soc.* **107,** 5016 (1985).
188. D. R. Wilson, J.-Z. Liu, and R. D. Ernst, *J. Am. Chem. Soc.* **104,** 1120 (1982).
189. E. W. Abel and S. Moorhouse, *J. Chem. Soc., Dalton Trans.,* 1706 (1973).
190. D. Seyferth, E. W. Goldman, and J. Pornet, *J. Organomet. Chem.* **208,** 189 (1981).
191. D. R. Wilson, A. A. Di Lullo, and R. D. Ernst, *J. Am. Chem. Soc.* **102,** 5928 (1980).
192. M. Paz-Sandoval, P. Powell, M. G. B. Drew, and R. N. Perutz, *Organometallics* **3,** 1026 (1984).
193. J. R. Bleeke and M. K. Hays, *Organometallics* **3,** 506 (1984).
194. J. R. Bleeke and W.-J. Peng, *Organometallics* **3,** 1422 (1984).
195. H. Lehmkuhl and C. Naydowski, *J. Organomet. Chem.* **240,** C30 (1982).
196. L. S. Hegedus and S. Varaprath, *Organometallics* **1,** 259 (1982).
197. A. Nakamura, P.-J. Kim, and N. Hagihara, *J. Organomet. Chem.* **3,** 7 (1965).
198. S. Otsuka and A. Nakamura, *Adv. Organometal. Chem.* **14,** 245 (1976).
199. P. Racanelli, G. Pantani, A. Immirzi, G. Allegra, and L. Porri, *Chem. Commun.,* 361 (1969).
200. W. Bailey, M. H. Chisholm, F. A. Cotton, C. Murillo, and L. Rankel, *J. Am. Chem. Soc.* **100,** 802 (1978).
201. L. N. Lewis, J. C. Huffman, and K. G. Caulton, *J. Am. Chem. Soc.* **102,** 403 (1980).
202. L. N. Lewis, J. C. Huffman, and K. G. Caulton, *Inorg. Chem.* **19,** 1246 (1980).
203. W. A. Herrmann, J. Weichmann, M. L. Ziegler, and H. Pfisterer, *Angew. Chem.* **94,** 545 (1982).
204. Y. N. Al-Obaidi, P. K. Baker, M. Green, N. D. White, and G. E. Taylor, *J. Chem. Soc., Dalton Trans.,* 2321 (1981).
205. R. Ben-Shoshan and R. Pettit, *Chem. Commun.,* 247 (1968).
206. R. E. Davis, *Chem. Commun.,* 248 (1968).
207. I. I. Kritskaya, *Russ. Chem. Rev.* **41,** 1027 (1972).
208. T. Beetz and R. M. Kellog, *J. Am. Chem. Soc.* **95,** 7925 (1973).
209. A. Nakamura and N. Hagihara, *J. Organomet. Chem.* **3,** 480 (1965).
210. W. Keim, *Angew. Chem.* **80,** 968 (1968).
211. R. P. Hughes and J. Powell, *J. Organomet. Chem.* **20,** P17 (1969).
212. M. Leyendecker and C. G. Kreiter, *J. Organomet. Chem.* **260,** C67 (1984).
213. C. G. Kreiter, M. Leyendecker, and W. S. Sheldrick, *J. Organomet. Chem.* **302,** 35 (1986).
214. M. R. Churchill, K. N. Amoh, and H. J. Wasserman, *Inorg. Chem.* **20,** 1609 (1981).
215. B. J. Brisdon, D. A. Edwards, J. W. White, and M. G. B. Drews, *J. Chem. Soc., Dalton Trans.,* 2129 (1980).
216. C. G. Kreiter, M. Leyendecker, and W. S. Sheldrick, *J. Organomet. Chem.* **302,** 217 (1986).
217. S. J. La Placa, W. C. Hamilton, and J. A. Ibers, *Inorg. Chem.* **8,** 1928 (1969).
218. O. W. Howarth, C. H. Mc Ateer, P. Moore, and G. E. Morris, *J. Chem. Soc., Chem. Commun.,* 506 (1981).
219. F. Timmers and M. Brookhart, *Organometallics* **4,** 1365 (1985).
220. T. V. Ashworth, A. A. Chalmers, E. Singleton, and H. E. Swanepoel, *J. Chem. Soc., Chem. Commun.,* 214 (1982).
221. K. H. Franzreb and C. G. Kreiter, *Z. Naturforsch.* **39B,** 81 (1984).

222. P. O. Nubel and T. L. Brown, *J. Am. Chem. Soc.* **106**, 3474 (1984).

223. P. O. Nubel and T. L. Brown, *J. Am. Chem. Soc.* **106**, 644 (1984).

224. K. H. Franzreb and C. G. Kreiter, *J. Organomet. Chem.* **246**, 189 (1983).

225. C. G. Kreiter, K. H. Franzreb, and W. S. Sheldrick, *Z. Naturforsch.* (in press).

226. C. G. Kreiter, K. H. Franzreb, W. Michels, U. Schubert, and K. Ackermann, *Z. Naturforsch.* **40B**, 1188 (1985).

227. R. Davis and I. A. O. Ojo, *J. Organomet. Chem.* **110**, C39 (1976).

228. R. F. Gerlach, D. N. Duffy, and M. D. Curtis, *Organometallics* **2**, 1172 (1983).

229. J. A. Iggo, M. J. Mays, P. R. Raithby, and K. Hendrick, *J. Chem. Soc., Dalton Trans.*, 205 (1983).

230. R. B. King, P. M. Treichel, and F. G. A. Stone, *J. Am. Chem. Soc.* **83**, 3600 (1961).

231. M. D. Fryzuk, T. Jones, and F. W. B. Einstein, *Organometallics* **3**, 185 (1984).

232. P. Caddy, M. Green, L. E. Smart, and N. White, *J. Chem. Soc., Chem. Commun.*, 839 (1978).

233. J. B. Keister and J. R. Shapley, *J. Organomet. Chem.* **85**, C29 (1975).

234. A. J. Deeming, S. Hasso, and M. Underhill, *J. Organomet. Chem.* **80**, C53 (1974).

235. J. J. Guy, B. E. Reichert, and G. M. Sheldrick, *Acta Cryst.* **B32**, 3319 (1976).

236. A. D. Clauss, M. Tachikawa, J. R. Shapley, and C. G. Pierpont, *Inorg. Chem.* **20**, 1528 (1981).

237. W. Fellmann and H. D. Kaesz, *Inorg. Nucl. Chem. Lett.* **2**, 63 (1966).

238. M. R. Churchill and R. Bau, *Inorg. Chem.* **6**, 2086 (1967).

Index

Cumulative List of Contributors

Abel, E. W., **5,** 1; **8,** 117
Aguilo, A., **5,** 321
Albano, V. G., **14,** 285
Alper, H., **19,** 183
Anderson, G. K., **20,** 39
Armitage, D. A., **5,** 1
Armor, J. N., **19,** 1
Atwell, W. H., **4,** 1
Baines, K. M., **25,** 1
Barone, R., **26,** 165
Behrens, H., **18,** 1
Bennett, M. A., **4,** 353
Birmingham, J., **2,** 365
Blinka, T. A., **23,** 193
Bogdanović, B., **17,** 105
Bradley, J. S., **22,** 1
Brinckman, F. E., **20,** 313
Brook, A. G., **7,** 95; **25,** 1
Brown, H. C., **11,** 1
Brown, T. L., **3,** 365
Bruce M. I., **6,** 273; **10,** 273; **11,** 447; **12,**
    379; **22,** 59
Brunner, H., **18,** 151
Cais, M., **8,** 211
Calderon, N., **17,** 449
Callahan, K. P., **14,** 145
Cartledge, F. K., **4,** 1
Chalk, A. J., **6,** 119
Chanon, M., **26,** 165
Chatt, J., **12,** 1
Chini, P., **14,** 285
Chisholm, M. H., **26,** 97
Chiusoli, G. P., **17,** 195
Churchill, M. R., **5,** 93
Coates, G. E., **9,** 195
Collman, J. P., **7,** 53
Connelly, N. G., **23,** 1; **24,** 87
Connolly, J. W., **19,** 123
Corey, J. Y., **13,** 139
Corriu, R. J. P., **20,** 265
Courtney, A., **16,** 241
Coutts, R. S. P., **9,** 135
Coyle, T. D., **10,** 237

Craig, P. J., **11,** 331
Cullen, W. R., **4,** 145
Cundy, C. S., **11,** 253
Curtis, M. D., **19,** 213
Darensbourg, D. J., **21,** 113; **22,** 129
Deacon, G. B., **25,** 237
de Boer, E., **2,** 115
Deeming, A. J., **26,** 1
Dessy, R. E., **4,** 267
Dickson, R. S., **12,** 323
Eisch, J. J., **16,** 67
Emerson, G. F., **1,** 1
Epstein, P. S., **19,** 213
Erker, G., **24,** 1
Ernst, C. R., **10,** 79
Evans, J., **16,** 319
Evans, W. J., **24,** 131
Faller, J. W., **16,** 211
Faulks, S. J., **25,** 237
Fehlner, T. P., **21,** 57
Fessenden, J. S., **18,** 275
Fessenden, R. J., **18,** 275
Fischer, E. O., **14,** 1
Forster, D., **17,** 255
Fraser, P. J., **12,** 323
Fritz, H. P., **1,** 239
Furukawa, J., **12,** 83
Fuson, R. C., **1,** 221
Gallop, M. A., **25,** 121
Garrou, P. E., **23,** 95
Geiger, W. E., **23,** 1; **24,** 87
Geoffroy, G. L., **18,** 207; **24,** 249
Gilman, H., **1,** 89; **4,** 1; **7,** 1
Gladfelter, W. L., **18,** 207; **24,** 41
Gladysz, J. A., **20,** 1
Green, M. L. H., **2,** 325
Griffith, W. P., **7,** 211
Grovenstein, Jr., E., **16,** 167
Gubin, S. P., **10,** 347
Guerin, C., **20,** 265
Gysling, H., **9,** 361
Haiduc, I., **15,** 113
Halasa, A. F., **18,** 55